AF733128

Psychology Revivals

Information, Natural Law, and the Self-Assembly of Rhythmic Movement

Originally published in 1987, the introduction states: "the authors have successfully accomplished their program – to explain, based on physical representations, the observed relations among various parameters of wrist-pendulum oscillations. Thereby a set of new ideas and concepts, including those developed recently by the scientific school to which the authors belong, are introduced to biology. These concepts are closely related to the experimental data. This accomplishment makes the book especially attractive and demonstrates once more the productivity of applying physics to biology."

"Clear language, simple figures, and physical examples illuminate rather complicated problems. These attractive features should make the book intelligible to a variety of investigators in the field of motor control, not only to the specialists with physical and mathematical education."

From the foreword: " Kugler and Turvey have written strategic physical biology, and shown that, after all, dynamics (including both kinetics and kinematics) may support a unitary physical view of some of the profound operations of our brains... This is a grand start on what I hope is a larger program of demystifying behaviour."

Information, Natural Law, and the Self-Assembly of Rhythmic Movement

Peter N. Kugler and Michael T. Turvey

Routledge
Taylor & Francis Group
LONDON AND NEW YORK

First published in 1987
by Lawrence Erlbaum Associates

This edition first published in 2016 by Routledge
27 Church Road, Hove BN3 2FA

and by Routledge
711 Third Avenue, New York, NY 10017

Routledge is an imprint of the Taylor & Francis Group, an informa business

© 1987 Lawrence Erlbaum Associates, Inc.

The right of Peter N. Kugler and Michael T. Turvey to be identified as authors of this work has been asserted by them in accordance with sections 77 and 78 of the Copyright, Designs and Patents Act 1988.

All rights reserved. No part of this book may be reprinted or reproduced or utilised in any form or by any electronic, mechanical, or other means, now known or hereafter invented, including photocopying and recording, or in any information storage or retrieval system, without permission in writing from the publishers.

Publisher's Note
The publisher has gone to great lengths to ensure the quality of this reprint but points out that some imperfections in the original copies may be apparent.

Disclaimer
The publisher has made every effort to trace copyright holders and welcomes correspondence from those they have been unable to contact.

A Library of Congress record exists under ISBN: 0898596408

ISBN: 978-1-138-12357-1 (hbk)
ISBN: 978-1-315-64846-0 (ebk)

INFORMATION, NATURAL LAW, AND THE SELF-ASSEMBLY OF RHYTHMIC MOVEMENT

PETER NOBLE KUGLER
Crump Institute for Medical Engineering
University of California,
Los Angeles

MICHAEL T. TURVEY
University of Connecticut
and
Haskins Laboratories

LEA LAWRENCE ERLBAUM ASSOCIATES, PUBLISHERS
1987 Hillsdale, New Jersey London

Copyright © 1987 by Lawrence Erlbaum Associates, Inc.
All rights reserved. No part of this book may be reproduced in any form, by photostat, microform, retrieval system, or any other means, without the prior written permission of the publisher.

Lawrence Erlbaum Associates, Inc., Publishers
365 Broadway
Hillsdale, New Jersey 07642

Library of Congress Cataloging in Publication Data

Kugler, Peter Noble.
Information, natural law, and the self-assembly of rhythmic movement.

Bibliography: p.
Includes indexes.
1. Movement, Psychology of. 2. Psychology, Experimental. I. Turvey, Michael T. II. Title.
BF295.K84 1987 599'.018 86-13541
ISBN 0-89859-640-8
ISBN 0-8058-0055-7 (pbk.)

Printed in the United States of America
10 9 8 7 6 5 4 3 2 1

To

Arthur Iberall

for the inspiration he has provided by his efforts to establish a physics of living systems, we dedicate this volume in the series, Resources for Ecological Psychology.

RESOURCES for ECOLOGICAL PSYCHOLOGY

Edited by
Robert E. Shaw, William M. Mace, and Michael Turvey

This series of volumes is dedicated to furthering the development of psychology as a branch of ecological science. In its broadest sense, ecology is a multidisciplinary approach to the study of living systems, their environments, and the reciprocity that has evolved between the two. Traditionally, ecological science emphasizes the study of the biological bases of *energy* transactions between animals and their physical environments across cellular, organismic, and population scales. Ecological psychology complements this traditional focus by emphasizing the study of *information* transactions between living systems and their environments, especially as they pertain to perceiving situations of significance to planning and execution of purposes activated in an environment.

The late James J. Gibson used the term *ecological psychology* to emphasize this animal-environment mutuality for the study of problems of perception. He believed that analyzing the environment to be perceived was just as much a part of the psychologist's task as analyzing animals themselves, and hence that the "physical" concepts applied to the environment and the "biological" and "psychological" concepts applied to organisms would have to be tailored to one another in a larger system of mutual constraint. His early interest in the applied problems of landing airplanes and driving automobiles led him to pioneer the study of the perceptual guidance of action.

The work of Nicolai Bernstein in biomechanics and physiology presents a complementary approach to problems of the coordination and control of movement. His work suggests that action, too, cannot be studied without reference to the environment, and that physical and biological concepts must be developed together. The coupling of Gibson's ideas with those of Bernstein forms a natural basis for looking at the traditional psychological topics of perceiving, acting, and knowing as activities of ecosystems rather than isolated animals.

The purpose of this series is to form a useful collection, a resource, for people who wish to learn about ecological psychology and for those who wish to contribute to its development. The series will include original research, collected papers, reports of conferences and symposia, theoretical monographs, technical handbooks, and works from the many disciplines relevant to ecological psychology.

Series Dedication

To James J. Gibson, whose pioneering work in ecological psychology has opened new vistas in psychology and related sciences, we respectfully dedicate this series.

CONTENTS

FOREWORD
Physics and Biology

F. Eugene Yates
Crump Institute for Medical Engineering
University of California at Los Angeles
Los Angeles, California

"Nothing is more practical than a good theory" (Kurt Lewin, cited in Marrow, 1969), yet in matters biological it seems more practical to make models of data rather than to fashion models on theory. As has been emphasized by Patrick Suppes (1969), the most mature programs in science try to connect the two kinds of models by some formal structure. Models of theory specify the theory and fit it for tests against data. Models of data may sometimes suggest theory, or causal relationships, but all too often they are little more than demonstrations of statistical regularities or ad hoc postulations of convenience. They rarely go deep unless they connect to theory. This book by Kugler and Turvey joins theory and data, and exemplifies the mature scientific program.

The formidable discoveries of molecular biology seem to me to have regrettably weakened appetite for theory. By the middle of the 1960s many leading biologists actually believed that the central problem of explaining living systems had been solved, and that a linear array of nucleotides was the most essential feature of the living state. The rest was—as Dirac might have said also in this context—"mere chemistry." No appeal to a strategic physical theory was involved in recent spectacular advances in biological science. Though physicists participated, their contributions were not of the reflective sort to be found earlier in Schrödinger's *What is life* (1945)—rather, the contributions represented tactical applications of physical methods such as X-ray crystallography to biological materials. The successes were stunning, but only a minority of biologists correctly suspected that this rush of discovery would slow when it encountered the problem of coordination, competition, and cooperativity which are present at all levels in the hierarchies within and among living organisms.

They thought that the next round of great advance in biological science would occur when DNA and RNA were once again seen as molecules, not spirits, or the elan vital. To this minority, molecular biology appeared rife with vitalism. Far from showing that life had been reduced to physics, molecular biology worsened the gap between them; DNA looked too "intentional."

Intentionality is one of the most elusive of all notions. Is it physical, or always only biological? Does it even exist? Is it an illusion? If a lighted candle is covered by a tall cylinder, the flame will waver, flicker, gutter, and die, smothered in its own exhaust. However, if just before the death of the flame, a baffle is lowered down the center of the cylinder until it is just above the flame, the flame stabilizes, leans steadily to one side of the baffle and survives, having organized a sustaining downward airflow delivering oxygen on one side of the baffle, and an exhaust of hot gases up the other side. The flame creates a draft and a chimney. It acts as if it intends to stay alive. What are we to make of this? The candle adapts without a genetic program. Perhaps because the metaphor of flame as life has been so poetically expressed, we feel especially close to it. ("And all our yesterdays have lighted fools the way to dusty death. Out, out, brief candle!"—*Macbeth*, Act V, Scene 3, Line 19.)

The relationship between physics and biology has not been easy in this century. In fact, the relation between theoretical physics and biology has, historically, never been close. According to R. Rosen (1986):

> For a long time, theoretical physics has concerned itself with the articulation of universal and general laws. From that perspective, biology seems limited to a rather small class of very special systems; indeed, *inordinately* special systems. Clearly, then, organisms are not the sort of thing that physicists seeking universal principles would look at. To a physicist, what makes organisms special is conceived as a plethora of constraints, initial conditions and boundary conditions which must be superimposed upon the true, general physical laws, and which must be independently stipulated before those laws bear directly upon the organic realm. Physicists rightly felt that the determination of such supplementary constraints was not their job. But no physicist has ever doubted that, since physics is the science of material nature in all of its manifestations, the relation of physics to biology is that of general to particular.

One of the hopes of the Enlightenment, as seen especially in the works of Saint-Simon and Comte resting on the physics of Galileo and Newton, was to account for phenomena of man according to the laws and regularities of natural science. This program failed—chiefly because there was then no physics for truly complex systems. Such systems have noisy interactions, historical and evolutionary aspects, and informational or linguistic features that seem to lie outside physics. Complex processes seen in the geosphere (Lyell), in biology (Darwin), in neurology and psychology (Freud), and in social processes (Spencer, Marx and Engels) crushed the hopes that the physics of the Enlightenment might comprehend man. But now that disappointment is being dispelled by a new synthesis.

Although our technological experience concerns systems that become organized by commands from the outside, as when human intentions lead to the building of machines, the attention of many scientists has now turned to natural systems that become structured by a flow of energy through their own internal processes: these are the self-organizing systems. The emergence of order within them is a complex phenomenon that intrigues physicists, biologists, differential topologists, computer scientists, linguists, psychologists, geologists, civilizationists, meteorologists, economists, embryologists, paleontologists, anthropologists, cosmologists. It would be hard to find a more comprehensive topic. Similarities among problems of self-organization and of dynamics in complex systems may be thought of as the agenda for science in our age. It raises the question of whether or not there is a parsimonious set of scientific principles adequate to account for all the varieties of motion and change in the universe. Can there be a strategic physics rich enough to serve as the foundation of the science of *all* complex systems? Kugler and Turvey in this book seem to answer "yes," and imaginatively point the way. But the way is not just another round of Newtonian mechanics.

The Newtonian program was so overwhelmingly successful that it still dominates almost every mode of system description even today. His ideas represented the supreme embodiment of the concept of natural law and causal order. The epitome of the application of the Newtonian program to biological systems can be found in the work of Ross Ashby (1952). For example, Ashby wrote: "It will be assumed throughout that a machine or an animal behaved in a certain way at a certain moment because its physical and chemical nature at that moment allowed it no other action. . . . Our purpose is to explain the origin of behavior which *appears* to be teleologically directed." Robert Rosen reviews Ashby's claims and finds them permeated by unduly restrictive assumptions arising straight out of the Newtonian picture of the universe. In its reductionistic aspects, that picture shows every material system as consisting of mass points whose velocities and accelerations contain the solution to every scientific problem. To understand any material system, we merely have to characterize its particles and the forces acting on them, formulate the necessary equations of motion, and integrate them. But Rosen notes that there is also a paradigmatic aspect to Newtonian mechanics in the present day; as he said in the Ashby Memorial Lecture in 1984, it supposes that:

> the *language* in which Newton described his theory of systems of mass points is the universal language for talking about systems in general, even if they have not been, or cannot be, reduced to systems of mass points . . . The essence of this language . . . is that systems have states, and that their behaviors are represented by dynamical laws superimposed on these states. These states are the cognates of mechanical phases; the dynamical laws are the cognates of mechanical forces . . .

Kugler and Turvey, accepting and understanding Newtonian physics, go beyond it by introducing the notion of complexity and reexamining a question that has always been fundamental to physics—one that has defined a large part of physical endeavor since Aristotle: Why do things flow? In answering, they show that complexity consists of interactions of several kinds. On one hand, complex systems can have force-field descriptions that belong to kinetics. These are their mass-dependent, Newtonian aspects. On the other hand, much of the seemingly "intentional" behavior of such systems arises out of kinematic interactions in which masses count for little or nothing, and velocities (flows) and their interactions provide an alternative description of system behavior. It is in this latter, kinematic description that intentional and linguistic aspects of systems arise and can be connected to physics. (But not to a restricted Newtonian physics. This book is very much in the spirit of the studies of James J. Gibson [1979] that began with foundations of "ecological optics," carried through to problems in event perception and ended by establishing his "ecological realism" as the freshest and strongest account of perception in terms that did not abandon connections to natural science.)

Attempts to relate one branch of science to another, even when one feels they *must* be related, can lead either to new insights or to pernicious mischief. The perversions of Social Darwinism that followed upon the works of Herbert Spencer are a case of the mischief. It would seem that evolutionary processes and selection principles *ought* to have some bearing on the social phenomena of man, just as on the origin of species, but political action based on such an assumption has overwhelmed us with disasters. Less disastrous, but even sillier, is a notion arising in a field sometimes called "strong AI" (for artificial intelligence) that man's mind is a "program" that in principle can run on a Turing machine, and that it is merely an inconvenient accident of history that it runs on a neural substrate (see Churchland, 1982, for criticisms of strong AI). In such assertions we see a complete confounding between relations among inferential propositions in logic, which cannot possibly be causal, and the causal order of events. In the world of logic, mathematical systems permit us to study implications formally and extract conclusions from assumptions (theorems from hypotheses). Much of the work in theoretical physics has attempted to bring mathematical systems into correspondence with some causal nexus in the universe. When the correspondence is established, the theorems of the mathematical system thereby become agencies of prediction about events in the dynamical universe. The successes of mathematical, theoretical physics have been so great that, as in the case of the successes of molecular biology, they have beguiled us into supposing that theirs is the only "right stuff" of scientific thought or practice.

In this book, detailed applications of physical thought are made to the extremely difficult problem of self-organization in complex systems. The use of physics is careful and respectful, but there is no apotheosis. The threads with which the authors weave are drawn from Newton, Huygens, Einstein, Ehrenfest, and, more recently, Prigogine for the underlying mechanics; from Gibson for the precision of formulating the psychological themes; from Pattee (1977) for reflections on the relationships between information and dynamics. The resulting fabric is strong, and the tapestry new. But in spite of the many sources drawn upon for this work, it is not eclectic; rather, it is systematic. The reader is carefully coaxed toward the realization that information has a physical basis (other than in the trivial sense that a "bit" of selective information, if stored, must occupy a physical degree of freedom in some real system). Throughout, the authors insist upon identifying principles that connect to natural science as they explain characteristics of rhythmic movement. They stand on "ye shoulders of giants" (Newton), but their balance is good and their perspective farsighted.

Motor actions of human beings, even when directed strongly by commands from the outside, have to be regarded as intentional. After all, one might not comply with a command, regardless of the consequences. Kugler and Turvey have taken on a problem of coordination of intentional motor action by human subjects in an unusual experimental design. This design exposes how rhythmic actions may become cooperative through a kinematic description involving the nervous system as the only linkage between two mechanical engines (the forearms, wrists, and hands). The experimental results are surprising, and the authors show great insight in discovering regularities in their data, and in establishing an interpretive context from theory. The result is narrowly interesting in the illumination provided about specified human motor behaviors. But it is also more generally appealing because it gives confidence that this style of research—that of interacting models on physical theory with models of experimental data—can actually crack the problems of high dimensionality, complexity, cooperativity, and self-organization and make them comprehensible through the use of terms and concepts already well established in natural science. It is all the more remarkable that the authors accomplish this synthesis without neologisms. Theirs is a triumph of physical biology that does justice to both the physics and the biology. They take us beyond the obvious and the familiar and confront our fragile explanations and illusions.

The beguiling suppleness of natural language often belies serious flaws in our understanding. We have no trouble in our day-to-day talk with the concepts of information, order, organization, levels, intentions, awareness, or freedom of action. They seem familiar and obvious. Yet so little are we aware of the internal constraints that shape and limit our possibilities that it takes an act of concentration to believe that we are related to other animals, or that we are not free to do anything we want to the surface of this planet (without risking the stability of that thin veneer, the biosphere). Even worse, many of us feel threatened by the successive scientific "demotions" of man, the highlights of which are usually said to be the Copernican-Galilean relocation of the earth from its previous position in the center of the universe; the connecting of man to more primitive hominid/primate ancestors by the work of Darwin; Freud's alerting us to the possibility that we don't always say what we mean or mean what we say, or know at the level of our prized consciousness why we do what we do; the demonstrations of molecular biology that we store our past in a rather dull chemical; recurring proofs that our attempts at governance seem inevitably to bring upon us a new wave of disasters very reminiscent of old ones (Tuchman, 1984; Jacobs, 1984)—recognition that we do not learn or see clearly from our own social and political affairs; we do not have any theory of macroeconomics that provides correct predictions for economic behavior even over very short intervals.

All such scientific scrutiny exposes man as being somehow less than he thinks he is, but fails to satisfy us that he is described as he is. Is he, after all, a natural kind? If so, can we synthesize the many different perspectives with which we view ourselves? Philip Rhinelander has asked (1973): "Is man incomprehensible to man?" We seem to be. We are stuck with pluralistic accounts, complementarities where we seek unities, and with paradoxes (the opposite of every profound statement is another profound statement, e.g., God is dead). It is not clear where wisdom lies. Voltaire's Zadig was wise because he saw differences where others saw only uniformity; but, the power of invariances in scientific explanations suggests that he might have been wise also to see uniformities where others saw only differences. At least in science the drive for monism, the search for unity and unification in explanation, remains strong despite repeated failures. We have a hunger for synthesis. Pluralisms, even just as dualisms (Yin/Yang) are only reluctantly accepted in science, and then only when a single frame of reference dramatically fails to account for all the relevant experimental data bearing on the answer to an important question such as the nature of electromagnetic radiation.

Perhaps the relationship between information and dynamics is of a complementary kind, as Pattee (1977) suggests. If so, then explanations posed in the dynamical frame of reference cannot answer or contradict those posed in the informational frame of reference. Each offers only a partial account of larger phenomena not reachable by a unitary approach. But striving against all these confusions and doubts that arise on every side (complementarity, incompleteness, undecidability, and the loss of certainty of mathematics) there is a constructive sculpting of a single form of explanation that is physical, but still robust enough to address man and his intentional acts. To do that, information must be brought within the reach of dynamics. In the present work that important connection is made through examining the consequences of suppressing the mass dominance of kinetic systems and showing that out of that one change a new class of interactions (kinematic, flow-field, informational) arises that can provide alternative dynamic descriptions of complex natural systems. The class of kinematic interactions has singularities, invariances, and symmetries that play back out causally into the kinetic world through constraints. This is a comprehensive and powerful notion.

It may seem strange that a book addressing movement through a very particular (some would say extremely limited) experimental protocol should inspire in the authors (or in me) grand visions of the relationships of physics to biology. But this book cannot be fully appreciated or understood without recognition that the authors take a strong stand precisely on that deep issue. This book contains within it a synthesis for that issue. It condenses informational aspects of biology that have proved heretofore intractable to physical accounts. I see in this book an extension of physical biology that goes far beyond the tactical program announced by Lotka (1924) 60 years ago. We need strategies; the tactics can follow. In my opinion, Kugler and Turvey have written strategic physical biology, and shown that, after all, dynamics (including both kinetics and kinematics) may support a unitary physical view of some of the profound operations of our brains—those committed to sensory-motor mappings and coordinated movements. This is a grand start on what I hope is a larger program of demystifying behavior.

REFERENCES

Ashby, R. (1952). *Design for a brain*. London: Chapman and Hall.

Churchland, P. M. (1982). Is *thinker* a natural kind? *Dialogue*, *21*, 223–238.

Gibson, J. J. (1979). *The ecological approach to visual perception*. Boston: Houghton Mifflin.

Jacobs, J. (1984). *Cities and the wealth of nations*. New York: Random House.

Lotka, A. J. (1924). *Elements of physical biology*. (Reprinted in 1956 as *Elements of mathematical biology*). New York: Dover.

Marrow, A. J. (1969). *The practical theorist: The life and work of Kurt Lewin*. New York: Basic Books.

Pattee, H. H. (1977). Dynamic and linguistic models of complex systems. *International Journal of General Systems*, *3*, 259–266.

Rhinelander, P. H. (1973). *Is man incomprehensible to man?* San Francisco: W. H. Freeman and Company.

Rosen, R. (1984). "The physics of complexity," Ashby Memorial Lecture, (preprint).

Rosen, R. (1986). Some epistemological issues in physics and biology. In *Quantum theory and beyond*, B. J. Hiley and F. T. Peats (Eds.). Oxford, England: Routledge and Kegan.

Schrödinger, E. (1945). *What is life*? London: Cambridge University Press.

Suppes, P. A. (1969). *Studies in the methodology and foundations of science*. New York: Humanities Press.

Tuchman, B. (1984). *The march of folly*. New York: Knopf.

INTRODUCTION
Physical and Neurophysiological Perspectives on Movement

M. B. Berkinblit
A. G. Feldman
Institute of Problems of Information Transmission
Academy of Sciences
Moscow, USSR

This book is concerned with the field of theoretical biology in which biological phenomena are interpreted through explanations proceeding from natural, primarily physical, laws. Borelli (1680, 1681) was perhaps the first to develop such an approach to living organisms. His book *On the movements of animals* focused attention on the problems of statics or equilibrium of an animal's body using mechanical representations of levers developed, in particular, by Archimedes. The problems of dynamics of living movements could not be considered scientifically in that book because necessary prerequisites—provided by Newton's classical work—did not appear until seven years later. Even the study of the kinematics of living movements was poorly represented, being limited by the crude methods of recording movements available during Borelli's time. This situation was radically changed after the appearance of photography which remains an effective method of movement study today.

As a rule, natural biological movements (e.g., locomotion) involve a great number of degrees of freedom of the body so that the resolution of the respective dynamic equations is hardly possible by improvised means. This obstacle was overcome with the advent of computers. Thus, the progress in biomechanics essentially depended on technical advances, many of which were borrowed from physics. The penetration of physical ideas into biology was of no less importance. The discovery of the law of conservation of energy provided a basis for understanding bioenergetics. Such physical concepts as stiffness, viscosity, and others appeared in biomechanics as well. This stream of ideas and methods was to a considerable extent unidirectional from physics to biology.[1]

[1]Remember, however, that the concept of Brownian motion and osmotic pressure came to physics from plant physiology.

The methodological difficulties of biomechanics led some investigators to conclude that the direct use of physical approaches and concepts could not generally be employed as an effective strategy for explaining biological movements: Unlike the movements of physical bodies, the movements of humans and animals are purposeful and execute certain functional goals. In addition, the nervous system is capable of using many muscles and joints and it varies their coordinations from time to time, even though the movement goal remains the same. In other words, living movements are controllable.

Bernstein (1947) formulated a number of control problems confronting the movement physiology and suggested solutions to some of them, such as the principle of afferent feedback for movement regulation. It has become evident that for the explanation of living movement the laws of classical mechanics must be supplemented at least by the principles of control theory. The first attempts to apply control theory to biological movements were rather straightforward. Wiener's suggestion that the cerebellar tremor is similar to the instability of a feedback-controlled regulator is relevant here. The reader may be less well-acquainted with the work of Korenev (1977), who attempted to build a theoretical biomechanics by introducing "aiming equations" that described goals of movement "goals," as auxiliary equations for classical dynamics. To overcome the problem of nonuniqueness of the possible solutions of the dynamic equations for living movements, the theory of optimal control was often used. Such efforts, despite their elaborate formalisms, favored the development of theoretical biology.

The history of biomechanics warns that one must not neglect the specifics of the biological objects while using the principles of physics, communication or control theories. It is true that natural laws are the same everywhere, but it does not follow that their forms are everywhere identical. Thus, two main problems confront modern biomechanics. On one hand, it is necessary to show that the achievements of physics, control theory, and other sciences can be effectively used for explanation of biological phenomena. On the other hand, it is necessary to show how the theories must be modified to reach this goal. These problems are not simple. Although the physical and mathematical principles are formulated clearly, additional biological principles in biomechanics are usually less distinct and often introduced without a substantiation in biology itself (as the previously mentioned Korenev's "aiming equations" or various principles of optimization). As an example of biomechanic studies in which biological data are used in an adequate manner, one should see the book by Beletskii, *Biped walking* (1984).

We have offered this brief historical account to place the Kugler and Turvey volume in the context of modern theoretical biology.

Locomotion and other rhythmic movements in vertebrates, including humans, are traditionally associated and compared with the oscillations of pendulums—classic objects of study in mechanics and nuclear physics. The use of the elucidated general physical laws of oscillations in the analysis of animal rhythmic activity seems to be quite natural and necessary. It is clear, however, that living organisms not only obey natural laws but also have developed, in the process of evolution, a unique ability of using them effectively. This ability achieves its highest form in voluntary motor acts. Exposing the principles underlying this special property of biological systems—*using* natural laws instead of being in passive subordination to them—is one of the complex but fundamental problems confronting the physiology of motor activity. This book contains essential investigations in exactly this direction. Employing physical representations combined with comparatively simple experimental tools, the authors elucidate general principles underlying the pendular, clocking mode of the organization of rhythmic limb movements in humans. It should be noted that a significant part of the book is focused on the consideration of modern concepts of various branches of mathematics, physics, and engineering—especially qualitative dynamics, nonequilibrium thermodynamics, and similarity analysis—with special reference to biological movements. This part of the book is of interest not only in relation to the analysis of experimental data, but also on its own as one of the few attempts to outline the fundamentals of physical biology. Clear language, simple figures, and physical examples illuminate rather complicated problems. These attractive features should make this book intelligible to a variety of investigators in the field of motor control, not only to the specialists with physical and mathematical education.

Rhythmic movements in living organisms are executed by neuronal oscillators that have many degrees of freedom and are extremely adaptive to various peripheral (afferent) signals and mechanical conditions. The theory of dynamic systems, especially of dissipative systems, seems to be the most adequate for the description of such oscillators. However, this theory is still only feebly elaborated, and neurophysiologists are compelled to content themselves chiefly with the accumulation of experimental data without knowing their true "worth" which could be established by a theory. Opposing the common tendency to regard neural oscillators as "hard-molded" central pattern generators, the authors regard them as "soft" structures. This means that "a biokinematic system with periodic behavior is assembled temporarily and for a particular purpose from whatever elements are available and fitting the task" (Chapter 1, p. 3). Such an understanding of the nature of rhythmicity suggests that one ought to see something more in neuronal oscillators than a likeness to electronic systems. This approach envisages, in full measure, the fact that biological systems are open, self-sustaining, nondeterminate, and guided by information. But the concept of information is used in a nontraditional sense—as something specifically reflecting the forms of dynamic processes in the biological system (e.g., "Haptic Information as the Abstracted Morphology of Movement Dynamics," Chapter 7).

A popular interpretation holds that biological rhythm is supported by an impulsive, phase-dependent offset of the energy lost per cycle, as is the case in the escapement used in the design of manufactured clocks. For a number of reasons, this analogy to rhythmic behavior is rejected by the authors (see Chapter 5). They argue that rhythmic behavior of living systems is a consequence of the self-assembly of dissipative structures: When the amount of energy flowing continuously into the system exceeds a critical value, a reorganization of the operational components of the system results, so that they can now support its stable oscillations by dissipating free energy. In such systems, the parameters of rhythmic processes (e.g., natural periods or amplitudes) can easily be controlled, as is the case for most rhythmic motions of organisms.

The experimental part of this book is devoted to the analysis of oscillations executed by subjects holding one or two pendulums of specific masses and lengths, and involving a single wrist joint in isolation or two wrist joints in combination. For a pair of hand-held pendulums, the notion of a virtual system that is qualitatively equivalent to the double system is defined (see Chapter 6). This notion symbolizes the idea that the two hands are constrained, in the given experimental conditions, to act as a single functional unit or coordinative structure. Based on the experimental data, the authors identify invariant structurally stable relations—natural laws—among period, amplitude, mass, and length. From these it is suggested that the subjects make use of an intrinsic metric and coordinate space that relate biological times to biological masses and lengths. Since mass is one of the coordinates of this space, the latter appears, in essence, to be dynamic. This idea is worth emphasizing, since "inner coordinate spaces" are usually associated in physiological literature only with kinematic variables. In this respect the authors develop more universal representations. Both metrics and the degree of curvature of the dynamic coordinate space are suggested to vary from subject to subject, and these properties affect the natural law that relates the previously mentioned parameters. The contribution of these properties is measured to reveal the invariant form of the law. This law relating periodic time to the mass and length of a wrist-pendulum system is theoretically identified through the use of similarity analysis, dimensional analysis and allometry, and is corroborated by the experimental data (see Chapter 8).

Chapters 10, 11, and 12 of this book are the most attractive and elegant. Here attention is focused on a law relating amplitude to the frequency of oscillations of the wrist-pendulum system. It is known that these parameters in mathematical pendulums can vary independently of each other. The experimental data relating amplitude to frequency of oscillation in the wrist-pendulum systems appears at first glance to have an independence (sometimes amplitude increases with increases in frequency; sometimes amplitude is constant with increases in frequency; and still other times, amplitude decreases with increases in frequency). However, having analyzed these data, the authors put forward the hypothesis that these changes of the amplitude of oscillation are a consequence of an adiabatic transformation of the system.

Chapter 10 begins with the analysis of conditional periodicity in adiabatically transformed conservative systems. In a conservative system an adiabatic change of state is to be regarded as a transformation through a sequence of intermediate reversible states that are near, or at, equilibrium. This is the case, for instance, when the length of a mathematical physical pendulum is slowly decreasing. Thereby, the average kinetic energy of the system increases in the same proportion as the frequency (Ehrenfest's theorem). In Chapter 11 the authors modify, and extend, Ehrenfest's theorem to encompass open chemical-mechanical-thermodynamic systems similar to wrist-pendulum systems. The theorem is used as a theoretical basis for constructing a theory of soft-molded escapements that gate energy flows responsible for (a) assembling and (b) sustaining biomechanical oscillations. And finally, Chapter 12 provides analyses of experimental data relating amplitude changes to frequency, mass, and length of the oscillator. The experimental results are in full accord with the proposed theory.

On the whole, the authors have successfully accomplished their program—to explain, based on physical representations, the observed relations among various parameters of wrist-pendulum oscillations. Thereby a set of new ideas and concepts, including those developed recently by the scientific school to which the authors belong (Kelso, Holt, Kugler, & Turvey, 1980; Kugler, Kelso, & Turvey, 1980; Saltzman, 1979), are introduced in biology. These concepts are closely related to the experimental data. This accomplishment makes the book especially attractive and demonstrates once more the productivity of applying physics to biology.

This book may give rise to some feeling of dissatisfaction, however, since the authors' attention is chiefly focused on general natural laws underlying the rhythmic phenomena, whereas the relevant physiological processes (the work of muscles, receptors, motoneurons, and other nervous cells) remain, to a certain degree, hidden.

This problem—the interfacing between physical and neurophysiological descriptions—has recently been addressed by Feldman (1979). He proposes that the nervous system executes movements through shifts in equilibrium operating points defined at the level of motor apparatus. The proposal incorporates physical concepts in combination with neurophysiological concepts specific to the biological motor system. The concept of *invariant characteristic* is also introduced to describe the behavior of muscles in combination with local reflexes and more central neural mechanisms subsuming its activity. In addition, the concept of *central command* as an independent measure of the influence of higher brain-level influence is also proposed. The model incorporates both physical and neurophysiological concepts. The strategy guiding the formulation is the identification of invariant characteristics of movement regulation that can be instantiated in a wide variety of neural designs. The analysis focuses on the identification of the functional invariants underlying movement production, rather than on neuronal detail that can vary dramatically across instantiations.

The conceptual gap associated with the interfacing strategy linking physical concepts with neurophysiological concepts, however, remains large. This gap will need to be addressed before a comprehensive theory of motor control can be realized. The conceptual bridging will require not only a consistent theoretical formulation, but also a translation statement of terms (from physical to neurophysiological and vice versa, depending on the uniqueness of terms) to allow a mutual coordination of physical and neurophysiological data bases. The theme of this book focuses essentially on the elucidation of general physical principles. To address the neurophysiological issues, other kinds of studies will be required.

Thus, the authors deliberately do not treat the question of how the neuronal structures that execute rhythmic behaviors must be organized to realize the observed biomechanical forms and flows of information and energy. The gap between the concrete, neuronal analysis of motor systems and an analysis directed toward exposing general principles[2] constitutes a central issue for the scientific agenda. Without underestimating the importance of the first approach (neurophysiological), this book by Drs. P. N. Kugler and M. T. Turvey successfully demonstrates the necessity of the second approach (physical).

REFERENCES

Beletskii, V. V. (1984). *Biped walking*. Moscow: Nauka (monograph in Russian).

Bernstein, N. A. (1947). *O postroenii dvizheniy*. (Concerning the construction of movements). Moscow: Medgiz.

Borelli, G. A. (1680, 1681). *De motu animalum*. (On the movements of animals). Rome: Bernabo, Vol. I–II.

Feldman, A. G. (1979). *Central and reflex mechanisms in motor control*. Moscow: Nauka (monograph in Russian).

Kelso, J. A. S., Holt, K. G., Kugler, P. N., & Turvey, M. T. (1980). On the concept coordinative structures as dissipative structures. II. Emperical lines of convergence. In G. E. Stelmach and J. Requin (Eds.), *Tutorials in motor behavior*. New York: North-Holland.

Korenev, G. V. (1977). *Introduction to human mechanics*. Moscow: Nauka (monograph in Russian).

---

[2]An approach based on the search for general design principles must be carefully tempered so as not to lose too much of the system's detail (such as its individual identity). Theories based on general principles must inevitably make decisions on the saliency of facts, throwing out some facts in favor of others. This problem is exemplified in the case of weather forecasting: The kinetic theory of gases provides very little help in the forecasting of weather. The strength of an approach based on the use of general principles, however, provides a more direct methodology for identifying biological invariants.

Kugler, P. N., Kelso, J. A. S., & Turvey, M. T. (1980). On the concept of coordinative structures as dissipative structures. I. Theoretical lines of convergence. In G. E. Stelmach and J. Requin (Eds.), *Tutorials in motor behavior*. New York: North-Holland.

Saltzman, E. (1979). Levels of sensorimotor representation. *Journal of Mathematical Psychology*, *20*, 91–163.

"You could always say, 'When you come to living things, anything can happen?' If you accept that you will never understand living things." (Fenyman, 1967 p. 165)

"Biology . . . is characterized by cycles in systems which are approaching steady states of intermediate stability . . . such cycles are a general consequence of the physics of systems undergoing energy flux and need not be introduced as special biological phenomena." (Morowitz, 1968, p. 34)

". . . it is easy to understand how a simple change in a single variable can result in very complicated changes in a large system of particles. This is the normal physical situation. It is not easy to explain how complicated changes in a large system of particles can repeatedly result in a simple change in a single variable." (Pattee, 1970, p. 225)

"The rules that govern behavior are not like laws enforced by an authority or decisions made by a commander: behavior is regular without being regulated. The question is how this can be." (Gibson, 1979, p. 225)

"Local thermodynamic engine cycles are capable of coming into existence by self-organization. A spinning, falling leaf, or an Aeolian harp, or the tides in the atmosphere illustrate such processes. However, if many such processes develop into or around a formal system, and become self-entraining, self-serving and persistent (e.g. an erupting geyser or a tornado), then the system may ultimately become viable, that is, it may be able to persist beyond the range of the local 'workshop' in which it was created, and may be able to seek out and draw from scattered and even randomly-appearing energy sources, in a fully autonomous fashion." (Iberall, 1978, p. 531)

"The role of geometrical representations in classical physics is well known. Classical physics is based on Euclidean geometry, and modern developments in relativity and other fields are closely related to extensions of geometric concepts. But take the other extreme: the field theory used by embryologists to describe the complex phenomena of morphogenesis . . . (In embryogenesis) . . . we see a progressive organization of a biological space in which every event proceeds at a moment and in a region that make it possible for the process to be coordinated as a whole. The space is functional, not geometrical. The standard geometrical space, the Euclidean space, is invariant with respect to translations or rotations. This is not so in the biological space. In this space the events are localized in space and time and not merely trajectories." (Prigogine, 1980, xiv)

"Symbolic representation of qualitative entities is doomed to its rightful place of minor importance in a world where flowers and beautiful women abound." (Einstein, 1937, Hyperbolic Aesthetic)

PREFACE

The work reported here is directed at understanding the physical basis of a hallmark characteristic of animals—the generation of rhythmic movements of the kind that occur most commonly in terrestrial locomotion. Central to this effort is the notion of information. Orthodox construals of information fail to come to terms with the sense of information that guides the assembling and sustaining of comparatively simple and basic acts, such as locomotion or the rhythmic swinging of an individual limb. Gibson's (1979) construal of information—as qualitative, macroscopic properties of structured energy distributions lawfully generated by, and thereby specific to, events and actions—is more successful. This specificational sense of information seems to be especially important to a physical biology that aims to embrace characteristic phenomena of living things through the careful extension of scale-independent, natural principles. Living things are nondeterminate. In conventional physical terms, knowing their initial conditions at a given point in time, and the laws of motion that they (like other natural things) obey, is insufficient for predicting their trajectories. Yet, their acts can be, and most frequently are, regular and reproducible. Information, it seems, is the basis of this reliability, and it can be so because information in the specificational sense is lawfully determined. Behavior is nondeterminate but information is determinate. In part, the success of a physical biology depends on understanding the laws that condense out, sustain, and dissolve patternings of low energy distributions specific to the facts of animals, of environments, and of their relations. It depends also on the availability of a means for describing and quantifying these lawfully generated distributions or patternings of energy. A "morphological" perspective on information is intimated.

Stated more particularly, the goal of the present work is to identify the natural laws that shape reliably the biomechanical flow fields that inform people how to assemble repeatedly and to maintain rhythmic movements defined over one or two variously loaded limbs in controlled laboratory settings. It is an attempt to account for rhythmic movements without any "self-actional explanatory terms" as Dewey and Bentley (1948) might say, or without any "loans on intelligence" as Dennett (1978) would phrase it. The enterprise proves to be far-ranging. The conceptual tools of various branches of mathematics, physics, and engineering—especially qualitative dynamics, nonequilibrium thermodynamics, and similarity analysis—are brought to the task of interpreting the data. These tools are not commonplace in movement research. Consequently an effort is made throughout the book to bring as many as possible of these conceptual tools into play gradually, as and when the data analysis demands.

Despite its physical trappings, the research reported in this book is a fairly standard piece of human experimental psychology. To begin with, the problem under study is most disconcertingly psychological for one who would take a physical tack. Even a meager description of what is entailed is shot full of psychological idioms: A person producing a rhythmic movement of a certain kind must *intend* that movement; *information about* the formation and maintenance of that movement must be available; and the person must be able to *detect* and *use* this information in order to successfully *perceive* the facts of the movement and thereby fulfill his or her *intention*. Furthermore, the problem under study involves measures of behavior and the manipulation of independent variables to which these measures are somehow related. As with a typical piece of human experimental psychology, a rationalization of the independent variable-dependent variable linkage is sought. The ideal is a rationalization that includes a minimum of explanatory concepts, where each concept is as principled in its individual derivation as is allowed by the current state of the sciences of living things. The present text does little more than pursue this ideal. If it appears eccentric in the light of much of contemporary experimental psychology it is only because the analysis is biased toward what Gibson liked to call "radical solutions to scientific problems" (Reed & Jones, 1982, p.14). As he expressed it: "General explanations are always preferable to piecemeal explanations ("models" as they are nowadays called) and this is all that is meant by a radical theory" (Reed & Jones, 1982, p. 14).

One final, related point: The experimental task has a noteworthy scientific pedigree. The task, basically, is that of swinging a hand-held pendulum back and forth through the action of the wrist. In the seventeenth century, pendular motion laid the basis for classical mechanics. Certain facts had been established about pendulums (that the period of a swing varied with length but not with mass and not with small amplitude variations) and Newton proceeded to derive an equation that described them. On the assumption that gravitational force always operated in a downward direction on the mass of the pendulum, he arrived at the equation for the acceleration of the pendulum, namely, *force* = *mass* × *acceleration*. The mass term dropped out when this equation was solved to determine the pendulum's period, consistent with observation. It soon became apparent that the equation was consistent with very many diverse observations. The equation captured a lawful regularity of immense scope.

The pendular motion of interest to the present monograph is governed by gravity only in part. There is an additional (elastic) force structure that is assembled repeatedly through an irreversible thermodynamic process. (Swinging a hand-held pendulum requires that a person systematically degrade chemical fuel into mechanical energy.) At issue is the principled assembling of this force structure in the gravitational context. The content of this monograph, therefore, can be looked at as continuing a centuries old tradition—that of detailing and analyzing the properties of pendular motion to reveal nature's hand.

REFERENCES

Dewey, J., & Bentley, A. F. (1948). *Knowing and the known*. Boston: Beacon Press.

Dennet, D. C. (1978). *Brainstorms: Philosophical essays on mind and psychology*. Montgomery, VT: Bradford Books.

Gibson, J. J. (1979). *The ecological approach to visual perception*. Boston: Houghton Mifflin.

Reed, E., & Jones, R. (1982). *Reasons for realism: Selected essays of James J. Gibson*. Hillsdale, NJ: Lawrence Erlbaum Associates.

ACKNOWLEDGMENTS

The work reported here was supported in part by the following sources: Basic research funds of the Crump Institute for Medical Engineering, University of California at Los Angeles; NIH Grant for Basic Medical Research RR5596 and office of Naval Research Grant N0014-83-c-0083 awarded to Haskins Laboratories; and, NSF Grant BNS 76 22943 awarded to the Center for Advanced Study in the Behavioral Sciences, Stanford, California.

This manuscript would never have reached fruition without the technical assistance of Barbara Wasneski, Terry Cobb and Art Lizza who efficiently directed the assembly of the book in its various stages of production: Multitude of thanks for being so patient with revisions. We would like to thank: Barbara Zeman for editorial assistance; Tova Clayman, Linda Reiser, L. Rosenbaum, and R. C. Schmidt for helping with data collection and analysis; Dave Truesdell and Phil Rubin for programming assistance; Joe Paccia, Dan Hayashi, Claudia Carello and Agnes McKeon for preparation of figures. Special thanks to Joe Paccia for preparation of the insect nest-building series (Chapter 4) and to Claudia Carello for preparation of the cover design and editorial assistance in all phases of preparation and production.

The facts and ideas expressed in this book originated and evolved through conversations and correspondences with the following individuals: E. Deland, S. Dudevoir, C. Fowler, S. Giszter, B. Kay, L. Katz, J.A.S. Kelso, T. Lombardo, G. Lukatela, W. Mace, J. Milstein, G. Moore, K. Newell, H. Pattee, T. Poston, R. Remez, L. Rosenbaum, R. Rosen, P. Rubin, S. Runeson, E. Saltzman, R.C. Schmidt, R.E. Shaw, R. Thom, M. Wade, and W. Warren. The final form of the ideas was carefully tempered by many discussions with A.S. Iberall and F.E. Yates on the topic of physical strategies as they manifest themselves in complex systems. We are grateful for their insistence on, and commitment to, a unitary view of science.

Our personal thanks to M. Berkinblit and A. Fel'dman for reading an early version of the manuscript and writing an introduction and to F.E. Yates for writing a foreword about "physics and biology". We would like to thank R.E. Shaw and W. Mace for their invitation to include this book in the Resources for Ecological Psychology Series.

P.N.K. is indebted to Gene Yates and Mike Gold for providing financial and intellectual resources at the Crump Institute that permitted the occasion to "dare to know" (*sapere aude*, Kant, 1784): to take the risk of discovery, to exercise the right of unfettered criticism, to accept the loneliness of autonomy. P.N.K. extends thanks to Charles Boer, Lee Kugler and Dena McLerran for their continued encouragement and support. And finally, P.N.K. offers his most endearing thanks to Lois Rosenberg Kugler for her love, sense of being, and imagination; and to Michael T. Turvey for his friendship, sense of scholarship, and persistence.

M.T. Turvey wishes to thank Nell Turvey, Georgia Turvey and Claudia Carello for their love and support. To Claudia a very special thanks for her endless good humor, patience and affection. The task would have been so much harder, perhaps impossible, without her. M.T. Turvey also wishes to acknowledge his debt to his dear friend and colleague Robert E. Shaw for the inspiration and vision (of psychology) that he has provided over the past dozen years. And finally M.T. Turvey would like to thank Peter Noble Kugler for letting him share the fruits of his creativity.

1 PERSPECTIVES ON RHYTHMIC MOVEMENT

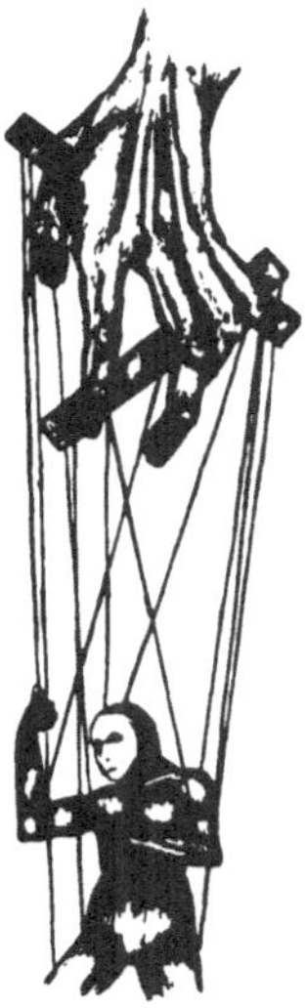

ABSTRACT

Rhythmic activity, such as bipedal and quadrupedal locomotion, obviously has a neural substrate. It is not clear, however, that a neurally oriented analysis of rhythmic movement will reveal its mechanism. Nor is it clear that the closely related motor-program perspective, with its emphasis on symbol strings, can provide much help in this matter. Arbitrariness plagues both of the orthodox perspectives on rhythmic movement, given their mutual lack of principled constraints. The first chapter sketches a law-based approach to animal movement, intimating how an account of biological phenomena might be fashioned that is methodologically continuous with the physical sciences. It identifies a central claim that natural phenomena at all scales are organized by a common set of (physical) strategies, although each scale may exhibit unique regularities. At their own scale, living systems are open to exchanges of matter and energy, they are self-sustaining and nondeterminate and they are dominated by information more than they are by forces. These characteristics of living things are discussed as factors that shape the regularities defining a physics of animal movement.

1.1. Rhythmic Phenomena in Biology

Rhythmic phenomena abound in biology at all levels of analysis (Aschoff, 1979; Berridge, Rupp, & Treherne, 1979; Iberall, 1972; Winfree, 1980). At the behavioral level, rhythms can be observed in which the entire body or parts of the body move in a cyclic, repetitive fashion. Running, swimming, flying, breathing, chewing, and grooming are the most cited examples, but rhythmic behaviors need not be so stereotypic, a fact attested to by the musical accomplishments of humans.

Although they are commonplace, behavioral rhythmicities are not well understood. To date, the efforts of neuroscience to explain behavior such as locomotion have been shaped in part by the issue of whether the behavior is generated centrally, by properties intrinsic to the central nervous system, or whether it is achieved by sensory feedback from moving parts of the body (Delcomyn, 1980; Evarts, Bizzi, Burke, Delong & Thach, 1971). Conventional wisdom and the weight of evidence favor the view of central pattern generation: A repetitive rhythmic output is said to be the product of a neural "oscillator" conceived either as a single pacemaker neuron (e.g., Maynard, 1972) or as a network of neurons (e.g., Wilson & Waldron, 1968).

The pioneering work of von Holst (1935/1973), an early proponent of central-pattern generators, addressed the problem of the coupling of two or more neural oscillators. This problem arises as soon as the focus of inquiry into locomotion shifts from the rhythmicity of a single limb or body segment to the rhythmicity of the locomotory act in full. Examination of the coupled rhythmic movements of the pectoral and dorsal fins of fish that swim with the main body axis immobile led to the identification of two major classes of mutual influence. One of these influences, termed "superposition," is manifest (primarily) in the amplitude of the rhythmic movements, while the other influence, termed the "magnet effect," is manifest in their frequencies. Treating the generators of individual fin rhythmicity as automatisms, von Holst (1935/1973) regarded the magnet effect as the tendency of automatisms to attract each other to their respective preferred tempos. von Holst regarded the magnet effect as general: Not only did it operate in the coupling of individual automatisms but it operated in the coupling of the cells that comprise an individual automatism. Further, he observed that when two or more automatisms converged on a common, stable frequency, they lost nothing of their individuality (see also Arsharvskii, Kots, Orlovskii, Rodionov & Shik, 1965; Shik & Orlovskii, 1965). Individual preference for this or that frequency remained, suggesting that the coordinated rhythmic states exhibited by the collection of rhythm generators comprised continuously competing tendencies among components. Although some of von Holst's empirical claims have been questioned (e.g., Roberts, 1967) the foregoing observations remain secure (Stein, 1977). In retrospect, they are observations that place locomotion squarely in the class of cooperative phenomena (Haken, 1977). How is this locomotory instance of cooperativity to be understood?

1.2. Hard-Molded and Soft-Molded Rhythmic Movements

Where explanations of rhythmic behavior such as locomotion have been neural in their focus, the tendency has been to regard the behavior in question as *hard molded* (*hard wired*, *hard geared*, *hard coupled*, *hard guided*, and *hard algorithmed*). These terms, borrowed from engineering, refer, in general, to the directing of "signals" along rigidly determined kinematic paths. The commonly proposed neural mechanisms, as noted, are single cells or ensembles of cells *specialized* for the generation of behavioral rhythmicities. Of the two, the specialized ensemble or network is thought to be the more general mechanism. A central pattern generator works, presumably, by guiding the chemico-electrical field along hard constraints (specific neural pathways linking specific neural elements) to produce periodic tensile states in the associated musculature. Selverston (1980) has argued that understanding how a hard-molded central pattern generator works for any given instance of rhythmicity is equated with identifying all of the ensemble's neural elements, all of the membrane and synaptic properties of those elements, and all of the connections among them. Putting aside doubts about the success of such a program (see Selverston, 1980, and commentaries) there are questions of its propriety. Rhythmic motions of the body can be, and frequently are, *soft molded* (*soft wired*, *soft geared*, *soft coupled*, *soft guided*, and *soft algorithmed*), which is to say, loosely speaking, that a biokinematic system with periodic behavior can be assembled temporarily and for a particular purpose from whatever neural and skeletomuscular elements are available and befitting the task.

Both hard and soft construction of rhythmic movements must follow from principles that govern the cyclic mode of biological organization in general. An account of hard-molded periodic behavior might be pursued without identifying these principles, but a satisfactory account of soft-molded periodic behavior cannot be attempted without them. In their basic format these principles are not likely to be unique to biology. An educated guess is that they are to be numbered among the few universally generalizable principles of physics (Morowitz, 1968, 1978; Yates & Iberall, 1973). Understanding the production of rhythmic movements that are softly molded from biological materials must rest on an understanding of how a specific set of strategies can be realized in distinct morphological settings subject to very general laws (cf. Iberall & Soodak, 1986). A program for understanding soft-molded behavior contrasts with that identified above for hard-molded rhythmic behavior in that it must give priority to physical law.

On at least two counts, it would seem that a law-oriented program is essential to the success of a neural-oriented program, regardless of whether the latter's target is an understanding of hard, or hard and soft, instances of rhythmic behaviors. First, there is the issue of how to constrain rationally the choice of *relevant properties* to be studied. Ideally one wishes to manipulate those parameters of the central pattern generator that govern its operation. Unfortunately, the parameter set of a neural ensemble contains, by some counts, 46 entries that could be relevant to a neurophysiological explanation (Bullock, 1975). Principles beyond those of neurophysiology are required to guide the selection of parameters. Second, there is the issue of how to explain the *characteristic quantities* of a rhythmic behavior—for example, its period, amplitude, and energy per cycle. These quantities cannot be rationalized on neural considerations alone.

1.3. The Concept of Motor Program

It is a small step from the concept of central pattern generator to that of motor program. Often the two concepts are used interchangeably. Like the concept of central pattern generator, the motor-program concept is a "centralist" counterpoint to the "peripheralist" treatment of rhythmic movement (as a concatenation of reflexively mediated, afferent-efferent connections). At the core of the centralist view is the recognition that the spatiotemporal details of a cycle of locomotion appear, in very large part, to be *anticipated*. By the peripheralist account, the successive and adjacent order of the skeletomuscular variables and their magnitudes are determined during the cyclic movement on a moment-to-moment basis. In contrast, the centralist account regards the successive and adjacent order of the skeletomuscular variables and the ballpark magnitudes of these variables as being determined prior to the cyclic movement. There are moment-to-moment influences, but these are reflected mainly in the magnitudes of the variables (as a fine tuning) and minimally in their ordering. Some proponents of the centralist view (e.g., Keele, 1968) have ruled out fed-back influences. But as others have pointed out, this prohibition is unwarranted (e.g., Reed, 1982; Schmidt, 1982) and inconsistent with the concept of program (MacKay, 1980). By definition, a program must receive (negations and affirmations) as well as give (instructions).

An example of a programming formulation of a step cycle in locomotion is provided by MacKay (1980) and reproduced in Figure 1.1. As can be seen, the kinematic and kinetic details are ordered by a sequence of commands and queries. The commands and queries are symbol strings that guide the performance of the step cycle: They function to constrain the continuous dynamical process by which the step cycle is realized. The contrast between rate-independent symbol strings and rate-dependent dynamical processes is said by Pattee (1973, 1977, 1979) to identify a complementarity principle. For Pattee, this particular complementarity is the hallmark of living systems. His argument

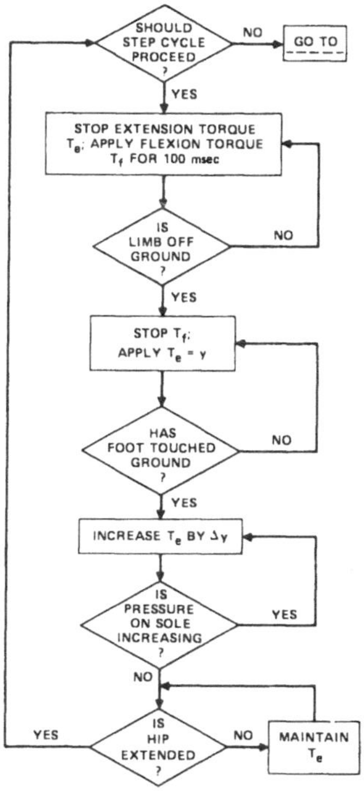

FIGURE 1.1. A MOTOR PROGRAM PERSPECTIVE ON THE LOCOMOTORY STEP CYCLE (adapted from MacKay, 1980).

is that living systems operate in two modes, the symbolic and the dynamic, which are incompatible and irreducible. Consequently, for Pattee, understanding biological phenomena (such as rhythmic movements) rests with the elaboration of this complementarity. By Pattee's view, the motor-program approach is flawed to the extent that it emphasizes the discrete symbolic mode to the neglect of the continuous dynamical mode. The phenomena in question are the result of a mutuality between the two modes.

There is, however, an inequality between the discrete symbolic and continuous dynamical modes that has to be appreciated. Nature uses the symbolic mode—nonholonomic, nonintegrable constraints (see Pattee, 1983, 1986; and Appendix A)—sparingly. Dynamics are used to the fullest, wherever and whenever, to achieve characteristic biological effects (see also Gould, 1966; Gould & Lewontin, 1979). Symbol strings are used, now and then, to direct dynamical processes and to limit their complexity—in other words, to trim the dynamical degrees of freedom. In Figure 1.1 the opposite strategy is at work. Very many nonholonomic constraints are exploited to achieve ("to explain") the kinetic and kinematic regularities of a locomotory step cycle. The question of how the dynamics, properly construed for the biological scale, might fashion the phenomenon is ignored. Also ignored is the question of how the symbol strings interface with the dynamics.

Pattee's analysis is an important one (see Carello, Turvey, Kugler, & Shaw, 1984) for the motor-program perspective (given that perspective's emphasis on the symbolic mode) because it suggests that only in the working out of the physics of a regularity can one identify the nature and type of symbol strings (nonholonomic constraints) needed to complete the explanation. Beginning with the symbolic mode, and adhering strictly to it, invites an account (of coordinated movement, for example) that will be plagued by arbitrariness. Beginning with the dynamical mode—that is, taking a law-based approach—and pursuing it earnestly promises an account that will be principled.

1.4. Physical Biology

Broadly construed, a law-oriented program is an attempt to give an account of biological phenomena (such as rhythmic movement) that is continuous with physical methodologies. A physical approach to biology views living systems as ordinary physical systems of great complexity; all biophysical transformations must satisfy the usual principles concerning the types of physical events and changes of free energies. The fact that problems may arise in the analysis due to the complexity of the system should not imply that the basic principles of physics must be altered. Rather, the implication is that biological systems have developed extraordinary means of making use of these principles.

Physical interpretations of the characteristic properties of living things are often viewed as an impermissible reductionism to simple physical properties at some reduced scale of analysis, a striving to convert complicated biological processes into simpler physical ones. The traditional (and questionable, see von Bertalanffy, 1952; Koestler & Smythies, 1969; Weiss, 1967, Fodor, 1975; Bunge, 1977; Bohm, 1980) reductionism to "fundamental properties" at a "fundamental scale" is being challenged, however, by a very different reductionism, one that seeks to define a universal set of organizing strategies that apply with equal emphasis across the various scales and disciplines of the natural sciences (Haken, 1977; Iberall & Soodak, 1986; Prigogine, 1980; Soodak & Iberall, 1978; Yates, 1982). As a reductionist program, physical biology is not an attempt to subordinate biology to physics, but rather an effort to ascertain the unity of animate and inanimate nature. In terms of fundamental strategies, physics as the science of matter and field is by no means simpler than biology and psychology as the sciences of living things. Moreover, to attempt a reductionism of phenomena in a test field (e.g., biology) to scale-independent strategies is not to jeopardize the independence or significance of that test field. The promise is that the test field of interest will be provided with a much more solid foundation (and science as a whole with more parsimony).

The test field of interest in this monograph is animal movement, and the focus is self-sustained rhythmic movement, especially stable and reproducible rhythmic movement. By the preceding statement of a level-independent physics, a putative physics of movement that is as parsimonious as a physics of motion is possible, and its pursuit is encouraged. But, by the same token, the details of a putative physics of movement should not be expected to look like the details of the established physics of motion. In physically nontrivial ways, living systems in action are not point masses in motion. The organizing strategies may be common, but the forms of the regularities will be different. The following properties should give a unique shape to the physics of biological movement.

1.4.1. Living Systems are Open. The isolated, closed physical system is not a model of living systems. Living systems are open to the flow of energy and matter. This openness, when subject to physical scrutiny guided by traditional physical principles, gives rise to a heterodox view of causal dynamics (von Bertalanffy, 1952; Prigogine, 1969; Glansdorff & Prigogine, 1971; Nicolis & Prigogine, 1977; Yates, 1986). Opening a physical system to the flow of energy and matter generates internal constraints that curtail the system's internal degrees of freedom. In reducing the internal degrees of freedom, new spatiotemporal geometries arise, accompanied by new modes of interaction. Because of these evolved internal constraints and their interactive consequences, an open system's behavior is not fully accounted for by initial and boundary conditions. In a word, openness incurs autonomy (to a lesser or greater degree) and this autonomy can be (and usually is) partitioned over a number of process scales (macro and micro). Importantly, this autonomy is sovereign at each scale. The points that follow elaborate this autonomy and sovereignty, underscoring that it does not deny natural law.

1.4.2. Living Systems are Self-Sustaining. Internally, living systems maintain a potential. It arises from delaying the transport of energy into and out of the interior and from the ability to sustain this exchange. Living systems must be served energy by an external agency, or serve themselves. Given a sustained on-board source of potential energy (usually in the form of chemical fuel) that can be drawn upon, capabilities such as self-reparation and self-propulsion are possible. The facility to produce generalized external work cycles—to produce, for example, periodically propulsive forces—has the significant consequence of giving a system a certain amount of flexibility with respect to external gradient fields. The system's trajectories need not follow minimum energy trajectories, defined relative to the external gradient field, if the system has the capability to generate forces comparable to those generated externally. (In comparison, the trajectories of the classically conceived particles are committed to the field's geodesics.) If these departures from and returns to the externally defined minimum trajectories (however brief) are systematic, then a sensitivity to gradients is implied. An ability to discriminate selectively low-energy potential gradients is likely to accompany the ability to self-sustain (see Section 1.4.4).

1.4.3. Living Systems are Nondeterminate. Whether addressing closed or open systems, physics is committed to natural laws. More elaborately, whatever its focus, physics is committed to a natural continuity in energy states. These, ideally, are reducible to symmetry statements (equations) that are defined on conservations (typically, mass, momentum, energy). Inhomogeneous distributions of conserved quantities (that is, potentials) give rise to forces that, in turn, give rise to motion. The laws of motion, whether they are expressed in the vocabulary of forces or, more profoundly, in the vocabulary of conservations, are inferred from trajectories. The guiding assumption behind the classical strategy for defining a law of motion has been that in the trajectories of particles are revealed the causal actions of potential fields. The conditions on the assumption are these: (a) a potential field external to a particle is constant in both space and time and (b) the particle has no internal means for introducing and absorbing forces. If these two conditions are satisfied, then the trajectory of a particle is fashioned solely by the potential and the inference from a system's trajectory to the field's causal action on the system—the strategic move on which the definition of a law of motion is classically based—is legitimate.

Living things, however, do not meet the second of the two conditions. They are "particles" with on-board potentials. They can, as noted, resist the minimum of the immersing field. The stable states with which they are associated are fashioned from the forces generated by the external potential *and* the forces generated by the internal potential, linked in a circular causality (Iberall, 1977/8; Yates, 1982): external→internal→external→. . . .

The point being made is actually twofold: first, living systems are *nondeterminate*. Nothing extraordinary is intended by this term. Its proper context is given by the physical convention for labeling phenomena "determinate" and "indeterminate." A system is determinate if (a) its trajectory strictly conforms to the minima and gradients of the potential, (b) its initial conditions can, in principle, be known precisely, and (c) the disposition of the external potential can, in principle, be known precisely. A system is indeterminate if (a) and (c) are satisfied but (b) is not and it is nondeterminate if (a) fails to hold even though (b) and (c) are likely. Second, the trajectories of living systems cannot be used to develop equations *of movement* analogous to the classical equations *of motion*. To reiterate, the equations of motion follow from observations of the trajectories of closed, isolated particles; they describe how the surrounding field acts on the isolated particles to cause their motion.

Does this twofold point deny a causal account of living systems? Evidently nature follows some order of cause and effect, but the significance of causality has undergone change and development in the growth of natural science (see Bohm, 1958; Margeneau, 1950). The attitude here is that causality means no more and no less than that nature (including living systems) is comprehensible in terms of laws that are discoverable (cf. Bergman, 1962; Gibson, 1967/1982; Margeneau, 1950).

1.4.4. Living Systems are Guided by Information. It will be argued in this monograph that information is implied whenever and wherever trajectories of a system are stable and reproducible in a field but are not the result of the field acting directly on the system through the mass dimension (conventionally, through forces). By this implication, the trajectories of nondeterminate systems are information-guided. Ordinarily, however, stable and reproducible trajectories point to laws. There need be no contradiction between these two implications. All that is required for agreement is that information be based in natural law (behavior is nondeterminate but information is determinate). This is Gibson's (1966, 1979) treatment of information. At the ecological scale of animals and their environments, information is a term that refers to macroscopic patternings of lower-energy fields (e.g., optical, acoustical, chemical, etc.) that are generated lawfully by higher-energy fields (e.g., the layout of potentials and reflective surfaces) and by the displacements of living systems relative to these fields. Because they are generated lawfully, these macroscopic spatial/temporal patternings will, under ordinary conditions of animal life, specify the layout of potential sources and reflective surfaces and the displacements relative to them.

Gibson's specificational sense of information can be contrasted with the more commonplace indicational/injunctional sense of information (Reed, 1981; Turvey & Kugler, 1984 a,b). The latter sense likens information to symbol strings (see Section 1.2) that indicate a state of affairs (e.g., "road work ahead") or direct a state of affairs (e.g., "slow down"). Information conceived of as a symbol string is incomplete. This is because it is an alternative description of a state of affairs that is useful only to the extent that it is less detailed than that to which it refers (Carello, Turvey, Kugler & Shaw, 1984; Pattee, 1973). Because a symbol string cannot, by definition, specify trajectories, it cannot provide a lawful basis for a trajectory. This is not to say that information in the indicational/injunctional sense cannot be responsible for a trajectory in part, only that it cannot constrain a trajectory in full. Without information in Gibson's specificational sense, information in the indicational/injunctional sense is impotent (e.g., to comply with the injunction "slow down" requires that information *specific to* the retarding of forward motion be available).

The macroscopic, lawfully generated patterns envisaged by Gibson carry, in their topological form, properties that are specific to components of change and components of persistence in the living system-surround relation, and they are meaningful because they define gradient values with respect to the living system's internal potentials. The conceptual promise is that the information carried in the *evolving geometry of structured energy distributions* can be information *about* the living system's dynamics (internal potential layout) relative to the environmental dynamics (external potential layout). As interpreted here, Gibson's conception of information is close to Thom's conception of information as geometric form (cf. Kugler, Kelso & Turvey, 1980, 1982). For Thom (1975 p. 145), geometric form "*can be the carrier of information, and in the set of geometric forms carrying information of the same type the topological complexity of the form is the quantitative scalar of the information.*"

1.5 . Preview

The chapters that follow develop a law-based perspective on rhythmic movement. The perspective grows out of an analysis of data collected in the laboratory and is molded by concepts and methods from (a) the mathematical domains of qualitative dynamics (Abraham & Shaw, 1982; Thom, 1975), similarity analysis (Rosen, 1978), and stability theory (Lyapunov, 1947); (b) the physical domains of nonequilibrium thermodynamics (Glansdorff & Prigogine, 1971; Iberall, 1977/8; Morowitz, 1978; Nicolis & Prigogine, 1977; Prigogine, 1980; Soodak & Iberall, 1978; Yates, 1982) and nonlinear mechanics (Andronov & Chaikin, 1949; Haken, 1977; Minorsky, 1962); and (c) Gibson's ecological approach to perception and action (Gibson, 1979; Lee, 1976, 1980; Michaels & Carello, 1981; Reed, 1982; Shaw & Turvey, 1981; Turvey & Kugler, 1984a).

Some of the laws under study here are not conventional. They might be called information laws, in view of the discussion in 1.4.4. Although the understanding of information laws expressed in these pages is limited, this much seems certain: Their antecedent conditions can be kinetic, kinematic, geometric, or temporal descriptions of a material layout. Their consequent conditions, in comparison, can only be kinematic, and/or geometric, and/or temporal. More exactly, the consequent conditions are large-scale kinematic, and/or geometric, and/or temporal properties that can function as soft constraints on a system's kinetics. Because they are based in law, and because laws are invariable under relevant transformations, these kinematic, and/or geometric, and/or temporal properties make possible an additional repertoire of stable and reproducible functions for any system able to establish a linkage (cf. Rosen, 1978) between these properties and its own kinetic properties.

Although information laws may be unconventional, they can be approached and identified through a fairly conventional methodology. Consonant with the strategy of the physical sciences, the methodology to be pursued emphasizes singularities, similarity transformations (that is, symmetries), and invariants. These notions form the core of conventional, physical theories of measurement and modeling.

REFERENCES

Abraham, R. H., & Shaw, C. D. (1982). *Dynamics: The geometry of behavior, Part one—periodic behavior*. Santa Cruz: Aerial Press.

Andronov, A., & Chaikin, C. E. (1949). *The theory of oscillations*. Princeton, NJ: Princeton University Press.

Arsharvskii, Yu. I., Kots, Ya. M., Orlovskii, G. N., Rodionov, I. M., & Shik, M. L. (1965). Investigations of biomechanics of running by the dog. *Biophysics, 10*, 665–672.

Aschoff, J. (1979). Circadian rhythms: General features and endocrinological aspects. In D. Krieger (Ed.), *Endocrine Rhythms*. New York: Raven Press.

Bergman, P. G. (1962). *Basic theories of physics*. New York: Dover Publications.

Berridge, M. J., Rupp, P. E., & Treherne, J. E. (1979). *Cellular oscillators*. Cambridge: Cambridge University Press.

Bertalanffy, L. von. (1952). *Problems of life*. New York: Wiley.

Bohm, D. (1958). *Causality and chance in modern physics*. London: Routledge.

Bohm, D. (1980). *Wholeness and the implicate order*. Boston: Routledge.

Bullock, T. H. (1976). In search of principles in neural integration. In J. C. Fentress (Ed.), *Simple networks and behavior*. Sunderland, MA: Sinauer Associates.

Bunge, M. (1977). Levels and reductions. *American Journal of Physiology: Regulatory, Integrative and Comparative Physiology, 23*, R75–R82.

Carello, C., Turvey, M. T., Kugler, P. N., & Shaw, R. E. (1984). Inadequacies of the computer metaphor. In M. S. Gazzaniga (Ed.), *Handbook of cognitive neuroscience*. New York: Plenum Press.

Delcomyn, F. (1980). Neural basis of rhythmic behavior in animals. *Science, 210*, 492–498.

Evarts, E. V., Bizzi, E., Burke, R. E., Delong, M., & Thach, W. T. (1971). The central control of movement. *Neuroscience Research Program Bulletin, 9*, 1–170.

Fodor, J. A. (1975). *The language of thought*. New York: Thomas Y. Crowell.

Gibson, J. J. (1966). *The senses considered as perceptual systems*. Boston: Houghton-Mifflin.

Gibson, J. J. (1979). *The ecological approach to visual perception*. Boston: Houghton-Mifflin.

Gibson, J. J. (1967/1982). James J. Gibson autobiography. In E. Reed and R. Jones (Eds.), *Reasons for realism: Selected essays of James J. Gibson*. Hillsdale, NJ: Lawrence Erlbaum Associates.

Glansdorff, P., & Prigogine, I. (1971). *The thermodynamic theory of structure, stability and fluctuations*. New York: Wiley.

Gould, S. (1966). Evolutionary paleontology and the science of form. *Earth Sciences Review, 6*, 77–119.

Gould, S. J., & Lewontin, R. (1979). The spandrels of San Marco and the Panglossian paradigm: A critique of the adaptationist programme. *Proceedings of the Royal Society* (London), *B205*, 581–598.

Haken, H. (1977). *Synergetics*. Heidelberg: Springer-Verlag.

Holst, F. von (1935/1973). *The behavioral physiology of animal and man*. Coral Gables, FL: University of Miami Press.

Iberall, A. S. (1972). *Towards a general science of viable systems*. New York: McGraw-Hill.

Iberall, A. S. A field and circuit thermodynamics for integrative physiology. I, II, and III. *American Journal of Physiology: Regulatory, Integrative and Comparative Physiology, 2*, R171–R180, 1977; *3*, R3–R19, 1978; *3*, R85–R97.

Iberall, A. S., & Soodak, H. (1986). A physics for complex systems. In F. E. Yates (Ed.), *Self-organizing systems: The emergence of order*. New York: Plenum Press.

Keele, S. W. (1968). Movement control in skilled motor performance. *Psychological Bulletin*, *70*, 387–403.

Koestler, A., & Smythies, J. R. (1969). *Beyond reductionism*. Boston: Beacon Press.

Kugler, P. N., Kelso, J. A. S., & Turvey, M. T. (1980). On the concept of coordinative structures as dissipative structures: I. Theoretical lines of convergence. In G. E. Stelmach and J. Requin (Eds.), *Tutorials in motor behavior*. New York: North-Holland.

Kugler, P. N., Kelso, J. A. S., & Turvey, M. T. (1982). On the control and co-ordination of naturally developing systems. In J. A. S. Kelso and J. E. Clark (Eds.), *The development of movement control and co-ordination*. New York: Wiley.

Lee, D. N. (1976). A theory of visual control of braking based on information about time to collision. *Perception*, *5*, 437–459.

Lee, D. N. (1980). Visuo-motor coordination in space-time. In G. E. Stelmach and J. Requin (Eds.), *Tutorials in motor behavior*. New York: North-Holland.

Lyapunov, M. A. (1947). *Probleme general de la stabilite du mouvement*. Princeton, NJ: Princeton University Press.

MacKay, D. (1980). The motor program: Back to the computer. *Trends in Neuroscience*, *3*, 97–100.

Margeneau, H. (1950). *The nature of physical reality*. New York: McGraw Hill.

Maynard, D. M. (1972). Simple networks. *Annals of the New York Academy of Sciences*, *193*, 59–72.

Michaels, C. F., & Carello, C. (1981). *Direct perception*. Englewood Cliffs, NJ: Prentice-Hall.

Minorsky, N. (1962). *Nonlinear oscillations*. Princeton, NJ: Van Nostrand.

Morowitz, H. J. (1968). *Energy flow in biology: Biological organization as a problem in thermal physics*. New York: Academic Press.

Morowitz, H. J. (1978). *Foundations of bioenergetics*. New York: Academic Press.

Nicolis, G., & Prigogine, I. (1977). *Self-organization in nonequilibrium systems: From dissipative structures to order through fluctuations*. New York: Wiley-Interscience.

Pattee, H. H. (1973). Physical problems of the origin of natural controls. In A. Locker (Ed.), *Biogenesis, evolution, homeostasis*. Heidelberg: Springer-Verlag.

Pattee, H. H. (1977). Dynamic and linguistic modes of complex systems. *International Journal of General Systems*, *3*, 259–266.

Pattee, H. H. (1979). The complementarity principle and the origin of macromolecular information. *Biosystems*, *11*, 217–226.

Pattee, H. H. (1983). Cell psychology: An evolutionary approach to the symbol-matter problem. *Cognition and Brain Theory*, *5(4)*, 325–341.

Pattee, H. H. (1986). Instabilities and information in biological self-organization. In F. E. Yates (Ed.), *Self-organizing systems: The emergence of order*. New York: Plenum Press.

Prigogine, I. (1969). Structure, dissipation and life. In M. Marios (Ed.), *Theoretical Physics and Biology*. Amsterdam: North-Holland, 23–52.

Prigogine, I. (1980). *From being to becoming*. San Francisco: Freeman and Co.

Reed, E. S. (1981). Indirect action. Unpublished manuscript, Center for the Study of Human Learning, University of Minnesota.

Reed, E. S. (1982). An outline of a theory of action systems. *Journal of Motor Behavior*, *14(2)*, 98–134.

Roberts, B. L. (1969). Spontaneous rhythms in the motorneurons of spinal dogfish (Seyliorhirus caricula). *Journal of Marine Biology Association* (U.K.), *49*, 33–49.

Rosen, R. (1978). *Fundamentals of measurement and representation of natural systems*. New York: North-Holland.

Schmidt, R. A. (1982). More on motor programs. In J. A. S. Kelso (Ed.), *Human motor behavior: An introduction*. Hillsdale, NJ: Lawrence Erlbaum Associates.

Selverston, A. I. (1980). Are central pattern generators understandable? *Behavioral and Brain Sciences*, *33*, 535–571.

Shaw, R. E., & Turvey, M. T. (1981). Coalitions as models for ecosystems: A realist perspective on perceptual organization. In M. Kubovy and J. Pomeranz (Eds.), *Perceptual organization*. Hillsdale, NJ: Lawrence Erlbaum Associates.

Shik, M. L., & Orlovskii, G. N. (1965). Co-ordination of the limbs during running of the dog. *Biophysics*, *10*, 1148–1159.

Soodak, H., & Iberall, A. S. (1978). Homeokinetics: A physical science for complex systems. *Science*, *201*, 579–582.

Stein, P. S. G. (1977). Application of the mathematics of coupled oscillator systems to the analysis of the neural control of locomotion. *Federation Proceedings*, 336, *7*, 2056–2059.

Thom, R. (1975). *Structural stability and morphogenesis*. Reading, MA: Benjamin.

Turvey, M. T., & Kugler, P. N. (1984a). An ecological approach to perception and action. In H. T. A. Whiting (Ed.), *Human motor actions: Bernstein reassessed*, Amsterdam: North-Holland.

Turvey, M. T., & Kugler, P. N. (1984b). A comment on equating information with symbol strings. *American Journal of Physiology: Regulatory, Integrative Comparative Physiology*, *15*, R925–R927.

Weiss, P. (1967). 1 + 1 ≠ 2 (When one plus one does not equal two.) In G. C. Quarton, T. Melnechuck and F. O. Schmitt (Eds.), *The neurosciences: A study program*, New York: Rockefeller University Press.

Wilson, D. M., & Waldron, I. (1968). Models for the generation of the motor output pattern in the flying locust. *Proceedings of the Institute of Electrical and Electronics Engineers*, *56*, 1058–1064.

Winfree, A. T. (1980). *The geometry of biological time*. New York: Springer-Verlag (series in biomathematics, Vol. 8).

Yates, F. E. (1982). Outline of a physical theory of physiological systems. *Canadian Journal of Physiology and Pharmacology*, 60, *3*, 217–248.

Yates, F. E. (Ed.) (1986). *Self-organizing systems: The emergence of order*. New York: Plenum Press.

Yates, F. E., & Iberall, A. S. (1973). Temporal and hierarchical organization in biosystems. In J. Urquhart and F. E. Yates (Eds.), *Temporal aspects of therapeutics*. New York: Plenum Press.

2 PROBLEMS OF MEASUREMENT AND MODELING

ABSTRACT

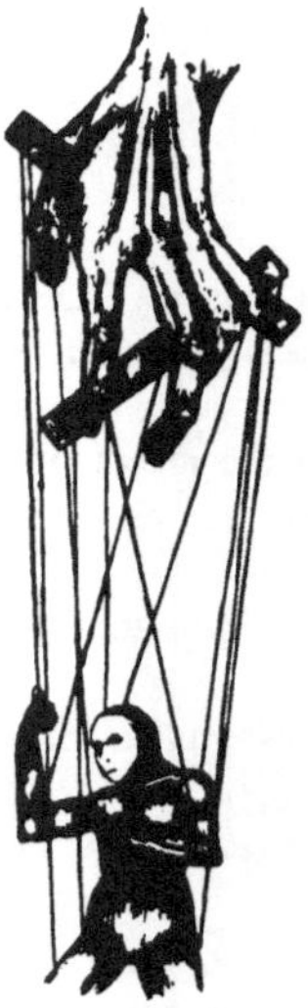

This chapter and the next two address the view that the methodologies of physics can be cast into a very general form that is applicable to all systems. As noted, this is not a reductionistic view that seeks to express all phenomena in terms of a standard set of elementary physical laws and boundary conditions. Rather it is the belief that the methodological strategies of the physical sciences can be extended to encompass the events connected with complex (biological) systems as well as simple (physical) systems. As will become apparent, the decisive idea is that of "flow." The path taken here from a simple system to a complex system begins with an understanding of mass-dependent causes of flow, passes through an appreciation of the stable geometric properties to which flows give rise, and ends with the observation that mass-independent characterizations of flow function as the informational basis for animal activity. The path is cluttered, however, with issues of measurement and modeling. Some of these issues are the concern of the present, prefatory chapter.

Any biological system succumbs to analysis at a number of scales, where the scale of any given analysis refers to its ordinal position in an extensive dimension of magnitude. The most commonly cited scales of analyses are the macroscopic and microscopic, but more specific distinctions in scale can be drawn. For example, with regard to a locomoting animal it is commonplace to note a scale at which the body segments, their associated torques, and their motions are the variables to be measured, and an immediately lower scale at which the muscles, their oxygen perfusion, and their tensile states are the objects of analysis. Scales of analyses contrast in terms of the number and type of components to which they refer; muscles outnumber body segments, for example, and the lower scale's states of perfusion are not duplicated at the upper scale. This contrast can invite a number of others, most notably contrasts in metrics (roughly, real valued "distance functions"), unit quantities (those adopted as standards of measurement), and levels of abstraction (degree of specificity or generality in the measurement of a scale's properties).

All natural systems conform to this image of a progression of scales of analyses. Further, and relatedly, any natural system, whether it is the system of interest in full or in part, can be regarded alternatively as an individual system (e.g., an atomism) or as a set of systems (e.g., a continuum), as Chapter 3 underscores. Whatever the regard, issues will arise concerning the selection of metrics, units, and level of abstraction.

Another class of problems that arises for a multiple-scale analysis of a system, assuming that the division into scales is principled, is that of how to construct models that relate different scales of analysis. With respect to biological systems, there is no single agreed-upon answer, although it is popular to construct models that use computational/representational linkages: The state of affairs at one scale of analysis is the basis for computing the state of affairs at the next scale of analysis (which is usually an ordinally lower scale when the system produces output and an ordinally higher scale when the system receives input). This approach to modeling is continuous with arithmetic formalisms. It favors a symbolic interpretation of a scale of analysis and of the relation among scales, and it begs many questions. Is there a more principled basis for interpreting and relating scales? The strategy pursued in physical science is to link scales of analyses through the symmetries associated with conservations. These symmetries do not change with changes in a system's description that are incurred by analyses at multiple scales. The model for biomechanical organization to be pursued here rests on a methodological commitment to symmetries and conservations as the common thread linking scales of organization in complex (biological) systems.

2.1. Aspects of the Problem of Measurement

2.1.1. Systems of Coordinates (or Mesh-Systems). The description of a thing in motion rests, in part, on the ability to locate points, and to measure distances and directions. Consider a point in space. To define its position is to refer it to some system of reference. Only relative to such a system can points be defined. For an unlimited plane the customary system of reference is that advanced by Descartes. It consists of two kinds of straight lines, vertical and horizontal, that enclose equal areas (in the Euclidean sense) to form a network or a *mesh* of equal Euclidean squares. By assigning consecutive whole numbers (indefinitely from zero) to each vertical and horizontal line of the mesh-system it becomes apparent that a point of the plane coincident with an intersection of any two particular lines can be described by numbers assigned to those lines. These numbers define the *coordinates* of the point.

The Cartesian system of coordinates is of limited application. A two-dimensional mesh-system of equal squares, while appropriate to a plane, is inappropriate for any surface that is curved. The Cartesian strategy for creating a mesh-system easily generalizes, however. Two families of intersecting *curves* replace the two families of intersecting straight lines. The curves, like the lines, can be numbered consecutively, thereby providing a mesh-system for unambiguously defining points in the curved space. Broadly speaking there are indefinitely many mesh-systems (Cartesian and others) applicable to plane surfaces and indefinitely many mesh-systems (ordinarily termed Gaussian) applicable to curved surfaces. Mesh-systems of either category are in large measure arbitrary. The distinction between the categories is not completely arbitrary, however; it is determined by the nature of the surfaces. Plane mesh-systems with equal areas enclosed by the meshes cannot play proxy for curvilinear mesh-systems with unequal areas enclosed by the meshes.

A *metric* is a function that associates with any two points (any two pairs of coordinates) a real nonnegative number that represents the distance between them. From one two-dimensional mesh-system to another the appropriate metric or distance function maintains a constant general form, viz.:

$$ds^2 = g_{11}du^2 + 2g_{12}dudv + g_{22}dv^2, \tag{2.1}$$

where ds is the (infinitesimal) distance between any two (infinitesimally close) points A and B, du is the distance between these points on one of the two kinds of lines or curves, and dv is the distance between them on the other (see Figure 2.1). Figure 2.1 suggests that a change in mesh-system incurs a change in the coordinate numbers of points in the surface and in the magnitudes of du and dv. It also indicates that the distance from A to B is a quantity that need not be affected by the mesh-system. Distance is, in fact, a scalar—that is, a magnitude that transcends systems of coordinates. This latter fact allows for an intuitive understanding of the g terms, or *metrical coefficients*, in Equation (2.1). They refer to magnitudes whose role is essentially corrective, specifically, counterbalancing the variations in du and dv (d'Abro, 1950). Obviously the metric coefficients of Equation (2.1) will vary with variations in mesh-system, in common with the variations in coordinates and the du and dv terms. Most importantly, to the extent that variations in du and dv are uneven from place to place in the mesh-system, the values for the g terms will similarly be inconstant; there can be, in short, a *distribution* of g values. Only in the special Euclidean

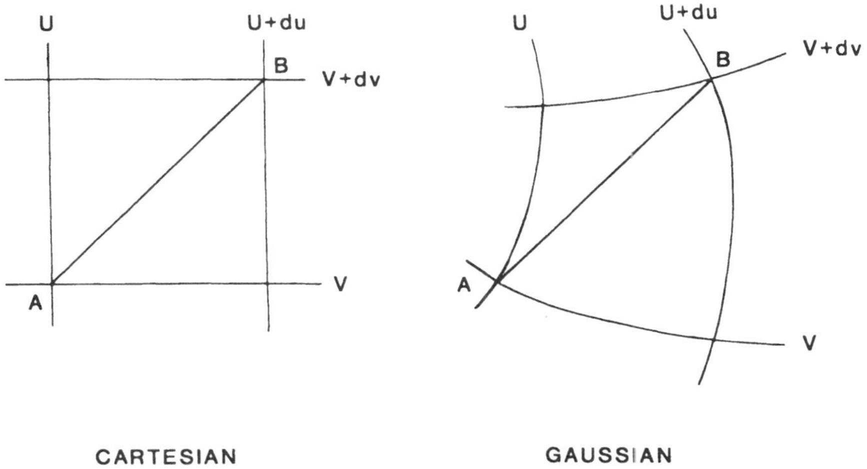

FIGURE 2.1. CONTRASTING PERSPECTIVES ON MESH-SYSTEMS.

(a)—Cartesian mesh

(b)—Gaussian mesh

cases of the Cartesian mesh-system and a mesh-system with diamond-shaped mesh are the g values constant throughout the surface. Because a Cartesian and diamond-shaped mesh-system are among the mesh-systems that can be imposed on a plane surface, it follows that plane surfaces can give rise to a constant *g-distribution*. Conversely, since a Cartesian and diamond-shaped mesh-system cannot be imposed on a curved surface, it follows that curved surfaces can never give rise to a constant *g-distribution*.

For the g values to be constant or variable from place to place in a mesh-system means that the metric, identified by Equation (2.1), is constant or variable from place to place. On the basis of the immediately preceding comments it would therefore appear that the metric of a system of coordinates is shaped by two rather different factors: the system of coordinates itself and the geometry of the surface to which the coordinate system is applied. Whereas the first factor is a matter of convention, the second factor has to do with *properties intrinsic to the situation*. Consequently, whereas the first factor is time independent—a mesh-system, once chosen, persists until replaced by another choice—the second factor can be time dependent—in principle, the intrinsic properties could alter over time, systematically or otherwise. In sum, a metric can vary from place to place (as a function of spatial variations in the intrinsic curvature of the surface), and it can vary from time to time (as a function of temporal variations in the intrinsic curvature of the surface).

2.1.2. Physically Interpreting Metrical Coefficients. The points just made, though mathematical in their origin, are not restricted to that domain. Their extension to and fusion with physical concerns are celebrated (Einstein, 1917/1920) and guide much of the discussion that follows. Of special and immediate importance is the physical interpretation given to *intrinsic curvature*. To be brief: Cartesian mesh-systems correspond to the physical notion of Galilean (constant velocity) frames of reference; Gaussian mesh-systems correspond to the physical notion of non-Galilean (accelerated) frames of reference. These correspondencies come about as follows.

In a Galilean frame, inertial force is absent (at most) or of constant value throughout (at the very least). Because force can be expressed often in terms of a potential (see Chapter 3), a Galilean frame is one in which the potential is homogeneous throughout, whether zero or of some definite value. A Galilean frame can be likened to a surface in which the ease of moving between any two points in either direction is the same throughout the entire surface. That is, it can be likened to the geometric notion of a surface without any curvature—a plane surface. A plane surface, and now by analogy, a Galilean frame, can be "captured" by a Cartesian mesh-system. Recalling that the *g-magnitudes* in the metric of a Cartesian mesh-system are constant throughout suggests a parallel between them and the physical notion of potential. The parallel is strengthened by a consideration of non-Galilean frames. Inertial forces are present in such frames. For these forces to arise, the potential can be neither nonexistent nor constant throughout the frame. The situation is akin to a surface that is not uniformly flat. There is curvature. Such surfaces, and now by analogy non-Galilean frames, can be "captured" only by Gaussian mesh-systems. And characteristic of Gaussian systems of coordinates is variation from place to place in the metrical coefficients. Thus, inhomogeneities of potential and inconstancies in the g terms go hand in hand. The implication is that the g terms, the metrical coefficients, define potentials and, conversely, that *potentials shape metrics*.

The spatial stratification of a biological system, such as a locomoting animal, will yield a succession of frames of reference or physical settings for events. A detailed discussion of how they might be conceptualized is deferred until Chapter 3, but some preliminary remarks are required here. The open character of biological systems means that each scale will most likely involve a flow of materials. Matter-energy will not be distributed evenly within the reference frame that the scale of analysis defines, although it could approximate closely a uniform distribution. The implication is that reference frames produced by an ideally principled stratification (Iberall, 1977/8) of a biological system will be non-Galilean in the significant sense of exhibiting variation in one or more potentials. Consequently, the mesh-systems that suit them are Gaussian, not Cartesian. Relations among physical quantities, whose measurements are referred to a biological frame of reference, may be obscured, therefore, by ignorance of the "curvature" of the reference space—a problem compounded by the fact that the "curvature" may be time varying rather than fixed (see the foregoing).

2.1.3. Symmetries and Invariants. Whereas a length has meaning without a prescribed mesh-system, a direction does not. In a two-dimensional Cartesian mesh-system, the direction of a given straight-line trajectory is provided by the two components of the two projections of the trajectory on the two straight lines of the Cartesian mesh-system that pass through its origin. The two components give the trajectory's length in addition to its direction. A directed magnitude of the kind just described is, of course, a vector.

A change in the mesh-system (for example, rotating or stretching the mesh) results in a change in the vector components. How does one evaluate whether it is the same vector or not? Ordinarily, one does so by identifying that mathematical rule that transformed the original mesh-system into the new one and determining whether its application to the original vector components yields the new vector components. If the components satisfy the mathematical rule, then the vector per se is seen to be independent of the mesh-system, unlike the numerical values of its components. One is inclined to refer to the vector as an *objective* magnitude and its components as *relative* magnitudes. It remains true, nevertheless, that where a vector is measured in reference to a particular mesh-system it is always the vector's components that are measured, not the vector itself.

Transformations of the mesh-system that do not affect the directed magnitude—that is, those that leave it invariant—capture the *symmetry* of the quantity. Quantities can be symmetrical (they can possess symmetry) in the way that geometric configurations are symmetrical. A set of points in a two-dimensional Cartesian mesh-system that is unaffected by integer multiple rotations of $\pi/2$ and by reflections about vertical, horizontal and diagonal axes (but which *is* affected by other transformations) constitutes a square. The reflections and rotations are symmetry operations and they define, in combination, the square's symmetry.

A square is a relation among points and it is this relation that has (rotational and reflectional) symmetry. Let a relation be defined between two vectors in a two-dimensional mesh-system; for simplicity, let both be situated at the same point and be equal to each other. There is, therefore, a vector equation which can be expressed through the respective equalities of the two components of the first vector with the two components of the second vector:

$$X_1 = X_2 \quad ; \quad Y_1 = Y_2. \tag{2.2}$$

With a transformation of the mesh-system, the *Xs* and *Ys* will assume different numerical values: X_1 becomes X_1' and X_2 becomes X_2', and so on. Obviously, for the relation between the vectors (the vector equation) to subsist through the transformation, the vector components of both vectors must change in exactly the same way—that is, they must covary. It is a small leap, therefore, to the understanding that the symmetry of a vector equation is defined by those transformations (symmetry operations) for which the equation is *covariant*. The symmetry is given by the group of transformations over which the components of the vectors change but over which the form of the equation relating the vectors does not change.

A vector is an instance of a more general mathematical magnitude commonly referred to as a tensor. Whereas a vector possesses only as many components as there are coordinate dimensions, a tensor's components can exceed the number of coordinate dimensions. And whereas a vector can be represented by an arrow, a tensor is more difficult to image, corresponding physically to the state of compression or tension at each point in a distorted elastic medium. It is true, nevertheless, that the notion of covariance applies equally to vector and tensor equations. Tensor equations (like vector equations) possess symmetry.

At any scale of analysis, the form of a relation deserving of the label "law" must be impervious to the various physical frames of reference and corresponding mesh-systems that characterize the scale. Only vector and tensor equations can satisfy this requirement. This is not to say that all vector and tensor equations will preserve their form across all relevant frames of reference; rather, it is to say that only equations linking vectors or tensors can be candidates for objective (mesh-systems free) status. Not surprisingly, the customary form of the laws of nature is that of vector or tensor equations.

In practice, there are two approaches to physical symmetries (cf. Rosen, 1978). One approach begins with a group of symmetry operations and seeks to isolate the associated invariant quantity or the covariant vector or tensor equation. The other approach begins with the invariant quantity or covariant equation, usually a conservation (see following) or a law, respectively, and proceeds to reconstruct the symmetry operations (more precisely, the different frames of reference) to which it belongs. With respect to a biological system like a locomoting animal, the frames of reference to which measures are referred are largely ill defined. But if it is known that the measurements are of lawfully related quantities, then a commitment to the lawful relation promises an understanding of the frames of reference. A law can function as a benchmark for assessing the properties of a physical frame of reference: Deviations of the components (of vectors or tensors) from the ideal values that they would assume in a Galilean frame of reference point to properties of the non-Galilean frame of reference that actually houses the law.

2.1.4. Conservations as Invariant Quantities. Conservation principles in physics refer to the fact that certain measurable quantities (for example, electric charge, mass-energy, and linear and angular momentum) remain constant over changes in frames of reference when defined within an isolated physical system. All conservation principles are based on observed geometric symmetries. In macro physics, the physics of bulk matter, the conservation principles are observed in and reducible to geometric symmetries of space and time. In a closed system, for example, energy is invariant over temporal change; to be more exact, it is invariant over observations of the system on different occasions. In microscopic physics, the physics of phenomena at the atomic scale and smaller, the same symmetry-based conservation principles apply, with additional temporal asymmetries playing a role; they account for the transport processes and propagations of such quantities as energy and electric charge. In the theory of subatomic particles a mathematical description is given of systems undergoing rotations, reflections, and translations, not only in a physical space-time reference frame but also in reference frames of a more abstract nature. At the subatomic scale, the application of symmetry operations leads not only to a classification of the particles, and of their interactions with each other, but to a number of new conserved quantities that have no macroscopic counterparts.

Table 2.1 identifies some of the previously mentioned symmetries and their related conservations. The first four conservations are termed *continuous* because the changes can be infinitesimal; a finite change can be made increment by increment. The resulting constants that relate to the motion are classical quantities and are additive. The other symmetries and conservations, unlike the classical ones, are *discrete*: the changes cannot be arbitrarily small.

2.1.5. Natural Units and Intrinsic Measurement Structures. The identification of symmetries in physical events is the basic strategy for defining natural laws. This is so because any natural law amounts to a (continuous) statement about one or more conservations. Fundamental to this strategy is the recognition of the symmetries associated with the frame of reference in which the physical event is measured and, in consequence, the mesh-system used to represent it. These symmetries should be identified prior to the measurement of the physical event. If this requirement is not met, then symmetries unique to the physical event will be confounded by symmetries unique to the reference frame. (To fix

TABLE 2.1
Symmetry Operations and Conservations

Symmetry Operations	*Conservations*
Form of laws invariant under translations of origin	Linear momentum, p
Form of laws invariant under rotations (orientations of axes) in space	Angular momentum, L
Form of laws invariant under translations in time	Energy, E
Form of laws invariant under Lorentz transformation of special relativity (space–time is isotropic)	Velocity v of the center of mass of a system
Form of laws invariant over similar particles	Indistinguishability of similar particles
Form of laws invariant over certain operations on the quantum mechanical wave function	Electric charge, Q

this idea, a square's symmetry is identical to that of a Cartesian mesh-system of equal squares but different from that of a polar mesh-system of equally distanced concentric circles; conversely, a circle's symmetry is different from that of the Cartesian mesh-system but identical to that of the polar mesh-system.) Furthermore, accuracy in the determination of the symmetries associated with physical events is strictly conditioned by the constancy of the symmetries defining the reference frame/mesh system. Constancy of the reference frame/mesh system is conveniently guaranteed by assuming that it approximates an ideal mathematical space, defined by a fixed set of symmetry operations. A system of coordinates defined independently of the physical events being measured constitutes an *extrinsic system of measurement*. The price paid for the use of an extrinsic system of measurement is relativity, that is, law statements that fail to hold across all frames of reference.

As foreshadowed in Section 2.1.2, Einstein's theory of general relativity avoided the problems of Galilean and special relativity by defining an intrinsic measurement structure for space, time, and matter. In general relativity the metrical character (curvature) of a geometric (four-dimensional space-time) continuum is defined at every point by the matter at that point and the state of motion of that matter. In a consistent (general) relativity, inertial mass has no reference to absolute space; instead, inertial masses are defined relative to one another. The physical geometry defined by the distribution of masses determines the inertia of the system and identifies a mesh system in which physical laws are invariant. Space and time are defined strictly in terms of *relations* among physical objects. Changes in the relative positions of masses give rise to changes in the defining geometry. Matter curves geometry (space-time), and curved geometry (space-time) moves matter: Matter and space define a (canonical) conjugate pair of mutually dependent variables. (The term *canonical conjugate* originates in Hamiltonian mechanics. The coordinate q, an extensive quantity, and momentum p, an intensive quantity, were described by Hamilton as canonically conjugate quantities. In general relativity, matter-energy is the intensive quantity conjugate with the extensive quantity of space-time.)

The general theory of relativity does not merely replace the Euclidean space of traditional science, or the Euclidean space-time of the special theory, with Riemannian space-time. Rather, it reveals that all such systems of relations are relative, as they have their (geometric) basis defined solely by the relation of the absolute masses and motions to the particular mass that has been arbitrarily chosen as a reference body for measuring. The general theory of relativity has a more thoroughgoing physical application to the problem of measurement than is customarily assumed. General relativity is a theory addressing the general problem of coordinating the measurements of physical systems for the purpose of observing the invariance of laws. It states that *the geometry of the coordinate space used in the measurement system must be defined in terms of the subject matter under inquiry regardless of the nature of the subject matter*. Einstein was concerned with mechanics, and its subject matter is (macro) masses and their motions. The mathematical expressions used in general relativity are shaped by the distribution of (macro) masses.

In an *intrinsic* measurement system the geometric structure of the measurement space is dictated by the physical properties of the reference frame, and not vice versa. Physical systems whose reference frames and motions relate as conjugates can have no single, absolute measurement structure in terms of which all of the behaviors of all such systems can be analyzed. That is, the classical reductionist view of a single, ultimate measurement structure (defined by a fixed set of units, particles, interactions, and primitive variables) is inappropriate for systems whose motions relate conjugately to the coordinate space. For "conjugate" systems, the classical reductionistic view of measurement must be replaced by a view of "natural" measurement units, that is, units defined intrinsically for each system of interest (cf. Rosen, 1978). While the mathematical tools necessary for intrinsic measurement are more complicated than those required for extrinsic measurement, the physical assumptions (and the physical explanations to which they ultimately lead) are simpler.

The term "natural system" is applied here to identify those systems in which intrinsic geometries are used to define measurement units. "Natural" identifies a class of systems whose motions relate conjugately to a coordinate space; it refers to an analytic relation defined by the conjugacy property. "Natural" is not to be confused with "real" or "physical" as qualifiers of systems. "Natural" identifies a general class of systems whose motions, qua *generalized actions*, are *not* independent of their coordinate spaces, qua *generalized context*. All measurement analysis must entertain some a priori commitment (explicit or implicit) to the type of relationship that exists between the measurement coordinate space and the motion of the system. The appropriate strategy will depend on whether a conjugate or nonconjugate relationship is defined between the behavioral actions of a system and the context in which the actions are embedded.

2.2. Aspects of the Problem of Modeling

2.2.1. Modeling and Induction. Modeling complements measurement; it explores a system's behavior by abstracting a set of observed (measured) properties from a subset population of events and using them to predict the existence of similar properties for a more general population of events. The technical term for the *abstraction* and *generalization* of properties from subsets to supersets is "induction." Models are inductive statements about observable properties. In physical science the properties of interest are generally relations (laws) defined over a set of fundamental units (mass, length, and time). The relations can be represented abstractly using a set of symbols and mathematical operations. In essence the abstracted properties are (ideally) lawful relations or symmetries that relate token members of the observed subset of events. These symmetries are used to generate statements—which constitute the model—about the *type* of which the *tokens* are members. Models generally predict linear continuities (symmetry-preserving transformations) or sudden, nonlinear discontinuities (symmetry-breaking transformations) for a superset of events.

The accuracy of the model is tested by using the symmetries to predict *type* properties about *token* members that are outside the originally observed token set. The success or failure of the model centers in large part on the generalizability of locally observed symmetries. The observed symmetries are matched continuously with the symmetries predicted by the model construct in accordance with some (a priori) nominated criteria for "goodness of fit." Modifications in the model are made by changing the abstract set of properties so as to improve "goodness of fit."

A necessary though not sufficient condition for all physical models is the inclusion of conservational symmetries. Conservational symmetry is an a priori model criterion that is independent of the subset of physical events under inquiry. It is the hallmark of physical models. Physical models start with a conservation assumption and proceed to identify the necessary boundary conditions that are required to yield predictions that concur with observed events. Put succinctly, physical models start with a commitment to the existence of conservational symmetries and proceed to converge on uniqueness through the addition of boundary conditions.

2.2.2. Extrinsic and Intrinsic Perspectives on Physical Modeling. Physical models are like measurement structures; they can be classified according to the extrinsic/intrinsic distinction. An extrinsic physical model is based on the following general assumption: An account (e.g., a set of laws) of physical events can be identified that has universal applicability. The account does not depend on the frame or context to which the events are referred. Extrinsic physical modeling is central to the formalisms of classical mechanics used to represent the behaviors of dynamical systems. This modeling perspective, however, has proven to be less than ideal for problems that involve the coupling of noninertial frames.

An intrinsic physical model, in contrast, is based on the following general assumption: An account (e.g., a set of laws) of physical events must contain a statement of the coupling relations that link the laws to the reference frames or contexts in which the events reside. An intrinsic physical model rejects the assumption that there are universal "units of analysis" or "geometries" that apply to all systems being modeled. Instead, it assumes that each system defines uniquely its own set of "natural" units and interactions (see Rosen, 1978, for an informed discussion of this issue). There are no a priori specifications for the reference frame nor, therefore, for the coordinate space geometry, to which the events are referred. Similarly, there are no a priori specifications for the fundamental units.

According to the intrinsic perspective on physical models, nothing is fundamental with respect to anything else. Rather, everything is fundamental. Things differ from one another only in terms of their symmetry characteristics. Consequently, the coordination of different frames of reference is realized through intrinsically defined field symmetries. Exemplary of this method of coordinating distinct frames is Einstein's general relativity. Noninertial frames are coupled continuously using intrinsic symmetries defined over matter distributions. In general relativity both the inertial properties and the gravitational properties derive their metrics from the distributional symmetries of matter fields.

2.2.3. Gödel's Proof and Turing's Machine: Modeling Limitations of Arithmetic Formalisms. Computer-like models are popular instantiations of the trajectory-cum-equations-of-state motion descriptions of conventional physical theory. The most general computer model, in a logical and mathematical sense, is the Turing machine. This machine was originated by Turing to clarify, in part, the implications of Gödel's incompleteness proof. Gödel's proof had shown that in any logical formalism complicated enough to include arithmetic, there were true theorems which could not be proved, and that if a proof could be formulated within the formalism then the formalism was inconsistent! Turing's machine was in effect a "universal" machine capable of processing data in a manner simulating the Gödelian proof process in any given formal system. The universal machine, given a properly encoded description of an arbitrary machine, and an encoded version of the data on which the latter was to operate, would produce an encoded version of what the arbitrary machine would produce. The universal machine could simulate any arbitrary machine processing arbitrary data.

In addition to the theoretical issue of incompleteness, Gödel's proof implied a number of practical limitations on the use of a Turing machine as a generalized model construct. The most important of these is the algorithmic unsolvability of the halting problem. A Turing machine, given data to process, will either eventually halt, giving a "result" or "decision," or (neglecting wearout or power failure) run on forever. The proof states that it is impossible to build a Turing machine that, given an encoded description of any arbitrary machine and its input, will always decide correctly whether that machine will eventually halt after processing its input data. Real computers, of course, differ from Turing machines. For example, they are subject to practical and economical limitations. Nevertheless, the limitations of the theoretical Turing machine form a subset of the limitations of real computing machines.

The existence of undecidable questions and the corresponding impossibility of constructing some machines (e.g., those that have halting solutions) are limitations on the Turing construct that derive from the arithmetic formalism.

The generalizability of these limitations from the model construct to the system being modeled will depend ultimately upon whether the assumptions that are fundamental to the arithmetic formalism can be carried over validly to the system being modeled. At all events, if the Turing machine is to be used as a universal model construct, capable of modeling any set of physical phenomena, then the question must be raised as to whether physical systems suffer from the same limitations: Are the problems of a Turing machine unique to the arithmetic formalism or are these problems universal and realizable in all physical processes? The strong "explanatory" version of the Turing construct rests on the assumption that the arithmetic formalism maps isomorphically onto the set of physical events. In the weak version of the construct, the relationship between arithmetic and physical events is not an identity. The Turing construct reduces to a means or a method for "simulating" physical events through logical extensions of continuous (analytic) functions; that is, the model is merely a *description* of the physical events in terms of (i.e., in the language of) analytic functions.

2.2.4. Token Physicalism/Turing Reductionism. Many duties can be and are assigned to the strong explanatory version of the Turing construct. Perhaps the most notable duty is that of transforming logical formulae into causes. Phenomena of biological systems, especially those customarily studied by psychologists and to a lesser (but nevertheless significant) degree by physiologists, do not readily reduce to physical accounts. It is difficult (even infeasible, some are inclined to believe) to refer the predicates of psychological and abstract (see following) physiological explanations to physical processes. The standard version of type physicalism construes the physical types to which psychological and physiological types are to be reduced as orthodox physical properties. In the standard form type physicalism has not fared well, giving the impression that it may not fare well in any form (Fodor, 1975). Token physicalism, which is less demanding, has fared better. It merely claims that any particular psychological or abstract physiological state is a token physical state. Significant among the implications of token physicalism is the indifference of psychological and abstract physiological states to the materiality of the system that expresses them (Fodor, 1981a,b). To hasten this discussion, think of psychological states and physiological states (like those governing the locomotory step cycle depicted in Figure 1.1) as software in the sense that computer science uses that term. Then it is roughly apparent that noncarbon-based (e.g., silicon) systems could share psychological and abstract physiological states with carbon-based (that is, biological) systems. The difference in materiality is not a difference that counts for very much. Looking at Figure 1.1, the point to be made is that the software description of the step cycle does not *logically* require neurons and the physico-chemical field processes that they mediate.

The foregoing illustrates nicely how token physicalism is interpreted by the Turing construct. To summarize: According to Turing reductionism, psychological states and abstract physiological states (such as those states depicted in Figure 1.1, that are meant to steer macroscopic mechanical processes and microscopic physiological processes) can be expressed in terms of the program states of Turing machines. According to token physicalism, psychological states and abstract physiological states are physical states. They are, therefore, potentially causal. What Turing reductionism putatively shows is how it is that these states can be causally efficacious despite their physically heterodox nature. It shows that they can be expressed as the states of a (particular kind of) physical machine.

There is a question of some importance to be asked: Is the causal status ascribed to logical formulae by token physicalism/Turing reductionism *real*? Causal realism holds that causation is a category of connections and determinations corresponding to the actual traits of the physical world. The causes ascribed by token physicalism/Turing reductionism are more nearly Humean, that is, categories of relations among ideas. There is a closeness between token physicalism/Turing reductionism and the tradition of logical positivism that merits consideration.

As an orientation to scientific explanation, logical positivism rejected causal realism. It viewed the scope of science as limited to the collection of data and the linking of these data together in a formally closed system. In the logical positivists' scheme of things it is logic that provides the connections and it is logic that enables the scientist to make predictions about new data. The ultimate antecedents in the explanatory systems advocated by logical positivists are the logical predicates postulated in advance by the scientist. Conceived as abstract logical entities, independent of observables, these "causal" antecedents need not refer to anything physical.

Token physicalism/Turing reductionism's attitude toward scientific explanation continues the tradition of logical positivism in its emphasis upon explaining phenomena through logical formulae that are connected logically in a formally closed system. It is, at the same time, discontinuous with that tradition in that it ascribes ontological status to the logical formulae. The Turing construct allows that if a set of events (say, those composing the locomotory step cycle) can be expressed as logical formulae, then those events can be simulated by a Turing machine. Referring again to Figure 1.1, the software both "explains" the phenomenon of the step cycle and is physically realizable. It is in this sense that token physicalism/Turing reductionism goes beyond logical positivism: It aspires to raising the status of explanation from merely logical to real.

There are different kinds of realisms. Token physicalism/Turing reductionism is an instance of representational realism. The antecedents of a phenomenon to which causation is ascribed by token physicalism/Turing reductionism are representations or logical descriptions of the actual physical states of affairs antecedent to the phenomenon. The style of scientific explanation advocated by token physicalism/Turing reductionism, therefore, is not truly a causal realism.

This fact has unwelcome implications with regard to the issues of the uniqueness and testability of explanation. One of the major criticisms of the original positivistic orientation to explanation was that it did not promise uniqueness: Whereas physical events are said to follow uniquely from their causes, internally consistent, logical descriptions of the causal process are multiple. The same problem that confronted the positivist confronts the advocate of explanation based on token physicalism/Turing reductionism: How does one get from the existence of multiple (logical) descriptions to a unique (causal) description? Dressing up logical formulae in machine instantiable programs does not resolve the uniqueness problem. Many programs can give rise to the same sequence of machine outputs (Marr, 1977).

The testability problem reads as follows: If (a) all logically specifiable events can be simulated by a Turing machine and (b) all events are logically specifiable, then (c) there are no events to be considered that cannot be simulated, by definition. Under token physicalism/Turing reductionism, the notion of testability is limited to whether or not the scientist can logically specify the set of events. The test is of the scientist, not of the model construct.

There is one further consideration that makes the token physicalism/Turing reductionism perspective on explanation suspect. Natural systems stratify into multiple levels of embedded structures or grains of analyses. The levels are not logically independent. For any two "adjacent" levels, the more microscopic level defines the coordinate space for the events at the more macroscopic level (see Section 3.1 that follows). From the intrinsic measurement stance, the measurement space provided by the more microscopic level is defined by the *physical properties of that level*. This means that the defining of events at the more macroscopic level must depend significantly on the materiality of the underlying microstructure, a conclusion that is potentially counter to the cornerstone of token physicalism/Turing reductionism. Micro-levels penetrate macro-levels and macro-level properties materially implicate micro-level processes. (Examples will be forthcoming.) The larger point to be made, by way of a concluding remark, is that type realizations of psychological and abstract physiological states should not be ruled out. Realizations in terms of some physical properties may fail (as in biochemistry, for example). But the type of realization that psychological and abstract physiological states may have in biological systems may be describable in more functional higher-order ways. The crucial feature of types is that what counts is always relative to a description (Searle, 1983).

2.2.5. Physical Theory and Conservational Closure: Model Constructs Based on a "Completeness" Assumption. The critical assumption upon which physical theory rests is not arithmetic in its origin. The assumption is drawn from observations about physical states of affairs. It constitutes a metastatement about physical states and relations, not arithmetic ones, and, in this regard, the explanatory power of the model is with reference to physical properties. In contrast, the explanatory power of the Turing construct is with reference to arithmetic properties. The foundation for all physical theories rests fundamentally on the completeness assumption concerning *conservational closure in isolated physical systems*. This is a physical assumption about the invariance of conservations under geometric transformations in space and time. This assumption provides an intrinsic completeness requirement for the modeling of physical events (cf. Benacerraf & Putnam, 1964; Suppes, 1969a,b).

The physical completeness required of conservational systems stands in sharp contrast to the formal incompleteness characteristic of arithmetic systems. In physical models the formalism of arithmetic is used as a tool in the analysis. It is not used to derive assumptions about the metastructure of the model construct. The mathematization of a physical model is not to be confused with the complete reduction of the physical model to an arithmetic model that includes all of the formalism's intrinsic characteristics. Arithmetic is the servant of the physical assumptions and constraints, and not vice versa.

The physical theories of quantum mechanics, classical (and relativity) mechanics, statistical mechanics, and thermodynamics are all examples of model constructs formed using the assumptions of intrinsic conservational completeness. These theoretical constructs all rest on the assumption that symmetry transformations can be derived that invariantly map conservations through space and time and across multiply embedded frames of reference. The assumption of conservational closure in isolated systems can be extended to open systems if the states of interest are at equilibrium (no entropy production, $P[S] = 0$) or steady-state (entropy production, $P[S] > 0$). In these states a conservational closure can be defined on that portion of the conservations that is being transported into and out of the system. The "art" of physical modeling is the derivation of sets of transformations that satisfy the assumption of conservational closure.

Some of the strategies associated with the theories of thermodynamics and statistical mechanics are outlined in Chapter 3. The focus is on identifying the book-keeping strategies that map the conservations across macro and micro descriptions of field processes. While the strategies can be mathematized, the role of mathematization is pragmatic, not ontological. Even if a physical set of strategies can be modeled using the formal tools of arithmetic, it can not be concluded that the ontological nature of the modeled physical events is to be "explained" in terms (i.e., in the language and connecting rules) of arithmetic. This conclusion is an abuse of the role of mathematics in physical modeling. The strength of physical models is in their minimal set of physical assumptions (such as conservational invariance), not in the mathematics used to instantiate these assumptions. Physical models are based on a physical assumption of completeness (conservational invariance under transformations) and this completeness is not jeopardized by the incompleteness arguments associated with arithmetic. The physical nature of events under inquiry is not to be confused with the computational nature of the arithmetic used to model the physical assumptions. The formal, mathematized model is *not* the ontological physical model, and physical events are *not* computing (in the arithmetic sense) state transitions.

REFERENCES

Benacerraf, P., & Putnam, H. (1964). *Philosophy of mathematics: Selected readings*. Englewood, NJ: Prentice-Hall.

D'Abro, A. (1950). *The evolution of scientific thought from Newton to Einstein*. New York: Dover.

Einstein, A. (1917). *Relativity, the special and general theory*. Braunschweig: Vieweg, 1917; New York: Holt, translated from the German edition by R. W. Lawson, 1920.

Fodor, J. A. (1975). *The language of thought*. New York: Thomas Y. Crowell.

Fodor, J. A. (1981a). The mind-body problem. *Scientific American*, *244*, 114–123.

Fodor, J. A. (1981b). *Representations*. Cambridge: MIT Press.

Iberall, A. S. (1977/1978). A field and circuit thermodynamics for integrative physiology. I, II, and III. *American Journal of Physiology: Regulatory, Integrative and Comparative Physiology*, *2*, R171–R180, 1977; *3*, R3–R19, 1978; *3*, R85–R97, 1978.

Marr, D. (1977). Artifical intelligence—a personal view. *Artificial Intelligence*, *9*, 37–48.

Rosen, R. (1978). *Fundamentals of measurement and representation of natural systems*. New York: North-Holland.

Searle, J. R. (1983). *Intentionality*. New York: Cambridge University Press.

Suppes, P. (1969a). Models on data. *Studies in the methodology and foundations of science*. New York: Humanities Press.

Suppes, P. (1969b). A comparison of the meaning and uses of models in mathematics and the empirical sciences. *Studies in the methodology and foundations of science*. New York: Humanities Press.

3 WHY THINGS FLOW: AN OUTLINE OF PHYSICAL THEORY FOR SIMPLE ATOMISMS

ABSTRACT

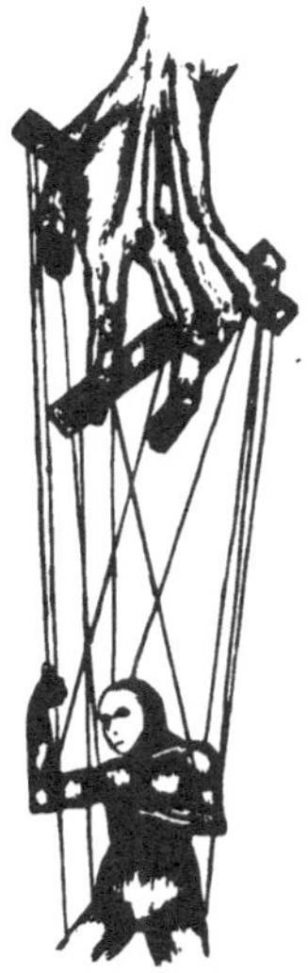

The purpose of this chapter is to sketch the physical nature of flow processes involving atomisms that interact through mass-dominated conservations (i.e., energy and momentum). These flow processes identify global tendencies toward uniform distributions of energy and momentum. Once the equipartitioning of conservations is achieved, the field is globally stable, and all flow processes disappear. This state is defined as equilibrium. A concentration of a conserved entity in any portion of a field identifies a potential, that is, a local storage depot. Potentials are assembled and disassembled by changing the distributional characteristics—the geometrical "layout" of the conservation. A layout or distribution of a conservation constitutes a thermodynamic field. In closed systems, flow processes identify global tendencies toward the stable, equipartitioned state. The second law of thermodynamics associates these tendencies with the "breaking of constraints"—the annihilation of organizational states. In open systems, however, flow processes can develop that are associated with tendencies toward nonequipartitioned states. The result is stable states exhibiting macroscopic distributional inhomogeneities. Under these conditions new macroscopic cooperative units can emerge. The breaking of constraints is referred to as entropy production, and the making of constraints is called negentropy production.

The second law of thermodynamics identifies the fact that constraints, regardless of their material nature, have a limited "lifetime." One interpretation of the second law is the prediction that the number of constraints in the universe decreases with time, ultimately converging on an equilibrium state, defined by no constraints and exhibiting no flows (e.g., "heat death"). The causal dynamics associated with the second law, however, can lead to a dramatically different interpretation if the system of interest is sufficiently complex to define a stratified (macro/micro) field structure. Under these conditions the role of the second law in dictating organizational constraints can vary, depending on which level of organization (macro or micro) is observed.

The physical principles that fashion states within and between upper and lower, macro and micro, frames of reference have to do with the flow of conserved quantities (mass, energy, momentum, etc.). Particular attention is given to the fact that kinetic descriptions (mass-related, e.g., MV^0, MV^1, MV^2*) provide the connections among different levels of a system composed of simple atomisms. In a stratified field the couplings that relate the micro force structures to the macro flow structures can evolve physical nonlinearities that serve as sources for "new" constraints. These constraints* ***restrict*** *or* ***classify*** *both the macroscopic and microscopic trajectories of the system. Brownian motion is typical of how such a nonlinear force/flow structure evolves.*

3.1. Stratified Frames of Reference and Space-Time Metrics

At any spatial scale of analysis a physical system will exhibit a "graininess" that is characterized by a large number of discrete, rather similar or identical entities that can be referred to as "atomisms" (Iberall, 1972). At some larger spatial scale these entities can interact, forming a coherent, continuum-like ensemble or pattern. Whenever the atomistic entities interact, the larger scale of analysis defines a *field*. The stratification into "grains" of field structure provides a physical basis for defining a number of natural coordinate spaces. As remarked in Chapter 2, the ordinally lower scale of analysis is the context or reference frame for the events occurring on the next ordinally higher scale of analysis. The reference frames and corresponding mesh-systems defined accordingly at each scale constitute intrinsic measurement systems for physical events.

There will be principled unit divisions for each frames' dimensions where the size of a unit will selectively determine the detail revealed in analysis. Large unit metrics reveal macroscopic details and small unit metrics reveal microscopic details. The concepts of macroscopic frame of reference and microscopic frame of reference are defined relatively, with respect to one another; they are not defined with reference to any absolute extrinsic criteria. The concepts of mechanical frame of refence and thermodynamic frame of reference can be defined in this manner.

3.1.1. Mechanical Energy and the Macroscopic Frame of Reference. The energy of a particle in a macroscopic inertial frame of reference is invariant and consists of the sum of two terms, the potential energy and the kinetic energy (for reviews, see, Goldstein, 1950; Landau & Lifshitz, 1976; Sommerfeld, 1950). The potential energy is a function of the coordinates only; it is considered to be the stored energy that a particle possesses by virtue of the work that had to be done to bring it to its present position. The kinetic energy is a function of the velocity only and is therefore associated with motion. It is the amount of energy possessed by a particle by virtue of moving at a certain velocity. As a particle moves along a trajectory, potential energy may be converted into kinetic energy and vice versa, but the sum of the two quantities must remain invariant.

The conservation of energy is a physical property that is independent of the frame of reference. With respect to the general class of inertial frames of reference (e.g., frames of constant velocity), it is possible to consider frames of reference that differ in size from macroscopic (continuum) to microscopic (atomistic). In extending the conservation laws to the microscopic level, a difficulty arises concerning the large number of degrees of freedom associated with the atomistic constituents. An ordinary piece of matter contains an enormous number of atoms, so that any detailed accounting of the dynamic variables in terms of mechanics becomes impossible. There are, however, three related methods of analysis used by physicists to address physical problems involving systems with large numbers of degrees of freedom. These methods are thermodynamics, statistical mechanics, and kinetic theory.

Thermodynamics uses only macroscopic parameters and derives certain equations that must hold between experimentally determined quantities; these equations are called phenomenological relations (see Prigogine, 1967; Zemansky, 1937). Thermodynamics requires no detailed knowledge of atomic theory or the mechanism of reactions or processes. Its laws and concepts are independent of any theory of matter or mechanism of interaction. Statistical mechanics considers average (or mean state) properties and probability distributions associated with atomistic ensembles and their interactions (see Tolman, 1938; Gibbs, 1960). These distributions and averages have, in general, been worked out only for systems at equilibrium, so that statistical mechanics has not been very useful in the study of (nonequilibrium) transport properties. Kinetic theory deals with mechanical model systems at a molecular level and the appropriate averaging over the molecules to predict macroscopically measurable properties. While it can be used for transport properties, it suffers from the difficulty that for all real cases the mathematics becomes exceedingly complex and often impossible. Statistical mechanics and kinetic theory have limited application, thus far, outside the equilibrium state. Thermodynamics, however, has had more success in its extension into the nonequilibrium region.

3.1.2. Thermal energy and the microscopic frame of reference. The first law of thermodynamics predicts that energy within an isolated physical system will always remain invariant, and the second law of thermodynamics predicts the failure of the conservation of *mechanical* energy for all physical events. The consistency of the first law with the second law of thermodynamics is maintained by the fact that energy can assume forms other than mechanical. The first law is preserved intact by noting that the missing mechanical energy that is predicted by the second law is appearing as kinetic energy in the microscopic-molecular frame of reference. The microscopic kinetic energy is referred to as thermal energy when viewed from the macroscopic frame of reference. The loss of mechanical energy is always associated with a corresponding increase in thermal energy (e.g., temperature) somewhere in the system. For example, if a car is traveling along a road, it has a certain macroscopic kinetic energy ($MV^2/2$). If on a level surface the brakes are applied and the car comes to a stop, the mechanical energy goes to zero (i.e., the macroscopic kinetic and potential energy disappear). The "lost" energy is to be found in the heating of the brakes. The kinetic energy at the macro-mechanical level has been transformed into thermal energy at the micro-thermodynamical level. Because temperature does not enter into the formalism of mechanics, a new strategy must be introduced that provides a method for conservational bookkeeping in processes involving multiple (macro/micro) frames of reference. Processes that transfer energy from macro to micro frames of reference are referred to as "frictional" or "dissipative." The methods used for the study of these processes constitute the analytic framework of thermodynamics.

The loss of mechanical energy, with the subsequent rise in temperature, was the most important observation in the historical development that led to the distinction between the macroscopic-mechanical and microscopic-thermodynamical frames of reference. The realization of the role played by the microscopic-thermodynamical frame of reference led to the development of techniques for measuring the dynamic properties associated with the microscopic coordinate space. These techniques complemented the tools of classical mechanics. Energy in the macroscopic modes was accessible to analysis and measurement by the methods of classical and relativistic mechanics, while energy in the microscopic modes was accessible to analysis through the methods of thermodynamics.

3.1.3. Thermodynamics and the Equilibrium State. When confronting a physical analysis entailing more than one frame of reference, it is essential that the coordinate space (geometric measurement structure) be defined relative to a common intrinsic metric (see Chapter 2). The space-time symmetries of the equilibrium state provide the metric for thermodynamic systems. Thermodynamic processes are defined in terms of whether they sustain these symmetries (reversible, equilibrium processes) or annihilate these symmetries (irreversible, nonequilibrium processes).

Equilibrium states are defined as those states in which the conserved

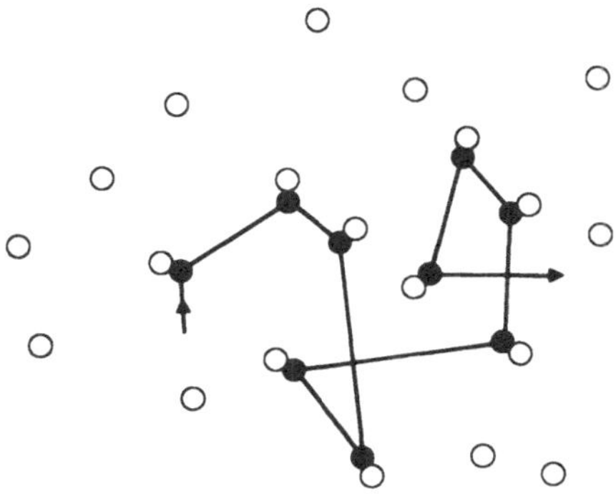

FIGURE 3.1. MEAN FREE PATH AND RELAXATION TIME. The mean free path is the average distance traveled between collisions. The mean relaxation time is the average time elapsed between collisions.

properties are uniformly distributed in space and time relative to boundary conditions. The distribution of these entities within the field forms the physical basis for an intrinsic space-time geometry. The equilibrium state defines a *softly assembled (and maintained), fluid-like, geometric structure*. The fundamental units of space and time are dynamically defined at equilibrium by the mean free path (δ) and mean relaxation time (τ). A free path is the distance traveled by an atom between two consecutive collisions with other atoms, and the average distance between collisions is the mean free path. A few free paths are indicated in Figure 3.1, where one particular atom, shown by the black dot, is followed as it collides with other atoms, shown by open circles. The relaxation time is the time elapsed between two consecutive collisions and the mean relaxation time is the average time elapsed between collisions. The initial measurement of the free paths and relaxation times requires the use of an extrinsic measurement structure (see Figure 3.2). Once the mean free path and mean relaxation time have been determined, their magnitudes and the symmetries of the equilibrium state can be used to define an intrinsic measurement structure. Figure 3.3 identifies an example of an intrinsic coordinate space so defined.

3.2. Physical Processes: Transformations on Conservations

Physical processes refer to transformations on conservations within a system relative to its equilibrium state. Two types of transformations are possible: (a) those in which the form of the conservation is transformed (e.g., potential energy transformed into kinetic energy, electric energy transformed into magnetic energy, etc.), and (b) those in which the frame of reference of the conservational distribution is transformed (e.g., a macro-mechanical frame of reference is

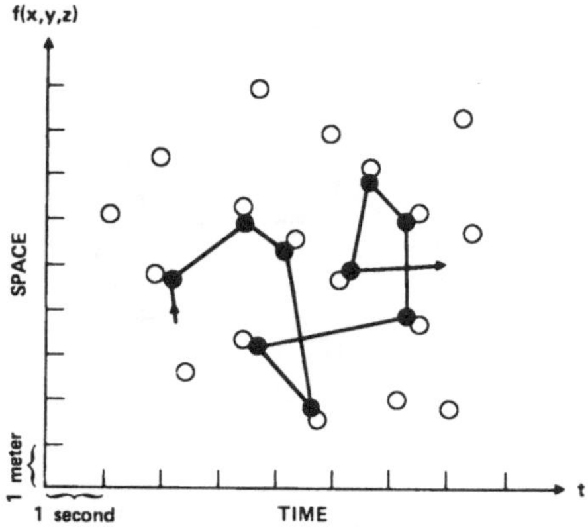

FIGURE 3.2. EXTRINSICALLY DEFINED (COORDINATE SPACE) METRICS. The free paths and relaxation times between collisions are measured using a Cartesian mesh-system. The coordinate space metrics (meters and seconds) are extrinsically defined.

transformed into a micro-thermodynamical frame of reference). Processes that involve transformations in the form of the conservations are referred to as reversible processes and processes involving transformations in the frame of reference are termed irreversible processes (see Figure 3.4).

3.2.1. Reversible Processes, Symmetry-Preserving Transformations, and Single Frames of Reference. Reversible processes refer to those physical transformations (events) that preserve the symmetry of the equilibrium state.

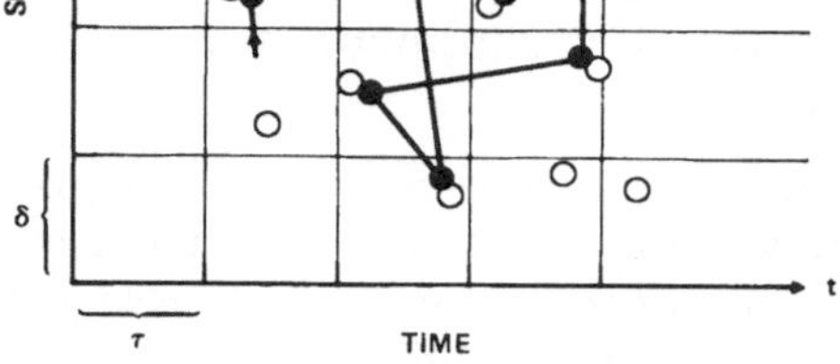

FIGURE 3.3. INTRINSICALLY DEFINED (COORDINATE SPACE) METRICS. The mean free path and mean relaxation time between collisions are used to define the space and time metrics.

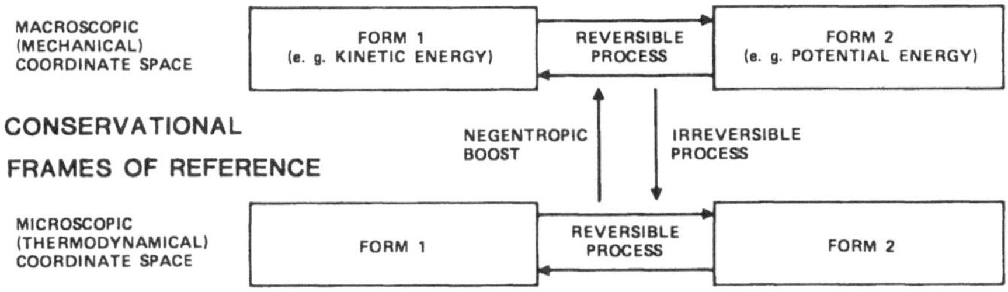

FIGURE 3.4. PHYSICAL PROCESSES: REVERSIBLE AND IRREVERSIBLE TRANSFORMATIONS ON CONSERVATIONS. Energy transformations can be classified according to (a) the [space, time, and matter] scale of the frame of reference in which the energy bookkeeping is measured (e.g., macro or micro), and (b) the "forms" (e.g., kinetic, potential, etc.) that the energy assumes. A reversible process is a symmetric set of transformations in energy forms that occur within a single frame (macro or micro) of reference. Reversible transformations are fundamental to Newtonian mechanics and equilibrium thermodynamics. An irreversible process is an energy transformation that involves changes in the scale of the frame of reference responsible for the energy bookkeeping measurements. Irreversible transformations are fundamental to nonequilibrium thermodynamics.

These symmetry-preserving transformations involve changes only in the form assumed by the conservation. Generally the conserved quantity (energy) is transformed from a space-dependent (potential) form to a time-dependent (kinetic) form. Reversibility identifies the fact that there is no temporal preference in the transformational ordering. Kinetic energy can be transformed into potential energy as readily as potential energy can be transformed into kinetic energy.

If the field boundary of the system is isolated from the flow of conservations, then the space-time dynamics of the equilibrium state (i.e., mean free path and mean relaxation time) define a fixed coordinate space for measuring the conservations associated with the reversible processes. Conservational closure can always be satisfied using the equilibrium frame of reference. For this reason the term *conservative* has been applied to systems exhibiting this characteristic. Fundamentally, the methodology for identifying reversible processes consists of detailing systematically—within a single frame of reference—a closed, symmetric set of oscillations (symmetry-preserving transformations) in the form assumed by the conserved entity. The important feature of a reversible process is that the transformation does not affect the intrinsic equilibrium geometry of the coordinate space.

3.2.2. Irreversible Processes, Symmetry-Changing Transformations, and Multiple Frames of Reference. From a measurement perspective irreversible processes identify symmetry-changing transformations of the equilibrium geometry that defines the coordinate space. Physically, irreversible processes identify flows of conserved quantities from one frame of reference (macroscopic mechanical) into another (microscopic thermal). Whereas reversible physical processes require only a single frame of reference to provide conservational closure, irreversible processes achieve closure only through the use of multiple frames of reference. The label *nonconservative* has been applied to systems involving symmetry-changing transformations in recognition of the failure to satisfy the summational invariance within a single measurement frame. The flow of conservations from one frame of reference to another exhibits a temporal asymmetry that is predicted by the second law of thermodynamics. In isolated physical systems the flow of conservations is irreversibly ordered in the direction of increasingly more micro frames of reference.

3.2.3. Second Law of Thermodynamics and Evolution of Thermal Symmetry. The first law of thermodynamics states that while energy can be transformed into many different forms, the overall sum of energy remains invariant, regardless of the nature of the transformations. The second law states that the possible transformations (interconversions) of energetic forms are subject to certain restrictions. Specifically, the second law applies to the ordering of the transformations of energy forms. Gradients (asymmetries) of temperature, concentration, and mechanical forces (such as pressure) tend to disappear in isolated systems, but the reverse never occurs (i.e., gradients never arise spontaneously). All these tendencies are examples of the second law of thermodynamics: Systems in nature undergo spontaneous changes in the direction of increased microscopic (thermal) symmetry, where entropy serves as the measure of the symmetry. Entropy is a quantitative measure of the number of available degrees of freedom in the micro field. It is an "extensive" parameter that summationally relates the micro constraints of a system to the system's ability to do macroscopic (mechanical) work.

Decreasing the number of system constraints results in (a) a loss in the system's ability to do (mechanical) work (extensive variation—*Δentropy*) and (b) an increase in thermal energy (intensive variation—*Δheat*)[1]. The breaking of

[1]The mathematical difference between extensive and intensive quantities is that where extensive quantities are measures in the measure-theoretic sense, intensive quantities are not (Bunge, 1973). If *a* and *b* are spatially separated bodies, and their physical addition (e.g., juxtaposition, concatenation) is denoted by "$a + b$," then the composit mass (M) is given by the sum of the individual masses, viz.,

$$M(a + b) = M(a) + M(b) .$$

Duration is extensive, like mass; temperature and density are not. Temperature and density, for example, do not satisfy an addition law. Temperature can be added and densities can be added but their resultant does not represent a temperature or a density, respectively, of a composite thing.

In truth, not all extensive quantities are strictly additive, just nearly so (for example, entropy).

constraints is associated with entropy production. The present definition of entropy is in terms of constraints. However, an alternative, and equivalent, definition can be stated in terms of the system's available degrees of freedom. From the perspective of degrees of freedom, the increase of entropy identifies an increase in the system's degrees of freedom.

According to the second law, all constraints have a finite temporal duration ("life time"). Or, stated differently, the second law predicts that configurational constraints for all physical (noncrystalline) systems are time dependent. This leads immediately to the conclusion that the only stable field symmetry (as time approaches infinity) is the microscopic state of thermal symmetry. Entropy is the measure of the "death" of these constraints. An entropic or thermal death is the state of thermal equilibrium.

The concept of entropy is typically associated with a statistical (probability) measure on ensemble fields (see Section 3.2.4). The ensemble size, however, need not be very large when entropy is defined from the perspective of a configurational constraint (see Section 3.2.5). The minimum requirements for a configurational constraint are two particles and a single constraint—where a constraint is defined as a loss of a degree of freedom. According to the second law, the constraint constitutes a temporal as well as a spatial boundary condition. The breaking of the constraint in time results in an increase in the system's entropy. The loss of the constraint results in an increase in the system's (macroscopic or microscopic) degrees of freedom.

Figure 3.5 provides a diagrammatic summary of the relationships linking the concepts of intrinsic geometries (coordinate spaces), ordered frames of reference (microscopic and macroscopic), and physical processes (reversible and irreversible). Trajectories parallel to the space-time axis identify reversible processes, transformations in spatial constraints that have temporal symmetry. The number of configurational constraints remains invariant under transformations. Nonparallel trajectories identify irreversible processes that are associated with the making and breaking of configurational constraints. These transformations have a temporal asymmetry associated with changes in constraints. Trajectories toward the micro frame of reference identify the spreading of the thermal symmetry (i.e., the breaking of constraints) into the macroscopic frame of reference. These are the trajectories predicted by the second law for isolated systems. The trajectories that project from the micro frame to the macro frame are associated with the assembling of constraints associated with open systems. These are trajectories predicted for systems that are evolving toward greater structure.

3.2.4. Thermodynamic Probability and Entropy. The global equilibrium state of a system can be used to define a state from which deviation can be

---

Further, it is the intensive quantity rather than the extensive quantity that is basic. By integration intensive quantities (e.g., density) generate extensive quantities (mass).

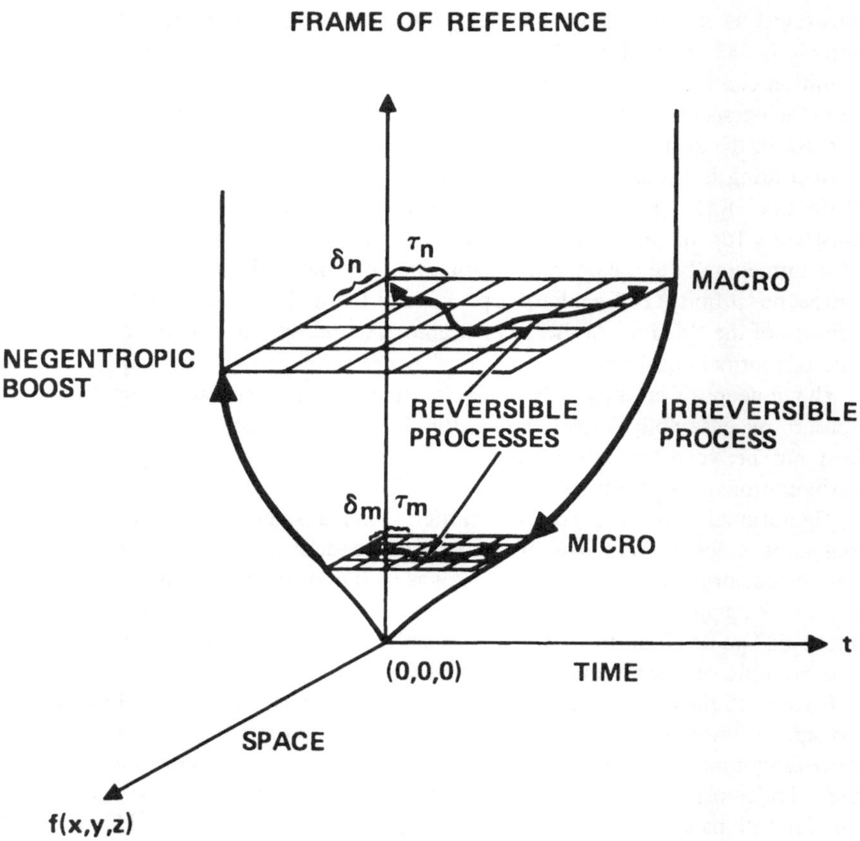

FIGURE 3.5. INTRINSIC GEOMETRIES, FRAMES OF REFERENCE, AND PHYSICAL PROCESSES.

associated with a certain probability. The second law predicts that the most probable state is the equilibrium state and successive departures from the equilibrium state are associated with less probable states. By comparing the probabilities of two states of a thermodynamic system, the direction of a process can be determined. All processes in isolated systems move in the direction of the most probable state—the equilibrium state. The calculation of the thermodynamic probability is a function of the number of different combinations of molecules for some state of the system. Entropy is a physical quantity that can be defined in terms of thermodynamic probabilities. Boltzmann showed that entropy is proportional to the logarithm of the thermodynamic probability

$$S = K \log W \tag{3.1}$$

where S is entropy, W is the thermodynamic probability and K is a proportionality factor ($K = 2.3k$, where k is Boltzmann's constant). If a system goes from one state to another, entropy changes (dS) as a function of variations in the thermodynamic probabilities

$$dS = S_2 - S_1 = K\log W_2 - K\log W_1 = K\log(W_2/W_1) \tag{3.2}$$

where the subscript *1* refers to the first thermodynamic state and the subscript *2* refers to the second thermodynamic state.

Consider a system in a state of global thermodynamic equilibrium. Assume that the state of the system changes as the result of local fluctuational instabilities leading to local nonequilibrium states. Further assume that the change in the state of the system can be characterized by a change in a certain internal parameter ξ, whose value depends on the geometry of the entire conservational distribution. The parameter ξ might identify the relative density of mass, energy, or momentum. In a global equilibrium state the parameter ξ has a value $\xi = \xi_O$, whereas in a local nonequilibrium state the parameter value differs from ξ_O. In a global equilibrium state the density of the conservation ξ_C is uniform over the entire field. As a result of a fluctuation, the system can spontaneously move locally into a nonequilibrium state with a variable density ξ.

The entropy S of the system can be defined as a function of the parameter ξ, so that $S = S(\xi)$. In an equilibrium state $S_O = S(\xi_O)$. The probability (dW) that a system will move into a state characterized by a value of the parameter ξ that lies in the interval between ξ and $\xi + d\xi$ can be found using Boltzmann's formula:

$$dW = \text{const.}\ \exp\left[\frac{S(\xi) - S(\xi_o)}{K}\, d\xi\right] = \text{const.}\ \exp\left[\frac{dS}{K}\, d\xi\right], \tag{3.3}$$

where the constant (const.) is determined by the normalization condition (the requirement that the sum of the probabilities equals *1*, $\Sigma dW(\xi) = 1$). (Strictly speaking, the constant also depends on the parameter ξ. It can be shown, however, that for a system containing a sufficiently large number of particles the dependence on ξ of the factor that stands in front of the exponential is insignificant in comparison with the dependence on the exponential.) The value of the change in the entropy is negative.

The entropy of a system has several characteristics that should be noted. The first is that the entropy of a system is determined completely by the current state of the system and does not depend on the nature of the process by which the system reached that state. The reason for this is that (a) each state of the system is determined by a certain number of combinations of molecular states; (b) each state corresponds to a definite thermodynamic probability; and (c) entropy is a function of the thermodynamic probability. The summational characteristic of entropy over the entire field identifies it as a state variable and, in this regard, it is similar to energy (see Sections 3.3.3–3.3.5 for its similarity to potential energy as a variational function). The second characteristic is that the total entropy of a system consisting of several independent parts is equal to the sum of the entropies of these parts. The final characteristic is that entropy always *increases* in irreversible processes, such as diffusion or the conversion of mechanical energy into internal energy.

3.2.5. Entropy and Constraints. Stated in the language of constraints, entropy is a measure of the number of constraints that couple the atomisms of a field. Entropy production ($P[S]$) is a measure of the *changes* in atomistic degrees of freedom due to the making or breaking of organizational constraints (i.e., couplings). The second law states that all organizational constraints are time dependent (on some scale of analysis) and that the breaking of constraints is associated with changes in the atomistic (internal) degrees of freedom. The second law predicts the increase of atomistic degrees of freedom with time.

The number of degrees of freedom in a system is equal to the number of atomisms multiplied by the dimension of the space in which the motion of the atomisms occurs, minus the number of constraints that couple the atomisms. The equation for determining the atomistic (internal) degrees of freedom for the continuum is:

$$F_{cont} = N(D) - C \tag{3.4}$$

where F_{cont} is the degrees of freedom, N is the number of atomisms, D is the dimension of the coordinate space in which the atomistic motion is occurring, and C is the number of constraints that couple the atomisms. For example, a single atomism in a three-dimensional space has three degrees of freedom. If there are two atomisms in the same space, there are six degrees of freedom. If a constraint couples the two atomisms, then there are five degrees of freedom for the composite, continuum system (composed of the two atomisms and the constraint).

The entropy of a system can be defined in terms of the difference between the isolated atomistic degrees of freedom and the internal (atomistic) degrees of freedom of the continuum. Let $(D_i)_{atom}$ denote the degrees of freedom associated with the ith atomism and $(D_1, D_2, \ldots)_{cont}$ denote the set of atomistic degrees of freedom associated with the continuum. If all of the atomistic degrees of freedom are independent of one another then the continuum's degrees of freedom are equal to the sum of the atomistic degrees of freedom:

Continuum composed of completely independent atomisms

$$(D_1, D_2, \ldots)_{cont} = (D_1 + D_2 + \ldots)_{atom} \tag{3.5}$$

If there are constraints that couple the motions of atomisms, then the continuum degrees of freedom are fewer than the independent atomistic degrees of freedom:

Continuum composed of coupled atomisms

$$(D_1, D_2, \ldots)_{cont} < (D_1 + D_2 + \ldots)_{atom} \tag{3.6}$$

The entropy production *dS* of a system is a measure of the gain of atomistic degrees of freedom. As noted in Section 3.2.4, the entropy production of a system can be associated with a change in probabilities associated with potentially available states uniquely composed of atomistic degrees of freedom. Entropy can be viewed, therefore, as a function *S* of the available atomistic degrees of freedom:

$$dS = S(D_1, D_2, \ldots)_{atom} - S(D_1, D_2, \ldots)_{cont} \tag{3.7}$$

where

$$S(D_1, D_2, \ldots)_{atom} = S(D_1) + S(D_2) + \ldots \tag{3.8}$$

and

$$S(D_1, D_2, \ldots)_{cont} \leq S(D_1) + S(D_2) + \ldots \tag{3.9}$$

The quantity *dS* is zero for systems of completely independent atomisms, positive for systems composed of atomisms and coupling constraints, and attains its maximum value for the pure case in which there are no internal degrees of freedom. The state of maximum entropy *S* is associated with the state in which all the coupling constraints are dissolved. In contrast, condensation, crystallization, polymerization, and other forms of aggregation are associated with state transitions entailing the formation of coupling constraints resulting in evolutionary transitions away from the maximum entropic state.

3.2.6. Time and Levels of Description. Figure 3.5 identifies three levels of description for time (cf. Prigogine, 1976). The first level is the *mechanical* level and it is fundamental to classical or quantum mechanics. Time is associated with locations in space. The trajectories of motion are in coordinate spaces in which the configurational constraints are unchanging. At this level of description there are *no* events associated with the making or breaking of constraints. Motions are symmetric (reversible) in both space and time. In Figure 3.5 these trajectories have their projections parallel to the space-time axis. They are the data base for a time-symmetric, conservative, "flatland" physics (Iberall & Soodak, 1986).

The second level of description is the *thermodynamic* level. This description originated in classical thermodynamics. At this level the breaking of constraints plays an active role in defining an "arrow of time." For this level the second law of thermodynamics prescribes the direction of the arrow: All processes proceed in the direction of decreasing the number of micro constraints. This level is associated with the continual breaking (e.g., "death") of constraints. In Figure 3.5 the level is associated with nonparallel trajectories that are evolving in the direction of increasingly more microscopic frames of reference.

The third level identifies thermodynamic situations that are irreversibly associated with the "assembling" or "making" of constraints. In this level the arrow of time points in the direction of increasing configurational order. In Figure 3.5 this level is identified by the nonparallel trajectories projecting from the micro frames into the macro frames. This level of description is associated with self-organizing systems.

3.3. Conservational Distributions, Potentials, and Force Fields

Equilibrium states define uniform distributions of conserved entities (e.g., mass, energy, momentum, charge). A system at equilibrium is defined by a symmetry in configurational constraints. The trajectories are dominated by reversible (time-symmetric) flow processes. Constraints are not broken or created, they are merely transformed in space-time orientations. This motion is defined as mechanical. If, however, one of the conserved entities becomes concentrated at a local region of the field, destroying the symmetry of the configurational constraints, then a flow process will result that acts to irreversibly redistribute the conservation into a symmetrical distribution. The transformation process will involve the breaking of local constraints and the increase of entropy. The gradient instabilities leading to the breaking of constraints stabilize when the field's conservations are symmetrically distributed relative to the boundary conditions. All irreversible flows involve instabilities (i.e., the breaking of constraints). These instabilities transform energy from macro (e.g., mechanical) modes into micro (e.g., thermal) modes. The instabilities associated with irreversible processes stabilize as the field's conservational distribution becomes symmetric (uniform). Asymmetry begets symmetry as instability begets stability.

A "potential" is defined as any *local* concentration of a conservation. Potentials relate to irreversible processes in that irreversible processes are instabilities (i.e., transport processes) that redistribute the conservations into uniform (symmetric) distributions. Any potential can be labeled abstractly a *source*. A region of lower concentration is termed a *sink*. Irreversible processes are instabilities that arise when sources are physically connected to sinks. The global flow instabilities are annihilated as the source concentrations converge onto their respective sink concentrations—equilibrium is defined when all gradients are annihilated. At equilibrium local fluctuational instabilities can arise that generate *random walk* diffusion processes (see Section 3.4).

Reversible and irreversible processes are the result of the global field distributions (symmetries and asymmetries) being manifested locally. Transport processes define *cooperativities* that link or couple atomisms of a field to the distribution characteristics of the field's conservations. Flow processes result from global field geometries defined with reference to the conserved entities.

This "strategy" for producing flows is independent of the conservation involved.

The concept of a *force* can be derived from the field's concentration gradients (where concentration is used in its *generic* sense). Gradients (asymmetries) give rise to forces that act in the direction of uniform conservational distributions. Forces are completely determined by the intrinsic geometry of the conservational distribution: The magnitude of a force is directly proportional to the rate of change of the potential (i.e., gradient); and the direction of a force is always such that the flow accelerates toward the place where the potential is lowest. The equilibrium state defines that state in which forces are minimal. It is worth noting that the dimension of the force field is always one greater than the dimension of the corresponding potential field. A scalar potential, conceived of as a scalar function defining a spatial rate of change, gives rise to a vector force field; a vector potential gives rise to a tensor force field, and so on.

3.3.1. Sources, Sinks, and Irreversible Flow Processes. A generalized irreversible flow process can be assembled out of any of the system's conservations. As examples, (a) a concentration of mass M can give rise to an irreversible mass transport dM (e.g., a drop of ink released into a field of water); (b) a concentration of momentum MV can give rise to an irreversible transport of momentum dMV (e.g., a metal ball dropped onto the surface of a fluid field); (c) a concentration of energy E can give rise to an irreversible transport of energy dE (e.g., a hot metal ball released into a container of cold water). The behavior of a dynamical system is causally linked to the state of the conservations—in the form of an *equation of state* (Equation 3.10)—and to the state of the transport processes—in the form of an *equation of change* (Equation 3.11).

$$\mathrm{f}[\Sigma \mathrm{M}, \Sigma \mathrm{MV}, \Sigma \mathrm{E}] = 0 \qquad \text{Equation of State} \tag{3.10}$$

$$\mathrm{g}[\Sigma \mathrm{dM}, \Sigma \mathrm{dMV}, \Sigma \mathrm{dE}] = 0 \qquad \text{Equation of Change} \tag{3.11}$$

Or, stated more generally,

$$\Sigma \mathrm{f}(\text{conservations})_i = 0 \tag{3.12}$$

$$\mathrm{d}\Sigma \mathrm{g}(\text{conservations})_i = 0. \tag{3.13}$$

If, for a given system, the values of the state variables are independent of time, the system is in thermodynamic equilibrium and the states are completely determined by the equation of state. If the state variables vary with time, then the system is in a nonequilibrium state and requires the equation of change to determine its dynamics. The number of state variables required to describe a nonequilibrium state may be larger than that required to describe the system at thermodynamic equilibrium.

As noted above, an irreversible flow is a function of the sink/source concentration ratio. In an isolated system, the irreversible flow always asymptotes onto a state dominated by reversible processes: the equilibrium state. The magnitudes of the sink and the source concentrations provide a relative measure of the local departure from the equilibrium state. The "figure of merit" scaling the magnitudes of the source and the sink is the concentration metric at the equilibrium state. While the equilibrium concentration is only physically observable at equilibrium, the equilibrium state has a *virtual* status for nonequilibrium states. The virtual state results from a symmetry that relates the source concentrations to sink concentrations in isolated systems. The asymptotic approach to the equilibrium state is characterized by a symmetric reduction in the source and sink concentrations. In this regard the symmetry defining the equilibrium state is the metric against which the field processes are both scaled and defined; sources and sinks are strictly defined relative to the (global) equilibrium state. The following dimensionless π-*numbers* can be used to define source, sink, and equilibrium states.

Definitions:

(1) The relative state condition for a source is defined by:

$$\pi_{\text{source}_i} = \left[\frac{\text{local state concentration}}{\text{global equilibrium state concentration}}\right] > 1 \tag{3.14}$$

(2) The relative state condition for a sink is defined by:

$$\pi_{\text{sink}_i} = \left[\frac{\text{local state concentration}}{\text{global equilibrium state concentration}}\right] < 1 \tag{3.15}$$

(3) The relative state condition for an equilibrium state (i.e., no sources or sinks) is defined by:

$$\pi_{\text{equil}_i} = \left[\frac{\text{local state concentration}}{\text{global equilibrium state concentration}}\right] = 1 \tag{3.16}$$

where i indexes all relevant conservations.

The aforementioned definitions can be expressed in the form of a phase diagram (see Figure 3.6). Combining Equations (3.14) and (3.15) yields an additional π-*number* that identifies the phase complexions of flow processes:

$$\pi_{\text{flow}_i} = \frac{\pi_{\text{sink}}}{\pi_{\text{source}}} \tag{3.17}$$

where

$$\pi_{\text{flow}_i} < 1 \quad \text{identifies irreversible flow processes} \tag{3.18}$$

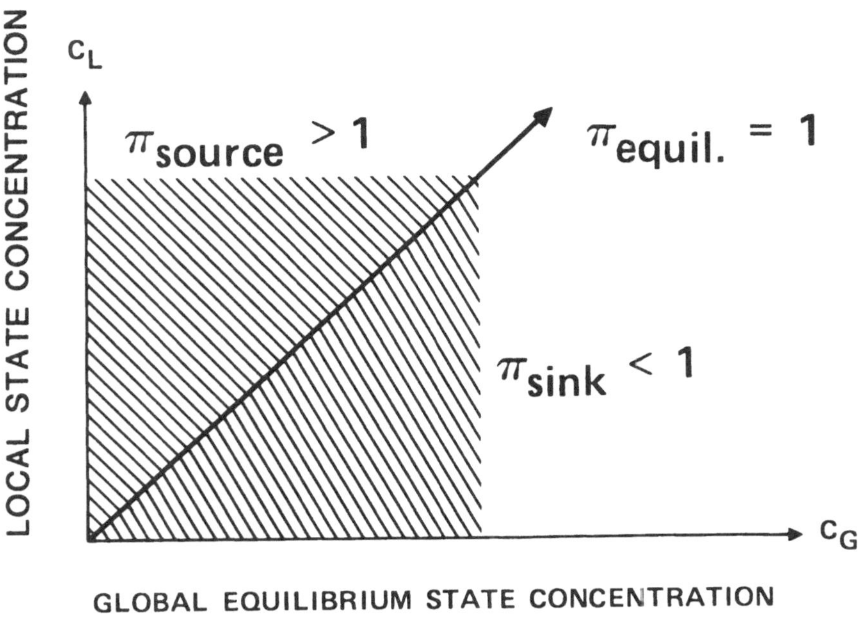

FIGURE 3.6. PHASE DIAGRAM OF SOURCE, SINK, EQUILIBRIUM, AND NONEQUILIBRIUM STATES. Regions of concentration where $\pi > 1$ identifies "sources." Nonequilibrium flow processes orient *away* from these regions. Regions of concentration where $\pi < 1$ identifies "sinks." Nonequilibrium flow processes orient toward these regions. Irreversible transport processes orient the distribution of conserved entities from sources toward sinks. Regions of concentration where $\pi = 1$ identifies equilibrium states. At equilibrium all irreversible flow processes cease, leaving only reversible processes.

and

$$\pi_{\text{flow}_i} = 1 \qquad \text{identifies reversible flow processes.} \tag{3.19}$$

3.3.2. Irreversible Flows and Onsager's Phenomenological Relations. For all irreversible flows, two types of characteristic parameters can be identified (see DeGroot & Masur, 1962). The first parameter is an "intensive" measure of the concentration gradient that is associated with the driving forces (thermodynamic or generalized). The second parameter is an "extensive" measure of the corresponding flows (rates, fluxes, or currents). A number of empirical generalizations have been established for classes of irreversible processes that linearly relate concentration gradients to flows. For example, Fourier's law for heat flow in an isotropic medium states that the heat flow is proportional to the gradient of the absolute temperature. Fick's law identifies a linear relation between the diffusion flow of matter and the concentration gradient. Ohm's law for electrical conduction and Newton's law for fluid motion are further examples of linear force-flow laws. These laws can be written in the general form

$$J_i = L_{ii}X_i \tag{3.20}$$

where J_i identifies the rate of flow, X_i the force, and L_{ii} the phenomenological coefficients (of heat conductivity, electric conductivity, diffusion, etc.). At equilibrium all coefficients become zero.

The theoretical motivation for the linear form of these laws was first offered by Onsager in 1931. In addition to deriving the linear laws, Onsager formulated the crosscoupling relations that predict when two transport processes take place simultaneously. As a general rule, any force is able to couple to any flow creating a symmetric matrix of force-flow couplings. These crosscouplings allow for the possibility that some step(s) of a reaction may progress in a direction contrary to that prescribed by its own affinity. An example is the coupling of diffusion and heat transport, which gives rise to the Soret effect and the Dufour effect. The Soret effect or thermal diffusion is the maintenance of a concentration gradient as a result of a temperature gradient. This effect may be described mathematically by adding to Fick's law a term that is proportional to the temperature gradient. The Dufour effect is the temperature difference arising from a concentration gradient. Mathematically this can be derived by adding to Fourier's law terms that are proportional to each of the concentration gradients. Other examples are the interference of heat conduction with electrical conduction, diffusion with electrical conduction, and heat conduction in various directions in isotropic crystals. In each of these cases the empirical equation is a linear, homogeneous relationship between a flow J_i and several forces X_j

$$J_i = L_{ij}X_j \tag{3.21}$$

where the phenomenological coefficients L_{ij} are independent of the forces. The coefficients L_{ii} in Equation (3.20) relate conjugated flows and forces, while the coefficients L_{ij} $(i \neq j)$ in Equation (3.21) give rise to crosscoupling.

The Onsager relations are valid for systems with constant phenomenological coefficients. In general this applies to near equilibrium systems that exhibit small space and time gradients of the local thermodynamic variables. Systems far from equilibrium are not candidate systems for the Onsager relations, because they characteristically exhibit steep gradients of the thermodynamic variables and changing phenomenological coefficients. The closer a system is to equilibrium, the more closely the Onsager relations approximate the system's state.

3.3.3. Causal Mechanics and Variational Mechanics[2]. The Onsager rela-

[2]Physical laws are generally expressed in one of two forms. The first form causally maps initial states forward in time into final states. The mathematical form of the "forward" projecting law is

tions identify causal laws that continuously relate the initial force conditions X_j to the final flow conditions J_i according to (symmetric and asymmetric) phenomenological coefficients L_{ii} and L_{ij}. Using the Onsager relations, a continuous flow trajectory can be defined from a continuous force input. The mapping function is linear if the coupling coefficients L_{ii} and L_{ij} are constants.

An alternative approach to the force-flow coupling can be sought using the tools of variational mechanics. The general strategy is to focus on the relationship between a potential function (the derivative of which is force) and the observed flow rate. The strategy is based on the assumption that the observed flow rates identify the minima of the potential functions that generate forces that drive the flows. The minima are defined relative to local boundary conditions. The minimum of the potential function is referred to as an *attractor* state. From the variational perspective, observed flows can be used to identify *trajectories of attractor states*, each of which defines a local minimum relative to some potential function.

In a conservative system, a local minimum is defined in terms of *spatial coordinates*; the force gradients that surround the local minimum, defining a ''potential well,'' scale as a function of spatial displacement from the minimum (e.g., ''attractor'' or ''singular'') state. In contrast, for a nonconservative (dissipative) system a local minimum is defined in terms of a *spatio-temporal coordinate space*. In a dissipative system ''attractor'' states are defined over minima carried in time-dependent dissipative energy functions. These time-dependent, nonconservative, *dissipative energy wells* stand in sharp contrast to space-dependent, conservative, *potential energy wells*. When the tools of variational mechanics are applied to dissipative field processes, the mimima, around which the processes stabilize, are defined in terms of time-dependent energy functions such as entropy production, excess entropy production, etc. The geometric ''form'' of the potential or dissipative well serves as a boundary condition for restricting or classifying the dynamic trajectories assumed by the

---

that of a differential equation. The second form causally maps final states ''backwards'' in time into initial states. The mathematical form of the backward projection is that of an integral equation.

The laws of classical mechanics were exclusively differential in form, while ''variational'' mechanics was the name of the discipline that was devoted to the development of laws that were integral in form. These two approaches to mechanics expressed two distinct philosophical orientations. Briefly, their differing motivations can be explained as follows: Since causes necessarily precede effects, the differential form, as used by classical (Lagrangian) mechanics, was considered the appropriate function for expressing ''causal'' laws. In classical mechanics, latter states are caused by former states. By contrast, variational (Hamiltonian) mechanics expressed its laws in the integral form so that final states (usually optimal) could influence (i.e., constrain) the trajectory of approaching states. In variational mechanics, former states are influenced by latter states. Mathematically, these two forms of expressing dynamic laws yield identical quantitative results.

field processes. Variational analyses identify "geometric wells" that constrain field processes around stable states.

3.3.4. Lyapunov Stability, Attractor States, and the Second Law of Thermodynamics. The basic concept of a stability analysis for dissipative systems is that of Lyapunov stability (Minorsky, 1962). Lyapunov stability is defined by the assertion that solutions (final states), related to neighboring initial states, evolve in the neighborhood of each other. This notion of stability explicitly refers to the limit ($t \to +\infty$) and implies the notion of a direction of time. The concept of asymptotic stability implies that any finite perturbation will eventually be damped back onto the original state through dissipative processes. An asymptotically stable state defines an *attractor state* in the state space evolution of the system's dynamics. Physically this refers to the asymptotic convergence of the source and sink concentrations (for $t \to +\infty$) onto a singular point defining the equilibrium state (no entropy production, $P[S] = 0$) for an isolated system or onto a steady state (where, $P[S] > 0$) in the case of a thermodynamically open flow system. In either case, the convergence of the trajectory onto the attractor state is time dependent. The geometrical meaning of the attractor state and this asymptotic convergence with time is illustrated in Figure 3.7 .

The second law of thermodynamics can be restated using the language of stability theory: For all physical processes a Lyapunov function exists with global (thermodynamic) equilibrium or steady states that act as *attractors* for nonequilibrium states (see Prigogine, 1980). The time-dependent dynamic trajectories engendered by the second law distinguish among various classes of concentration states (e.g., source, sink, equilibrium, and steady state), some of which act as attractors for others (Planck, 1932). In an isolated system the gradient-free field dynamic of the equilibrium state identifies an attractor for all

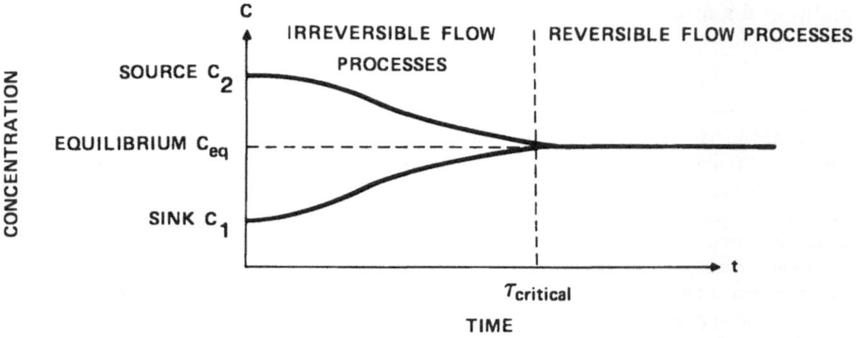

FIGURE 3.7. TIME EVOLUTION PLOT OF CONSERVATIONAL DISTRIBUTION. The second law of thermodynamics identifies a single time-dependent "attractor" state for isolated systems. The attractor state is equilibrium. At equilibrium, the conserved entities are distributed uniformly (according to the equipartition theorem) over all the available degrees of freedom.

other states. In an open (or closed) system a steady state identifies the attractor state. In either case, these attractor states are assembled from any, or all, of the field's conservations, resulting in forces that tend to drive the system toward various equilibrium or steady states: thermal (uniform distribution of temperature), mechanical (uniform distribution of momentum and mass), or chemical (uniform distribution of chemical potential), etc.

3.3.5. Evolution Criteria and Variational Characteristics. Classical thermodynamics provides a macroscopic description of equilibrium situations in terms of state variables that are minimized or maximized. For example, in an isolated system the equilibrium state is defined when the entropy S achieves its maximum possible value for the given boundary conditions. If, however, the volume and the temperature of the system are held constant, then the equilibrium state is defined relative to the Helmholtz free energy $F = E - TS$, which achieves a minimum at the equilibrium state. The variables around which the stability of the equilibrium states is defined identify the thermodynamic potentials. These potentials play the organizing role for thermodynamic trajectories similar to the organizational role played by potential energy in (Hamiltonian) mechanics.

Prigogine (1967) has proposed a set of variational arguments for the trajectories assumed by nonequilibrium (dissipative) states. These states identify processes (a) that have not reached equilibrium but are on a trajectory toward equilibrium; or (b) that are prevented from going to equilibrium because of boundary conditions. Prigogine has proposed the theorem of minimum entropy production for defining stable nonequilibrium states. The theorem, however, is valid only within the linear thermodynamic range of the Onsager relations. More recently Glansdorff & Prigogine (1971) have had some success in applying a generalization of the theorem of minimum entropy production (*min P[S]/dt*) to regions beyond the linear range. As a system is driven further away from the local equilibrium states (chemical, mechanical, thermal, etc.) through new conservational flows (energy, momentum, mass, etc.), new attractor states can suddenly emerge that are associated with dramatic reductions in the entropy production. These states are carried within the *excess entropy production function.*

3.4. Statistical Mechanics

3.4.1. Relationship Between Statistical Mechanics and Thermodynamics. In thermodynamics, the laws governing the equilibrium state of an ensemble system are derived from continuum parameters such as pressure, volume, entropy, and temperature. In contrast, statistical mechanics derives the identical equilibrium laws using atomistic parameters such as velocity, displacement, energy, and momentum. The theory of statistical thermodynamics provides a formalism for relating continuum, thermodynamic parameters to atomistic, mechanical parameters. The methods of statistical mechanics are drawn from probability theory and classical mechanics. The program of statistical mechanics is similar to thermodynamics in that it is concerned with the study of properties of atomistic fields at global, equilibrium states. It derives these properties from the known or assumed atomic structure of matter—defined by quantum mechanics—and the laws of interaction—defined by classical mechanics.

Statistical mechanics provides a theoretical connection between the observable behavior of large material objects and the dynamics of the invisibly small atomisms constituting these objects. The strategy is to derive ensemble state descriptions that are consistent with the mean dynamical states of every atomism in the system. Because the macro and micro descriptions are state equations defined in terms of conservations, and conservations are invariant over state descriptions, the two equations (macro and micro) can be set equal to one another. Macroscopic and microscopic state equations are alternative descriptions of the same conservational fields.

Statistical mechanics differs from thermodynamics in that it makes explicit reference to the nature of the micro structure of the field processes. Whereas thermodynamics provides a bookkeeping strategy for tracking the global equilibrium states of ensemble behavior in terms of continuum variables, statistical mechanics provides a bookkeeping strategy for tracking global equilibrium states—at the macroscopic, continuum level and the microscopic, atomistic level—in terms of atomistic variables. The methods of both statistical mechanics and thermodynamics provide continuum field descriptions in terms of state equations involving very few state variables.

3.4.2. Small Fluctuations in Continuum Fields. There are three important facts to be considered when addressing continuum fields composed of interacting atomisms: (a) all constraints that eliminate atomistic degrees of freedom become time dependent at some scale of analysis; (b) at global equilibrium the stability of the continuum field is a function of the uniformity of the distribution of conserved entities; and (c) when global stability is defined on the continuum level there are local instabilities at the atomistic level that show up as fluctuations. The first two facts derive their substance from thermodynamics. They are the consequence of the second law. The third fact derives its substance from statistical mechanics. It is the consequence of the fact that continuum fields *cannot* simultaneously be both globally and locally stable. Or, put differently, continuum fields *cannot* simultaneously be both globally and locally at equilibrium. The relationship between global stability and local instabilities is the key to dissipations (thermodynamics) and fluctuations (statistical mechanics) and provides a continuous up-down bootstrapping for all dynamic processes.

A uniform conservational distribution defines a thermodynamic criterion relative to the continuum force-field level of description. The evolution of systems operating at global equilibrium is such that local degrees of freedom change as a function of the breaking of local constraints, according to the second law. The second law states that all local constraints are susceptible to fracture at some time scale of analysis. If a system is at global (thermodynamic) equilibrium, then locally the force field is unstable, resulting in atomistic (statistical mechanical) fluctuations. The local instabilities generate atomistic fluctuations that are continually damped out by the global stability requirements defined by the second law. Under certain conditions the local instabilities can amplify, resulting in global instabilities and the emergence of new organizational regimes (see Chapter 4). Statistical mechanics provides a methodology for predicting the details of the local fluctuations.

Thermodynamically, a fluctuation constitutes a spontaneous deviation from a globally defined equilibrium state. When viewed globally, the fluctuation appears as a transition from a more probable into a less probable state. When viewed locally, from the perspective of the instability, the fluctuation is in fact moving in the direction of a more probable local state. The laws of statistical mechanics (Fowler, 1936; Landau & Lifshitz, 1959; Tolman, 1938) examine the nature of these fluctuations in systems exhibiting large numbers of interacting atomisms, that is, systems of high dimension. The tendency of fluctuations to depart locally from the global equilibrium state identifies a natural process (i.e., instability) that competes with the entropic process defined by the second law of thermodynamics (c.f. Levich, 1971): *The second law of thermodynamics predicts a probabilistic decrease in large-scale (global) gradient-dependent dynamics, while the statistical mechanical law of fluctuation predicts a probabilistic increase in small-scale (local) gradient-dependent dynamics.* Local fluctuations are the consequence of stability defined at the continuum level. The system cannot simultaneously satisfy both global and local stability requirements. If the system satisfies global stability requirements, it *must* exhibit local instability—resulting in fluctuations. If the system satisfies local stability requirements, it must exhibit global instability—resulting in phase transitions, steady-state transitions, etc. Thermodynamic factors drive the continuum field toward globally stable states and statistical mechanical factors drive the atomisms toward locally stable states—i.e., the system's overall behavior is bootstrapped by competitions between global and local stabilities.

3.4.3. Brownian Movement. Local atomistic (statistical mechanical) instability in the context of global continuum (thermodynamic) stability ordinarily results in only small fluctuations relative to the size of the continuum field processes. Under certain conditions, however, the local atomistic fluctuations can become greatly amplified so as to enter into active competition with the macroscopic field processes. If these conditions persist, then the local instabilities can spread, resulting in a global instability. An example of the physical strategy behind the amplification of atomistic fluctuation is Brownian movement.

Brownian movement is the "zigzagging" motion observed in a macroscopic particle when it is suspended in an atomistic liquid or gas medium. The medium is composed of a micro structure of fluctuating atomisms that continually collide with the particle and transfer momentum to it with each collision. As an ensemble, the atomisms of the medium exert a pressure on the surface of the particle. The collisions of the atomisms with the surface of the particle are completely random and from all directions. If the surface of the particle is sufficiently large relative to the size of the atomisms, so that a large number of atomisms impinge on it in a very short time interval, then it can be assumed that the momenta that are transferred to the particle from all directions are, on the average, exactly balanced.

The foregoing assumption breaks down, however, if the macroscopic particle is of a critical size relative to the size of the atomisms and the rate of atomistic collisions. For these *Brownian* particles (whose radius is of the order of $10^{-4}cm$), it can no longer be assumed that the microscopic collisions are creating a balanced (symmetric) force structure on the surface of the particle. The surface of the Brownian particle is so small that in a short time a limited number of atomisms impinge on it. There is an increasing probability that the resultant forces exerted by the colliding atomisms on the surface of the particle will not balance. Over very small time intervals, more atomisms hit the particle from one side than from the other sides, resulting in particle motion (see Figure 3.8). In the next instant more atomisms will hit it from another side, and it will take off in a new direction. The random pattern followed by the particle is called *random walk*.

The quantitative theory of Brownian motion, simultaneously proposed by Einstein and Smoluchowski, was derived from the theoretical claim that the *mean* kinetic energy associated with the macroscopic motion of the Brownian particle is exactly equal to the *mean* kinetic energy associated with the microscopic collisional motion of a molecule:

$$\left[\frac{M\bar{V}^2_{macro}}{2}\right] = \left[\frac{m\bar{V}^2_{micro}}{2}\right]. \tag{3.22}$$

This is true because the average kinetic energy of a particle does not depend upon the kind of particle (macro or micro), but only on the temperature of the field. Therefore, at a given temperature the most probable speed of a particle is always inversely proportional to the square of its mass. As a Brownian particle and the colliding molecules are in a thermal field at a uniform temperature, and the mass of the Brownian particle is considerably greater than that of the colliding molecules, the speed of the Brownian particle must be considerably less than that of the molecules. From Equation 3.22 it follows that

$$(\bar{V}^2_{macro} / \bar{V}^2_{micro}) = (m_{micro} / M_{macro}) \tag{3.23}$$

and

$$\bar{V}_{macro} = \bar{V}_{micro}(m/M)^{1/2} \tag{3.24}$$

where the bar denotes the mean speeds, and *m* and *M* are the respective masses of the molecule and the Brownian particle. Given that a Brownian particle is composed of many millions of molecules, $M/m = 10^{10}$, and that the average predicted speed of gas molecules is several hundred meters per second (according to assumptions of the kinetic theory of gases), the resulting speed of the Brownian particle is predicted to be about several millimeters per second, which can be observed.

3.4.4. An Elementary Dissipation-Fluctuation Theorem. A Brownian particle can be regarded as a macroscopic object from the point of view of the force transaction. The total effect of the atomisms' colliding with the Brownian

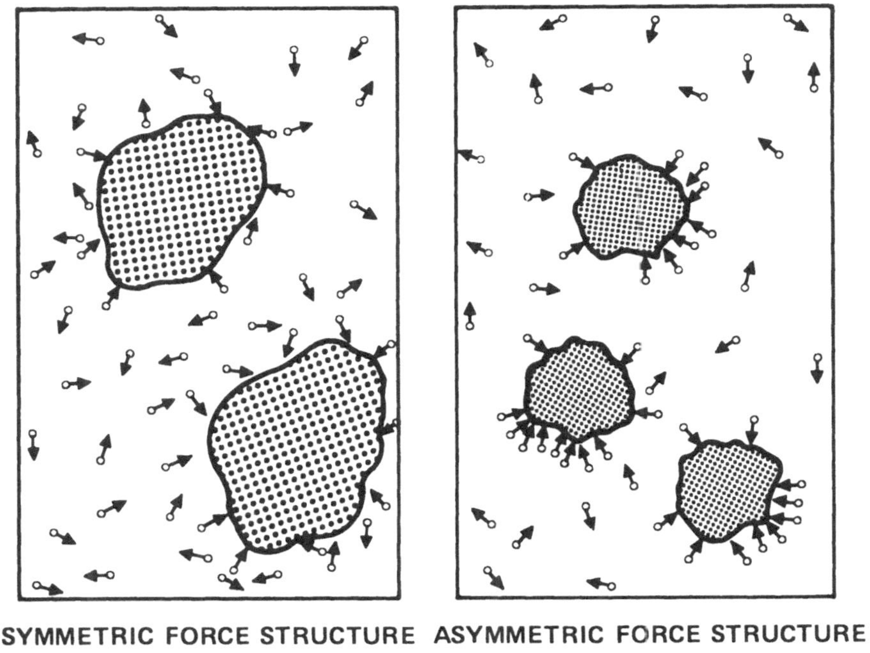

FIGURE 3.8. BROWNIAN FORCE STRUCTURE. For a large macro particle (left panel) the distribution of atomistic collisions on its exterior surface is uniform. The result is a symmetric force structure with a net work output of zero. For smaller macro particles (right panel) the collisional distribution has a significant probability (*10%*) of becoming nonuniform. The result is the development of an asymmetric force structure on the surface of the macro particle which leads to the production of a work cycle (a Brownian "hop").

particle can be equated with a viscous force. Usually viscous forces tend to retard motion; however, in the case of the Brownian particle the viscous forces tend to *both* accelerate and retard the particle's motion. The atomistic collision first provides the *source* of momentum that accelerates the particle over a distance δ and then provides the *sink* that absorbs the momentum that results in the retarding of the motion over a similar distance δ. The process is then repeated in a cyclical fashion.

The viscous force that drives and retards the motion can be approximated using Stoke's hydrodynamic law for fluid viscosity:

$$F = 6\pi r \eta v \tag{3.25}$$

where r is the radius of the Brownian particle, v is the velocity of the particle and η is the coefficient of viscosity. The work $WORK_{viscous}$ done by the average force F_{av} of viscous friction over the spatial displacement δ is approximately twice the kinetic energy of the atomistic motion $K.E._{micro}$:

Microscopic viscous work

$$\text{WORK}_{\text{viscous}} = F_{av}\delta = 6\pi r \eta v \tag{3.26}$$

Microscopic thermal energy

$$\text{K.E.}_{\text{micro}} = (mV^2_{\text{micro}})/2 \tag{3.27}$$

Microscopic viscous work equals twice the thermal energy

$$\text{WORK}_{\text{viscous}} = 2(\text{K.E.}_{\text{micro}}). \tag{3.28}$$

Since the kinetic energy of the atomistic motion is equal to the kinetic energy of the motion of the macroscopic particle from Equation (3.22), then the macroscopic kinetic energy of the particle can be used to determine the microscopic dissipative forces:

$$2(\text{WORK}_{\text{macro-mechanical}}) = MV^2_{\text{macro}} = \text{WORK}_{\text{viscous}}$$
$$= F_{av}\delta = 6\pi r \eta v \tag{3.29}$$

The relation expressed in Equation (3.29) constitutes an elementary form of a fluctuation-dissipation theorem. The size of the fluctuation is proportional to the size of the dissipation. The importance lies in the fact that the conservational relationship allows for the calculation of the amount of dissipation by means of the amount of fluctuation and vice versa. In Brownian motion the dissipation could be calculated from the macroscopically observed mean free path and relaxation time. Using this energy relationship, a theoretical prediction was derived for the mean distance through which such a particle should drift in a given time, through a given medium, solely as a function of microscopic fluctuations. The mean distance could be directly observed and compared with the theoretical prediction. Thus, the observable macroscopic (mechanical) displacements δ of the particle could be noted and compared with the displacement times τ so as to predict properties of the microscopic (thermodynamic) domain. The Einstein-Smoluchowski theory of Brownian motion was, in a rigorous and strong sense, the limiting theory for infinite friction. From this starting point it was possible to develop the theory of statistical mechanics and, thereby, the molecular-kinetic theory of thermodynamics.

Using the above theoretical construct, the kinetic energy of the observable macroscopic motions can be used to predict the microscopic dissipative properties. The macro/micro—mechanical/thermodynamic—relationship constitutes a fundamental (Brownian) similitude that has application for all systems operating at the frictional limit. It can be used to derive statements about unobservable microscopic (thermodynamic) characteristics of a system from observable mechanical characteristics. The significance of the Brownian similitude is that typically a nonconservative system requires a detailed analysis of at least two frames of reference before closure on the conservational invariants can be achieved. While the details of the macro frame of reference can be given, the micro frame is usually unobservable or highly complex, and thus defies a detailed analysis. If a macro/micro Brownian similitude exists, then conservational closure can be achieved based solely on the detailing of motions in a single (macro) frame of reference. Once the macro motion is detailed, the microscopic energy flow structure is completely determined by the similitude.

3.4.5. Q-factors and Engine (Work) Cycles. An engine is a set of constraints that systematically converts various forms of energy into mechanical energy or work. Brownian motion is an engine process: Thermal energy is converted into mechanical energy through the constraints provided by the particle surface. The engine process is driven by a momentum concentration (i.e., pressure) that is periodically (self) assembled in accordance with the distributional characteristics of the microscopic momentum fluctuations and depleted according to the viscosity characteristics of the medium. The engine cycle is characteristic of a "limit cycle" (e.g., relaxation oscillator). The cycling process consists of: (a) the fluctuational (self) assembly of a potential; (b) the conversion of the pressure into mechanical work; and (c) the dissipation of mechanical work into the momentum reservoir. The operational characteristics of the Brownian engine are limited strictly by the fluctuational properties of the momentum reservoir and the size of the particle.

A dimensionless number referred to as a *Q-factor* (quality factor) is used to identify the relative amount of energy carried by the mechanical and thermal energy phases of an engine cycle. The number is derived by dividing the average energy carried by the macro (mechanical) work phase by the average energy carried by the micro (thermodynamic-dissipative) heat phase:

$$\pi_Q = \left[\frac{\bar{E}_m}{\bar{E}_t}\right] \tag{3.30}$$

A *Q-factor* measures the flow of energy between macro and micro frames of reference. It is commonly used to measure the ability of the system to store energy from cycle to cycle within the mechanical frame of reference. A system with a high *Q-factor* stores a large portion of the energy carried in the macroscopic work cycle (oscillation) within conservative force structures. Very little energy flows between macro and micro frames of reference within each cycle (that is, the system has little dissipation). In contrast, a system with a low *Q-factor* stores very little of the energy carried in the macro work cycle. Each work cycle is partially reconstructed out of new energy that must flow into the macro frame of reference from another frame of reference. The new energy is required to offset the energy flowing out of the macro frame through dissipative processes. Brownian motion represents the limiting case for systems operating with dissipative forces: Each successive work cycle or Brownian hop must be constructed afresh, resulting in a *Q-factor* of $\pi_Q = 1.0$. It is interesting to note that in the case of Brownian motion, the forces that power the motion and the forces that retard the motion *both* have their origin in the system's microstructure. This type of force structure is also characteristic of many biomechanical movement systems in which the force-producing fluctuations and the force-absorbing dissipations are both a function of the metabolic tissues that are distributed throughout the entire biomechanical machinery (see Chapter 11).

3.4.6. Brownian Motion and Macro/Micro, Symmetry-Breaking Instabilities. From the theory of Brownian motion a set of physical requirements can be identified that constitutes a basic strategy for the spontaneous creation of symmetry-breaking instabilities. The instabilities are associated with constraints that temporarily drive the system away from mechanical equilibrium. The strategy involves a generalized thermodynamic coupling of microscopic field processes to macroscopic field processes.

Since the Brownian system is an isolated system, the same momentum must be continuously recycled from micro to macro through fluctuations and macro to micro through dissipations. This isolation places a strong restriction on the size of the macroscopic work cycle. If the system is open to the flow of conservations (e.g., energy, matter, momentum, charge), however, then the size of the potential (negentropic) boost from the microstructure can increase dramatically. The study of the conditions that lead to the self-organization of macroscopic constraints in open systems through the amplification of microscopic fluctuations has been a primary topic of investigation for Prigogine (1976, 1980) and his Brussels colleagues (Glansdorff & Prigogine, 1971; Nicolis & Prigogine, 1977).

The essential requirements for self-organization can be identified in the context of the macro/micro structuring of the Brownian system:

1. A conservational reservoir from which (generalized) work can arise. (The thermal bath provides a closed reservoir of energy giving rise to macroscopic work cycles—the Brownian motions. If the system is open to the flow of energy through an external source of heat, then the Brownian motions can become more amplified.)

2. A microcosm of elements with a stochastic fluctuating nature. (The atomistic microstructure is randomly driven by the thermal energy.)

3. The presence of nonlinear components. (The transition from a symmetric to an asymmetric force structure on the surface of the Brownian particle is a nonlinear symmetry-breaking instability.)

4. A scale change occurs, such that an instability is critically amplified. (The transition in the qualitative characteristics of the force structure is a function of the ratio of the mean radius of the Brownian particle to the mean radius of the atomistic particles.)

When these requirements are present the possibility exists for the transition of the macroscopic motion from a stochastic steady-state condition to a time-dependent Brownian regime. The regime is stable by virtue of the dissipation of energy by the micro field, and is maintained by virtue of the amplified pressure instability that "squirts" momentum from the micro (atomistic) field into the macro (continuum-particle) field. The particle's trajectory, constrained by the successive "squirts" and "dissipations" of momentum into and out of the macro motions is that of a limit cycle.

3.5 Newton's Paradigm: The Primacy of Force Couplings

> Every body continues in a state of rest, or of uniform motion in a right line, unless it is compelled to change that state by forces impressed upon it.
>
> Isaac Newton
> *Principia Mathematica*

The coupling that links the micro structure of the Brownian system (i.e., molecules) to the macro structure (i.e., Brownian particle) is a force coupling. While a Brownian system involves dissipative and fluctuational components that were not a part of Newton's original mechanics program, it nonetheless involves sets of interactions that are all dominated by force exchanges reducible locally to Newton's paradigm. Briefly, Newton's paradigm reduces all phenomena to descriptions of states plus forces (dynamics). The state of a system identifies what the system is like now; only forces can precipitate a change of state. Classically, this paradigm has provided the fundamental strategy for approaching the problem of causality in physics: Present states plus dynamical laws are necessary and sufficient to determine the next state (and, in fact, all subsequent states). In the Newtonian paradigm a given state is related to a future or past state by a dynamical law that specifies the forces acting on the given state. The notion of cause is inextricably tied to the notion of force; and, the notion of force is inextricably tied to the notion of mass (e.g., *force* = *mass* × *acceleration*).

Returning to Brownian motion, the force structures that determine the states realized by the macro particle have statistical mechanical properties originating in the system's micro states. For a macroscopic motion to occur, a force structure must be generated by the micro field that is sufficient in magnitude to complete actively with the macroscopic forces (i.e., the mass of the Brownian particle times its acceleration). In sum, the causal basis of the Brownian (macro) motion is a force field description of the macro/micro interaction. And insofar as Brownian movements *are* completely dominated by force interactions with the micro field, they are consistent with Newton's force paradigm, and the more modern conservational paradigm.

In the *19th* century Newton's laws of force dynamics were restated in the form of conservation principles. Newton's laws of dynamics were shown to be the necessary consequences of systems operating under the conservational invariants of mass, momentum, and energy. In essence, Newton's laws are the conservation principles stated in the language of forces. But whether the emphasis is on forces or conservations, it is evident that the dimension of mass dominates particle interactions (at the scale of classical mechanics). The prototypical experimental protocol highlighting this fact manipulates the mass of one particle while holding the mass of the other particle constant. The velocities of the two particles that result from interactions (i.e., collisions) are observed; the prototypical result is that variations in the mass of a particle are always associated with variations in the resultant velocities. It is in this regard that the mass of a particle assumes the central role in the notion of causality in dynamics.

3.6. A Summary

Flow processes in systems composed of simple atomisms involve competitions between global (macroscopic) and local (microscopic) force distributions. Global stability patterns (symmetries) can be defined by thermodynamic criteria. The second law defines the symmetry condition (i.e., uniform distribution of conservations over the spatial field) for a global equilibrium state. Globally defined thermodynamic equilibrium, however, results in local nonequilibrium states. The local nonequilibrium states are associated with atomistic fluctuations. The atomistic fluctuations (i.e., instabilities) operate according to statistical mechanical criteria. The relationship between the global thermodynamic stability and the local statistical mechanical instability, upward and downward, is the key to a stability bootstrapping. The macroscopic stability will dominate the various processes as long as the macroscopic force field is sufficient to damp the local fluctuational instabilities. If the local instabilities amplify to the point of overcoming the global damping, then a new global regime may result that is more equitable locally. Oscillation is an example of one of these new regimes. All thermodynamic fields operate according to this principle: Stability is engendered through a coupling of thermodynamic processes and statistical mechanical processes.

In any thermodynamic-statistical mechanical coupling, steady states are defined in terms of geometric properties that are created by the dissipation of energy. These dissipative geometries, as they may be called, emerge as a function of the competition between the globally defined thermodynamic (equilibrium) criteria and the locally defined statistical mechanical (fluctuational) criteria. An example of a dissipation-based geometric property is a limit point attractor or a limit cycle attractor. These attractors are steady states that exhibit a degree of stability when confronted with perturbations. Importantly their stability is tied to the absorptive capabilities of the system's micro states, that is, to the system's dissipative ability. In the absence of dissipation the system is neutral; one state is as good as any other.

Prigogine (1980) and his colleagues (Glansdorff & Prigogine, 1971; Nicolis & Prigogine, 1977) have emphasized the role of dissipation for self-organization in a variety of field processes (e.g., hydrodynamic, chemical, and those associated with traffic patterns, the collective behavior of social insects, etc.). Systems whose organizational constraints are dissipative in their origin have been termed dissipative structures by Prigogine (1969). In the chapters to follow it is argued that the stability and reproducibility of oscillatory states (e.g., locomotion) in biological systems are members of the class of *dissipative structures* (see Kugler, Kelso & Turvey, 1980, 1982). As dissipative structures, however, rhythmic biological systems have characteristics that distinguish them from the more commonly discussed dissipative structures that are assembled out of collectives of simple atomisms and force dominated interactions. In complex (biological) systems the connections among different systemic levels are dominated more by *kinematic*, *geometric*, and *temporal* (*spectral*) descriptions than by orthodox force descriptions. Whereas kinetic couplings identify *mechanical interactions*, kinematic couplings (and the like) can be said to identify *informational interactions*. Where the present chapter outlines the physics of flow processes that are dominated by kinetic descriptions, the next chapter discusses the physics of flow processes that are dominated by kinematic descriptions and how these processes can become strategically organized to yield a self-assembling information system. At the core of this monograph is the thesis that biological movement systems are exemplars of self-assembling information systems.

REFERENCES

Bunge, M. (1973). On confusing "measure" with measurement in the methodology of behavioral science. In M. Bunge (Ed.), *The methodological unity of science*. Boston: Reidel.

DeGroot, S. R., & Mazur, P. (1962). *Non-equilibrium thermodynamics*. Amsterdam: North-Holland.

Einstein, A. (1905/1956). *Investigations on the theory of Brownian movement*. Edited with notes by R. Furth, translated by A. D. Cowper. New York: Dover.

Fowler, R. H. (1936). *Statistical mechanics* (2nd ed.). Cambridge, England: Cambridge University Press.

Gibbs, J. W. (1960). *Elementary principles in statistical mechanics*. New York: Dover.

Glansdorff, P., & Prigogine, I. (1971). *The thermodynamic theory of structure, stability and fluctuations*. New York: Wiley.

Goldstein, H. (1950). *Classical mechanics*. Reading, MA: Addison-Wesley.

Iberall, A. S. (1972). *Towards a general science of viable systems*. New York: McGraw-Hill.

Iberall, A. S., & Soodak, H. (1986). A physics for complex systems. In F. E. Yates (Ed.), *Self-organizing systems: The emergence of order*. New York: Plenum Press.

Kugler, P. N., Kelso, J. A. S., & Turvey, M. T. (1980). On the concept of coordinative structures as dissipative structures: I. Theoretical lines of convergence. In G. E. Stelmach and J. Requin (Eds.), *Tutorials in motor behavior*, New York: North-Holland.

Kugler, P. N., Kelso, J. A. S., & Turvey, M. T. (1982). On the control and co-ordination of naturally developing systems. In J. A. S. Kelso and J. E. Clark (Eds.), *The development of movement control and co-ordination*. New York: Wiley.

Landau, L. D., & Lifshitz, E. M. (1959). *Course of theoretical physics. Vol. 5, Statistical mechanics*. New York: Pergamon Press.

Landau, L. D., & Lifshitz, E. M. (1976). *Mechanics*. New York: Pergamon Press.

Levich, B. G. (1971). *Theoretical physics*. (Vol. 2.) New York: Wiley.

Minorsky, N. (1962). *Nonlinear oscillations*. Princeton: Van Nostrand.

Nicolis, G., & Prigogine, I. (1977). *Self-organization in nonequilibrium systems: From dissipative structures to order through fluctuations*. New York: Wiley.

Onsager, L. (1931). Reciprocal relations in irreversible processes, *Physical Reviews*, *37*, 405.

Planck, M. (1932). *Theory of heat* (Brose, Trans.). New York: Macmillan.

Prigogine, I. (1967). Evolution criteria, variational properties, and fluctuations. In R. J. Donnelly, R. Herman and I. Prigogine (Eds.), *Non-equilibrium thermodynamics, variational techniques and stability*. Chicago: University Press.

Prigogine, I. (1969). Structure, dissipation and life. In M. Marios (Ed.), *Theoretical physics and biology*. Amsterdam: North-Holland, 23–52.

Prigogine, I. (1976). Order through fluctuation: Self-organization and social systems. In C. H. Waddington (Ed.), *Evolution and consciousness: Human systems in transition*. Reading, MA: Addison-Wesley.

Prigogine, I. (1980). *From being to becoming*. San Francisco: Freeman and Co.

Sommerfeld, A. (1950). *Mechanics*. New York: Academic Press.

Tolman, R. (1938). *The principles of statistical mechanics*. London: Oxford University Press.

Zemansky, M. W. (1937). *Heat and thermodynamics*. New York: McGraw-Hill.

4 WHY THINGS FLOW: AN OUTLINE OF PHYSICAL THEORY FOR COMPLEX ATOMISMS

ABSTRACT

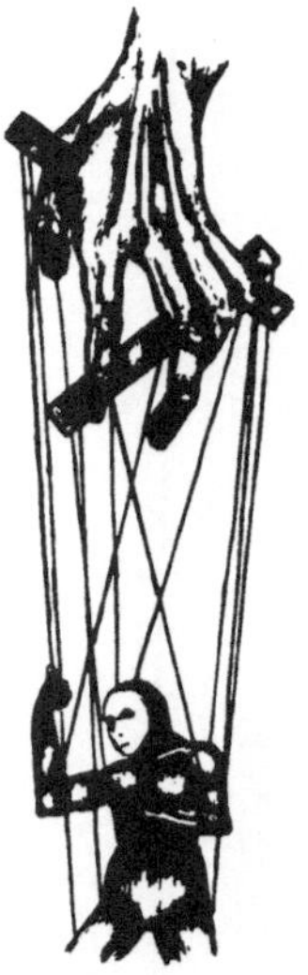

A locomoting animal is not a simple atomism in motion. Biological systems possess available on-board sources of potential energy. Their interiors are functionally rich. So how might the physical principles governing flow processes defined over simple atomisms be extended to complex biological atomisms? And if these principles are to be elaborated, then what direction must the elaboration take so that it does not impugn them? Answers are suggested, perhaps suprisingly, by the classical theory of collisions.

Interactions between simple Newtonian particles and their surroundings are dominated by descriptions that involve the mass dimension. Ordinary interactions are said to occur through forces (a force, dimensionally, is MLT^{-2}). In contrast, interactions between complex biological particles and their surroundings are dominated by descriptions in which the mass term is absent. These descriptions are kinematic (space/time), and/or geometric (spatial), and/or spectral (temporal). Indeed, interactions that do not involve mass-based descriptions or observables might be taken as the hallmark of systemic behavior at the ecological scale. At the ecological scale interactions are largely informational, not forceful. The significance of this fact is that suppression or removal of the mass dimension from the state description eliminates the "interactive violence" associated with the shock waves and impulses that strain *the internal constraints and ultimately lead to internal fracture and "organizational death."*

A historical challenge for the physically minded scientist has been the removal of vitalism from explanatory accounts of biological systems. A less heralded but no less important challenge is the removal of the interactive violence associated with mass-dominated interactions. A physical pursuit of this latter challenge brings to the forefront the centrality of the concept of information and puts a "non-Newtonian life" back into biology. This pursuit was anticipated by Gibson, and a paradigm for interactions or collisions that asserts the primacy of kinematic flow-field descriptions might be termed Gibsonian. An elaboration of Gibson's methodology focuses on the physical and functional significance of nonmass interactions in a manner that is continuous with the theory of collisions. The promise is of a natural transition from the physical theory of self-organization to a theory of self-organizing "information systems."

An example is given of a self-organizing information system—the periodic assembling of a nest by a population of social insects. A detailed analysis of this phenomenon focuses on the nature of various field couplings (macro/micro, flow/force) and the strategic role each assumes. More particularly, the analysis focuses on a coupling between the macroscopic locomotory processes of the insects and a microscopic, dissipative field of dispersing pheromones. The coupling is realized as a description that is not invariant over direction. The chemical diffusion field (i.e., the dispersing pheromones) interacts with the insects through a Newtonian ***force-field*** *description. In comparison, the insects interact in locomotion with the diffusion field through a non-Newtonian* ***flow-field*** *description.*

The significance of the embedded nest-building activities is that they exemplify a self-organizing system that is dominated macroscopically by "kinematic flow-field" properties and microscopically by "force-field" properties. This design may be very general. A presentation is given of Gibson's ecological optics and of the optical flow-field properties (kinematic, geometric, and spectral) that constrain locomotory activity. A case is made for a Gibsonian paradigm for interactions (asserting the primacy of flows) to supplement the Newtonian paradigm for interactions (asserting the primacy of forces) at the ecological scale of living things and their environments.

4.1. Interactions Involving Complex (Biological) Atomisms

An interaction between two things (particles, systems, and the like) is recognizable by the fact that, minimally, one of the things induces a dynamics (that is, a change of state) in the other. The prototypical interaction is that of two things coming into contact, as in the familiar example of colliding billiard balls. The theory of collisions developed to handle such matters is not, however, limited to contacting particles or systems. A collision is broadly defined in physics to include any temporally limited interaction. Thus, for example, molecules are said to collide, and the theory of collisions is brought to bear, even though molecules interact at a distance through their electric fields.

When one thing induces a dynamics in another thing there arises the question of the basis for this inducement: Through what observables or properties do the things in question interact? The Newtonian program answers this question with "forces." In Brownian motion, the exemplary case of a system that stratifies into micro and macro scales, the microstructure induces a behavioral change in the Brownian particle via forces applied asymmetrically to the surface of the particle. As emphasized in the preceding chapter, Brownian motion is an interaction between macroscopic and microscopic entities that takes place through kinetic observables or properties, that is, those that include the mass dimension. The micro/macro interaction can be referred to as a force coupling or, alternatively, as mass dominated.

The Newtonian program's answer to the preceding question about relevant observables does not seem well suited to the dynamics "induced," for example, in a locomoting animal by surrounding obstacles and apertures or in one rhythmically moving limb by another rhythmically moving limb. The problem—as this chapter tries to highlight—is that interactions such as these are nontrivially different from those to which the Newtonian program typically refers. In the interactions between animal and surround and between limbs the mass dimension is largely suppressed; that is, these interactions take place primarily through the observables or properties fashioned solely from the dimensions of length and time. Insofar as interacting systems are open to many descriptions, at issue is the particular description in terms of which they interact. Newton (or, more precisely, his program) ventured that one description, in terms of forces, would suffice for all interactions. The hypothesis entertained here is that a supplementary description, in terms of kinematic flows, is needed to extend the Newtonian methodology of collision analysis to complex (biological) atomisms. This hypothesis, to be fleshed out in the pages that follow, is consonant with (a) Rosen's (1978) thesis that the basis for physical interactions is much broader than has been considered to date and (b) Runeson's (1977/1983; Runeson & Frykholm, 1983) thesis that kinematic descriptions are informational (in the specificational sense) about the kinetic descriptions on which they are causally dependent.

The example through which the conception of flow-based interactions will be developed is that of the nest-building behavior of social insects. This example is chosen, in part, because of its similarity to Brownian motion. First, insect nest building involves a microstructure, namely, molecules (pheromones) diffusing in air. Second, insect nest building involves macroscopic entities, namely, insects, whose motions are linked to the aforementioned microstructure. As will become apparent, in this insect nest-building example the microstructure induces a dynamics in the macroscopic entities, and the macroscopic entities, in turn, induce a dynamics in the microstructure. The dynamics induced in the micro-to-macro direction occurs at the (microscopic) scale set by the size of the individual pheromone and at the (macroscopic) scale set by the individual insects. The dynamics induced on the insect's tissues and transducers by the transport of pheromones identifies an interaction that takes place through kinetic, mass-based observables such as energy and momentum. The dynamics induced on the insect's behavior by the global patterns of diffusing pheromones in the air identifies an interaction that takes place through kinematic and geometric, mass-independent, observables such as gradients and saddlepoint singularities. It is this latter kind of micro-macro interaction that the insect nest-building example is intended primarily to illustrate. A word of caution: Throughout the subsequent verbal and pictorial characterization of insect nest building, the field of diffusing pheromones is labeled "micro" even though the basis of the focal micro-macro interactions is to be found in coarse-grained, large-scale (qualitative) properties of the pheromone field.

4.2. Insect Nest Building: A Cooperativity of Complex Atomisms

The analysis that follows is of the nest-building activity of a population of social insects, such as African termites. The insects in the analysis relate to their surroundings largely in chemical terms and fly rather than crawl, although the arguments generalize readily to insects of the crawling variety. The principal notions expressed in the analysis derive from Grassé's (1959) and Bruinsma's (1977) observations and a thermodynamic treatment advanced by Deneubourge (1977).

The nest-building behavior consists of a number of qualitatively distinct phases of construction. In the first phase, building materials are carried into the site and deposited randomly. This phase comes to an end with the emergence of preferred sites that number far fewer than the number of original deposits. The second phase is associated with the material build-up of the preferred deposit sites until they take on the shape of pillars. When the pillars achieve a certain size and are separated by a critical distance, a third phase of construction emerges. The third phase is characterized by the mutual "curvature" of two neighboring pillars toward a "virtual midpoint." This phase is complete when the two pillars meet at the virtual midpoint, forming an arch. A final phase is marked by the construction of an arching dome that extends from the tops of the arches. The completion of the dome marks the end of the building cycle. The cycle can be repeated on the top of the dome beginning with the random deposit phase.

The primary questions to be addressed are: (a) how does an area peppered with deposits develop "preferred" locations; (b) what is the nature and origin of the constraints that restrict insect trajectories to those that converge on preferred locations; and (c) what is the nature of the description that links the behavior of the insects (a macroproperty) to the pheromone field (a microstructure)?

4.2.1. The Random Deposit Phase of Nest Building. The insects tend to follow two simple dictums: Move in the direction of the strongest smell and deposit where the smell is strongest. The smell in question is determined by the pheromones that the individual insects inject into the material used to construct the nest. If there are no material deposits in the area, or they are few and far between, then the locomotory path of an insect is "chemotactically" free, and the place at which a deposit is eventually made is virtually unrestricted. The earliest deposits, therefore, will be made at random (see Figure 4.1). However, the fact of a deposit means that pheromones diffuse into the air from that site, creating a field of attraction for insects in its vicinity. This local diffusion orients nearby insects toward the point of highest pheromone density. Such points, technically speaking, are singular states.

4.2.2. A Restriction on Macro-Micro Couplings. The attractant can be expected to diffuse widely from the site of a deposit. Fick's law (Section 3.3.2) demands that the pheromones diffuse at a rate in proportion to their density. Consequently, in the interval immediately following a deposit, a concentration gradient will be defined over the pheromone distribution in the air. An insect flying into the area of a deposit will intercept this gradient. An insect's sensitivity to the gradient, however, must be bounded and cannot be continuous throughout the pheromone field. A limitation on the sensitivity of insects to the graduated concentration of diffusing pheromones would mean that their macroscopic states, their trajectories, are not always shaped by the microscopic pheromone distribution in which they are immersed (see Figure 4.2). The existence of a perceptual limit would parse an insect's trajectories into those that are coupled to the pheromone field (see Figure 4.3) and those that are not (for example, the outlying trajectories to the right in Figure 4.2).

4.2.3. The Development of Preferred Sites. When there are only a few deposits in the building area, the pheromone concentration remains low and the probability that pheromone gradients will influence the behavior of an insect is very small. Under these conditions any locale is as attractive as any other; there are no preferred sites. These conditions can persist indefinitely if the number of insects involved in the nest-building activity remains small. If the number of insects increases to some marginal value, however, then the likelihood that any given insect will move into the vicinity of a recent pheromone deposit will increase (see Figure 4.4). The greater the number of random deposits within a given interval of time, the greater the probability that an individual insect will pass into a pheromone field defined by a previous deposit. The more attractive a site becomes because of its pheromone concentration, the more frequent are pheromones deposited (via material deposits) on that site, leading to the following sequence of events: (a) an increase in concentration level, followed by (b) an increase in probability of deposit, followed by (c) an increase in deposits, followed by *(a)*$\mapsto$*(b)*$\mapsto$*(c)*$\mapsto$*(a)*, etc. A cyclic sequence of the preceeding kind is referred to as an *autocatalytic reaction*: The rate of accumulation of material at X is in proportion to the material at X. The point is that below a critical number

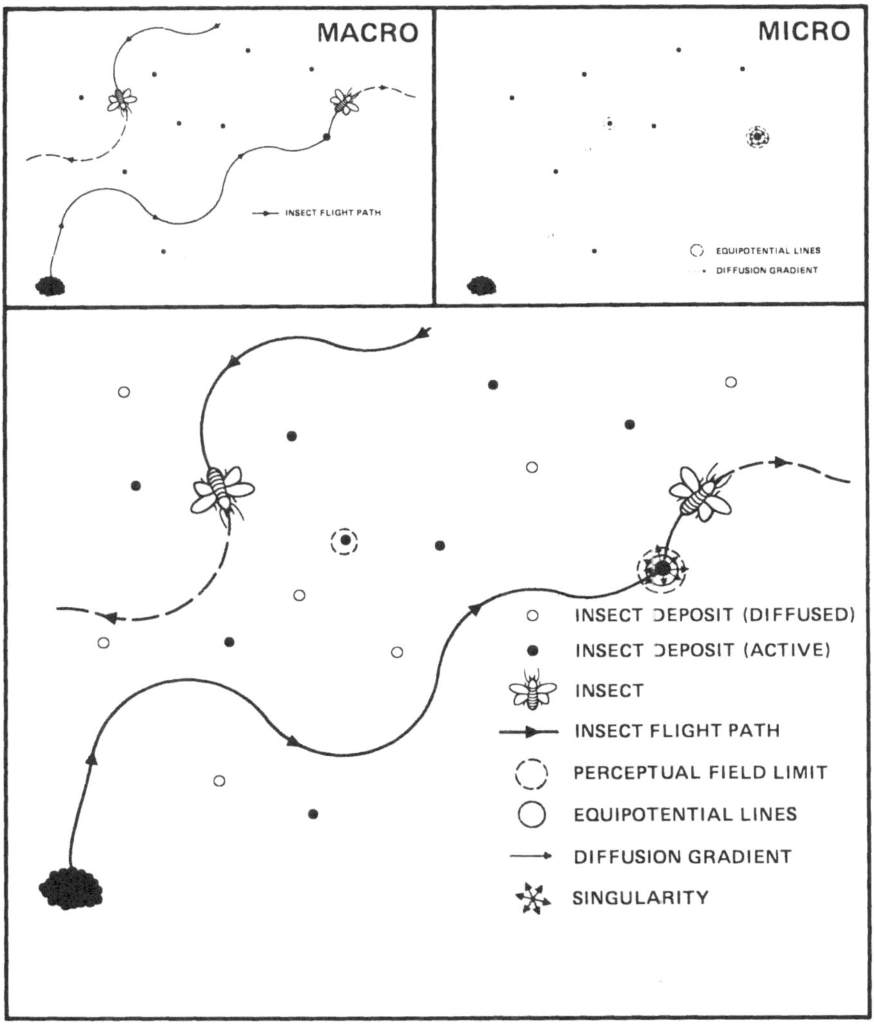

FIGURE 4.1. RANDOM FLIGHT PATTERN.

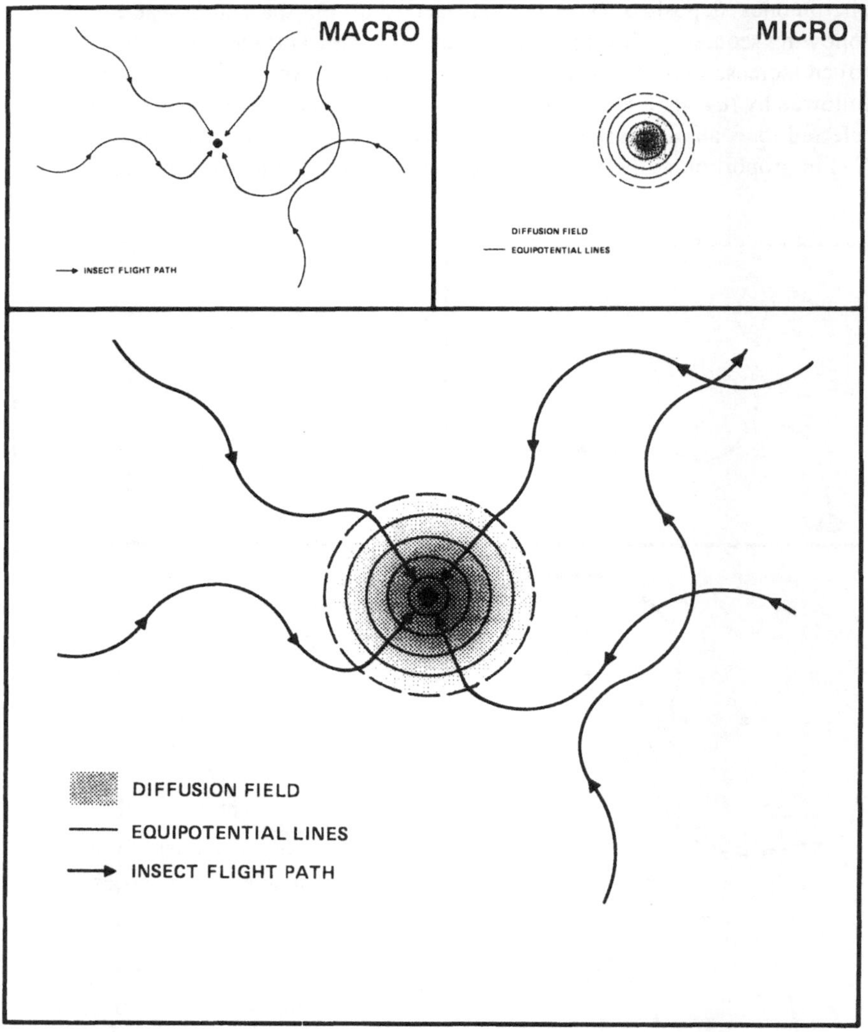

FIGURE 4.2. TRAJECTORIES IDENTIFYING "*PERCEPTUAL LIMIT*" OF THE FIELD.

of insects, the diffusive character of pheromones eliminates the possibility of any of the attractant persisting long enough to support the development of a preferred site. Above a critical number, the diffusive character of the pheromone field can amplify and take on a macroscopic space and time scale.

4.2.4. The Emergence of a Time-Dependent Structure as a Function of the Second Law of Thermodynamics. Thermodynamics' second law is behind the diffusion of the attractant throughout the nest-building area. The pheromones at a material deposit define a concentration of chemical potential, a conserved quantity. Accordingly, each deposit is a local source of instability driving the diffusion of pheromones in the neighboring medium (see Section 3.3). A globally stable distribution of pheromones is achieved when the chemical potential is partitioned uniformly throughout the medium enveloping the building area.

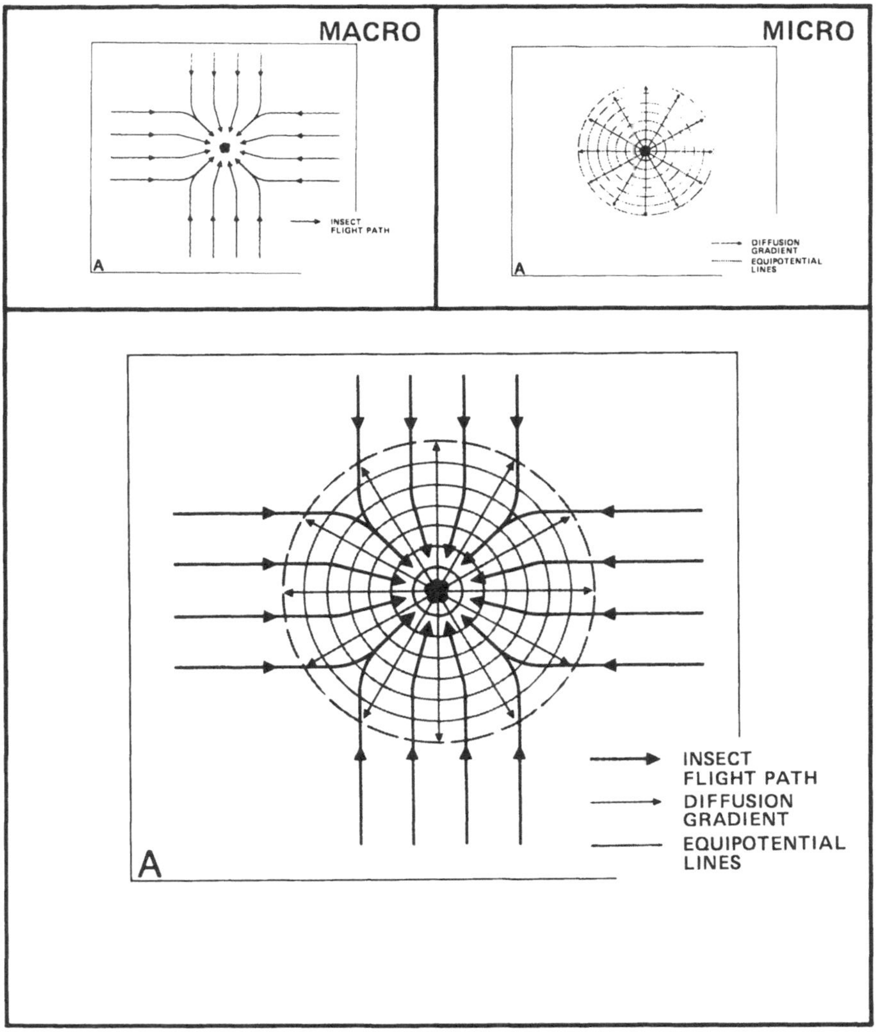

FIGURE 4.3. A MACRO/MICRO COUPLING OF INSECT TRAJECTORIES TO THE DIFFUSION FIELD.

The customary picture of the second law's role is that of dissolving constraints, of rendering structured organizations of atomisms into structureless aggregations (Chapter 3). For example, in the present context the second law eliminates the clustering of pheromones at a given site—it breaks up a particular spatial organization. The consequences of this dissolution of a relatively time-independent spatial arrangement, however, casts the second law in a less customary role, namely, that of a source of order. The consequent diffusive

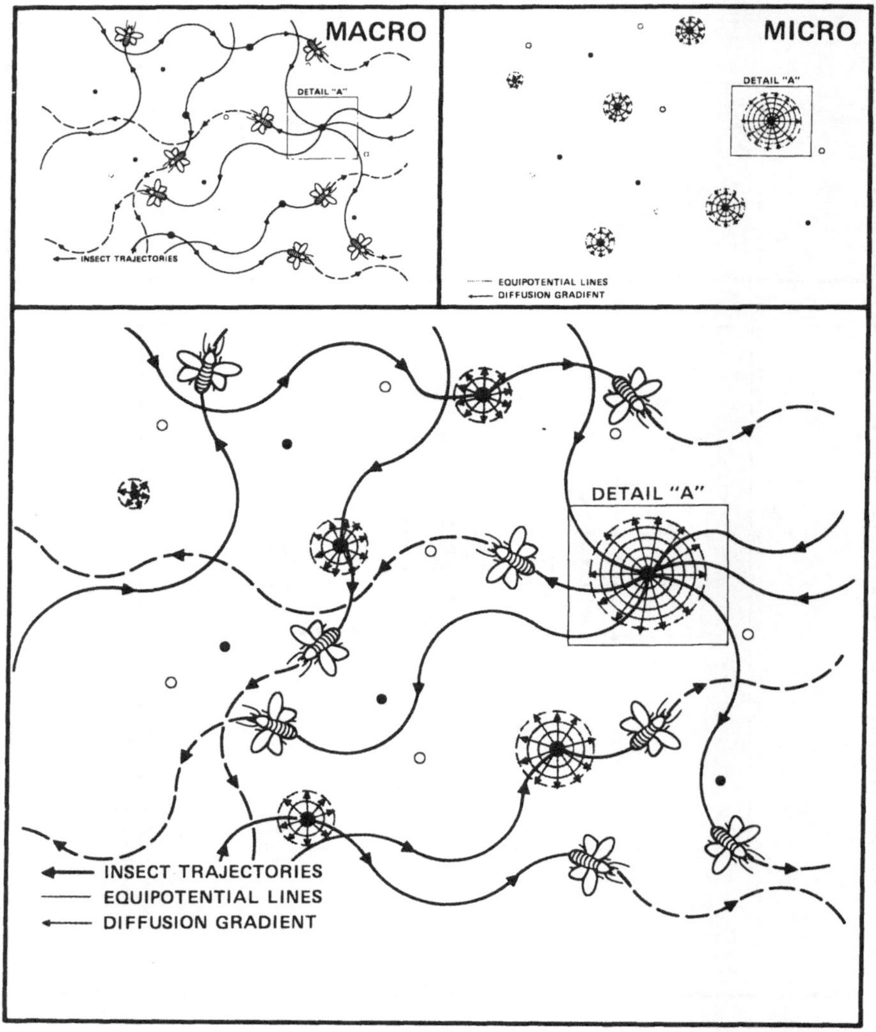

FIGURE 4.4. DEVELOPMENT OF PREFERRED SITES.

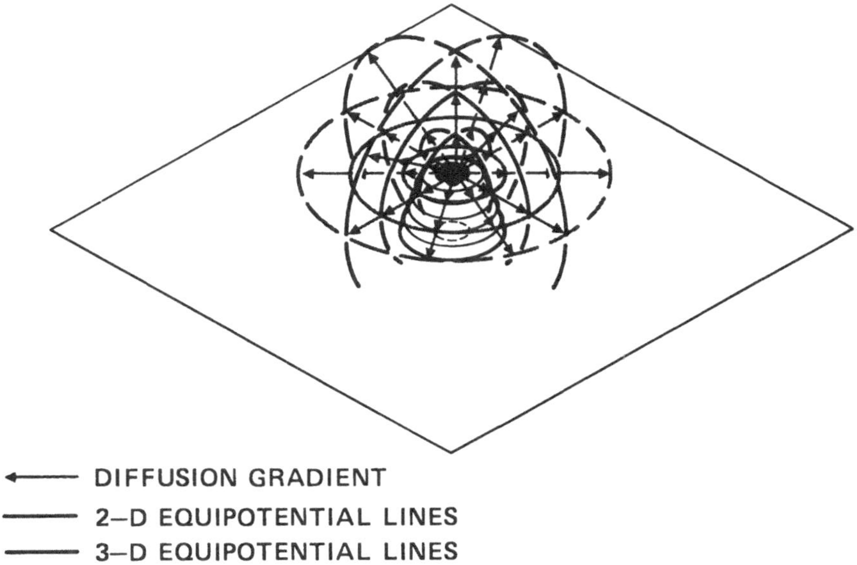

FIGURE 4.5. DEVELOPMENT OF A SINGLE PILLAR: TOP OF PILLAR IS THE ONLY ACTIVE SOURCE OF DIFFUSION.

flows constitute time-dependent configurations. The singularity and its associated gradient have been noted so far; richer configurations of diffusing pheromones are noted later as the analysis develops. Because the time-dependent configuration "induces" a dynamics in the insects (modulating their trajectories in the direction of preferred locations), and because the insects continuously bring energy and matter into the nest area and distribute them nonuniformly, the ordering tendency of the second law becomes especially apparent. As more sites, larger sites, and spatially proximate sites become established, the temporal orderings of diffusive flows become more numerable and their properties more varied. In a most important sense, the nest is a product of the second law. That is to say, this coherent time-independent structure (the nest) owes its origin and characteristic features in large part to the dissipative processes that the second law engenders.

4.2.5. Successive Point Attractors and the Development of Pillars. On the site of an original deposit a pillar may develop (Figure 4.5). The process by which it does so illustrates nicely the point about the second law as a source (ultimately) of time-independent structural organizations. It is noteworthy that as a pillar grows, its uppermost region at each new height acts as a point attractor for the insects while operating as a point repellor for the diffusing pheromones (see Section 3.3.3). (The pillar's top is always the most active site of pheromone diffusion.) This means that in the course of construction there is a continuous succession of point attractors or singularities, each characteristic of a relatively short-lived dissipative field. A pillar, therefore, can be viewed as a structural vestige (or, more abstractly, an alternative description) of properties defined temporarily in the geometry of pheromone diffusion fields. It is important to make the parenthetical remark about "alternative descriptions." It bears on the difficulty of understanding the origin of rate-independent structures in biology that constrain rate-dependent dynamics (see Pattee, 1972, 1979, and Chapters 1 and 13). The pillars of the termite nests are structures that restrict the trajectories of the insect builders. Similar to biological constraints, the pillars function to select one trajectory from among many virtual trajectories—to constrain the insect dynamics in a certain way at a certain time—but their physical origin is not mysterious; roughly, they arise from the play of the second law in the context of an open system, as outlined above.

To summarize: Pillars are included among the time-independent structures that constrain the dynamics of nest building. These structures, however, are erected on the basis of *time-dependent* geometries carried in diffusive flow fields. Is this a crude blueprint, a strategy, for the evolution of biological constraints in general? The advantage of the strategy sketched is that no extra-physical principles (consistent with but not reducible to physical principles) are implicated in the origin of biological constraints. A more detailed discussion of the strategy follows.

4.2.6. Saddlepoint Attractors and the Development of Arches. Figure 4.6 identifies a situation in which two pillars are being constructed independently. This situation occurs if the diffusion gradients do not interact within the insects' perceptual limit. If these conditions persist, then only individual pillars will be constructed. If, to the contrary, the two pillars are constructed sufficiently close to each other so that their diffusion fields interact within the perceptual limit of the insects, then a new type of singularity emerges in the flow-field geometry (see Figure 4.7). The interaction of the diffusion fields results in a saddlepoint singularity whose location is midway between the two pillars. Inspection of Figures 4.7 and 4.8 reveals that whereas there are indefinitely many flow lines (diffusion gradients, insect trajectories) focused on the singularities associated with the uppermost region of the individual pillars, there are only two one-dimensional flow lines focused on the saddlepoint singularity.

Figure 4.8 identifies the macroscopic paths of the insects as they couple to the saddlepoint in the pheromone field. Insects entering the pheromone field from certain angles will find the geometry of the field similar to that of a single, isolated pillar. Insects entering the pheromone field midway between the two pillars, however, encounter a flow gradient that takes them to a singular state marking a point roughly midway between the two pillars. After following the

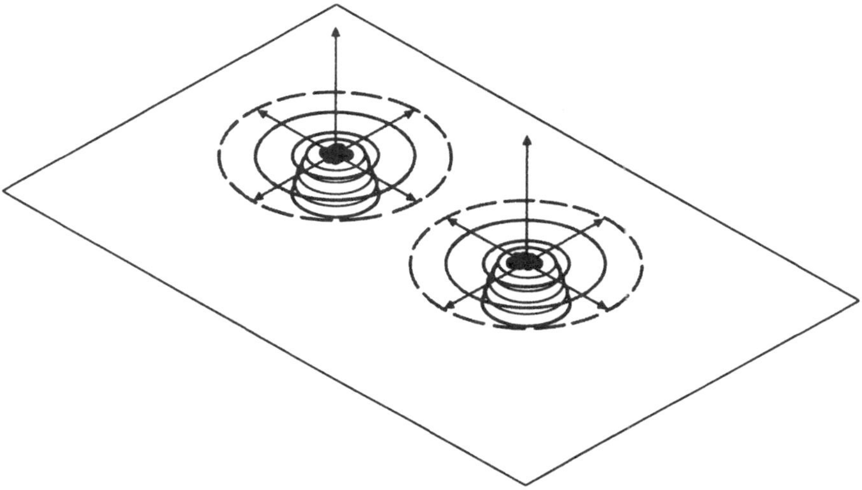

FIGURE 4.6. DEVELOPMENT OF A PAIR OF PILLARS WHERE DIFFUSION FIELDS INTERACT OUTSIDE OF THE PERCEPTUAL LIMIT.

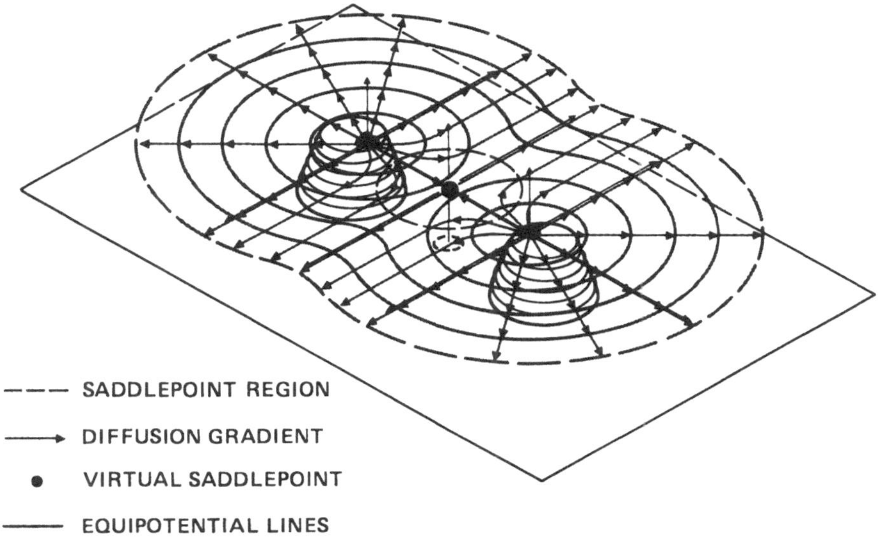

FIGURE 4.7. DEVELOPMENT OF A PAIR OF PILLARS WHERE DIFFUSION FIELDS INTERACT WITHIN THE PERCEPTUAL LIMIT.

gradient into this singularity, the insects encounter even steeper gradients that lead out of the singularity toward the two active pillar sites.

Singularities and their surrounding gradients are flow-field properties that bias the insects to make their deposits at particular sites. Patently, the singularities associated with the uppermost regions of the evolving pillars are geometric properties that have kinetic counterparts. Using more general terms, these

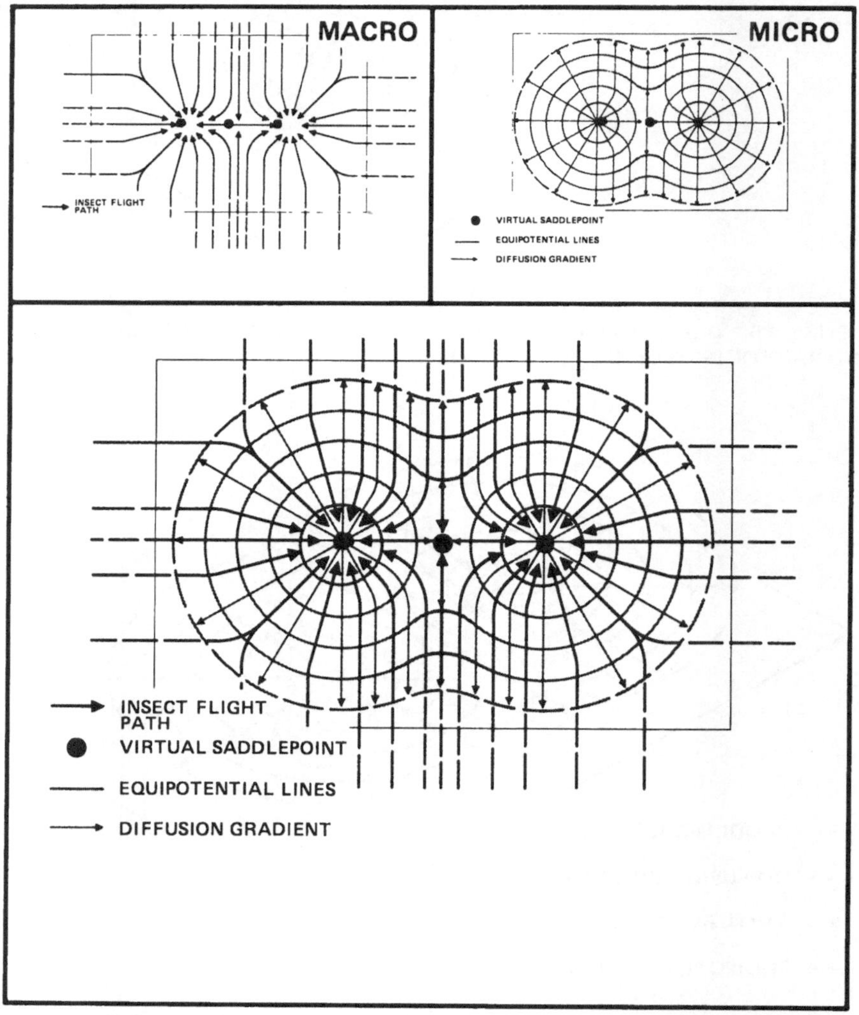

FIGURE 4.8. TRAJECTORIES OF THE INSECTS COUPLED TO THE DIFFUSION PATTERN'S FIELD.

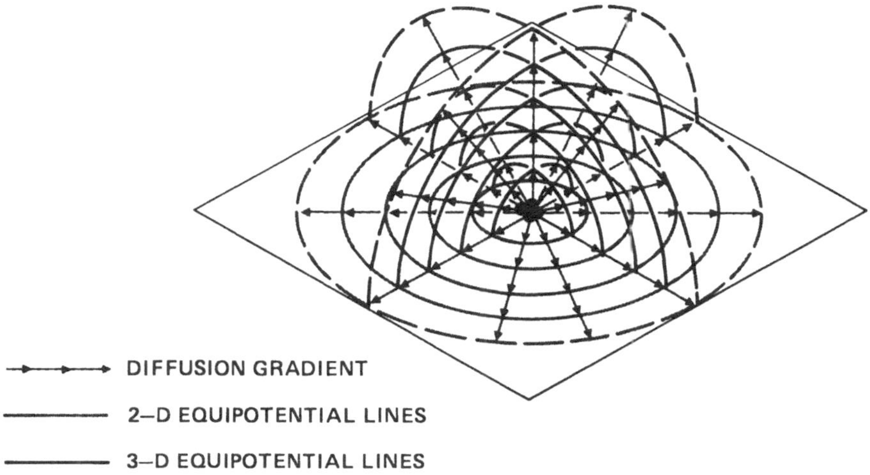

FIGURE 4.9. THREE-DIMENSIONAL DIFFUSION GRADIENT OF A PREFERRED DEPOSIT SITE.

flow-field properties specify kinetic properties, namely, concentrations of mass. The same thing, however, cannot be said of the saddlepoint depicted in Figure 4.7. It is a singularity carried in the flow field whose kinetic counterpart is *possible but not actual*. The saddlepoint is a flow-field property that specifies a *virtual* kinetic property, namely, a nonexistent (but incipient) concentration of mass lying between the two pillars at a distance above the ground surface that is roughly the average height of the two pillars. In strictly flow-field terms, the significance of the saddlepoint is that it introduces a new global symmetry that creates local asymmetries relative to the pillars. In strictly nest-building terms, the significance of the saddlepoint is that it introduces a depositing bias toward a region between the two pillars; loosely speaking, the saddlepoint encourages arch construction. And in strictly control-theory terms, the significance of the saddlepoint is that it is a low dimensional description, one that "alternatively describes," and one that plays back causally upon, a dynamics of high dimensionality.

The structure of the flow field induced by diffusion from two adjacent pillars can be studied in more detail. Three-dimensional descriptions of the flows resulting from a single pillar and from a pair of pillars are given in Figures 4.9 and 4.10, respectively. The saddle shape of the field created by the paired pillars is evident in Figure 4.10. It is also clear from inspection of Figure 4.10 that a plane perpendicular to the horizontal plane housing the saddlepoint singularity—an outset plane from the perspective of the microprocesses of diffusion (see Figure 4.11)—partitions the flow field generated by the pair of pillars. On either side of the outset plane are two one-dimensional spherical fields, each originat-

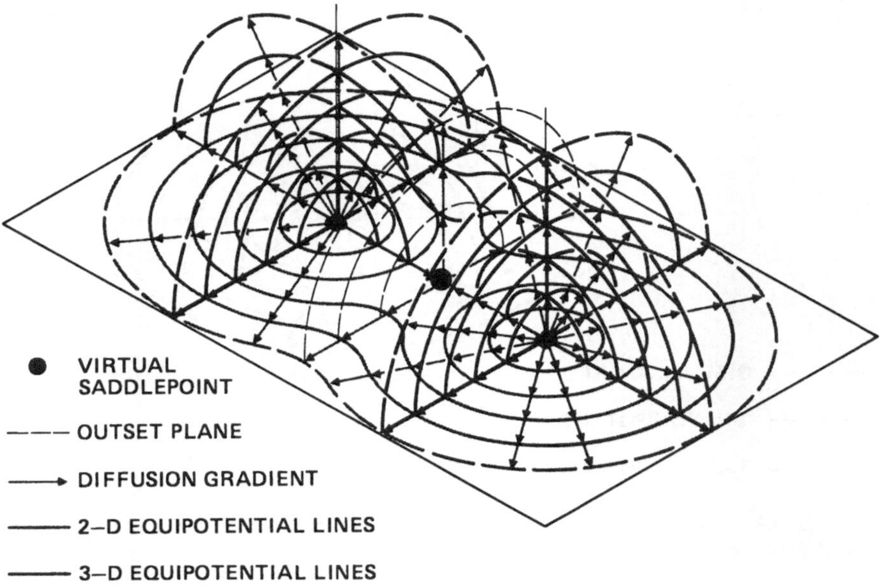

FIGURE 4.10. THREE-DIMENSIONAL VIEW OF EQUIPOTENTIAL LINES ASSOCIATED WITH INTERACTING PILLARS.

ing at the uppermost region of a pillar and flowing, in one hemisphere, radially toward the outset plane. At the outset plane the one-dimensional fields become two-dimensional, flowing away from the saddlepoint singularity. The coupling of the macroscopic insect trajectories to this microscopic diffusion pattern—the building of an arch—is shown in abbreviated form in Figure 4.12. Mirroring the diffusive flow lines, the possible insect trajectories are partitioned by a plane, technically an inset (see Figure 4.11), that is perpendicular to the plane in which the saddlepoint singularity lies.

4.2.7. The Final Phase: The Construction of a Dome. The completion of the arch is associated with the annihilation simultaneously of the two pillar singularities and the saddlepoint. The result is the emergence of a new point attractor (Figure 4.13) whose strength is a function of the combined strength of the two pillar singularities. The new singularity can interact with other local singularities, resulting in the emergence of an intricate pattern of new saddlepoints (see Figure 4.14). These saddlepoints and their associated gradient fields provide constraints that are used in the construction of a domed roof (see Figure 4.15). Upon completion of the roof a new cycle begins with random deposits on the upper surface of the roof (see Figure 4.16).

4.3. A Circular Causality of Self-Assembled Flows and Forces

The assembling of a pheromone flow field is kinetically based. Two forces are largely responsible: The force exerted muscularly by the individual insects in pursuing a given trajectory and in depositing material at a given site; and the ''generalized'' force sustained by the nonuniform distribution of chemical potential (see Section 3.3). The former force is responsible for the concentrations of matter within the building area; the latter force is responsible for irreversibly dispersing the pheromone attractant uniformly throughout the neighboring medium (force and flow vanish at chemical equilibrium). How forces that are sustained by the conservation gradients can drive flows is a question addressed by the Onsager relations identified in Section 3.3.2. Such force-to-flow couplings are obviously of significance to the nest-building activity, but they do not constitute the whole story.

The analysis laid out in the preceding reveals that the loci at which matter is deposited are the major determinants of the structure assumed by the field of diffusing pheromones. Individual insects flying in the nest-building area must generate particular forces in order to deposit at these select loci. How are the muscular forces of the insects constrained? The answer, of course, is ''by the

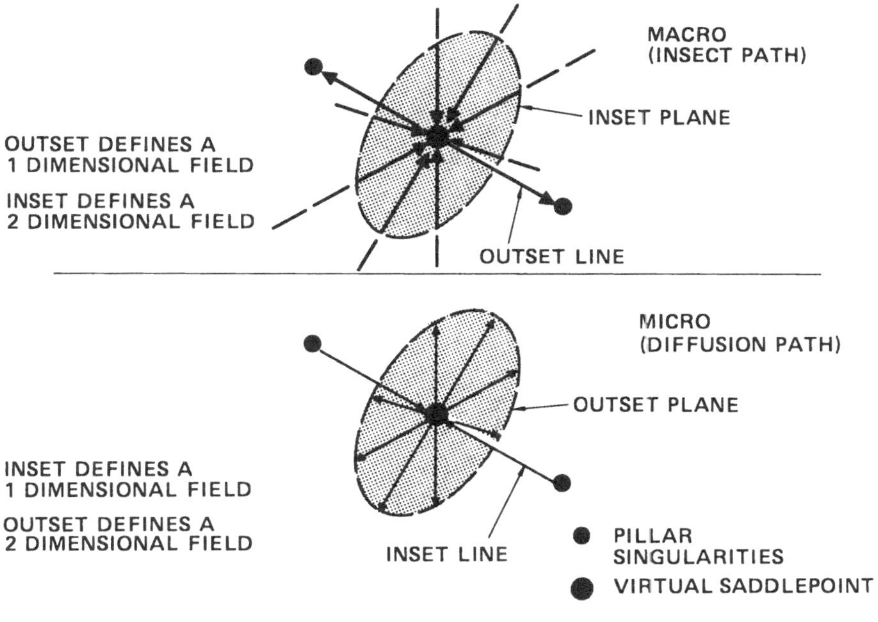

FIGURE 4.11. MACRO AND MICRO SADDLEPOINTS.

properties of the flow field.'' There is, therefore, a heterodox flow-to-(muscular) force coupling that complements the orthodox (generalized) force-to-flow coupling. This heterodox coupling is not addressed by the Onsager relations.

Understanding this heterodox coupling is primarily a matter of coming to terms with alternative descriptions of pheromone diffusion. To begin with, pheromone diffusion defines a chemical kinetic field and as such it can be assigned an energy value. The magnitude of this value is very small, however,

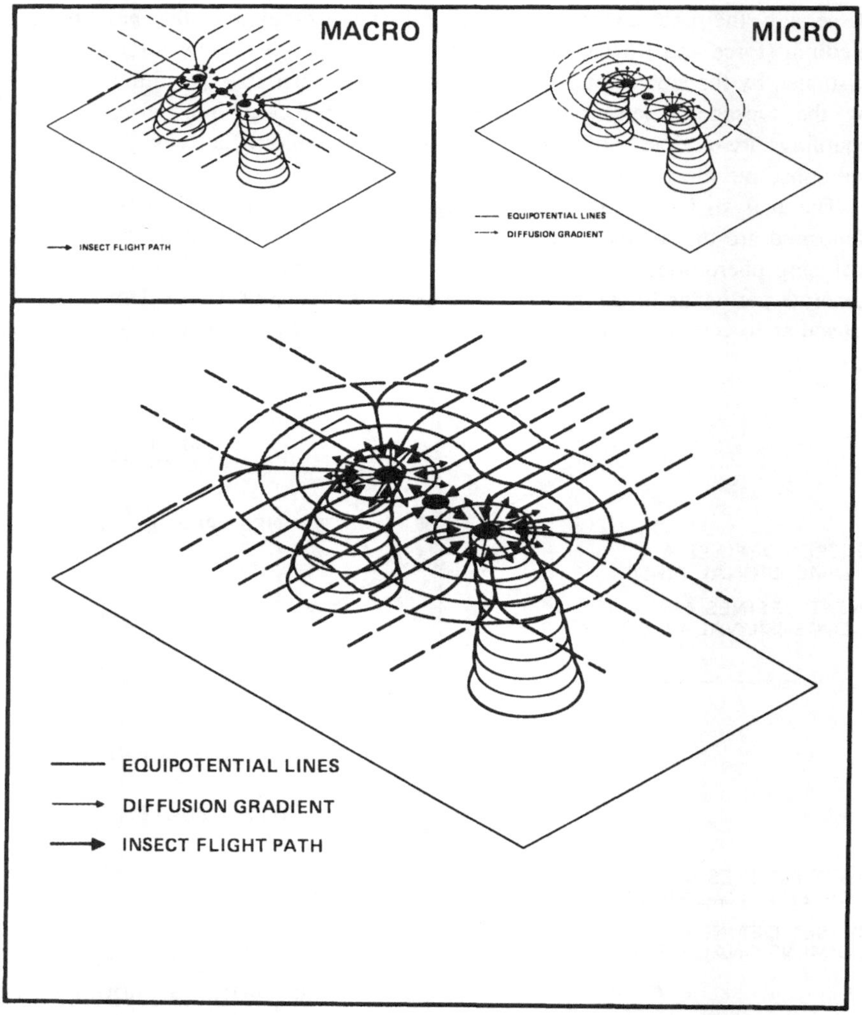

FIGURE 4.12. BUILDING OF AN ARCH.

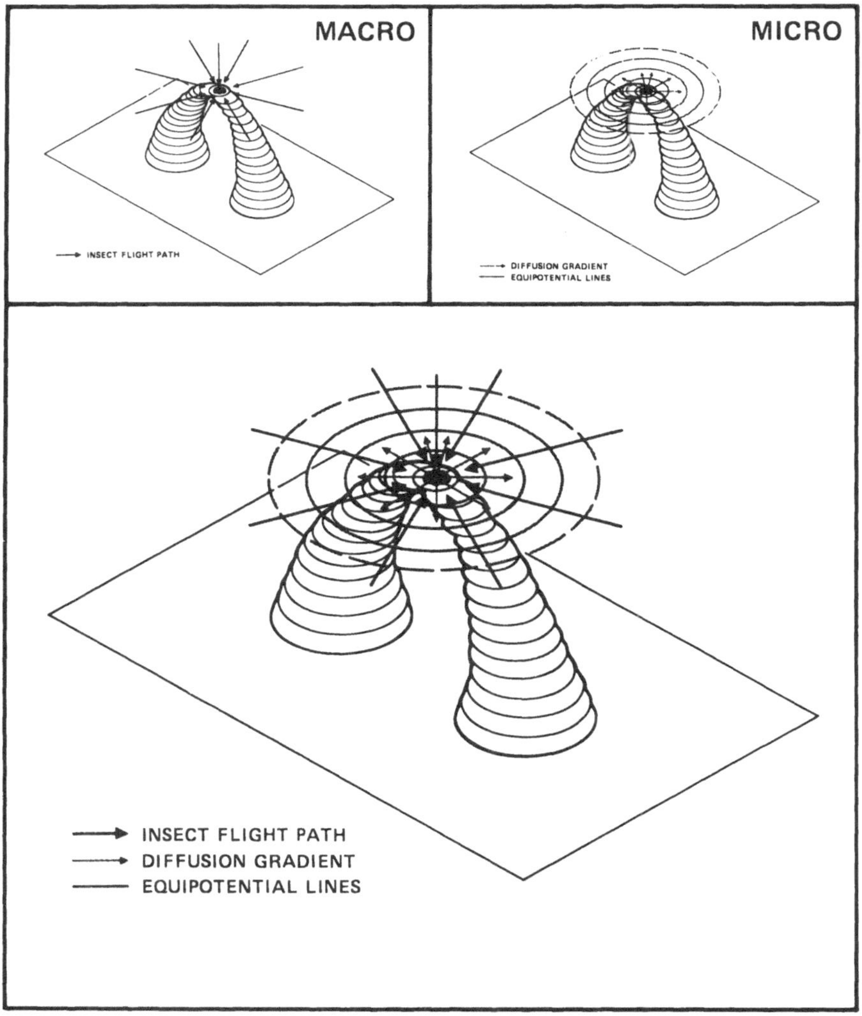

FIGURE 4.13. COMPLETION OF AN ARCH.

relative to the energy of the metabolic processes that drive the limbs and wings of the insect builders. Pheromone diffusion is a low energy field, relatively speaking. What is the implication of this fact vis-à-vis insect locomotion? It is simply that at the macroscopic scale of locomotory activity, the pheromone gradient cannot function as a force gradient. The approach by an insect to a deposit site is not a movement against a force gradient. Moreover, it is plainly *not* the case that departures of an insect from the chemical kinetic field's maxima

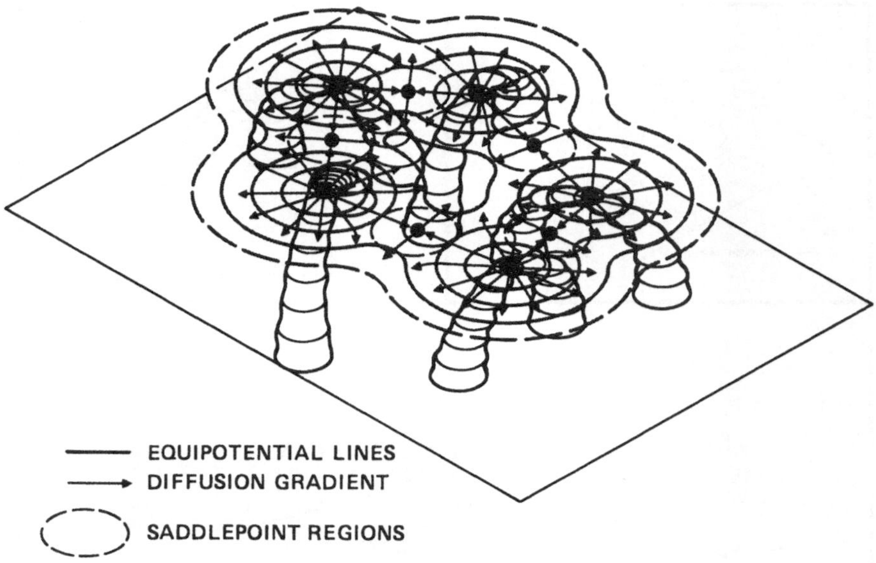

FIGURE 4.14. INTERACTIONS OF CHEMICAL (PHEROMONE) GRADIENTS EVOLVE NEW SADDLEPOINTS.

and minima give rise to forces that drive the insect back onto those states. In short, a locomoting insect's relation to pheromone diffusion is nothing like that depicted in Figure 4.17a, an insect is *not* driven by external forces (see Figure 4.17b).

What *is* significant about the pheromone diffusion, vis-à-vis insect locomotion, is the kinematic and geometric properties of the diffusion. Quite simply, what matters is an abstracted description of the properties of the chemical kinetic field—its pattern or morphology (see Section 4.5.1). In the act of locomoting, an insect's relation to pheromone diffusion is that depicted in Figure 4.18b; internally generated forces move the insect relative to the gradient (irreversible) and nongradient (reversible) regions.

The details of the heterodox coupling can be summarized as follows: It is a low-energy flow-force coupling. It might be termed a *perceptual coupling*, and the topological properties (i.e., singularities, gradients, saddlepoints, etc.) carried in the kinematic flow field might be termed *informational constraints* (cf. Runeson 1977/83; Runeson & Frykholm, 1983; Kugler, Turvey, Carello & Shaw, 1985; also Chapter 7). The intrinsic, dissipative geometry of the chemical flow field (that is, patterned layouts of singular and gradient states) provides a set of low-energy informational constraints that classify or restrict the high-energy force outputs generated by the insect's movement system.

Evidently, the behavior of the insects both contributes to and is constrained by the structural properties of the pheromone field. Insects contribute to the pheromone field through their frequent deposits and these deposits generate patterned diffusion fields that constrain, in turn, the depository activities of the insects. In this regard the evolving insect nest is exemplary of a "self-reading" and "self-writing" system. The system's "behavior" is both guided by—in the sense of "self-reading"—and a contributor to—in the sense of "self-writing"—the structure of the flow field. There is a "circular causality" of the following kind: *force field→flow field→force field→flow field . . . cycle*, which can be described, alternatively, as an *action→perception→action→perception . . . cycle* (see Figure 4.19). While the cycle is closed, by definition, in terms of complementary forces and flows, it is *open* in terms of the properties that constitute the descriptors of the flows, namely, the informational constraints. The primitives of the nest-building process (flow-field properties that govern its phases) arise *during* the force-flow cycles. Nest building is a self-complexing phenomenon (Kugler, Kelso & Turvey, 1982; Kugler, Turvey & Shaw, 1982); the governing properties of the diffusion field self-assemble and dissolve in conjunction with the expedient behaviors of the insects. The subsection that follows underscores the dual closure and self-complexing properties of nest building as an exemplary instance of circularly causal forces and flows. In the mean time the various ideas expressed in the present subsection are collected, by way of summary, in Figure 4.20.

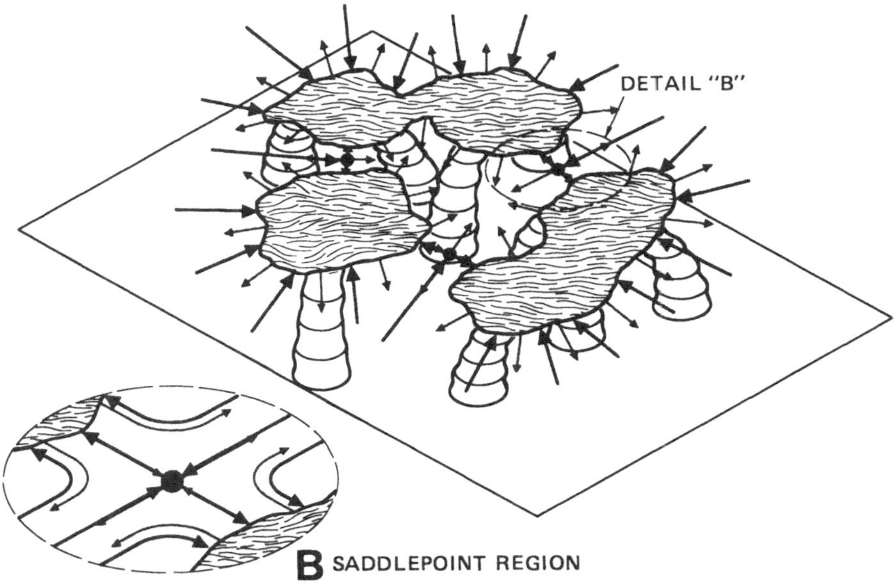

FIGURE 4.15. DEVELOPMENT OF A ROOF

4.3.1. A Closed Periodic Orbit Defined Over a Temporal Succession of Spatial Organizations. The insect nest-building activity defines a closed periodic orbit of macroscopic organizational modes. The closure is a function of ringing together successive stabilities and instabilities. A full cycle is completed when a set of (arbitrarily defined) initial conditions have been recovered. For instance, define the initial conditions as the random-deposit phase of insect

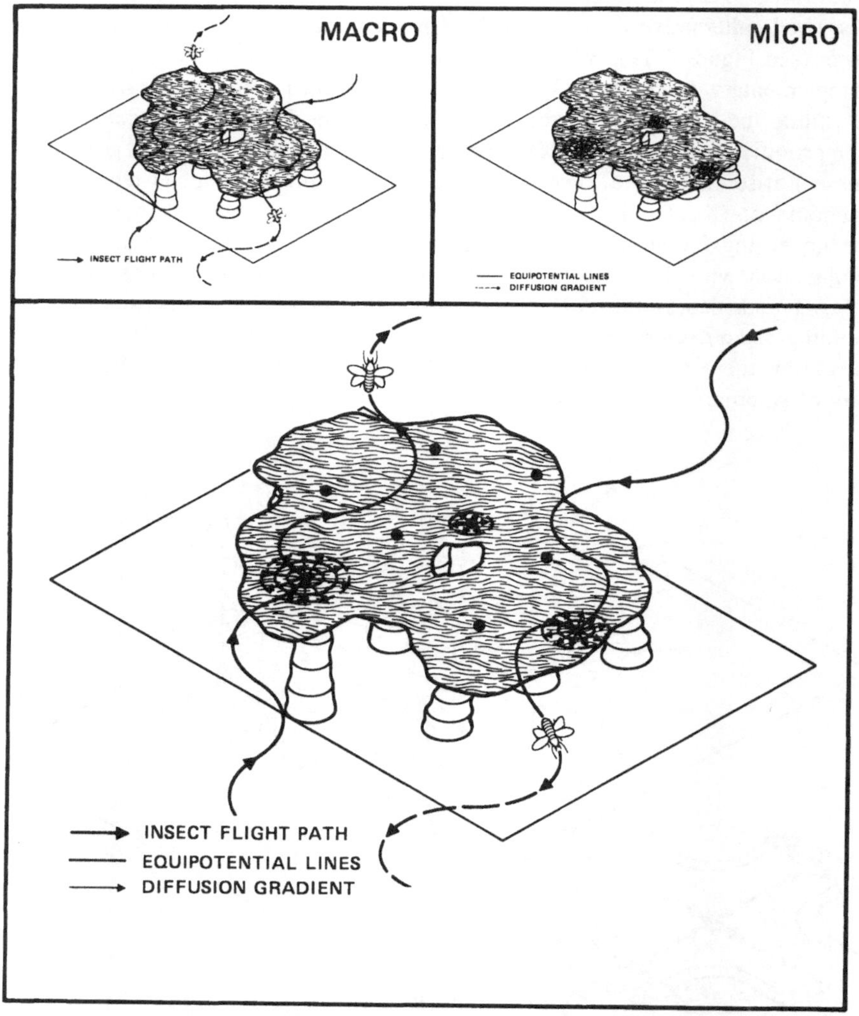

FIGURE 4.16. COMPLETION OF THE ROOF AND START OF A NEW CYCLE BEGINNING WITH RANDOM DEPOSITS.

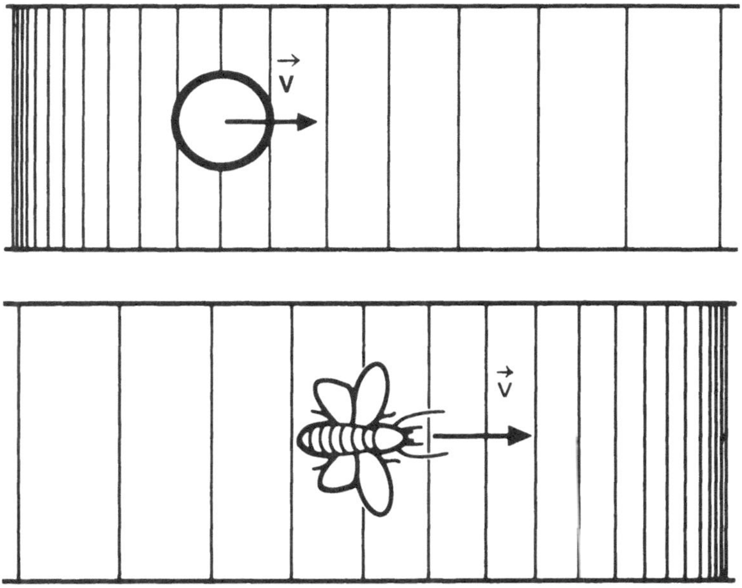

FIGURE 4.17.

Upper panel (a): IRREVERSIBLE FORCE FIELD GRADIENT.

The particle's motion is a *reaction* to the forces interacting with the particle. All changes in force arise externally.

Lower panel (b): IRREVERSIBLE FLOW FIELD GRADIENT.

The particle's (loco)motion is an *action* generated in response to (kinematic) flow gradients. Forces are generated from the interior of the particle. These forces can resist or complement external forces.

behavior. Then it is apparent that this set of initial conditions is recovered after the completion of the roof-building phase of construction. The recovery of the initial conditions, so defined, marks the closure of the building cycle.

The closure of the cycle is a function of (a) the topological constraints associated with the spatial and temporal processes and (b) the spatial (mean free path) and temporal (mean relaxation time) scales of the processes. As the building cycle evolves in time, from pillar construction to arch construction to dome construction, the spatial scale becomes successively smaller, completing a cycle when the building area is completely filled in. In this regard the temporal building cycle is a function of the spatial building constraints; the space scales the repetitive time course of insect nest building.

The successions of stabilities and instabilities are associated with the creation

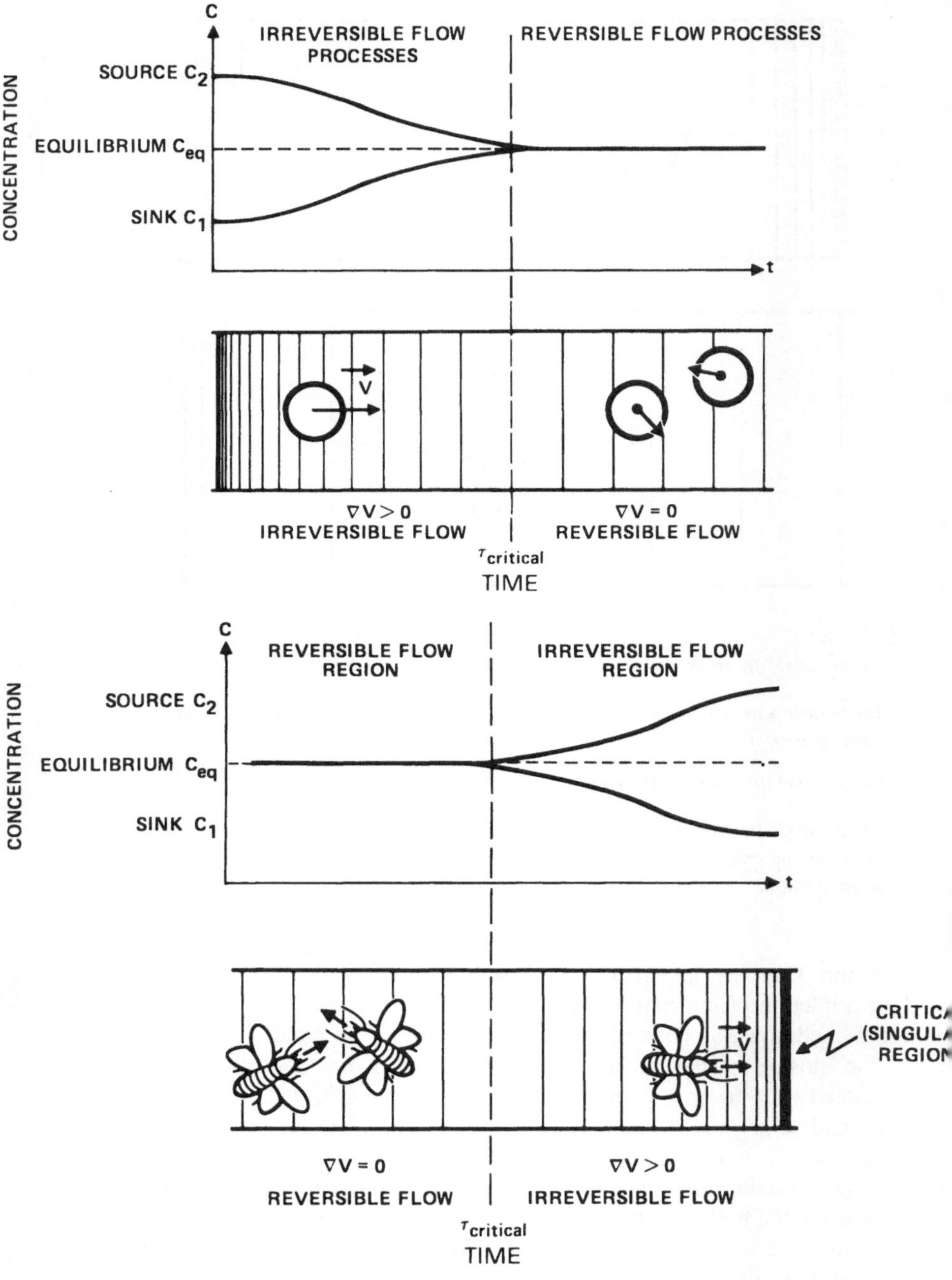
C
IRREVERSIBLE FLOW PROCESSES
REVERSIBLE FLOW PROCESSES
CONCENTRATION
SOURCE C_2
EQUILIBRIUM C_{eq}
SINK C_1
t
V
$\nabla V > 0$
IRREVERSIBLE FLOW
$\nabla V = 0$
REVERSIBLE FLOW
$\tau_{critical}$
TIME
C
REVERSIBLE FLOW REGION
IRREVERSIBLE FLOW REGION
CONCENTRATION
SOURCE C_2
EQUILIBRIUM C_{eq}
SINK C_1
t
V
$\nabla V = 0$
REVERSIBLE FLOW
$\nabla V > 0$
IRREVERSIBLE FLOW
$\tau_{critical}$
TIME

FIGURE 4.18

and annihilation of singularities in the topology of the pheromone flow field. Stabilities are associated with the processes in which the layout of singularities is invariant and instabilities are associated with the transition region in which singularities are created or annihilated. Each topologically distinct layout of singularities is associated with a qualitatively distinct, macroscopic mode of organization. Figure 4.21 identifies the circular ring of modes associated with the nest-building behavior.

4.3.2. A Comparison of Insect Architecture and Brownian Motion. At first blush, insect nest building and Brownian motion appear to have little in common. When examined from the perspective of micro-macro linkages and the types of nonlinear relations that can emerge, however, the two systems exhibit a good deal of "strategic similarity." The following strategies roughly parallel those identified for Brownian motion in Section 3.4.6:

1. A reservoir of energy from which (generalized) work can arise. (An energy reservoir is provided by the food that the insects consume during foraging. They draw on this reservoir to sustain their motions and to construct their nest.)

2. A microcosm of elements with a stochastic fluctuating nature. (The concentrations of pheromones constitute such a microcosm. Their appearance and active presence in the building area are stochastic properties to the degree that the locomotory activities and deposits of the insects in the initial stage are chemically unbiased.)

3. The presence of nonlinear components. (For example, the frequency with which deposits are made at a site and, therefore, the rate at which the site grows, are in direct proportion to the current size of the deposit at the site.)

FIGURE 4.18. *(Opposite page)* REVERSIBLE AND IRREVERSIBLE TIME EVOLUTION OF FIELD STRUCTURES. The time course (t) of evolution for an open system need not evolve toward a state of total time reversibility. Insect nest building provides an example of a time evolution process in which an initial reversible flow evolves into an irreversible flow. The initial reversible state is defined during the random flight phase of nest building. The emergence of preferred sites marks the transition *in time* from reversible (insect) flows to irreversible (insect) flows.

Upper panel (a): Transition of a simple atomism from a region of irreversibility to a region of reversibility. The "flow" of a simple (physical) atomism driven by an external force field generated by a potential (concentration) gradient ∇V.

Lower panel (b): Transition of a complex atomism from a region of reversibility to a region of irreversibility. The "flow" of a complex (insect) atomism "informed" by flow field properties generated by pheromone concentration gradient ∇V.

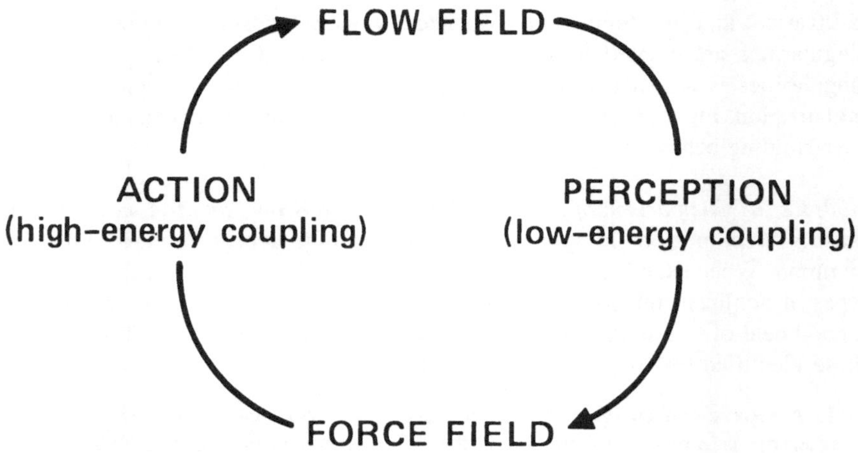

FIGURE 4.19. PERCEPTION/ACTION CYCLE.

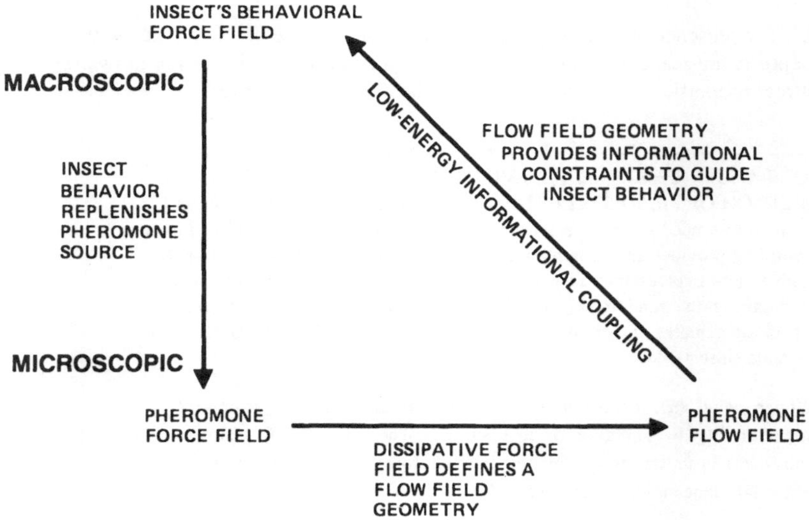

FIGURE 4.20. SELF-ORGANIZING INFORMATION SYSTEM.

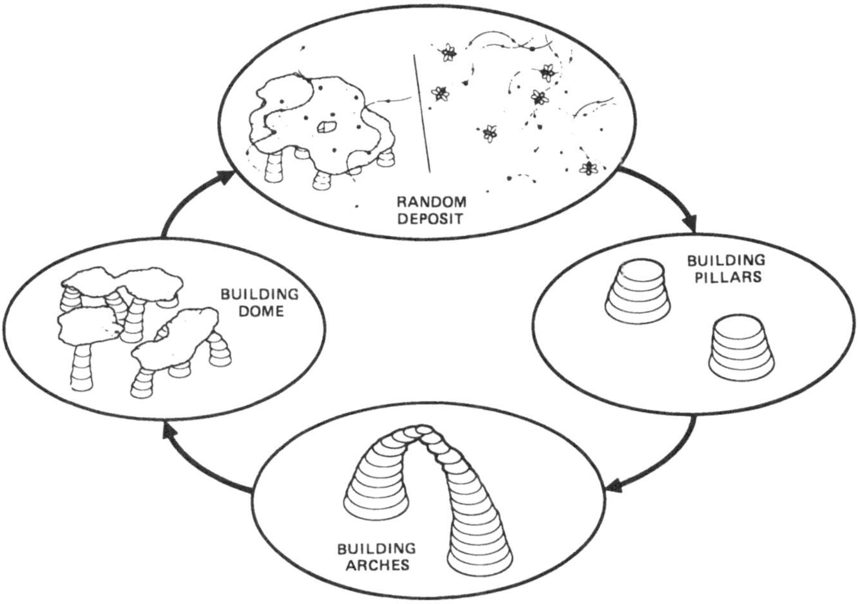

FIGURE 4.21. SELF-ORGANIZING RING OF NEST-BUILDING MODES. The evolving macroscopic morphologies, defined by the chemical diffusion fields, exemplify the self-organization of *living coordinate spaces*. These morphological patterns define intrinsic coordinates that orient the locomotory states of the insects. These *living coordinate spaces* can evolve and devolve in time. They stand in sharp contrast to *dead extrinsic coordinate spaces* that are time and space independent (see Section 2.2.2).

4. A critical quantitative change such that a nonlinear component is critically amplified. (For example, a qualitative transition in the macroscopic flow-field properties occurs as a function of the number of deposit sites which vary as a function of the number of insects participating in nest building.)

There are two major contrasts. The first is that the insect-nest system is *open* to the flow of energy (insect metabolism based on a food supply that is external to the nest) and matter (building materials, insects), whereas the Brownian system is *closed* to external flows. The macroscopic motions of a Brownian system are relatively small because of the continuous recycling of a fixed amount of momentum through the macro and micro field processes. In the case of insect architecture, however, the macroscopic organizations are not limited by internal reservoirs. The insects import and export conservations through the nest boundary. Because of this open flow of conservations, macrostructures such as pillars, arches, and walls can become amplified materially. The openness of the insect-nest system makes possible an elaborate evolution of geometric constraints and material structures.

The second contrast is with respect to the issue of causality. The behavior of a Brownian system is dominated by interactions that couple the macro motions of the Brownian particles to the micro field through kinetic properties. The behavior is in accord with the Newtonian paradigm of force causality. In contrast, the behavior of the insect system is dominated by interactions that couple the macro motions of the insects to the micro structure of the pheromone field through flow-field properties. In this regard the behavior of the insect system does *not* fall under the aegis of Newton's causal paradigm (see Figure 4.22). This contrast is sharpened in Section 4.4.

4.3.3. Self-Assembly and the Co-Mutation of Micro/Macro Field Constraints. Insect nest building is presented as an example of a self-assembling information system. The major lesson to be learned from this example may be the following.

It was remarked at the outset of this discussion that two perspectives can be taken on the pheromone field, one microscopic, the other macroscopic. What is happening from the micro perspective? Obviously a change is taking place. Constraints that bind the molecules together, that trap the molecules within a certain circumscribed area (viz., a deposit site), are continuously undergoing fracture in accordance with the second law of thermodynamics. When viewed microscopically, the field of pheromones is subject to entropic mutation. What is happening from the macro perspective? Again a change is taking place. But it is a change of a drastically different kind: Macroscopically, patterns are emerging, defined over the pheromone flow lines. From the macro perspective, with the emphasis on large-scale kinematic and geometric effects, the field is subject to negentropic mutation.

Traditionally the focus of physics has been microscopic mutations. This focus led to the formulation of thermodynamics' second law. In contrast, the traditional concern of biology has been macroscopic mutations. This concern spawned the formulation of a special theory of the evolution of form and function. However, if the nest-building example of a self-assembling information system generalizes, then it might be supposed that these two historically distinct concerns—entropic and negentropic change—may be intimately connected. That surely is one interpretation of current developments in nonequilibrium thermodynamics: A physics of self-assembly might rest fundamentally on a theory of *organization through co-mutation*, where co-mutation refers to an entropic microstructure and a negentropic macrostructure linked through circular causality.

4.4. Newton's Program and Nonkinetic Interactions

According to the arguments advanced thus far, nest-building behavior is grounded in the fact that the dynamics of individual insects is constrained by

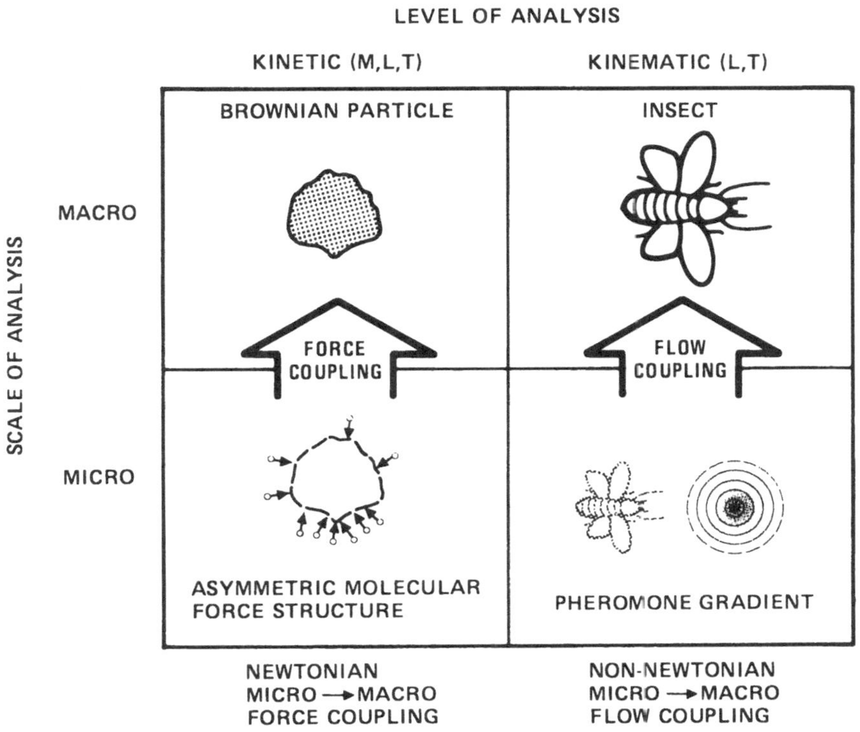

FIGURE 4.22. MICRO→MACRO COUPLING IN BROWNIAN AND INSECT SYSTEMS. In Brownian motion the forces at the micro scale can create symmetry-breaking instabilities. These lead to the creation of force structures that can participate actively at the force scale of the macro particle, resulting in a Brownian "hop." In the insect system, however, the forces at the micro scale can never assemble a force structure that is at the macro scale of the insect [because of the mass differential: *(pheromone molecule mass)/(insect mass)* << *1*]. Insect (loco)motion can, however, be influenced by the kinematic flow patterns generated by the micro force structure. In an insect system the forces responsible for orientation changes originate from the interior of the insect. These forces can be constrained or harnessed by the kinematic patterns of the pheromone diffusion field. The Brownian system is prototypical of a *micro→macro force coupling*. The insect system is prototypical of a *micro→macro information coupling*.

concentrations of matter, actual and virtual, that are (a) usually at a distance from the insects and (b) specified by patternings of pheromone distributions in the air. The decisive idea is that the material layout exerts its influence on the dynamics of insects through a field, where the relevant field properties are kinematic (and/or geometric, and/or spectral) rather than kinetic. What is to be made of interactions that occur through nonkinetic observables? As a partial response to this question, the present section addresses the commensurability of such interactions with Newton's program. A fuller response is provided

TABLE 4.1

LEVEL OF ABSTRACTION

	[1]KINETIC (M,L,T)	[2]KINEMATIC (L,T)	[3]GEOMETRIC (L)	[4]TEMPORAL (T)
MACRO	FORCE FIELD	FLOW FIELD	SPATIAL FIELD	SPECTRAL FIELD
MICRO	FORCE	FLOW	SPATIAL	SPECTRAL

ALTERNATIVE FIELD DESCRIPTIONS

Descriptions	*Dimensions*	*Properties*
[1]Kinetic (Newtonian Mechanics)	M, L, T	Forces—violent interactions; conservations such as mv^0, mv^1, mv^2; symmetries; potentials; constraints; singularities, X, $\dot{X}$ phase space
[2]Kinematic	L, T	Nonviolent interactions; singularities; constraints; symmetries; X, $\dot{X}$ phase space
[3]Geometric	L	Geometries, forms, boundary conditions
[4]Temporal	T	Spectra, frequencies; functions

The field descriptions are macroscopic. They do not include microscopic descriptions such as charge, spin and other attributes of interaction at the electromagnetic and quantum scale [adapted from Yates, F.E., & Kugler, P. N. (1984)].

by the section that follows. It details the characteristics of a paradigm in which flows rather than forces are primary.

It is suggested that the qualifier "informational" be applied to any interaction between two things (say, a system and its surround) that takes place through a description in which the mass dimension is suppressed. By definition, informational interactions are at odds with Newton's program, given that it seeks to account for all interactions in terms of mass-based descriptions (forces). There are, nevertheless, a couple of points of contact between the notion of informational (massless) interactions and the Newtonian program. To begin with, the Newtonian program exploits the interaction of one thing with another as a methodology for discerning relevant state descriptions. As remarked before, any system of interest can be described in numerous ways. Physical descriptions have two broadly defined degrees of freedom: level of abstraction and scale of analysis, as discussed in Chapter 2 and noted in Table 4.1. Ordinarily, however, when a system interacts with its surround, it does so through a small subset of possible descriptions. That is, interactions parse descriptions into relevant and irrelevant. Newton's interaction methodology was at work in the analysis of nest building and it revealed the hegemony of nonkinetic descriptions.

It has long been respected that not all mechanical interactions are identical. Where the interacting systems are ideal point masses (that is, they lack available internal degrees of freedom) the totals of mass, momentum, and energy of the two systems before and after the interaction are the same. This symmetry defines an elastic collision (Figure 4.23). An inelastic collision or interaction is defined by a different symmetry; total mass and total momentum are invariant but not total energy. Systems that participate in inelastic collisions are systems with interiors (Figure 4.24 and Appendix B). Energy is not conserved at the

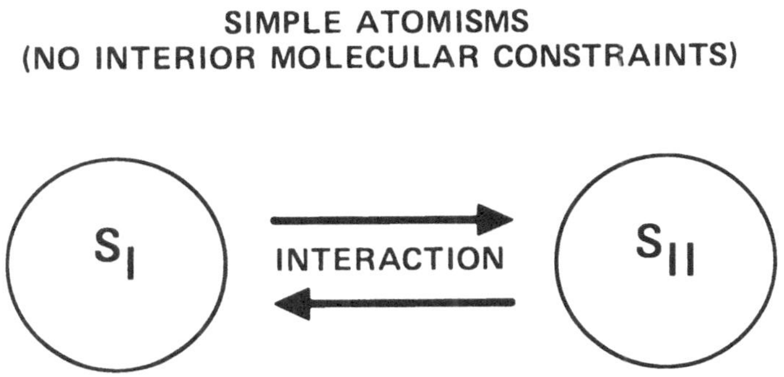

FIGURE 4.23. ELASTIC INTERACTIONS. Elastic interactions occur between particles that have no (interior) molecular constraints. In elastic interactions, all the energy and momentum appearing in the motion of the particles before the interaction also appears in the motion after the interaction.

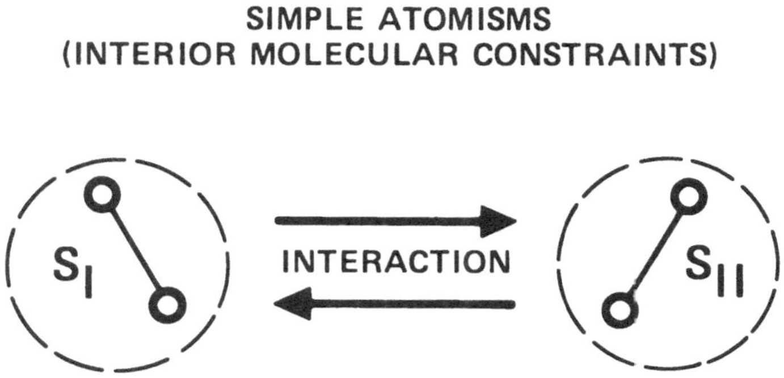

FIGURE 4.24. INELASTIC INTERACTIONS. Inelastic interactions occur between particles that have interior molecular constraints. In inelastic interactions, all the momentum that appears in the motion of the particles before the interaction appears in the particles' motion after the interaction. Energy, however, is not constant. The missing energy is absorbed by interior molecular constraints and is stored in vibrational or rotational modes.

macroscopic mechanical scale because some of the energy of interaction is absorbed by the internal degrees of freedom. Nevertheless, in common with elastic interactions, inelastic interactions take place through a force description.

Paradigmatic of systems that collide inelastically are oxygen molecules. These diatomic structures take up some of the energy of a collision in the atomic bond. But what of systems that are enormously more complex than oxygen, systems that (a) have a detailed, functionally rich internal microstructure and (b) persist as a kind only insofar as the interior is maintained intact? These systems (see Figure 4.25) cannot afford the expense of being restricted to force-based interactions with their surround. Yet interact they must (to garner fuel, for example) and their interactions must be frequent. By the present argument, complex atomisms and their surrounds interact primarily through informational interactions. The advantage of informational interactions, with their suppression of the dimension of mass, is that systems and surround can interact without violence; complex interiors need not be subject to shock waves. (See Appendix B for a discussion of collisions, fracture and internal complexity.)

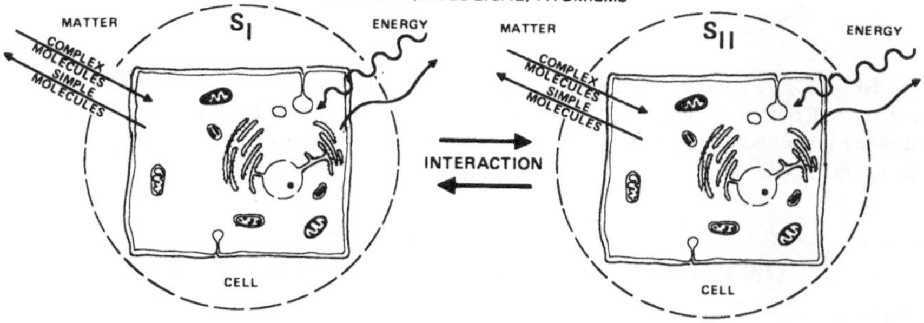

FIGURE 4.25. INFORMATIONAL INTERACTIONS. Particles with large numbers of interior (molecular) constraints can time-delay energy flows from the interior to the exterior. The result is the emergence of thermodynamic flow processes. These flow processes will vary as a function of the number of internal constraints. The time-delaying of *exterior→interior→exterior . . . energy flows* changes dramatically the particle's external mode of interaction; the particle is no longer a "slave" to external force fields. The time-delaying of internal (thermodynamic) energy flows provides a "particle-based" source of forces that can actively compete with the external forces. As the internal force fields increase with respect to the scale of the external force fields, the role of the mass dimension becomes successively less relevant for sustaining interactions. What remains relevant, however, are the fundamental dimensions of length and time. These nonforce dominated interactions are termed "informational interactions" and are viewed as methodologically continuous with elastic and inelastic interactions. Informational interactions are the physical consequence of particles with complex interiors that can time-delay large amounts of energy relative to external force fields.

The interactions participated in by biological systems as macroscopic entities are mainly "soft" and "controlled" (Kugler, Turvey, Carello & Shaw, 1985).

In sum, the intimation of the Newtonian program—that interactions may be typed by the internal complexity of the participating systems—is upheld by biological systems and their informational interactions. Nonkinetic, informational interactions (despite their oddness) are on a continuum with elastic and inelastic interactions.

4.5. The Gibsonian Paradigm: The Primacy of the Flow Description

A distinction has emerged with potentially far-reaching implications for understanding information's physical basis: There are interactions that take place through kinetic observables. These are orthodox, Newtonian interactions. And there are interactions that take place through kinematic (and/or geometric, and/or spectral) observables. These are heterodox, non-Newtonian interactions. As stated, they could be termed "informational."

It is self-evident that Newtonian interactions are causal; indeed, in the interactions addressed by particle mechanics are to be found the historical roots of the causal principle (Margeneau, 1950; Bohm, 1958). But wherein is the causality, if any, of informational interactions? Doubts about its presence are motivated, of course, by the fact that causal accounts are universally expressed in mass-dependent terms. In the remarks that follow, causality will be claimed for informational interactions, but the causality is with respect to the information, not the interactions.

Margeneau (1950) expresses the principle of causality as follows: Let X and Y be *complete* states of a specified system (say, particle and field) at times t_l and t_2, t_l being earlier than t_2. If X is realized, Y will certainly follow. The essence of this causal principle is the invariability of natural laws. The original mild requirement of invariance with respect to time was upgraded by Einstein to invariance with respect to transformations in which the observer passes from one system of reference to another (Chapters 2 and 3). In classical field theory, which addresses the juxtaposition of particles and field, there are two classes of laws—those that govern the action of the field on the material particles (or, conversely, the reaction of material particles to the field) and those that govern the internal relations of the field and the field's dependence on the presence of matter. Conventionally, the physical setting to which the classical field theory applies is that depicted in Figures 4.17a, and 4.18a. This conventional setting is permeated throughout by the causal principle—the particle's motions are causally determined and the field's properties are causally determined.

Figures 4.17b and 4.18b identify an unconventional setting for classical field theory. The particle has an on-board, deployable source of potential energy. Moreover, the energy level of the field is too low to function kinetically as a determinant of the particle's trajectory. If the particle moves, it is generally by virtue of forces that the particle itself generates, and these forces do not act on the field in question (the pheromone flow field) but on another field (e.g., air pressure). A state description of the particle/field system in terms of the momentum and position of the particle and the kinematic, geometric, and temporal properties of the pheromone field would not sustain a causal account of the motion of the particle depicted in Figures 4.17b and 4.18b. The reason is at least twofold: First, kinetic terms are absent from the relevant field description; and second, the particle and field descriptors given do not embrace the entire state of the particle/field system. (Note the italicized qualifier "complete" in the preceding statement of the causal principle.) The particle, a biological system, has a "history." Identifying the significant variables (variables of state) of any system to which hereditary equations might apply is notoriously difficult (cf. Shaw & Alley, 1985). Moreover, to compound the difficulty, the "history" is replete with successions of bifurcations.

This caveat, however, does not render the principle of causality irrelevant to the particle/field system of Figures 4.17b and 4.18b. It must be assumed, as a working scientific premise, that ignorance—of the proper observables and of the form (potentially heterodox) of the laws that relate them—obscures the causal principle. The point being made is not a metaphysical point: It is simply that an invariant causal law cannot be written over the kinematic, geometric, and temporal variables of the field, on the one hand, and the motion parameters of the particle, on the other hand. The particle's trajectory, from a consideration of properties of the field, is plainly not determined. When special boundary conditions (see Chapter 13) hold, the particle's trajectory is merely consistent with certain spatial/temporal properties of the field; when these conditions do not hold, even this modest relation is absent. To be redundant: The field's properties do not determine uniquely the particle's trajectory.

Nevertheless, it is true that reliable and reproducible effects follow from a juxtaposition of a particle and a field of the kind depicted in Figures 4.17b and 4.18b. After all, in the example under consideration, nests are repeatable from one occasion to another, from one setting to another, with similar features and design. What is the source of this reliability and reproducibility? The answer, it would seem, must lie with the field's properties. Relations within the field and the dependence of the field on the layout of matter must abide by the principle of causality. That is to say, there are invariable laws internal to the field, and there are invariable laws by which material layouts structure the field's properties. One might presume the existence of field equations, involving kinematic observables, wholly or in part, that play a role analogous to the kinetic field equations characteristic of classical field theory. The upshot is that in the juxtaposition of a complex atomism and a kinematic flow field, nondeterminacy of the atomism's behavior goes hand in hand with determinacy of the field's properties.

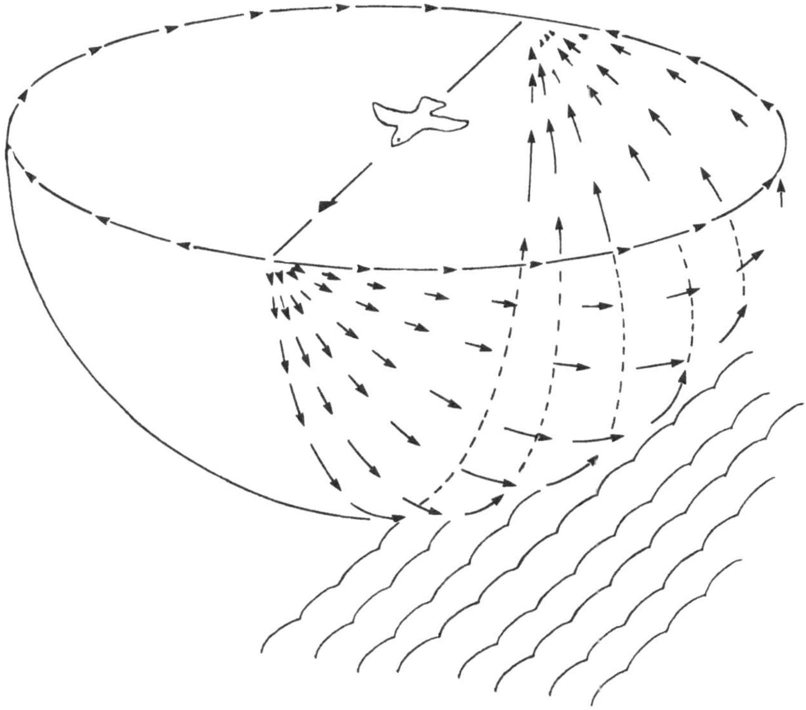

FIGURE 4.26. GLOBAL OPTICAL FLOW MORPHOLOGY SPECIFYING DISPLACEMENT (forward locomotion) relative to a textured medium (clouds or ground surfaces).
[adapted from Gibson, J. J. (1966)].

Three ideas are decisive in generalizing the preceding discussion to the locomotory activity of any animal, namely (a) alternative descriptions in macroscopic kinematic (and/or geometric and/or spectral) terms can be given of the low-energy distributions enveloping animals; (b) these alternative descriptions are law based and determinate; (c) these alternative descriptions do not cause the movements of animals but wholly constrain them. Together these three ideas characterize the perspective on the interactions of animals and environments that was advanced by Gibson (e.g., 1958, 1966, 1979; Reed & Jones, 1982) and previewed in Chapter 1. A conceptual framework exists, therefore, for understanding the interactions of complex atomisms (it should, perhaps, be termed the Gibsonian paradigm) that stands in contrast to the familiar Newtonian paradigm for simple atomisms (see also Weiss, 1963, for a related discussion, focusing on the limited applicability of Newton's program to the "cell as unit"). To close this circle of ideas, Gibson's standard descriptions of flow fields is presented with a few selected quotations.

Figure 4.26 is taken from Gibson (1966). It depicts a flow field generated by the movement of a complex atomism through a medium (air) filled with a reverberating network of light rays. The network is produced (ordinarily) by multiple reflections from a persisting layout of opaque, material surfaces. About the flow field's physical character Gibson (Reed & Jones, 1982) commented:

> I have been brought up on the idea that Isaac Newton's laws of motion are basic and so they are in a limited way. They are the basis of mechanics and of machine civilization. Newtonian motions are important, but as one investigates the laws of visual perception, one begins to realize that optical motion is altogether different from material motion. For example, the optical motions in a field of view when I am moving have no inertia. Maybe the photons of radiant energy, the stimuli for the photoreceptors, have inertia in some sense, but displacements or transformations in the field of view are noninertial. At any rate that's worth discussing. (p. 99)

Relatedly, he wrote (Gibson, 1966) about analytical depictions of optical flow fields, such as Figure 4.26, in these terms:

> There is something wrong with this graph, however. It is analytical, not topological. The vectors, of course, represent the velocities of points. Accordingly there is a temptation to think of the graph as a collection of stimulus motions analogous to the physical motions of particles in space, but this thought would be very misleading. The insubstantial motions of margins in an optic array are not analogous to the substantial motions of particles. Optical motions have no inertia and do not collide with one another; they are a set of relations, not of independent elements. The optical flow in the graph is not to be thought of, therefore, as a collection of separately moving parts, each perceptible by itself. The analysis of spot velocities is useful, but it should be remembered that the optic array does not consist of spots. If a terrestrial layout is analyzable by *faces* and *facets*, then its optical array is analyzable into *forms* within *forms*. Every form, even the smallest, undergoes a transformation, but this fact is not shown by the graph. (p. 198)

On the significance of kinematic flow fields to action he argued (Reed & Jones, 1982):

> No fact of behavior, it seems to me, betrays the weakness of the old concept of visual stimuli so much as the achieving of contact without collision—for example, the fact that a bee can land on a flower without blundering into it. The reason can only be that centrifugal flow of the structure of the bee's optic array specifies locomotion and controls the flow of locomotor responses. (p. 14)

He denied, however, that "there is ever a one-to-one correspondence between stimulation and perception," asserting instead that "perception is wholly constrained by . . . information" (Reed & Jones, 1982, p. 399). Finally, when chided by Boynton (1974) for the absence of "physical life" in his new sort of abstract geometric optics, Gibson (Reed & Jones, 1982) responded that his analysis:

> . . . is less concerned with seeing light than with seeing by means of light. Consequently I believe that it can *bracket* the disciplines of radiometry, photometry, and psychophysics and base itself on the *invariant* properties of an optic array. These geometric concepts may well prove to be more lifelike than the physics of photons. (p. 75)

Gibson distinguished his analysis of the optical support for perceiving and acting from the science of physical optics by referring to his enterprise as "ecological optics." "Ecological optics" is meant to emphasize that the focus is on the patterning of structured light fields that are at the ecological scale of animal/environment interaction (rather than on the wave/particle, emission/absorption properties at the quantum level). While quantum electrodynamics provides a theory of the low-energy photon field, of which the optical flow field is an abstraction, it provides no account of the systematic transformations of patterns sustained by the photon field. These patterns and their transformations are *scale sensitive* and their investigation must be conducted at the appropriate scales in space and time. Ecological optics is a theory of optics appropriately scaled, in space and time, to accommodate animal and environment interactions.

A limited presentation of ecological optics follows, one that will suffice to make the connection between Gibson's optical analysis and the field analyses provided earlier in this chapter. The limitations are not trivial, however. The presentation assumes open environments rather than object-cluttered environments. This assumption permits one the luxury of projective geometry; to be able to talk as if optical units map from preceding to succeeding arrays. In truth they do not. The clutter on the ground plane means that surfaces are concealed and revealed, optical units deleted and accreted, as an animal locomotes. In sum, the assumptions and tools of projective geometry apply only restrictedly to the transforming patterns of optical structure. In more general terms, Gibson doubted the propriety of available mathematics to the description of optical flow fields and encouraged the discovery and development of analyses tailored to the field's peculiar problems (Gibson, 1979; Reed & Jones, 1982). His skepticism ran deep: If optical events are not motions, then they resist description by changing coordinates. If optical structure comes into and goes out of existence in the course of an animal's locomoting, then efforts to apply a coordinate system will fail; nonexisting optical units have no coordinates. Gibson sensed that the analysis of light at the ecological scale would have to proceed without the conventional coordinate system notion that has served physics so well (Reed & Jones, 1982, p. 193).

4.5.1. Ecological Optics: An Overview of Optical Flow Fields. Physical optics is rarely applied in any thoroughgoing fashion to visual perception. It is more commonplace to describe the optical support for vision in the terms of geometric ray optics. In either case, however, the eventual description of the optical support will have very many degrees of freedom and the emphasis will be on the quantities that can be assigned to microscopic properties, individually or as aggregations. So described, the light that surrounds animals grossly underconstrains the details of their movements.

A very different description, however, shows that the ambient light is rich in action-related properties—properties that address directly the problems that animals must solve in orienting and adjusting their movements to the surrounding layout of surfaces (Gibson, 1979; Turvey & Kugler, 1984). This description is macroscopic and qualitative. Its construction is detailed briefly below (for more details, see Gibson, 1979; Koenderink & van Dorn, 1981; Lee, 1974, 1976, 1980).

All surfaces (except most mirrors and windows) are textured. An element of texture is a region that reflects the same amount of light from all points within its borders. At its borders, however, there is a marked transition in intensity. The light reflected from a surface, therefore, will have structure imposed upon it by the texture elements. And where surfaces differ in substance and textural detail there will be corresponding differences in the reflected light. Intuitively, one gets the impression of an infinitely dense nesting of structure; light is structured in the large by the faces of surfaces and in the small by the facets of surfaces. Accordingly, at any point of observation there will be a unique, regular arrangement—an array—of differences in intensity in different directions. It is particularly useful to think of this optic array as a bundle of narrow cones of light with their apexes at the point of observation (Gibson, 1979). Because each cone has as its base a distinct texture element, it will be distinguished from its neighboring cones by the intensity (and spectral composition) of the light it contains.

As intimated, a unique optic array is defined at each of the infinite number of points of observation. Now, imagine a trajectory such that many points of observation are intercepted successively. Such a trajectory occurs whenever a person or animal walks, runs, jumps, and so on. To keep things very simple, suppose that a person is walking forward in a straight line and none of the surrounding surfaces (the furnishings of a room, for example) is moving or changing. The person's movement could be represented by a velocity vector that has its origin in the person and points in the direction of movement. Suppose, however, that matters are reversed and the person is treated as *not* moving and the surrounding surfaces are treated as moving in unison. Then a zero velocity vector can be assigned to the point of observation that the person occupies and velocity vectors can be assigned to the many components of the optic array. By so doing a velocity vector field is created (see Figure 4.26). To generalize: How a person or animal moves relative to the surroundings, what these surroundings are like, and how they change relative to the person or animal define uniquely a velocity vector field—an optical flow field. It should be expected, however, that for movements (or surrounds, or changes) of the same type there will be similarities in the generated optical flow fields. That is to say, looked at *more macroscopically* the *form* or *morphology* of the velocity vector field should uniquely specify the kind of movement that is being made (or the type of surface layout that exists or the type of environmental change that is occurring).

The more coarse-grained, macroscopic perspective is built from the trajectories of the velocity vector field. A trajectory is a curve that connects velocity vectors under the stipulation that the curve's velocity vector agrees at each of its points with the connected velocity vectors. In short, a trajectory is a curve tangent to the vector field. A velocity vector field yields an indefinitely large number of trajectories. Nevertheless, the trajectory description sustains an even simpler description couched in terms of the *tendencies* of the trajectories: Are they tending to converge on a point? Are they tending to veer away from some region or other? Are they tending toward a closed orbit? Are the tendencies uniform throughout the field or are they different (yet symmetric) for different parts of the field? And so on. It is these tendencies, characterizing the optical flow field *qualitatively*, that define its form or morphology. The forms or morphologies of optical flow fields can be categorized roughly as global and nonglobal.

4.5.2. Global Symmetries (Morphologies). Returning to the activity of walking forward on a straight line in a room, the optical flow field that is lawfully generated by the walker's movement is describable as one in which *all* vectors in the field magnify concurrently. At one point in the field, the velocity is zero. This would be the point on the wall intersected by the walker's line of locomotion—literally, the point to which the person is walking. Moving forward, therefore, in a cluttered surrounding, produces a symmetric radial expansion, a global optical outflow with a singularity (see Figure 4.26). If one were to walk backwards, the optical flow would be characterized as a symmetric radial contraction as all of the vectors undergo minification and flow into a point. Again, it is the point where the line of locomotion of the point of observation intersects the wall, but this time the wall is the one from which the person is retreating.

Each of the very many ways an observer changes his or her relation to the surrounding layout of surfaces is characterized uniquely by a global morphology. That is the significant feature of the global morphologies: They constitute the information available to the visual system about the position, orientation, and movement of the body as a whole relative to the environment (Gibson, 1979). If one produces artificially a global optical change, then the observer should experience himself or herself moving in the way specified by the morphology of the optical flow field. A demonstration of this fact is provided by having a person walk forward in a room where the walls and ceiling are moving as a unit in the same direction but faster than the walker (Lishman & Lee, 1973). This circumstance generates global optical contraction. Consequently, the forward-moving walker experiences that he or she is actually walking backwards. [The same effect is to be expected if the same global morphology were defined over compression waves in the air (Lackner, 1977).]

4.5.3. Local Symmetries (Morphologies). Of course, there are many situations where the surrounding layout of surfaces is not completely static. Some substantial surfaces are displacing relative to the point of observation. These local movements give rise to localized disturbances in the optical flow field. Think about it this way: As a player moves forward, say, in a soccer game to get to a loose bouncing ball, there is *simultaneously* a symmetry (pattern) defined globally over the optical flow field specifying the player's forward movement and a symmetry (pattern) defined locally specifying an object moving in a certain way that is being approached by the player. Very many symmetries that are not defined globally over the optical flow field are to be expected. For example, there ought to be a particular symmetry of the flow pattern that is specific to approaching a brink in a surface (such as a curb or a wide gully).

A local symmetry that is particularly important is the local flow pattern produced by an approaching object. The object in question gives rise to one of the cones that has its apex at the point of observation. Let the object approach the point of observation at a constant speed. As it does so, the corresponding optical solid angle magnifies. The rate of magnification, the velocity of expansion, can be represented as dA/dt, where A is the area of the base of the solid angle or cone. Now it turns out that the *relative* rate of magnification is an especially important optical property. Its dimensions are T^{-1} (given that the units of area cancel, leaving a unit of time in the denominator). The *inverse* of the relative expansion velocity is an important temporal quantity. In fact, it is the time elapsing before the object contacts a plane located at the point of observation. After Lee (1976, 1980), this macroscopic optical property is designated as τ; thus

$$\tau = [(dA/dt)/A]^{-1}. \quad (4.1)$$

The significance of τ is that people and animals do not have to evaluate the speed and distance [or even acceleration in those situations where the velocity is not constant (Lee, Young, Reddish, Lough & Clayton, 1983)] of an approaching object in order to perceive when that object will be at their location. Rather, they need only detect τ, a macroscopic property of the optical flow field (see Todd, 1981).

The purpose of pursuing macroscopic descriptions of the optical flow fields is to get an understanding of what is invariant about the structure of light when the same action problem recurs in infinitely variable settings. Thus, in the preceding it is seen that for a huge number of environmental situations in which a person or animal must accurately synchronize actions to aspects of events—catching a ball or prey, stepping off a curb, returning a tennis serve, steering through a cluttered environment, and so on—there is, in principle, a macroscopic characterization of the optical flow field yielding a single property (τ) that makes this synchronization possible.

Consider now the invariant macroscopic form of the optical flow that specifies approach to a brink in a ground surface—a place where one will have to adjust one's locomotion, perhaps subtly, perhaps dramatically. Roughly and intuitively, the velocity vector field generated by approach to a brink can be described like this: There is a horizontal contour defining a discontinuity in the optic array that is generated by the occluding edge. At this discontinuity, the overall magnitude of the vectors changes sharply. Vectors above the horizontal are generally smaller than the vectors below the horizontal. With approach to the brink, all vectors magnify, but with important differences. Whereas the number of magnifying vectors below the contour decreases (as one approaches the boundary of an occluding surface, increasingly less of it is in front and increasingly more of it is behind), the number of magnifying vectors above the contour increases (as one approaches an occluded surface, increasingly more of it is revealed). In short, there is a distinct optical morphology associated with the approach to a brink where one ground surface, the one being walked upon, is occluding partially another ground surface, the one being approached. Regardless of the actual structure of these surfaces (their geometry and material composition), the rate and style of locomotion, the magnitude of the drop from one surface to the other, and so on, there is a form to the optical flow field that specifies invariantly a brink. The moral is that from a more microscopic perspective the velocity vector field may be terribly variable, but from a macroscopic perspective, one that underscores the qualitative form of the flow field, the velocity vector field is seen to be invariant.

4.6. A First Brief Summary: Flow Morphologies

The observations that flow morphologies are recurrent and invariant over media are not new (see Schwenk, 1976; Stevens, 1974; Thompson 1917/1942, and Weiss, 1967 for reviews). Nor is the observation that hard-molded material structures may owe both their design and their genesis to flows; that is, that flow morphologies may be archetypal and formative (e.g., Weiss, 1970; Vogel, 1981; Tributsch, 1982). Even the observation that flow morphologies might be exploited in the soft construction of stable reproducible behaviors has been made before. Schieferstein (1925; cited by Schwenk, 1976) observed that the wedge formation of birds in flight is the visible manifestation of a vibrating aerial form that surrounds and bears them. The trailing birds follow in the wake of the leading bird, riding the waves that it creates. But the trailing birds are not merely parasitic and the leading bird is not fated to make a disproportionate energetic contribution to the flock's flight. Each bird in the formation contributes to the persistence of the aerial flow field, with the bird at the apex able to draw on the field's energy just like the other birds because the spread of waves is faster (it is at the speed of sound) than the speed of the flock. Especially significant is the observation that the air that connects the birds functions much like an elastic medium; where one bird imparts more or less energy to the field, the others impart less or more, respectively.

What the present chapter adds to the preceding observations about flow morphologies is the following: Any flow morphology that can be defined reliably on a low-energy field (that is, a field whose kinetic terms are very small relative to those of the systems it envelops or connects) is potentially a source of information about the dynamics that gave rise to it. Chapter 7 presents the central nervous system, in particular, the haptic system, as such a field. Muscle-joint activities generate mass-based descriptions from which the ubiquitous mechanoreceptors "abstract" (Rosen, 1978) less detailed, but functionally sufficient, flow descriptions. The coordination of multiple muscle-joint complexes, it is suggested, is made possible by lawful interactions of these alternative descriptions. That is, high-energy muscle-joint complexes are *not* linked on the basis of Newton's mechanics. Their connectivity is not mediated by energy and momentum exchanges. Rather, the coordinated motions of muscle-joint complexes are made possible by flow morphologies defined over the low-energy neural field.

4.7. A Second Brief Summary: Perceptual Couplings

In the spirit of Chapter 1, and reiterating the major themes of the present chapter, a locomoting terrestrial animal can be thought of as a particle that is (a) in mechanical contact with a matter field (the layout of surfaces) and (b) possesses an on-board, deployable source of potential energy. Because of its inherent ability to produce suitably large forces, the particle is not guided solely (as are less well equipped, more conventional particles) by the reactive forces that result necessarily from its changing disposition to the matter field. Insofar as the particle's displacements do not adhere strictly to the force geodesics of the matter field, the interactive description must be nonmechanical (i.e., non-Newtonian). Nonmechanical couplings are commonly termed "perceptual." They are made possible, primarily, by the availability of a low-energy field (e.g., light). The latter is lawfully structured by the high-energy matter field (the layout of surfaces) and the particle's displacements (the animal's locomotion) relative to that matter field. As a result the macroscopic patternings and flows of the low-energy field (the optic array and its transformations) are specific to the properties of the matter field and to the displacements. To repeat some newly coined phrases: For systems that have an on-board source of potential energy and a means for deploying it, there are (relatively) low-energy based (non-Newtonian) *perceptual* couplings with the surround to be added to and contrasted with the (relatively) high-energy based (Newtonian) *mechanical* couplings with the surround.

REFERENCES

Bohm, D. (1958). *Causality and chance in modern physics*. London: Routledge.

Boynton, R. M. (1974). The visual system: Environmental information. In E. C. Cartarette and M. P. Friedman, *Handbook of perception, I*. New York: Academic.

Bruinsma, O. H. (1977). An analysis of building behavior of the termite *macrotermes subhyalinus*. *Proceedings of the VIII Congress IUSSI*, Wargeningen.

Deneubourge, J. L. (1977). Application de l'ordre par fluctuation à la description de certaines étapes de la construction du nid chez les termites. *Insectes Sociaux, Journal International pour l'ètude des Arthropodes Sociaux*, *24*, 117–130.

Gibson, J. J. (1958). Visually controlled locomotion and visual orientation in animals. *British Journal of Psychology*, *49*, 182–194.

Gibson, J. J. (1966). *The senses considered as perceptual systems*. Boston: Houghton Mifflin, 1966.

Gibson, J. J. (1979). *The ecological approach to visual perception*. Boston: Houghton Mifflin.

Grassé, P. P. (1959). La reconstruction du nid et les coordination interindividuelles chez *Bellicositermes natalensis* et cubitermes sp. La théorie de la stigmergie: essai d'interprétation des termites constructeurs. *Insectes Sociaux, Journal International pour l'ètude des Arthropodes Sociaux*, 6, 127.

Koenderink, J. J., & van Doorn, R. J. (1981). Exterospecific component of the motion parallax field. *Journal of Optical Society of America*, *71*, 953–957.

Kugler, P. N., Kelso, J. A. S., & Turvey, M. T. (1982). On the control and co-ordination of naturally developing systems. In J. A. S. Kelso and J. E. Clark (Eds.), *The development of movement control and co-ordination*. New York: Wiley.

Kugler, P. N., Turvey, M. T., & Shaw, R. (1982). Is the "cognitive impenetrability" criterion invalidated by contemporary physics? *Behavioral and Brain Sciences*, *(5)2*, 303–306.

Kugler, P. N., Turvey, M. T., Carello, C., & Shaw, R. E. (1985). The physics of controlled collisions: A reverie about locomotion. In W. H. Warren and R. E. Shaw (Eds.), *Persistence and change: Proceedings from the first conference on event perception*. Hillsdale, NJ: Erlbaum.

Lackner, J. R. (1977). Induction of illusory self-rotation and nystagmus by a rotating sound field. *Aviation, Space and Environmental Medicine*, *48*, 129–131.

Lee, D. N. (1974). Visual information during locomotion. In R. McLeod and H. Pick

(Eds.), *Perception: Essays in honor of J. J. Gibson*. Ithaca, NY: Cornell University Press.

Lee, D. N. (1976). A theory of visual control of braking based on information about time to collision. *Perception*, *5*, 437–459.

Lee, D. N. (1980). Visuo-motor coordination in space-time. In G. E. Stelmach and J. Requin (Eds.), *Tutorials in motor behavior*. New York: North-Holland.

Lee, D. N., Young, D. S., Reddish, P. E., Lough, S., & Clayton, T. M. H. (1983). Visual timing in hitting an accelerating ball. *Quarterly Journal of Experimental Psychology*, *35A*, 333–346.

Lishman, J. R., & Lee, D. N. (1973). The autonomy of visual kinesthesis. *Perception*, *2*, 287–294.

Margeneau, H. (1950). *The nature of physical reality*. New York: McGraw Hill.

Pattee, H. H. (1972). Laws and constraints, symbols and languages. In C. H. Waddington (Ed.), *Towards a theoretical biology, Volume 4*. Chicago, IL: Aldine.

Pattee, H. H. (1979). The complementarity principle and the origin of macromolecular information. *Biosystems*, *11*, 217–226.

Reed, E., & Jones, R. (1982). *Reasons for realism: Selected essays of James J. Gibson*. Hillsdale, NJ: Lawrence Erlbaum Associates.

Rosen, R. (1978). *Fundamentals of measurement and representation of natural systems*. New York: North-Holland.

Runeson, S. (1983). On visual perception of dynamic events. *Acta Universitatis Upsaliensis: Studia Psychologia* serial number 9 (originally published 1977 as dissertation).

Runeson, S., & Frykholm, G. (1983). Kinematic specification of dynamics as an informational basis for person and action perception: expectation, gender recognition and deceptive intention. *Journal of Experimental Psychology: General*, *112*, 585–615.

Schieferstein, H. (1925). Resonanzflug. *Zeitschrift fur Flugtechnik und Motorluftschiffahrt*, *5*.

Schwenk, T. (1976). *Sensitive chaos*. Translated by O. Whicher and J. Wrigley. New York: Schocken Books.

Shaw, R. E., & Alley, T. R. (1985). How to draw learning curves: Their use and justification. In T. D. Johnston and A. T. Pietrewicz (Eds.), *Issues in the ecological study of learning*. Hillsdale, NJ: Lawrence Erlbaum Associates.

Stevens, P. S. (1974). *Patterns in nature*. Boston: Little, Brown and Company.

Thompson, D. W. (1917/1942). *On growth and form*. London: Cambridge University Press.

Tributsch, H. (1982). *How life learned to live: Adaptation in nature*. Cambridge: MIT Press.

Todd, J. T. (1981). Visual information about moving objects. *Journal of Experimental Psychology: Human Perception and Performance*, *7*, 795–810.

Turvey, M. T., & Kugler, P. N. (1984). An ecological approach to perception and action. In H. T. A. Whiting (Ed.), *Human motor actions: Bernstein reassessed* Amsterdam: North-Holland.

Vogel, S. (1981). *Life in moving fluids: Physical biology of flow*. Boston: Willard Grant Press.

Weiss, P. (1963). The cell as unit. *Journal of Theoretical Biology*, *5*, 389–397.
Weiss, P. (1967). 1 + 1 ≠ 2 (When one plus one does not equal two). In G. C. Quarton, T. Melnechuck and F. O. Schmitt (Eds.), *The neurosciences: A study program*. New York: Rockefeller University Press.
Weiss, P. (1970). *Life, order and understanding*. The Graduate Journal, University of Texas, *Vol. VIII*, Supplement.
Yates, F. E., & Kugler, P. N. (1984). Signs, singularities and significance: A physical model for semiotics. *Semiotica*, *52*, 49–77.

5 WHY THINGS "TICK": PHYSICAL PREREQUISITES FOR SELF-SUSTAINED OSCILLATION

ABSTRACT

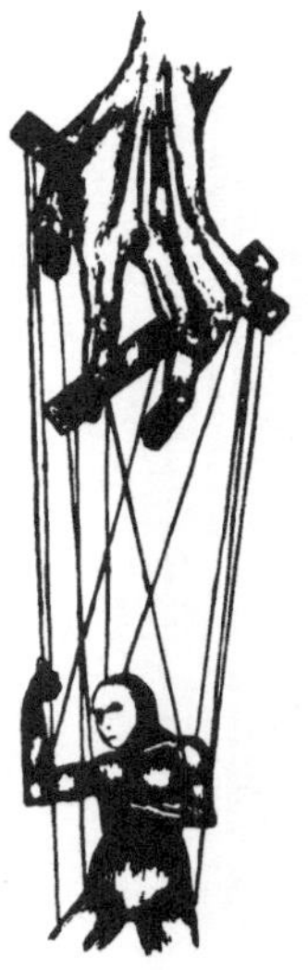

What kind of physical system is a terrestrial animal in the act of locomoting? A rationale will be provided in this chapter and the next for the claim that, in locomoting, a terrestrial animal assumes a pendular, clocking mode of organization. This mode of organization will become the topic of experimental study. The main experiment is described in Chapter 6, and its results and those of other related experiments are pursued in the chapters that follow. Of particular concern to the present chapter is the clocking aspect—the energy-flow process by which the organization that typifies locomotion can be sustained. The popular interpretation of the flow of energy that supports a biological rhythm as analogous to the escapement of a standard, man-made timepiece is rejected. This rejection is based on physical considerations that comprise the larger part of this chapter. In place of the clock escapement, an escapement is proposed that arises naturally when energy courses through structures with available macroscopic degrees of freedom whose energy-absorbing capacity is limited, whose persistence is independent of the coursing energy, and whose transition states are not associated with shock waves.

5.1. Introduction

The rhythmic movements of locomotion can be counted among auto-oscillatory phenomena on the grounds that the involved muscle-joint systems achieve sustained, repetitive motions by drawing on a chemical potential energy supply to offset the energy lost per cycle. The requisite physical conditions for self-sustained oscillation in muscle-joint systems can be brought into focus by contrasting a number of energy-flow models for persistent, rhythmic behavior. A persisting system that dissipates energy during motion requires some form of escapement to gate the flow of energy into its operational components. Two escapements will be contrasted: *Phase-dependent (in-flow) escapements* are used in the design of man-made clocks; *Sink-dependent (out-flow) escapements* are continuous with the design principles of self-organization in dissipative field processes.

5.2. The Theory of Clocks: Phase-Dependent Escapements

5.2.1. An Ideal, Harmonic Oscillator. A historically popular model of rhythmicity is the ideal, linear harmonic oscillator

$$m \frac{d^2\theta}{dt^2} + k\theta = 0 \tag{5.1}$$

where *m* and *k* are mass and stiffness coefficients, respectively, and θ is amplitude. The differential equation describing the ideal motion expresses the first law of thermodynamics: The total energy is conserved from cycle to cycle. An energy-flow system that forms a closed cycle relative to a set of fixed states constitutes a work cycle. These closed, cyclic energy-flow systems are referred to as *engines*. A cycle involving only reversible, conservative energy transformation, such as the system depicted in Equation (5.1) and Figure 5.1, is termed a *mechanical engine* (see Chapter 11 for a discussion of mechanical, thermodynamic, and chemical engines). The mechanical engine, with its repetitive sinusoidal behavior, is an ideal rather than a real entity. Its oscillations are sustained in the absence of energy dissipation. The exclusion of a dissipative term in Equation (5.1) violates the second law of thermodynamics.

5.2.2. A Dissipative Oscillator with Extrinsic Timing. Reality is more closely approximated by the linear damped harmonic oscillator

$$m \frac{d^2\theta}{dt^2} + \overline{B \frac{d\theta}{dt}} + k\theta = 0 \tag{5.2}$$

through the inclusion of a damping or frictional coefficient, $\overline{Bd\theta/dt}$ (the bar over the frictional term indicates a cycle average). Equation (5.2) models a thermodynamic system—one that loses useful energy—but it does not model an instance of *sustained* oscillation (see Figure 5.2a,b). What is required is some means of compensating for the energy loss brought about by the dissipation

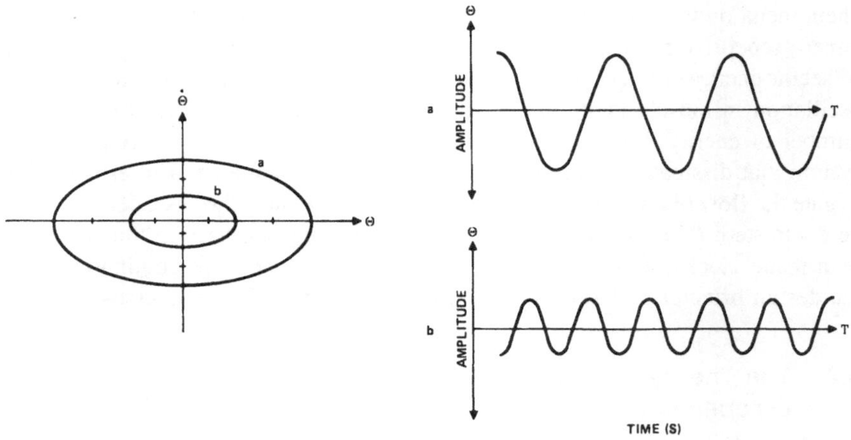

FIGURE 5.1. ENERGY TRANSFORMATIONS IN A CONSERVATIVE (MECHANICAL) OSCILLATOR. Trajectories *(a)* and *(b)* identify state changes in a conservative (mechanical) oscillator. A fixed amount of energy is periodically transformed from potential energy to kinetic energy (where $P.E. \propto \theta$ and $K.E. \propto d\theta/dt$). The left panel depicts the phase plane orbit. It is closed (infinitely continuous in time) and symmetric about the two axes. The right panel is the position versus time response plane.

coefficient B. There are two ways of achieving a sustaining compensation. The first is to introduce a forcing function $F(t)$, giving the following modification of (5.2):

$$m \frac{d^2\theta}{dt^2} + \overline{B \frac{d\theta}{dt}} + k\theta = F(t) \tag{5.3}$$

The left-hand term of (5.3) is a highly linear and isochronous unit, similar to a mass-spring system. The added right-hand term of (5.3) identifies a function that injects or squirts energy into the mass-spring like system independent of the system's state but dependent on time. As a persistent unit, (5.3) entails: (a) two logically independent components (given, respectively, by the left-hand and right-hand terms); (b) the notion of an extrinsic timing component (expressed by the right-hand term); (c) the nonautonomy of the system in the conventional mathematical sense of nonautonomy, namely, a forcing function that is time dependent; and (d) an unwelcome regress, in that a system homologous to that of (5.3) must be postulated to guarantee the persistence of $F(t)$ (in that insofar as the forcing function does work and dissipates energy, its losses must be compensated for).

5.2.3. A Dissipative Oscillator with Phase-Dependent, Intrinsic Timing. The second way of compensating for the energy losses, and thereby deterring the decay or relaxation process characteristic of (5.2), is to let energy be available in a constant, relatively *timeless* manner and to let the work accomplished by the constant energy source be occasional (rather than continuous), determined by the oscillatory component depicted by the left-hand term of (5.2). In short, a pulse of energy from a potential source is injected at a proper phase so that the resulting forcing function $E(\varphi)$ exactly offsets the energy loss averaged over each cycle (see Figure 5.3a,b). Thus the equation becomes

$$m\frac{d^2\theta}{dt^2} + \overline{B\frac{d\theta}{dt}} + k\theta = E(\varphi) \qquad (5.4)$$

which can be written as

$$m\frac{d^2\theta}{dt^2} + k\theta = E(\varphi) - \overline{B\frac{d\theta}{dt}} = \text{approximately zero}, \qquad (5.5)$$

to emphasize the near-zero net gains and losses in which the losses are essentially compensated for and the unit persists. Equation (5.4) models the ordinary dissipative clock. It exemplifies a thermodynamic engine. A system is labeled a *thermodynamic engine* if the work cycle persists through some form of energy degradation (Yates & Iberall, 1973).

As a sustained oscillation, Equation (5.4) (a) identifies a single entity decomposable into two logically dependent components; (b) incorporates the related notions of autonomy and intrinsic timing in that the forcing function is

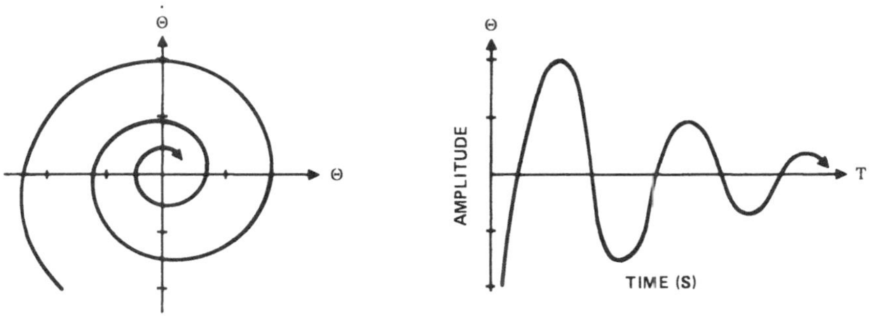

FIGURE 5.2. ENERGY FLOW TRANSFORMATIONS IN A CLOSED. NONCONSERVATIVE (DISSIPATIVE) OSCILLATOR. The dissipative nature of the oscillatory process is evident in the systematic reduction in the maximum amplitude (maximum potential energy) and maximum velocity (maximum kinetic energy) realized over successive cycles. Starting with an initial energy E_I (where $E_I = P.E. + K.E.$), the energy E_I is successively reduced in time to a final energy of $E_F = 0$. The second law of thermodynamics identifies the state E_F as an *attractor state*: All energy states $E_I > E_F$ will tend to converge on E_F as $t \to \infty$, independent of initial conditions. The oscillation becomes discontinuous at E_F.

Left panel (a): Velocity versus position.

Right panel (b): Position versus time.

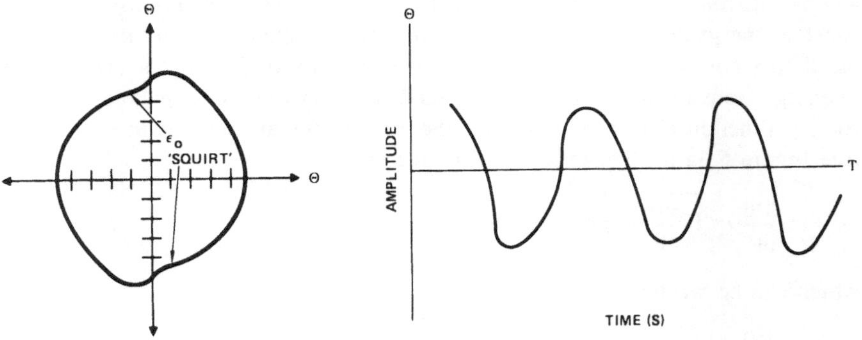

FIGURE 5.3. ENERGY FLOW TRANSFORMATIONS IN AN OPEN, (SELF-SUSTAINING) NONCONSERVATIVE OSCILLATOR. The dissipative losses associated with the oscillation are off-set by an escapement that injects new energy $\bar{\varepsilon}_0$ into the cycle at specific phases φ in the *P.E.→K.E.→P.E.* . . . transformation process. The $\bar{\varepsilon}_0$ exactly off-sets the dissipative losses, restoring the initial energy condition. The phase plane orbit defines a closed limit cycle that is infinitely continuous in time. Note the regions in the orbit associated with (i) the phase in which energy is injected into the orbit and (ii) the slow relaxational phase following the energy injection where the energy is transformed into (frictional) heat. The energy transformations associated with these two phases are not symmetric.

Left panel (a): Velocity versus position.

Right panel (b): Position versus time.

time independent, the *phasing* activity being determined by properties of the system itself; and (c) does not require additional mechanisms of the same type to guarantee persistence. Equations (5.4) and (5.5) model a *self-sustained* oscillation. The only known temporal stability for a system that sustains its periodic behavior by the means identified in (5.4) and (5.5) is the limit cycle (Minorsky, 1962). A limit cycle is a closed sequence of states in phase space (say, of velocity and displacement) that tends toward a relatively fixed frequency and amplitude indifferent to perturbations as long as they are below critical magnitudes and outside select phases. Below these critical magnitudes and outside these select phases, the frequency is preserved and the amplitude resists amplification even if the perturbations are at the operating frequency. Significantly, (a) each cycle in limit-cycle oscillation is ''initiated'' (or rather sustained) *de novo* and (b) a cycle in limit-cycle oscillation defines a period over which the energy bookkeeping is closed (as expressed in Equation 5.5).

5.2.4. Central Properties of the Phase-Dependent, Pendular, Clocking Mode of Organization. The foregoing requirements of self-sustained, limit-cycle oscillation are drawn from the elementary theory of clocks (Andronov &

Chaikin, 1949). A pendulum clock (see Figure 5.4a) permits the physical identification of the essential components of phase-dependent, self-sustaining clocking devices. The components are an *energy source* (the raised weight), an *oscillatory component* (the pendulum), an *escapement* that admits energy to the oscillatory component in an appropriately packaged amount and at a fixed phase (the anchor that allows the escape wheel to advance one tooth every half period), and a *coupling component* that controls the escapement, using the properties of

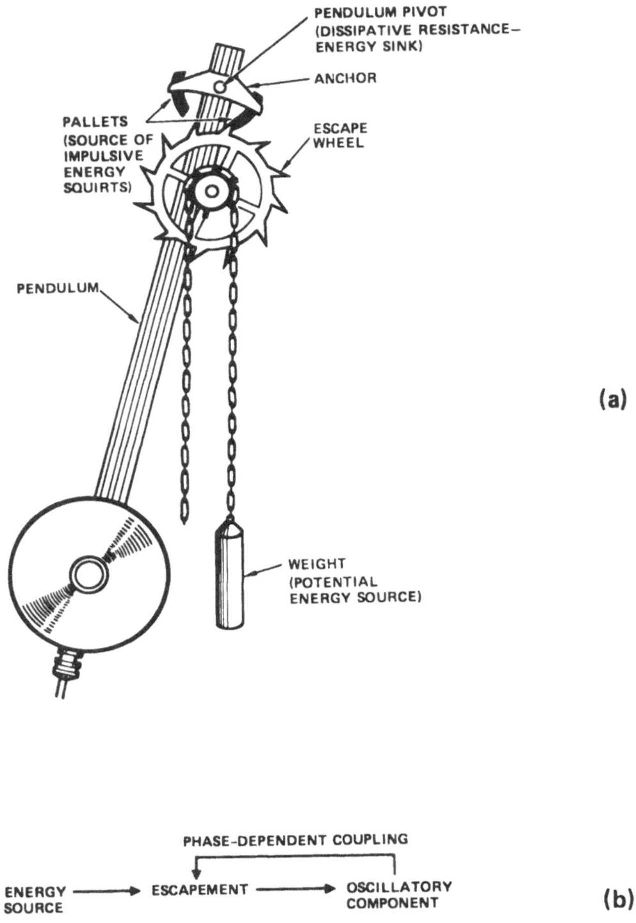

FIGURE 5.4.

Upper panel (a): Pendulum clock.

Lower panel (b): Components of a self-sustained oscillator.

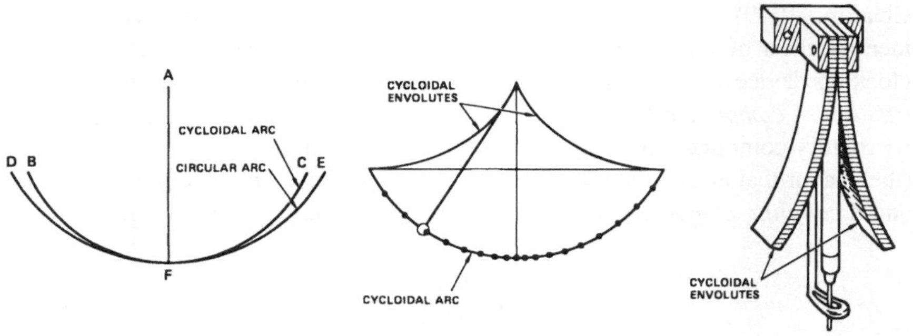

FIGURE 5.5.

Left panel (a): Examples of a circular arc *DE* and a cycloidal arc *BC*.

Center panel (b): Pendular motion structured by cycloidal envolutes.

Right panel (c): Huygens' pendular design for a cycloidal, isochronous clock.

the oscillatory system itself (the interaction of the anchor and the escape wheel). The four components are depicted in Figure 5.4b. Of central concern for the successful organization of an accurate phase-dependent escapement is the mechanical independence of the two quantities of amplitude and period.

5.2.5. The Independence of Amplitude and Period. The periodic time of a simple pendulum of length L for very small amplitude is given by $\tau = 2\pi\sqrt{L/g}$. As the amplitude increases, the period becomes longer according to:

$$\tau = 2\pi\sqrt{L/g}\,\{1 + \theta/16 + (11/3072)\theta^4 + \ldots\}, \tag{5.6}$$

where θ is amplitude in radians. The circular error (Equation 5.6, in braces), as it is called, arises from the circular potential well associated with the pendulum's motion. The circular potential well is forged from the rigid, single-degree-of-freedom design of the pendulum (which holds the pendulum bob to a circular trajectory) and the uniform gravitational field.

A pendulum clock will become a poor timepiece beyond a small range of amplitudes. Its reliability is ensured by keeping it within the small range of amplitudes to which the periodic time is indifferent. This is achieved through hard-geared connections that put an upper bound on the energy squirted cyclically into the pendulum. Such precautions would be unnecessary, however, if the potential well were cycloidal rather than circular (Figure 5.5a), as shown by Huygens in 1673. The cycloid is defined by the equations:

$$x = k(2\xi + \sin 2\xi) \tag{5.7}$$

$$y = k(1 - \cos 2\xi)/4, \tag{5.8}$$

where k is a constant and ξ is the direction of the pendulum bob (the angle to the horizontal). Under the conditions of oscillation in a cycloidal potential well, the periodic time is *always* independent of the amplitude of oscillation; that is, the oscillation is isochronous (Pippard, 1978). Huygens' design for an isochronous, cycloidal pendulum is depicted in Figure 5.5b. A simple pendulum is modified by suspending a weight from a cord placed between two pieces of metal shaped in the form of cycloids. As the pendulum swings, the cord wraps and unwraps on the cycloidal metal *envolutes*. The resulting trajectory of the suspended weight is transformed from a circular arc (characteristic of a simple pendulum) to a cycloidal arc (characteristic of a cycloidal pendulum). The period of the cycloidal pendulum is completely independent of the amplitude of the swing—it is isochronous.

There are two points to be emphasized. First, a necessary condition for an accurate pendular clocking mode of organization is that there exists a range of amplitude values in which amplitude can be regulated independently of period. Then variation in amplitude will not affect timing. Second, there are physical conditions under which the two quantities will always be independent. These conditions are probably few, limited perhaps to the continuous range of sheared, parabolic potential wells (Pippard, 1978).

5.2.6. The Packaging of Energy. The dissipative nature of a real pendulum clock requires that energy must be fed into the cycle for the oscillation to persist. The energy cannot be so great as to push it into the amplitude range in which its period is affected; but it cannot be so small as to fail to compensate for frictional losses. The energy cannot be supplied in a continuous, oscillatory fashion because continuous feeding would require another sustained oscillatory mechanism. Nor can it be pulsed at a phase or phases that would interfere with the periodic time. These conditions on the energy flow are met by a hard-molded, saltatory escapement—a set of mechanical arrangements that serves as a field boundary condition on the flow of energy. As a result, a fixed amount of energy is squirted discretely into each cycle of oscillation at a fixed phase. The amount of energy injected into the cycle is fixed by the design of the escapement, and its injection is strictly a function of phase. This means that any accidental or intended mechanical transformations of the pendulum of a clock (such as mass or length changes) that result in changes in the natural period of the pendulum will *not* be accompanied by variations in the amount of energy injected into each cycle. The energy flow characteristics of phase-dependent escapements are independent of transformations in the mechanical parameters associated with the oscillatory component (e.g., mass, length, period, etc.).

It is to be underscored that the period/energy dissociation in an artifactual, pendular, clocking mode of organization is not defined by a natural law; it is the outcome, rather, of an implicit set of instructions that identify how to assemble a sustained auto-oscillatory process, given the form of the potential well, the quantity of dissipation per period, and the dynamics of the oscillatory element. The theory of escapement advanced in the theory of clocks assumes a cleverly designed *dissociation* of the period and energy variables (see Figure 5.6).

5.2.7. The Inadequacy of a Phase-Dependent Escapement for Soft-Molded Locomotory Processes. There is an important difference between hard-molded, phase-dependent, pendulum clocks and the soft-molded, pendular clocking organization of a locomotory system. Hard-molded clocks are systems that cannot sustain, and have no need to sustain, stable oscillations at any period other than a single natural period (the criterion for a good timepiece, after all, is the invariance of period). In comparison, animals locomote at different periods, in part because they locomote at different gaits. Each gait is associated with a preferred period (Hoyt & Taylor, 1981; Pennycuick, 1975; see Section 8.7) and while animals tend most frequently to locomote in a given gait at the preferred period, they need not necessarily do so (Hoyt & Taylor, 1981; see Section 7.2.1). Animals can locomote, and commonly do locomote, for comparatively long intervals at periodic times other than the preferred. Given this fact of locomotion, namely, that locomoting occurs at *different* periods, two questions are posed: (a) Under what physical conditions can an oscillator of fixed mass and length oscillate at different periods in a stable fashion?; and (b) How does the period of oscillation relate to the energy (and, thereby, the amplitude) of the oscillation? According to the artifactual perspective on escapements, which equates escapements with mechanical constraints tied to phase as the control parameter, each different instance of the pendular clocking mode entails the construction of a new escapement. As noted, the construction of artifactual escapements is guided by an implicit set of instructions or rules. To assume that the conditions of energy flow for rhythmic movements, such as locomotion, are

---

FIGURE 5.6. *(Opposite page)* INDEPENDENCE OF ENERGY AND FREQUENCY AS A FUNCTION OF A PHASE-DEPENDENT ESCAPEMENT. The response characteristic of an oscillator with a fixed phase-dependent escapement is depicted. The frequency of the oscillator is conditioned by the length of the pendulum. The manner in which the length of the oscillator is changed is *independent of*, and *prior to*, the actual motion of the oscillator. (This method of changing oscillator length is contrasted in Chapter 11 with changes in oscillator length that are continuous with the motions of the oscillator.) The procedure is as follows: The oscillation is stopped, the pendulum replaced with one of a different length and, finally, a small perturbation (less than the escapement energy) is applied to the pendulum to start the escapement. The motion of the oscillator then spirals outward, asymptotically approaching the limit cycle orbit defined by the escapement energy. The limit cycle's maximum amplitude θ and maximum velocity $d\theta/dt$ values are *independent* of the oscillator's length. This is because (i) maximum amplitude and velocity scale to maximum potential and kinetic energy, respectively, and (ii) the total energy carried by the cycle is strictly a function of the escapement's energy.

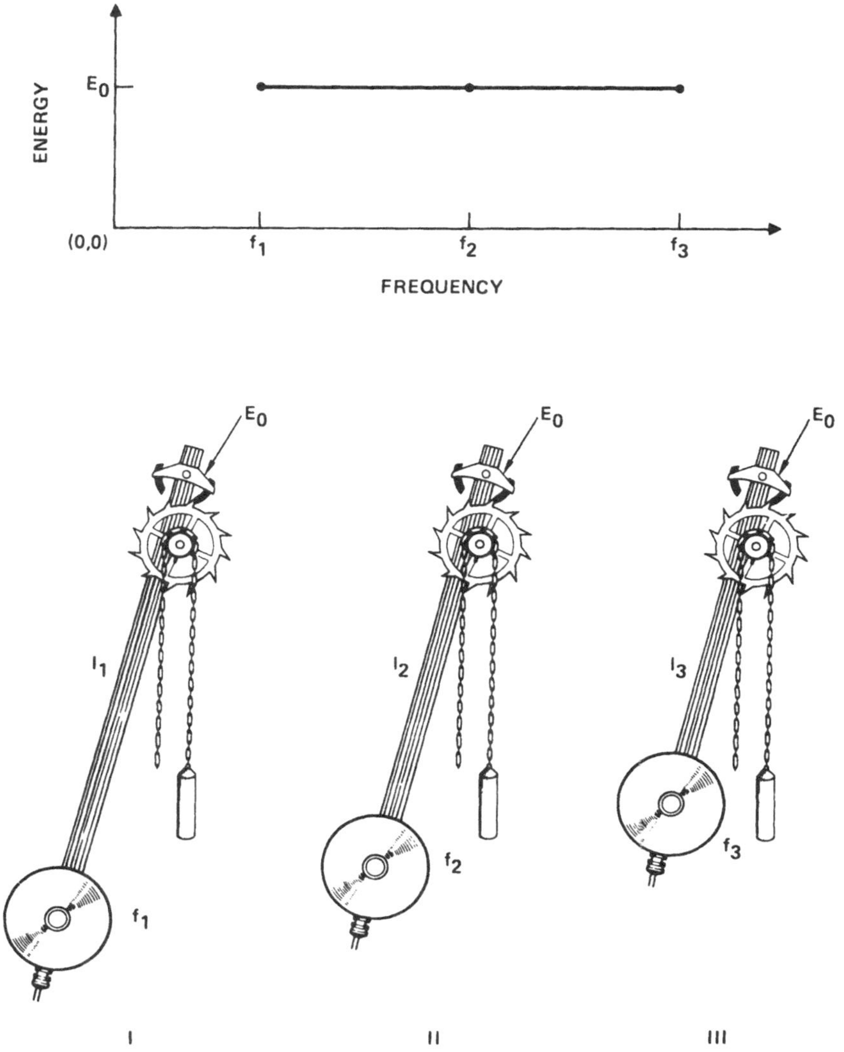

FIG. 5.6.

selected according to an analogous set of instructions is to take out a loan of intelligence (Dennet, 1969, 1971, 1978). The methodology pursued here recommends that such (controller-information) borrowing be avoided (Kugler, Kelso & Turvey, 1980; Turvey, Shaw, Reed & Mace, 1981). It must be supposed, therefore, that there is (a) a natural, principled basis by which escapements can be assembled—arising from lawful processes rather than imposed by arbitrary rules, and (b) a law-based coupling of period and energy so that as rhythmic components of the locomotory act are required (for whatever reason) to assume alternative, stable periodic times, the energy to assemble and sustain these alternative periodicities is specified. This latter supposition need not deny the separability of period and energy recognized in the theory of clocks, at least not in the following sense: The conditions that bring about a change in energy may be largely independent of those that bring about a change in period (even though energy and period may be systematically related).

The remainder of this chapter is devoted to exploring physical instances of naturally evolving escapements with an eye to their appropriateness for the biological phenomenon of rhythmic movement. Attention to the second supposition—the issue of a law-based coupling of period and energy—is given in Chapters 10, 11, and 12.

5.3. Finite-Sink Escapements

As will become apparent, an alternative design for escapements is to have the flow of energy into a system restricted by the ability of the system's internal operational components to absorb the flow of energy. Two escapements of this type will be considered: (a) hard-molded impulsive escapements that use the storage capabilities of conservative (elastic and gravitational) force structures to gate the flow of energy and (b) soft-molded nonimpulsive escapements that use the dissipative flow capabilities associated with entropy-producing field processes. Homeokinetics (Iberall & Soodak, 1978, 1986; Yates, 1982) emphasizes (a), whereas the theory of dissipative structures (Glandsdorff & Prigogine, 1971; Nicolis & Prigogine, 1977; Prigogine, 1978, 1980) emphasizes (b). Both of these escapements involve gating processes that are related to the finite capacity of the internal components of the system to absorb energy, conservatively or nonconservatively, in a single dynamic mode. Because of the dependence of these escapements on a limited ability to absorb energy, they are termed *finite-sink escapements*. Finite-sink escapements and phase-dependent escapements differ with respect to their causal dependence on the flow of energy. Phase-dependent escapements exist prior to, and independent of, the flow of energy, while finite-sink escapements are an *a posteriori* fact of, and causally dependent on, the flow of energy.

5.3.1. Resistive (Nonconservative) and Elastic (Conservative) Limits—Hard-Molded, Impulsive Escapements. A continuous flow of energy into a system can be transformed into a cyclic, impulsive out-flow of energy by constraining the energy to flow through an intermediate set of operational components that can temporarily store the energy. A conservative force structure (e.g., capacitor, *R-C* circuit, elastic structure, etc.) functions as a temporary storage depot. The energy stored is released impulsively when a storage (conservative) limit or resistive (nonconservative) limit is reached. The amount of impulse is a function of the degree of damping (i.e., the nonconservative forces) associated with the flow. The limited storage capacity is a function of: (a) the finite storage capabilities (i.e., elastic limit) of the conservative force structure and/or (b) the finite resistive capabilities (i.e., resistive limit) of the operational components that are in competition with the conservative force structure. In either case the energy is absorbed temporarily by the conservative force structure (e.g., storage device) until a limit (conservative or nonconservative) is reached, at which point the energy is suddenly ("impulsively") released, restoring the ability of the conservative force structure to function as an energy depot. Given that the energy flow through the operational components is continuous, impulsive release is followed by storage which, in turn, is followed by impulsive release, and so on. Which is to say that a continuous flow of energy through a conservative storage device, operating near a storage or resistive limit, can give rise to a self-sustained oscillation. Aeolian harps provide a simple example.

An Aeolian harp is an Ovidian instrument that is played by blowing a continuous stream of air into a chambered vessel. The geometry of the vessel is such that the exit passage for the air is covered by a set of strings. In a resting mode the strings assume a stable state defined by minimum potential energy. For small air flows, the strings offer little resistance. If, however, the (viscous) air flow is sufficiently increased, then the resistance can cause the strings to start storing (conservatively) potential energy in an elastic flow process. The storage of energy will continue until the amount of energy stored in the strings exceeds the viscous air resistance. When the resistive limit is reached, the elastic flow suddenly becomes unstable and the elastically stored energy is impulsively released. The result is a motion that transfers the potential energy stored in the elastic flow process into a mechanically oscillating pressure (sound) wave. With the sudden transfer of energy, the string returns—*snaps back*—to its original equilibrium position of minimum stored potential energy, and the cycle starts over again.

A second example of an impulsive finite-sink escapement is identified in Figure 5.7. The block of wood rests on a moving belt. Its motion is restricted to a one-dimensional trajectory by an externally attached spring (see Abraham & Shaw, 1982; Stoker, 1950). (The situation is similar to that of a Prony brake.) Because of the surface friction between the block and the belt, energy supplied to the block via the belt's motion results in a stretching of the spring. The stretching spring stores energy until the elastic forces exceed the frictional forces that bind the block to the belt's surface. When the (stored) elastic forces exceed

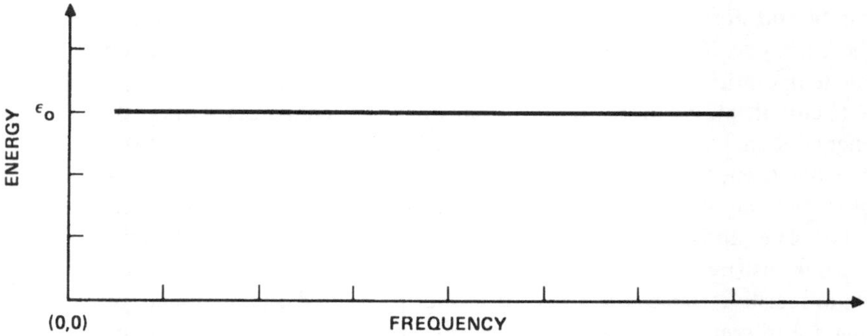

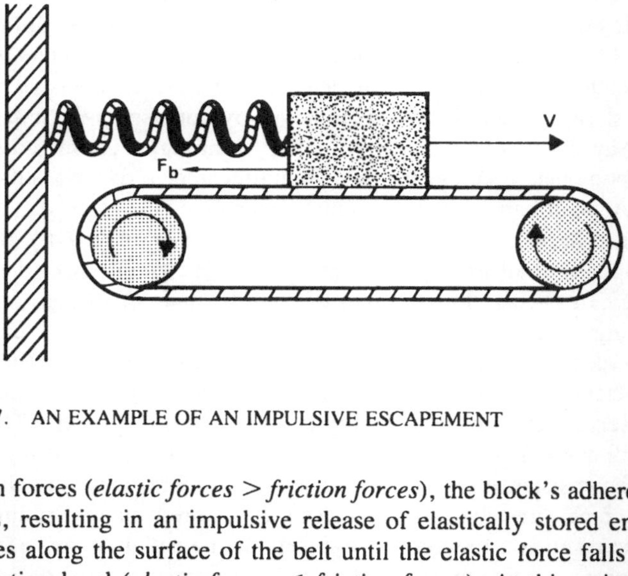

FIGURE 5.7. AN EXAMPLE OF AN IMPULSIVE ESCAPEMENT

the friction forces (*elastic forces* > *friction forces*), the block's adherence to the belt yields, resulting in an impulsive release of elastically stored energy. The block slides along the surface of the belt until the elastic force falls below the surface friction level (*elastic forces* < *friction forces*). At this point the block again adheres to the belt and begins to store energy elastically. The transition between force (elastic and frictional) structures is a function of the spring's displacement from its equilibrium state. A hysteresis results if the transition amplitudes are not symmetric in both directions. In either case, the forces governing the motion of the block oscillate between friction forces that tend to retard the block's motion and elastic forces that propel the block along the belt's surface. Because the critical transition points are amplitude dependent, the energy of the oscillation is independent of the velocity of the belt.

Now, it is true that some biological rhythms are marked by impulses, and for them the (impulsive) relaxation oscillator serves as a useful mathematical characterization. However, many biological rhythms, especially the soft-assembled rhythmic movements exhibited frequently by people, are not impulsive. There is a smoothness and symmetry to these movement cycles that cannot be duplicated by Aeolian harps and sliding blocks. Impulsivity in the foregoing examples derives from the energy returned by the conservative force structure (e.g., the elastic strain of the band). Obviously impulses would be eliminated if there were no return—that is, if the energy supplied to the operational components was completely dissipated. But can completely dissipative processes give rise to cycles?

There is a further aspect of the escapements and of the cycles arising from "natural" limits (e.g., resistive and elastic) that makes this class of finite-sink escapements less than ideal for an understanding of rhythmic movement. The microscopic material organization that supports the macroscopic frictional and elastic properties defining the escapement is present from the outset and fixed. That is to say, although an escapement founded on resistive and elastic limits is potentially natural (versus artifactual), it may not be sufficiently self-organizing and, therefore, sufficiently general for purposes of biology, physiology, and psychology. Can the microscopic material organization that yields the macroscopic detail of an escapement arise from lesser beginnings? Can escapements be truly soft molded?

5.3.2. Dissipative Structures: Soft-Molded, Nonimpulsive Escapements. An alternative design for a finite-sink escapement is suggested by the work of Prigogine and his colleagues on the formation of oscillatory regimes from dissipative field processes. The principal question being considered is: How can the flow of energy through a field of loosely coupled, fluid-like operational components give rise to a set of soft-molded constraints that temporarily organize the field? Prigogine (1967) has termed the emerging organizations *dissipative structures*, to emphasize the entropy-producing field processes inside the system that serve to create, maintain, and dissolve the organizational constraints. Speaking informally, within the regions of linear thermodynamic laws (see Chapter 3) *weak* processes operate that tend to *decrease* the organization of the field patterns. In regions far from equilibrium, however, *strong* processes can arise to generate instabilities that *increase* the field's organization. Fluid-like, irreversible energy flow cycles can emerge if certain conditions —identified in Chapter 3—are realized, namely:

1. A reservoir of potential energy must be available from which a (generalized) work cycle can arise.

2. The operational component(s) must have available more than one degree of freedom.

3. A competition must arise between the energy flowing into the operational components and the ability of the components to absorb the energy flow under their current configuration.

4. A critical scale change in the energy flow must be realized such that nonlinear interactions within the operational components are critically amplified.

The point is that if a flow of energy is passed through a stochastically assembled fluid-like ensemble, then the possibility exists for a sudden transition from a steady-state situation characterized by little spatio-temporal detail to highly structured spatial and temporal regimes (see Chapter 3). The precipitation and stabilization of these regimes are a function of the competition identified in the foregoing Point 3: As the amount of energy flowing into the operational components becomes excessive, an instability can ensue that drives the current field configuration into a new regime that is characterized by a more equitable energy flow relationship. This transition depends on the availability of degrees of freedom for absorbing the excess energy flowing into the system. A variety of stable states can emerge when the rate of energy flowing into the system is offset exactly by the rate at which the energy is being absorbed or equipartitioned by the system: (a) multiple steady states with systematic transitions from one to another; (b) rotation on a *limit cycle* around an unstable singular point, resulting in stable oscillation (several limit cycles are possible); (c) oscillation resulting in waves. The dissipative geometry is defined by the appearance in phase space of certain properties that can orient a system's trajectories. The appearance of the orienting properties (limit points, limit cycles, strange attractors, etc.) mark the transition from an ergodic phase space (exhibiting no preferred locations) to a nonergodic phase space (exhibiting preferred locations).

Not all advertised instances of dissipative structures will serve the understanding of rhythmic movements. Although the oscillatory states that can arise from entropy-producing field processes can be nonimpulsive and symmetrical, the transitions to such states are abrupt and generally construed as second-order phase transitions, similar in kind to those seen in hydrodynamic instabilities (e.g., the transition from laminar flow to turbulence). There are good intuitive reasons for supposing that the emergence of rhythmic movements and other biological oscillations does not engage second-order phase transitions and all that they entail. There are two aspects in particular that render second-order phase transitions unlikely components of self-organization in full-fledged biological systems. First, they are characterized by shock waves linked to the onset of the instability. These shock waves can be debilitating, weakening or eliminating whatever internal constraints already exist that bind cooperatively a field process. Second, they require high energy flows to sustain the instability that breaks the symmetry of the field's conservational distributions and the dynamical mode that results from the symmetry breaking. That is, whether or not the new mode resulting from a second-order phase transition persists depends on whether or not the high rate of energy flow persists. Said differently, the mode concomitant to a second-order phase transition is not *intrinsically* stable; there are no well-defined, stable operating points associated with the intrinsic *dissipative geometry* of the field dynamics (see Chapter 3). The stability is only by virtue of the stability of the boundary conditions defining the energy flow.

The transitions associated with the emergence of new biological organizations seems to contrast with second-order transitions in the following ways: (a) biological transitions are not typically characterized by shock waves at the instability regions; (b) the persistence of new biological modes is not tightly coupled to the rate of energy flowing into the field; and (c) the stability of the emerging modes is intrinsically defined relative to the geometry of the internal dynamics. Evidently, what needs to be understood for matters of biology is how different dynamical modes can be successively and *gently* organized—that is, without shock waves or high energy flows—through the interplay of macroscopic degrees of freedom and the field's dissipative geometry. The evolving modes of a spinning, falling leaf suggest that such gentle organizations can come about through the interplay of a system's macroscopically available degrees of freedom and the energy flowing into and out of the system.

A falling leaf is immersed in a field of gravitational potential energy defined by its position above the ground. As the leaf falls, potential energy is available to the leaf and is converted into motion. If the fall is very brief, then the absorbing motion may be translational only. If, however, the fall is of a longer duration, so that a critical ratio is realized between the leaf's velocity and the air resistance, then the possibility arises that the leaf's kinetic pattern will evolve a side-to-side yaw—that is, a vibration. The transition from a translational motion to a translational/yawing motion marks the emergence of a new *sinking mode*. The new (vibrational) degree of freedom appears when the capacity of the translational degrees of freedom to absorb energy has been exceeded. It might be said that the excess energy flowing into the leaf (as its motion accelerates) evolves a new sink. At still greater velocities, the leaf's motion evolves still a further sink—a rotational degree of freedom. The new pattern again results from the failure of the "old" motion, the translational/yawing motion, to absorb the energy flowing into the leaf. These particular *qualitative* patterns of motion (translational, yawing, and rotational) are tied to the particular geometrical properties that characterize the time-dependent sinks of the energy-flow system—namely, the leaf's surface structure in relation to the potential field. The new sinks evolve through interactions between macroscopic degrees of freedom and microscopic dissipative processes. The result of the interaction is the evolution of more efficient macroscopic dissipative geometries.

Although an infrequently advertised instance of a dissipative structure, the falling leaf exemplifies the characteristic properties: (a) a potential reservoir is available; (b) there are multiple degrees of freedom (translational, vibrational, and rotational) in the operational component (the leaf); (c) a competition arises between the amount of potential energy flowing into the leaf and the ability of the leaf's motion to dissipate the energy; and (d) a critical amount of energy flows into the leaf's motion so that new degrees of freedom are required to absorb the energy flow. Speaking more generally, and reiterating points made before, the new spatio-temporal regimes evolve (self-organize) as a function of a competition between the flow of energy into a field and the ability of the operational component's degrees of freedom to dissipate the energy. If internal degrees of freedom are available, then the dissipation of energy can be realized by a variety of configurations. If the flow of energy into the field exceeds the dissipative capacity of the current internal organization, then the excess energy can serve as an energy source for reorganizing the operational components into a new organization that can dissipate a greater amount of energy. The new internal organization can give rise to a new macroscopic oscillatory pattern.

An emerging self-sustained oscillation is a function of a soft-molded (dissipative) escapement that is assembled, maintained, and dissolved in accordance with the energy flow characteristics. The basis for the organizational constraints that gate the energy through the system is the entropic, dissipative flow of energy. In sum, macroscopic order evolves out of a more microscopic tendency toward disorder (see Chapters 3 and 4); in this regard *the second law of thermodynamics provides* ***both*** *a source of energy grades for powering the system and a source of constraints for organizing the system.*

The emergence of new patterns as a function of dissipative energy flows through sets of operational components identifies a universal design principle whose application is independent of scale size and material instantiation. Like falling leaves, transitions in locomotion suggest the gentle emergence of new organizations as a function of scale changes in the energy flow (associated with increasing locomotory velocity). Whereas the descent of a leaf evolves into successively different forms of organization due to the available translational, vibrational, and rotational degrees of freedom, the locomotion of an animal, at increasingly greater velocities, evolves into successively different gaits due to constraints on the degrees of freedom derived from the phase symmetries that relate the oscillating limbs (or so it may be conjectured). As instabilities arise with increased energy flows, new phase patterns relating the limbs must emerge, namely those that are successively more able to absorb the time-dependent, dissipative, energy flow (cf. Pedley, 1977). An account of the foregoing kind may rationalize the fact that at the speed at which the transition from the trot to the gallop occurs, an animal's stepping rate is scaled to its weight: *transition* $\propto$ *(body weight)*$^{-1/8}$ (Heglund, Taylor, & McMahon, 1974). Roughly interpreted, the conditions determining the shift in gaits accompanying increasing speed are the same for all quadrupeds.

Of course, an animal does not have to run faster in order to change gait. Intention can serve just as well to bring about a gentle (though discontinuous) gait transition. For biological systems, more macroscopically defined intentions as well as more microscopically defined fluctuations can give rise to structuralizations. Due recognition of this point is given in Chapter 13.

5.4. Autonomous, Clocking Mode of Organization: A Summary

An autonomous, clocking mode of organization is defined as the ability of a system to transform internally a continuously available energy supply into an oscillating out-flowing energy distribution. The design of the phase-dependent escapement, used in the building of timepieces, is motivated by the desire to produce an oscillation with a precisely repeated temporal unit. The phase-dependent escapement is an artifactual clocking mode. It does not self-assemble from local flow processes and material limits. A phase-dependent escapement typically requires an assembling agent. Natural clocking modes involve the use of finite-sink escapements. These escapements do not require an agent for their assembly; they can self-assemble from local flow processes and material limits. Not all such self-assembling escapements, however, are of equal relevance to biology.

Similar to the man-made clock, the oscillator assembled naturally from elastic and resistive limits sustains its periodic motion by impulses. As noted, impulsivity characterizes some biological rhythms, but it does not characterize all of them. Many rhythmic movements can be smoothly continuous and energetically symmetrical. Oscillators without saltations, that are sustained by a continuous flow of energy derived from an available energy reservoir, are prototypical of dissipative structures.

What is implied by the conclusion that the principles governing a sustained rhythmic movement are more akin to those governing dissipative structures than they are to those governing clocks, Aeolian harps, and sliding blocks? Two implications are noted. They are subtly but significantly felt throughout the analysis of rhythmic movement presented in this book.

All auto-oscillatory phenomena can be stratified into macroscopic, mechanical processes and microscopic, thermodynamic processes. In auto-oscillatory phenomena, exemplified by the clock, the Aeolian harp, and the sliding block, the macroscopic and microscopic processes are orthogonal and temporally discontinuous. In Pattee's (1973, 1977, 1979) sense they are complementary (see also Prigogine, 1978). Thus, for example, in the pendulum clock the origin of the friction that retards and the origin of the force that impels are quite separate and distinct. Moreover, in the pendulum clock the continuous dissipation of energy through friction contrasts with its (quasi) discontinuous replenishment through inertial force—the dissipative and inertial-squirt phases relate discontinuously. In auto-oscillatory phenomena assembled via dissipative processes, however, the two levels are connected intimately, and temporally they can be either discontinuous or continuous. One feature of this macro-micro intimacy is that the mechanisms that foster and the mechanisms that hinder sustained oscillations share a common origin. The first implication, therefore, of a dissipative-structure perspective on rhythmic movement is that the linkage between the inertial force that impels and the frictional force that retards is not arbitrary.

The second and related implication of a dissipative structure is that the neuromuscular escapement that packages the energy per cycle is conditioned by the dissipative geometry associated with the rhythmic movement. The presence of dissipation guarantees that the phase space of the movement system will be nonergodic. The nonergodicity is due to the presence of these dissipative geometries (limit points, limit cycles, strange attractors, etc.). These orienting properties generally identify preferred states in phase space that emerge whenever two or more flow processes, operating at different flow rates (for example, linear versus quadratic), are in competition. The orienting properties or preferred states are associated with the small set of solutions that satisfy simultaneously the flow problem for all competing (flow) processes. At all events, the quantity of chemical energy that is degraded into mechanical energy and gated into a rhythmic movement must be just that quantity that the organization of the movement can absorb. A change in the movement's organization—that is, a change in its energy sinks—must be matched by a corresponding change in the energy flow and, *a fortiori*, in the neuromuscular escapement. In a rhythmic movement the organizational constraints for absorbing energy are to be found locally and globally. Locally (or microscopically), a stable organization is to be found in the near-equilibrium perfusion/absorption states of the oxygen flow that provides the metabolic *choke* for the neuromuscular apparatus. Globally (or macroscopically), a stable organization is to be found in the spatiotemporal configuration of muscle-joint complexes defined in reference to the conservative (gravitational and elastic) force structures.

5.5. From Force-Dominated to Flow-Dominated Escapements

A final comment on the design of softly assembled escapements brings to the forefront the possible (and likely) energy-gating role of information as that term is defined in Chapter 4. Consider another perspective on the nest-building behavior of insects discussed in Chapter 4. Viewed from a relatively large space-time scale, the behavior is an instance of a work cycle that is assembled from the local activities of very many atomistic components. Arches are erected and a roof connecting them is constructed. The sequence then repeats itself (see Figure 4.20). The energy that fuels this work cycle is stored locally (the on-board chemical fuel of individual insects) and it is distributed fairly uniformly throughout the area in which the construction is taking place. The *directing* and *scaling* of the locally available and more or less uniformly distributed energy supply is done through the (topological) properties of the kinematic (flow field) abstraction of the pheromone diffusion field. As noted in Chapter 4, these properties orient the flight trajectories of the individual insects. They constrain the transformation or flow of energy in the nest-building area in the sense of constraining the very many local metabolic processes. Of course, the kinematic properties do not intrude upon the laws governing metabolism; rather, they articulate boundary conditions on these laws (see Chapters 11 and 13).

In sum, what makes the nest-building example interesting in the present context is the intimation that in dissipative structures, with their macro-micro couplings, a circularly causal process of forces and flows functions as an escapement. In the nest-building scenario, the kinematic flow-field properties modulate the on-board metabolic reservoirs to yield forces (conservative and nonconservative); these forces, in turn, modulate the flow-field properties. This continuously circulating causal process is embedded in a periodic process that plays out relatively slowly in a macroscopic work cycle, namely, the construction of arches and the roof that connects them.

The dissipative structures and periodicities of interest to the present monograph are those manifest in the limb movements of terrestrial animals. Chapter 7 generalizes the foregoing characterization of escapements in dissipative structures to the assembling of rhythmic movements in one or both wrist joints. The methodology to investigate the principles by which such movements are assembled is the topic of the chapter that follows.

REFERENCES

Abraham, R. H., & Shaw, C. D. (1982). *Dynamics: The geometry of behavior, Part one—periodic behavior*. Santa Cruz: Aerial Press.

Andronov, A., & Chaikin, C. E. (1949). *The theory of oscillations*. Princeton, NJ: Princeton University Press.

Dennet, D. C. (1969). *Content and consciousness*. London: Routeledge.

Dennet, D. C. (1971). Intentional systems. *Journal of Philosophy*, *68*, 87–106.

Dennet, D. C. (1978). *Brainstorms: Philosophical essays on mind and psychology*. Montgomery, VT: Bradford Books.

Glansdorff, P., & Prigogine, I. (1971). *The thermodynamic theory of structure, stability and fluctuations*. New York: Wiley.

Heglund, N. C., Taylor, C. R., & McMahon, T. A. (1974). Scaling stride frequency and gait to animal size: Mice to horses. *Science*, *186*, 1112–1113.

Hoyt, D. F., & Taylor, C. R. (1981). Gait and the energetics of locomotion in horses. *Nature*, *292*, 239–240.

Iberall, A. S., & Soodak, H. (1978). Physical basis for complex systems—some propositions relating levels of organizations. *Collective Phenomena*, *33*, 9–24.

Iberall, A. S., & Soodak, H. (1986). A physics for complex systems. In F. E. Yates (Ed.), *Self-organizing systems: The emergence of order*. New York: Plenum Press.

Kugler, P. N., Kelso, J. A. S., & Turvey, M. T. (1980). On the concept of coordinative structures as dissipative structures: I. Theoretical lines of convergence. In G. E.

Stelmach and J. Requin (Eds.), *Tutorials in motor behavior*. New York: North-Holland.

Minorsky, N. (1962). *Nonlinear oscillations*. Princeton, NJ: van Nostrand.

Nicolis, G., & Prigogine, I. (1977). *Self-organization in nonequilibrium systems: From dissipative structures to order through fluctuations*. New York: Wiley-Interscience.

Pattee, H. H. (1973). Physical problems of the origin of natural controls. In A. Locker (Ed.), *Biogenesis, evolution, homeostasis*. Heidelberg: Springer-Verlag.

Pattee, H. H. (1977). Dynamic and linguistic modes of complex systems. *International Journal of General Systems*, *3*, 259–266.

Pattee, H. H. (1979). The complementarity principle and the origin of macromolecular information. *Biosystems*, *11*, 217–226.

Pedley, T. J. (Ed.) (1977). *Scale effects in animal locomotion*. New York: Academic Press.

Pennycuick, C. (1975). On the running of the gnu (*Connochaetes taurinus*) and other animals. *Journal of Experimental Biology*, *63*, 775–799.

Pippard, A. B. (1978). *The physics of vibration, vol. I*. Cambridge, England: Cambridge University Press.

Prigogine, I. (1967). *Introduction to thermodynamics of irreversible processes* (3rd ed.). New York: Wiley.

Prigogine, I. (1978). Time, structure, and fluctuations. *Science*, *201*, 4358.

Prigogine, I. (1980). *From being to becoming*. San Francisco: Freeman and Co.

Stoker, J. J. (1950). *Nonlinear vibrations*. New York: Wiley-Interscience.

Turvey, M. T., Shaw, R. E., Reed, E. S., & Mace, W. M. (1981). Ecological laws of perceiving and acting: In reply to Fodor and Pylyshyn (1981). *Cognition*, *9*, 237–304.

Yates, F. E. (1982). Outline of a physical theory of physiological systems. *Canadian Journal of Physiology and Pharmacology*, 60, *3*, 217–248.

Yates, F. E., & Iberall, A. S. (1973). Temporal and hierarchical organization in biosystems. In J. Urquhart and F. E. Yates (Eds.), *Temporal aspects of therapeutics*. New York: Plenum Press.

6 AN EXPERIMENTAL METHODOLOGY FOR INVESTIGATING THE SOFT, SELF-ASSEMBLY OF THE PENDULAR, CLOCKING MODE

ABSTRACT

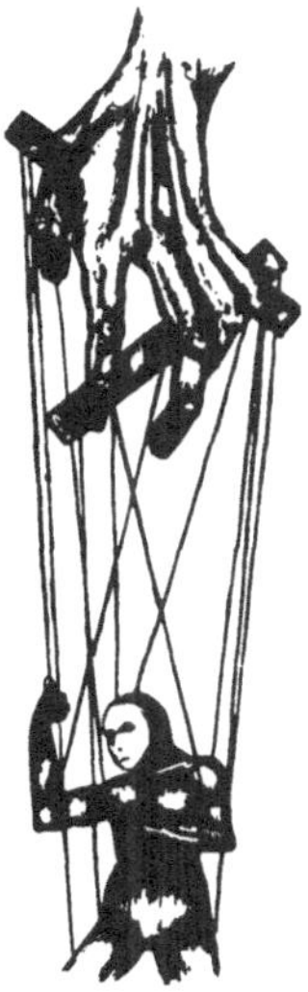

An experimental procedure to investigate the soft-construction of the pendular, clocking mode is described. In this procedure a person swings hand-held pendulums at the wrists, comfortably and rhythmically. This wrist-pendular, clocking mode can be assembled with one wrist-pendulum system (a pendulum held in either the left or right hand) or it can be assembled with two wrist-pendulum systems (one pendulum held in the left hand and another pendulum held simultaneously in the right hand). In the latter case the person's task is to swing the two pendulums comfortably together, at a common tempo and at a fixed phase relation. Several experiments using this procedure are reported in due course. The present chapter reports on one experiment, the main experiment on which much of the discussion in the book focuses.

The comfortable periodic times onto which a person settles vary with the magnitudes of the hand-held pendulums. The question posed is whether or not it is possible to describe the magnitudes of single wrist-pendulum systems and double (that is, coupled) wrist-pendulum systems in the same way through the same single quantities. To do so would lay the foundations for a general law-based account of wrist-pendular periodic motion.

A single wrist-pendulum system consists of three concentrations of mass—

those of the pendulum shaft, the added weights, and the hand—at different distances from a common axis of rotation. It is a compound pendulum. Through the method introduced by Huygens in the seventeenth century a compound wrist-pendulum system can be redescribed as a simple pendulum—that is, as a single concentration of mass at a single distance from the point of rotation—whose periodic timing is the same as that of the compound wrist-pendulum system. A double wrist-pendulum system is similarly conceivable as a common pendulum, albeit one in which mechanical forces are only partially responsible for the connections among the various masses. Neural processes, rather than mechanical forces, connect the total mass of the left wrist-pendulum system with that of the right wrist-pendulum system.

It is assumed herein that "Huygens' law" applies equally to single and double wrist-pendulum systems and that both, therefore, are describable through single quantities of mass, length, and time. The implication is that a double wrist-pendulum system can be characterized, in a principled manner, as a virtual single *wrist-pendulum system. The means by which this characterization is achieved, namely, Huygens' law, is a conservation based symmetry operation. In general terms, the operation transforms a system into a considerably simpler version of itself, leaving invariant the behavioral property of periodic timing.*

This physical notion of a virtual single system (or unitary process) assembled through the conservations pertains to very general aspects of the problem of how movements are coordinated and controlled. It is of particular relevance to the intuitive idea that in a coordinated activity such as locomotion, two or more biokinematic degrees of freedom are constrained to act as a single functional unit. As will become evident, the strategy of physically interpreting the behavior of two or more isochronous pendular systems as that of a (virtual) single system is fundamental to the analysis of the data and to the development of a law-based account of rhythmic movement.

6.1. Locomotion and the Pendular, Clocking Mode

Locomotion has the thermodynamical (dissipative) characteristics of a limit cycle, clocking mode of organization insofar as the limb motions in locomotion are auto-oscillatory. Further, locomotion has the mechanical (conservative) characteristics of a pendular-like mode of organization (interchanging kinetic energy and potential energy) insofar as the limbs in locomotion are rhythmically raised and lowered under the influence of gravity. Both of these modes, the (thermodynamical) clocking and the (mechanical) pendular, are simple dynamic regimes, and both can be soft molded. From a strictly physical perspective, the assembling of these modes raises the question of how a complex internal field (the interior of an animal) can self-organize an isolated segment of itself, or its

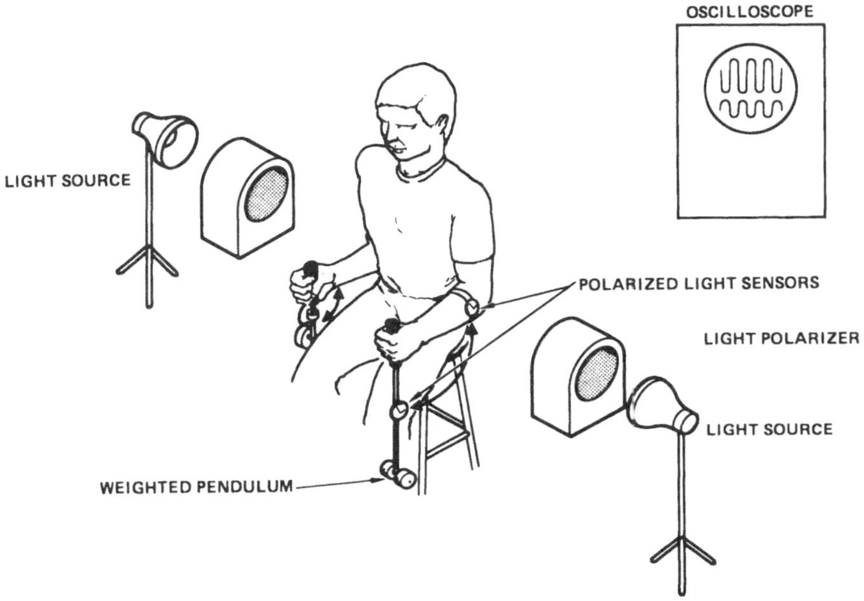

FIGURE 6.1. THE EXPERIMENTAL TASK FOR STUDYING THE SOFT, SELF-ASSEMBLY OF THE PENDULAR CLOCKING MODE.

entirety, into an effectively simple dynamical regime. Put differently, how does a complex system, dominated by plastic-elastic-gel processes (Iberall, 1972; Yates, 1982), simulate simple systems (Fowler & Turvey, 1978)?

Given the foregoing, a physical description of locomotion is that it closely approximates a pendular, clocking (limit-cycle) mode of organization. Consequently, understanding locomotion rests in part on understanding the physics that condenses out a simple pendular, clocking mode in one limb and in a pair of limbs. To this purpose the research to be reported used the task shown in Figure 6.1.

The figure depicts the case where a subject is swinging, out of phase by 180 degrees and at a common period, two hand-held pendulums of different lengths and masses. This can be construed usefully as a case of coupling two "oscillators" of different natural periods. As suggested by previous work on limb oscillators and their couplings, the natural periods of limb oscillations must be manipulated if the design principles of locomotion are to be uncovered (e.g., Graham, 1972; Shik & Orlovskii, 1965; Stein, 1973; von Holst, 1935/1973). In the research to be reported, variation in natural period is achieved by varying the length and the mass of hand-held pendulums.

6.2. Methodological Guidelines

The goal of the present research was to gain qualitative and quantitative physical insights into the soft construction of a motor organization akin to that of locomotion. Abiding by this goal the research was conducted according to the following methodological considerations. First, qualitative insights can be facilitated if the design focuses on equilibrium states rather than on transition states (Wesley, 1974). Second, it is necessary to identify the appropriate qualitative description before beginning a quantitative description (Poincaré, 1914). Third, and related, emphasis should be on the overall form of proposed equations and their physical significance rather than on the formal solution of differential equations (Stahl, 1962).

With respect to the first injunction, comprehending the changes in systematic behavior is hamstrung to the degree that the stable behaviors are not explicable. Understanding what determines a system's equilibria or steady states—the conditions that create them, maintain them, and dissolve them—is foremost in the understanding of a system. The equilibria or steady states and their governing conditions rationalize the transitions. In general, equilibrium and steady state descriptions involve fewer variables. (At equilibrium and in the steady state, variables are coupled through nonlinearities.) The equilibrium and steady state offer the simplest description of a system in the form of the equation of state (see Section 3.3.1).

With regard to the second injunction, initiating the study of any complicated psychological or biological process along strictly quantitative lines soon becomes a matter of the trees obscuring the forest. The details become unwieldy. A prudent tack is to pursue simplifying abstractions of the process as far as is permissible (that is, consistent with the underlying dynamics) and to develop an overall picture of how the states of the process relate. The first pass on the problem should entail the identification of the qualitative or modal characteristics of the organization. A good qualitative description can give direction to (subsequent) quantitative efforts.

Finally, with regard to the third injunction, it might appear inadequate to describe a physical or biological phenomenon in anything other than the calculus notation. In many instances, however, applying the tools of dimensionless numbers or criteria of similarity (Johnstone & Thring, 1957; Stahl, 1962) is the only practical route to take. Moreover, a description in terms of similarity criteria is neither just a matter of convenience nor simply a last-ditch effort. The form of such a description can be (and often is) suggestive of the underlying physical principles (Duncan, 1953).

6.3. The Experimental Task

The experimental setting consisted of a subject seated comfortably in a chair holding one or two pendulums of specific masses and lengths (see Figure 6.1).

The task confronting the subject was to perform a pendular oscillation involving a single wrist joint in isolation or two wrist joints in combination. The only restriction on the subject was that the forearm of the wrist-pendulum system be maintained in a position horizontal to the ground plane. From this orientation, a static or resting pendulum (i.e., freely hanging pendulum) formed a 90 degree angle with respect to the forearm. Dynamic oscillations occurred about this vertical axis. By maintaining the forearm in a relatively horizontal position, the pendular oscillation could be confined to the wrist joint(s).

6.3.1. The Design of Hand-Held Pendulums. The pendulums were constructed as follows. Rods of two different lengths (0.294*m* and 0.599*m*) but constant diameter were made from a lightweight rigid material (0.020*m* diameter wooden (ash) dowel). A lightweight tubular metal bar (0.006*m* diameter) was inserted through each rod approximately 0.040*m* from the lower end for the purpose of securing various weights. Aluminum washers and nuts were used to fasten metal weights to the bar and rod. A small magnetically attractive metal plate (0.030*m* × 0.060*m* × 0.001*m*) was attached approximately 0.150*m* from the upper-handle end of each rod for the purpose of securing a magnetic photo cell from the recording apparatus. Finally, a thin rubber grip was inserted over the upper, handle end of the rod, providing a firm and comfortable handle for controlling the pendular motions. In accordance with the above criteria masses of 0.1*kg* and 1.0*kg* were added to rods of 0.294*m* and 0.599*m* to produce four length(*m*)/mass(*kg*) conditions: 0.294/0.1, 0.599/0.1, 0.294/1.0, 0.599/1.0. These length/mass conditions are designated *A, B, C, D* respectively.

6.3.2. Goniometric Recording System. The angular motions of interest were the oscillatory movements occurring about the wrist joint(s). These movements were measured by (magnetically) attaching one photo cell sensor to the forearm of the subject and (magnetically) attaching a second photo cell sensor to the metal plate on the pendulum. Bathed in a field of rotating polarized light, the photo cells generated voltages proportional to angular rotation. The voltages were transferred to a digital representation via analogue-to-digital conversions at a sampling rate of 200 *hz*. A package of conversion routines (implemented on a PDP-11/45 computer) was then used to convert the digital signals and display them as time domain plots. Additional programs provided mean amplitude, frequency, and duration information. Importantly, the collection apparatus provided an accurate means for recording movement parameters, while maintaining a relatively unconstrained environment for the subject.

6.4. Procedure and Subjects

The four pendulum systems, constructed out of the two length conditions (0.294*m* and 0.599*m*) and the two mass conditions (0.1*kg* and 1.0*kg*) yielded eight single or isolated wrist-pendulum systems (four right wrist-pendulum systems and four left wrist-pendulum systems). Double or paired wrist-pendulum systems were formed by combining the various left and right single systems. Sixteen double systems were thus constructed. Single and double wrist-pendulum systems combined formed a total of 24 wrist-pendulum conditions.

Four subjects (two males and two females) with no reported motor disabilities participated in the experiment. The subjects were volunteers from a university population. Three of the subjects were right-handed and one subject was left-handed. The subjects are listed in the tables according to numbers 1-4. Subject 1 was male, age 32 years, right-handed, and weighed 65.8 *kg*; Subject 2 was female, age 29 years, right-handed, and weighed 63.6 *kg*; Subject 3 was male, age 26 years, left-handed, and weighed 68.1 *kg* and; Subject 4 was female, age 20 years, right-handed, and weighed 49.9 *kg*.

Each subject was exposed to the 24 wrist-pendulum conditions eight times, using a repeated measures design, for a total of 192 trials. The trials were distributed over two days, 96 trials on each day. The trials were presented in eight blocks of 24 conditions. Each block represented the complete set of single and double conditions. Within each block the single and double conditions were presented as separate sub-blocks. Within the sub-blocks the single and double systems were presented randomly. The experimental order began with the sub-block of eight single conditions followed by 16 double, eight single, 16 double, and so on.

For the single wrist-pendulum systems the subject was instructed to *search* through the range of possible rhythmical patterns (i.e., variations in frequencies and amplitudes) that could be assembled and locate the *preferred* set of oscillatory parameters. A "preferred" set of parameters was interpreted for the subject as that rhythmical motion that felt most "comfortable," "stable," and "natural"—a motion that he or she could sustain for a very long period if so required.

When the subject was presented with a pair of pendulums, the instructions were to assemble an oscillatory system in which the individual pendulum motions were *180 degrees out of phase* and at a *single common frequency*. For this organization, the subject's task was again that of "searching" through the range of possible rhythmical patterns that could be assembled (using the double pendulum system) to locate the "preferred" set of oscillatory parameters. A "preferred" set of oscillatory parameters was once again interpreted for the subject as that rhythmical motion that felt most "comfortable," stable and natural.

The subject was further instructed to signal verbally the experimenter when the "preferred" configuration had been assembled. Upon receiving the signal, the experimenter initiated a trial consisting of a 10-second goniometric recording of the rhythmical oscillatory movement patterns. Within each trial, the number of unit oscillations recorded ranged from 8 to 15 cycles, depending on the preferred cycle frequency. The data set for each pendulum condition ranged from $8 \times 8 = 64$ to $8 \times 15 = 120$ data points. This number of points was more than adequate to yield reliable data while avoiding the deteriorating effects of subject fatigue. If a subject had difficulty in sustaining a preferred oscillatory configuration over the 10-second trial, the trial was repeated. An oscilloscope was used to monitor continuously the subject's performance (output signal on the goniometric system) as a check for transient signals.

6.5. The Concept of a "Virtual" Single System

For purposes of analysis, it will be desirable to consider together the single wrist-pendulum systems and the double wrist-pendulum systems. The question is: How can the double systems be described in a way commensurate with the single systems?

A single wrist-pendulum system is made up of three masses rotating about a point in the wrist. The masses are the mass of the rod, the mass of the added weights, and the mass of the hand. It is, in short, a *compound* pendulum. A double wrist-pendulum system is made up of six distinct masses, three rotating about a point in the right wrist and three rotating about a point in the left wrist. It, too, is a compound pendulum, albeit more complicated. Compound pendulums can be represented as single simple pendulums. The mass of a single simple pendulum is located (concentrated) at a single point and its suspension from the point of rotation is massless. The sections that follow show how the transformation from single and double compound wrist-pendulums to single, simple pendulums is achieved. These sections (together with Appendix C) give the physics and the reasoning that make the transformation possible.

The upshot is that a single (compound) wrist-pendulum system can be identified by a single magnitude of mass M_i and a single magnitude of length L_i. A double wrist-pendulum system can be identified in like fashion. Drawing from the theory of compound pendulums, the concepts of virtual mass M_v and virtual length L_v are introduced to characterize a pair of wrist-pendulum systems coupled to the same period as a *virtual single system*.

The central problem is to determine the center of oscillation of a single wrist-pendulum system and of a pair of wrist-pendulum systems that are constrained to oscillate at a common natural period. The center of oscillation defines the point at which the total mass of a given compound system would have to be concentrated in order to obtain a simple, single pendulum system exhibiting the same natural period as the compound system. A solution to this problem is offered in Huygens' 1673 *Horologium oscillatorium* (see Bell, 1950, for a modern translation). It will be referred to, in this monograph, as Huygens' law.

6.5.1. A Generalized Derivation of the Center of Oscillation. In a simple pendulum the mass of the system is concentrated at a single point referred to as

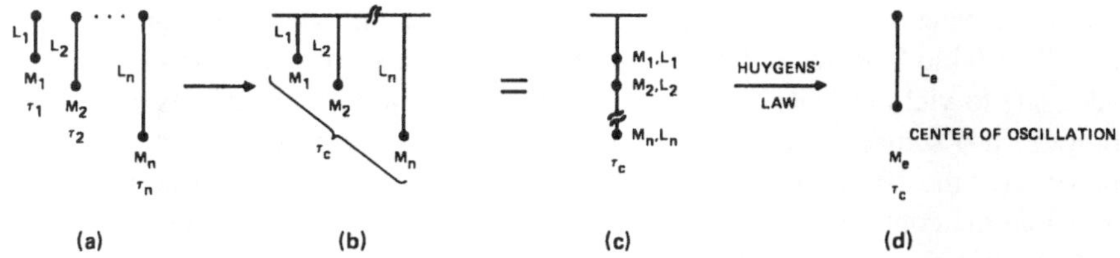

FIGURE 6.2. CENTER OF OSCILLATION OF COMPOUND PENDULUM.

Left panel (a): Ensemble of simple pendulums.

Center panels (b) and (c): Examples of compound pendulums.

Right panel (d): Equivalent simple pendulum.

the center of gravity (mass). In contrast, in a compound pendulum the distribution of mass extends beyond the point defining the center of gravity. The material (mass) particles that constitute the compound pendulum can be viewed as a rigidly connected ensemble of simple pendulums. Each point particle, and its length from the point of rotation, if independently suspended and allowed to oscillate freely, would exhibit a natural period of oscillation that is a function of its suspension length. As a compound pendulum, the individual point particles are rigidly connected so that the entire ensemble of particles oscillates at a single, common period. Figure 6.2a identifies an ensemble of simple pendulums of unequal lengths and masses. The natural periods associated with the shorter pendulums will be shorter than the natural periods associated with the longer pendulums (i.e., in Figure 6.2a $\tau_1<\tau_2< \ldots <\tau_n$). When the pendulums are coupled with a rigid constraint, the natural period of the resulting single compound pendulum will be shorter than the periods associated with the longer pendulums and longer than the periods associated with the shorter pendulums (i.e., $\tau_1<\tau_2< \ldots <\tau_c<\tau_n$, see Figure 6.2b). In a compound pendulum, the particles of the individual pendulums interact mutually in a manner that modifies each other's motions, resulting in the natural period defined over the ensemble.

Huygens proposed that the length of a single pendulum, whose natural period was identical to that of the compound pendulum, could be used to identify the center of oscillation of the compound pendulum (see Figure 6.2c). If the length of the simple pendulum was marked off against the compound pendulum, then a point would be identified that preserves the same period of oscillation in its connection with the other points that it would have if detached and left to oscillate in isolation. This point is the *center of oscillation*. Huygens' insight was the recognition that while the center of gravity (mass) and center of oscillation are coincident in a simple pendulum, they are not coincident in a compound pendulum.

The simplest example of a compound pendulum is presented in Figure 6.3. Two particles of mass M_l and M_2 are rigidly suspended at distances L_l and L_2, respectively, from an axis of rotation. Finding the center of oscillation requires identifying the length L_e of the equivalent simple pendulum that exhibits the same natural period as the compound pendulum. The pendulum defined by M_2 retards the motion of the pendulum defined by M_l, while M_l accelerates M_2, so that M_l oscillates more slowly and M_2 more rapidly than each would if separately suspended (i.e., $\tau_l<\tau_c<\tau_2$). Thus, the required L_e must lie somewhere between L_l and L_2 (i.e., $L_l<L_e<L_2$).

The solution offered by Huygens is derived from three physical claims. First, if any mass begins to move under the force of gravity, its center of gravity cannot rise higher than it was when the motion began. Second, the height to which any given mass of the pendulum rises above its lowest level is proportional to the square of the velocity that it acquires in falling to that level. And third, the velocities of the masses are proportional to their respective distances from the axis of suspension. The second and third claims derive from Galileo's laws of falling bodies. The first assertion, however, while implicitly recognized by Galileo, owes its formal origin to Huygens and marked the first explicit recognition of the conservation of energy principle.

Using the preceding physical claims as postulates, Huygens addressed the pendulum problem as if it were a problem in a branch of geometry. Relying heavily on the theorem of similar triangles, Huygens deduced a generalized solution for defining the center of oscillation for any rigid body undergoing rotational motion (see Appendix C for details of the derivation). The solution stated that the center of oscillation L_e is given by the sum of the product of the masses of the particles and the squares of their respective distances from the axis

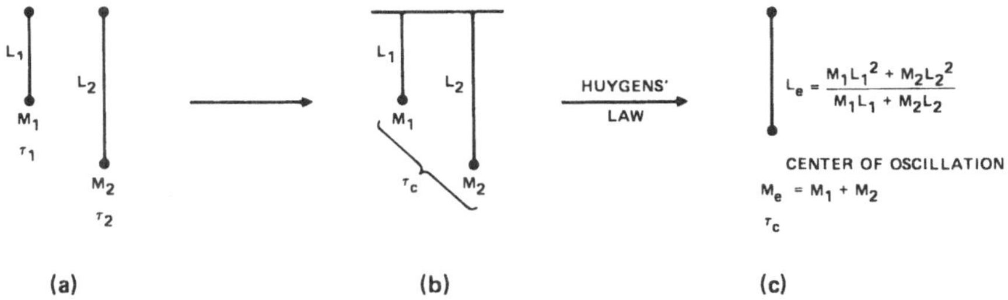

FIGURE 6.3. CENTER OF OSCILLATION OF MINIMAL COMPOUND PENDULUM.

Left panel (a): Independent pendulums.

Center panel (b): Compound pendulum.

Right panel (c): Equivalent simple pendulum.

of suspension, divided by the sum of the products of the masses and their respective distances. Stated in equation form,

$$L_e = \frac{M_1L_1^2 + M_2L_2^2 + \ldots}{M_1L_1 + M_2L_2 + \ldots} \tag{6.1}$$

or in more general form,

$$L_e = \frac{\Sigma M_iL_i^2}{\Sigma M_iL_i}. \tag{6.2}$$

The mass concentrated at the center of oscillation is equal to the total mass of the individual particle systems:

$$M_e = M_1 + M_2 + \ldots \tag{6.3}$$

or, stated more generally,

$$M_e = \Sigma M_i. \tag{6.4}$$

Hence, a single pendulum of length L_e and mass M_e could be defined that exhibited the same natural period as the compound pendulum (see Figure 6.3c). This solution, referred to as Huygens' law, forms the theoretical framework for the concept of *inertia*. Stated in terms of inertia, the law reads: The distance of the center of oscillation L_e of a compound pendulum from the axis of suspension is obtained by dividing the moment of inertia of the pendulum about that axis by its statical moment about the same axis:

$$L_e = \frac{\Sigma\,[\text{inertial moments}]}{\Sigma[\text{statical moments}]}. \tag{6.5}$$

In contrast, the center of gravity (mass) is defined as

$$L_g = \frac{\Sigma M_iL_i}{\Sigma M_i} = \frac{\Sigma\,[\text{statical moments}]}{\Sigma\,[\text{masses}]}. \tag{6.6}$$

For a simple pendulum of length L and mass M, the center of oscillation L_e and the center of gravity L_g are both equal to the pendulum length L:

$$L_g = \frac{ML}{M} = L \quad \text{and} \quad L_e = \frac{ML^2}{ML} = L. \tag{6.7}$$

For a compound pendulum, however, the center of oscillation and center of gravity differ because the distribution of mass varies as a function of L_i^2 for the center of oscillation and L_i for the center of gravity.

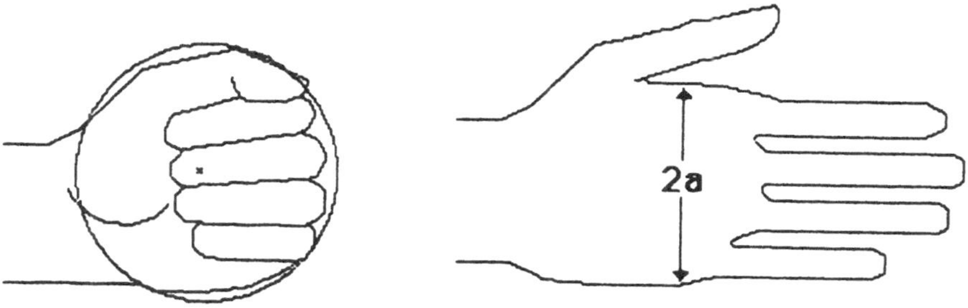

FIGURE 6.4. *Left panel (a)*: HAND APPROXIMATED AS A DISC.

Right panel (b): WIDTH OF HAND MEASURE.

6.6. A Simple Pendulum Equivalent of a Single Wrist-Pendulum System

An individual wrist-pendulum system is a compound pendulum replaceable by an equivalent simple pendulum. The length of an equivalent simple system can be derived as follows, through repeated use of the parallel axis (Steiner's) theorem (see, for example, Feynman, Leighton, & Sands, 1964):

1. The rods used in the construction of the pendulums were of uniform density and cross-sectional area. The constant product of density and area was found to be 0.248*kg-m*. Consequently the mass of any given rod was:

$$M_{rod} = (0.248)L, \tag{6.8}$$

where L was a rod's length in meters.

2. Let the clenched hand be approximated by a disc of radius a (Figure 6.4a), where $2a$ is the width of the hand measured from the second to the fifth carpometacarpal joint (Figure 6.4b).

3. Let b be the distance from the distal head of the radius bone in the forearm to the vertical axis of a rod held in the hand such that the top of the rod is flush with the top of the hand (Figure 6.5). The distances a and b will vary with the individual subjects. For the four participants in the main experiment the range of a was 0.033*m* to 0.048*m* and the range of b was 0.076*m* to 0.086*m*.

4. The center of mass of a rod lies at a distance of $[(L/2)-a]$ from the center of the hand which is at a distance b (on a line parallel to the ground plane) from the point of rotation O in the radio-ulnar joint (Figure 6.5). These distances are schematized in Figure 6.6. The distance from O to the center of mass of the rod is given by:

$$Q = [b^2 + ((L/2) - a)^2]^{1/2}. \tag{6.9}$$

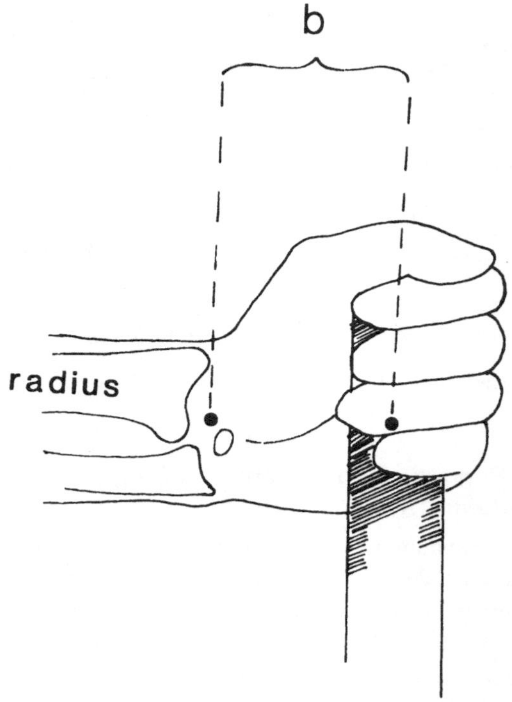

FIGURE 6.5. THE *b* PARAMETER.

By the parallel axis theorem the inertia of the rod about O is given by:

$$I_{rod} = M_{rod}Q^2 + (M_{rod}L^2)/12. \quad (6.10)$$

5. The added masses were discs attached to a bolt through the rod. The discs were evenly balanced on either side of the rod, allowing a construal of the discs as a single thick disc of radius $R = .03m$ at a distance c from the lower end of the rod. The center of mass of this "disc" is a distance P from the point of rotation O as shown in Figure 6.7:

$$P = [b^2 + (L - a - c)^2]^{1/2}. \quad (6.11)$$

By the parallel axis theorem the inertia of a disc about O is

$$I_{disc} = M_{disc}P^2 + (M_{disc}R^2)/2. \quad (6.12)$$

6. According to Bernstein's (1967) and Dempster's (1955) calculations the mass of a hand is approximately .006% of a person's total mass. For the participants in the main experiment the hand mass was: 0.395*kg* for Subject 1; 0.382*kg* for Subject 2; 0.408*kg* for Subject 3; and 0.299*kg* for Subject 4. Given that the distance of the hand's center of mass from *O* is *b* (Figure 6.5), the inertia of a closed fist (conceptualized as a disc, Figure 6.4a) is, by the parallel axis theorem,

$$I_{hand} = M_{hand}b^2 + (M_{hand}a^2)/2. \quad (6.13)$$

7. By Huygens' law, the length of an equivalent simple pendulum is given by the ratio of a compound system's total moment of inertia to its total static moment. Determining the latter quantity requires the distance from the point of rotation *O* to the compound pendulum's center of mass (*CM*). The *CM* of the compound pendulum is at a distance of L_{CM} from the handle end of a rod as given by:

$$L_{CM} = \frac{M_{rod}(L/2) + M_{disc}(L-c) + M_{hand}a}{M_{rod} + M_{disc} + M_{hand}}. \quad (6.14)$$

The distance *h* from the point of rotation *O* to the compound pendulum's center of mass (Figure 6.7) can then be calculated from:

$$h = [(L_{CM} - a)^2 + b^2]^{1/2}. \quad (6.15)$$

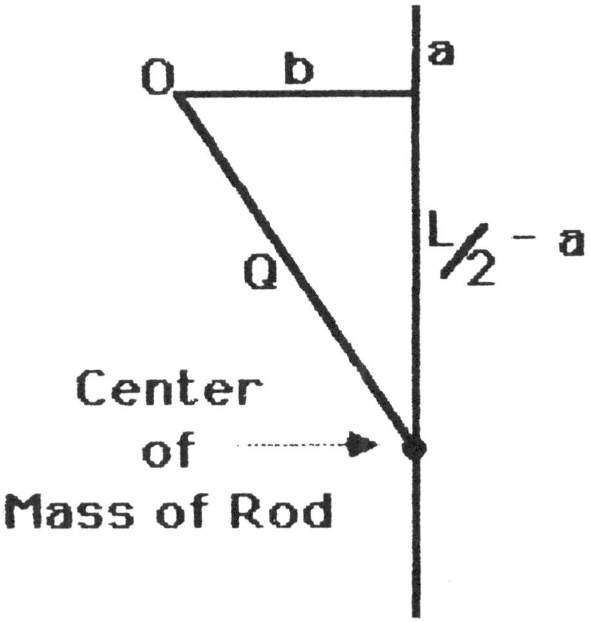

FIGURE 6.6. SCHEMATIC IDENTIFYING THE CENTER OF MASS OF THE ROD.

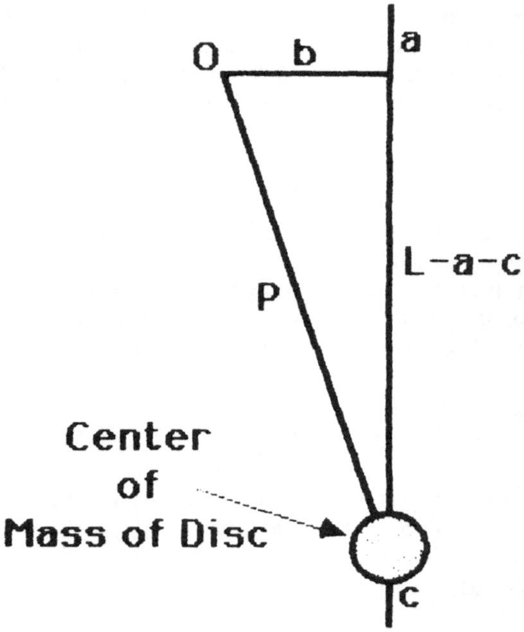

FIGURE 6.7. SCHEMATIC IDENTIFYING THE CENTER OF MASS OF THE DISC.

8. With h known, the length of an equivalent simple pendulum L_e is obtained from:

$$L_e = \frac{I_{rod} + I_{disc} + I_{hand}}{(M_{rod} + M_{disc} + M_{hand})h}. \tag{6.16}$$

9. The mass concentrated at L_e from the axis of rotation is given by the summed masses of the rod, disc and hand.

6.7. A Simple Pendulum Equivalent of a Double Wrist-Pendulum System

A compound pendulum is usefully thought of as two or more simple pendulums, rigidly coupled so as to restrict the systems' independent motions to a *single common natural period*. Figure 6.8 applies the concept of a compound pendulum

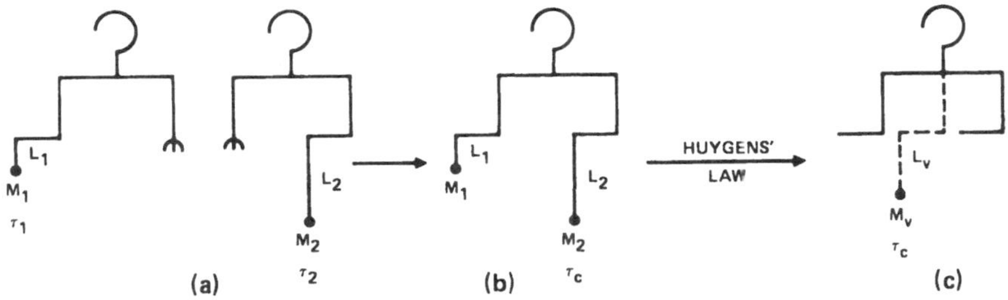

FIGURE 6.8. CENTER OF OSCILLATION FOR DOUBLE WRIST-PENDULUM SYSTEM.

Left panel (a): Independent wrist-pendulum systems.

Center panel (b): Double wrist-pendulum system.

Right panel (c): Equivalent "virtual" single simple pendulum system.

organization to the double wrist-pendulum system: Figure 6.8a identifies a pair of independent (isolated) wrist-pendulum systems and their associated natural periods; Figure 6.8b identifies the pair of pendulums coupled to form a double wrist-pendulum system with a single common period; and, Figure 6.8c identifies the equivalent "virtual" simple pendulum derived from Huygens' law. From Equations (6.5) and (6.3) the mass of a virtual system, M_v, is the sum of the masses of the two component systems and the length of the virtual system, L_v, is given as

$$L_v = \frac{M_1L_1^2 + M_2L_2^2}{M_1L_1 + M_2L_2} \qquad (6.17)$$

where M_1 and M_2 are the two individual system masses and L_1 and L_2 are the two individual lengths.

6.8. Period and Angular Excursion Data for Single and Double Wrist-Pendulum Systems

Table 6.1 identifies the length/mass conditions of the four single systems *A*, *B*, *C*, and *D* (see Section 6.3.1) for each of the four subjects. Tables 6.2 and 6.3 report the data on period and angular excursion, respectively, for each of the four subjects in each of the eight single wrist-pendulum conditions. Each datum is the mean of eight observations. Angular excursion refers to the magnitude of a full swing of a wrist-pendulum system, that is, the angle between the two extreme forward and back positions. This was the angle measured directly by the goniometers. The term *amplitude* as used in the analysis of oscillatory motion refers to one-half of the angular excursion. Most of the analyses are performed on amplitude, signified by θ.

TABLE 6.1
Single Wrist-Pendulum Systems' Length *(m)*/Mass *(kg)*/Inertia *(kg-m²)* Conditions

Subject	A	B	C	D
1	.194[a]	.449	.232	.515
	.568[b]	.644	1.468	1.544
	.021[c]	.130	.079	.410
2	.192	.454	.232	.518
	.555	.631	1.455	1.531
	.021	.130	.078	.411
3	.185	.452	.235	.516
	.581	.657	1.481	1.557
	.020	.134	.082	.394
4	.207	.471	.235	.514
	.472	.548	1.372	1.448
	.020	.122	.076	.383

[a]length *(m)*
[b]mass *(kg)*
[c]inertia *(kg-m²)*

TABLE 6.2
Periods of the Single Right and Left Wrist-Pendulum Systems

Length/Mass Condition	*Wrist-Pendulum System*	*Subject* 1	2	3	4
	Right	0.808*	0.791	0.735	0.883
A		0.026**	0.028	0.023	0.021
	Left	0.780	0.787	0.733	0.911
		0.020	0.029	0.021	0.019
	Right	1.166	1.155	1.163	1.393
B		0.029	0.028	0.021	0.024
	Left	1.160	1.161	1.171	1.344
		0.027	0.030	0.024	0.027
	Right	0.874	0.911	0.884	0.978
C		0.022	0.028	0.026	0.029
	Left	0.864	0.892	0.927	1.004
		0.023	0.035	0.023	0.032
	Right	1.321	1.362	1.359	1.453
D		0.029	0.025	0.029	0.036
	Left	1.337	1.370	1.379	1.453
		0.035	0.038	0.036	0.028

*The upper number is the mean period (seconds).
**The lower number is the mean standard cycle deviation.

TABLE 6.3
Angular Excursions of the Single Right and Left Wrist-Pendulum Systems

Length/Mass Condition	*Wrist-Pendulum System*	*Subject* 1	2	3	4
	Right	70.99*	54.37	84.05	78.78
A		3.97**	2.39	3.62	3.30
	Left	75.54	44.58	86.86	89.04
		3.75	2.74	3.11	3.47
	Right	83.14	58.96	116.08	80.73
B		3.49	2.39	3.62	3.20
	Left	91.04	55.75	111.04	81.70
		3.59	2.70	3.09	4.69
	Right	66.29	50.59	81.47	65.43
C		4.04	3.00	2.89	3.02
	Left	71.45	44.18	86.17	64.00
		3.74	3.09	3.82	3.27
	Right	75.17	55.92	102.93	65.26
D		3.60	2.23	3.71	2.86
	Left	79.02	50.42	102.04	67.84
		2.88	2.74	3.11	3.47

*The upper number is the mean angular excursion (degrees).
**The lower number is standard cycle deviation.

Tables 6.4 through 6.7 give the periods of the two wrist-pendulum systems composing a double system for each of the 16 double systems for each subject. Tables 6.8 through 6.11 give for each subject the angular excursions of the two wrist-pendulum systems for each of the 16 double systems. Unless stated otherwise, all of the analyses reported are on the data given in Tables 6.2 through 6.11. Subsidiary experiments were conducted and are referred to in due course.

Table 6.12 identifies the virtual mass and virtual length parameters derived from Huygens' law for the 16 double wrist-pendulum systems. Table 6.13 identifies a set of condition numbers that will be used to identify the respective virtual single lengths and masses.

Inspection of Tables 6.4 through 6.7 indicates that when coupled, the mean periods associated with the left and right systems (averaged over eight trials) differed on the average by less than 6*ms*. This approximate isochrony allows a single common period to be defined that is the average of the left and right periods. The common period, while a natural period for the pair of systems as a unit, is *not* the natural period of either of the individual systems, as is made

TABLE 6.4
Periods of the Double Wrist-Pendulum Systems[1]
Subject 1

Length/Mass Condition of the Left Wrist-Pendulum System	*Length/Mass Condition of the Right Wrist-Pendulum System*			
	0.194*m* 0.568*kg*	0.449*m* 0.644*kg*	0.232*m* 1.468*kg*	0.515*m* 1.544*kg*
0.194*m*	0.801*	1.100	0.851	1.233
0.568*kg*	0.025	0.051	0.029	0.044
	0.803**	1.100	0.853	1.238
	0.021	0.046	0.026	0.047
0.449*m*	1.107	1.220	1.167	1.285
0.644*kg*	0.045	0.034	0.039	0.067
	1.131	1.214	1.166	1.307
	0.036	0.031	0.036	0.055
0.232*m*	0.867	1.091	0.883	1.188
1.468*kg*	0.034	0.036	0.029	0.031
	0.863	1.091	0.882	1.188
	0.080	0.033	0.053	0.035
0.515*m*	1.299	1.330	1.253	1.364
1.544*kg*	0.080	0.033	0.053	0.035
	1.299	1.330	1.252	1.364
	0.035	0.029	0.036	0.029

[1]The upper number of each pair is mean period (seconds); the lower number is standard cycle deviation.
*The upper pair reports the values for the right wrist-pendulum system.
**The lower pair reports the values for the left wrist-pendulum system.

evident in the comparison of the natural period values in Tables 6.4 through 6.7 with the natural period values in Table 6.3.

It becomes evident in Chapter 8 that the significance of the virtual single system notion is that swinging together two pendulums with two hands at the natural period of the system given by their union is *dynamically identical* with swinging one pendulum with one hand at its natural period. This is so despite the greater number of neural elements and biokinematic degrees of freedom involved. Thus the notion of a virtual single system addresses the so-called problem of degrees of freedom (Bernstein, 1967; Gelfand, Gurfinkel, Tsetlin, & Shik, 1971; Greene, 1972; Turvey, 1977; Turvey, Shaw, & Mace, 1978). This problem, when expressed with reference to the human, is how control and coordination are manifest in a system of 100 or so joints and approximately 800 joint actuators (Turvey, 1980). The notion of a virtual single system implies that under the proper physical description, a given movement performed with n skeletal links may not be fundamentally different from the same type of movement performed, say, with $2n$ links. Put another way, performing a movement with $2n$ links may not necessarily pose a different (or more difficult) problem for the nervous system than performing the same movement with n links.

6.8.1. Distinguishing a Soft Molded Compound Pendulum from a Hard Molded Compound Pendulum. A wrist-pendulum system is a soft-molded auto-oscillatory system. The soft-molded constraints are functional and temporary, in contrast to hard-molded constraints, which are mechanical and permanent. For example, in a hard-molded compound pendulum the rigid coupling linking the individual pendulums is provided through a set of permanent mechanical constraints that form a force field by distributing globally the locally occurring

TABLE 6.5
Periods of the Double Wrist-Pendulum Systems[1]
Subject 2

Length/Mass Condition of the Left Wrist-Pendulum System	*Length/Mass Condition of the Right Wrist-Pendulum System*			
	0.192*m* 0.555*kg*	0.454*m* 0.631*kg*	0.232*m* 1.455*kg*	0.518*m* 1.531*kg*
0.192*m* 0.555*kg*	0.883*	1.152	0.972	1.327
	0.036	0.035	0.036	0.042
	0.881**	1.155	0.971	1.328
	0.044	0.057	0.055	0.087
0.454*m* 0.631*kg*	1.074	1.218	1.113	1.347
	0.040	0.034	0.040	0.027
	1.070	1.218	1.113	1.376
	0.038	0.025	0.030	0.039
0.232*m* 1.455*kg*	0.937	1.171	1.027	1.308
	0.042	0.028	0.035	0.030
	0.974	1.169	1.031	1.313
	0.038	0.065	0.031	0.056
0.518*m* 1.531*kg*	1.233	1.322	1.195	1.381
	0.062	0.035	0.064	0.034
	1.227	1.311	1.203	1.382
	0.042	0.039	0.037	0.028

[1]The upper number of each pair is mean period (seconds); The lower number is standard cycle deviation.

*The upper pair reports the values for the right wrist-pendulum system.

**The lower pair reports the values for the left wrist-pendulum system.

TABLE 6.6
Periods of the Double Wrist-Pendulum Systems[1]
Subject 3

Length/Mass Condition of the Left Wrist-Pendulum System	*Length/Mass Condition of the Right Wrist-Pendulum System*			
	0.185*m* 0.581*kg*	0.452*m* 0.657*kg*	0.235*m* 1.481*kg*	0.516*m* 1.557*kg*
0.185*m* 0.581*kg*	0.847*	1.046	0.952	1.203
	0.034	0.029	0.030	0.033
	0.834**	1.047	0.937	1.203
	0.034	0.036	0.041	0.054
0.452*m* 0.657*kg*	1.064	1.167	1.109	1.311
	0.050	0.030	0.040	0.031
	1.063	1.163	1.128	1.310
	0.028	0.026	0.026	0.033
0.235*m* 1.481*kg*	0.940	1.054	0.998	1.256
	0.034	0.030	0.028	0.029
	0.937	1.054	0.984	1.251
	0.029	0.034	0.033	0.061
0.516*m* 1.557*kg*	1.293	1.305	1.305	1.396
	0.049	0.038	0.050	0.033
	1.290	1.305	1.307	1.395
	0.031	0.028	0.028	0.033

[1]The upper number of each pair is mean period (seconds); the lower number is standard cycle deviation.
*The upper pair reports the values for the right wrist-pendulum system.
**The lower pair reports the values for the left wrist-pendulum system.

stresses and strains. The forces in the field are transmitted in a continuous fashion from one region to another through mechanical couplings. The realization of a single common period is due, therefore, to the mechanical transmission of stress/strain properties defined over the field. The global oscillatory motion of the hard-molded compound pendulum exhibits a characteristic natural period that is shared by all the composite subsystems. In addition to inducing a common period, the hard-molded constraints that link the subsystems further restrict the subsystems' behavior to a *single common amplitude*.

A soft-molded compound double wrist-pendulum system differs from a hard molded compound pendulum in four ways: (a) the compound double wrist-pendulum system is constructed out of a set of constraints that are only assembled *temporarily*; (b) the field defining the compound double wrist-pendulum system is constructed out of *functional* rather than mechanical constraints that span a

TABLE 6.7
Periods of the Double Wrist-Pendulum Systems[1]
Subject 4

Length/Mass Condition of the Left Wrist-Pendulum System	*Length/Mass Condition of the Right Wrist-Pendulum System*			
	0.207*m* 0.472*kg*	0.471*m* 0.548*kg*	0.235*m* 1.372*kg*	0.514*m* 1.448*kg*
0.207*m* 0.472*kg*	0.928*	1.378	0.990	1.460
	0.023	0.038	0.041	0.043
	0.933**	1.374	0.997	1.463
	0.028	0.086	0.034	0.056
0.471*m* 0.548*kg*	1.375	1.406	1.367	1.491
	0.060	0.034	0.060	0.046
	1.375	1.418	1.370	1.513
	0.032	0.031	0.037	0.045
0.235*m* 1.372*kg*	1.057	1.350	1.039	1.455
	0.039	0.026	0.051	0.040
	1.054	1.345	1.043	1.468
	0.041	0.053	0.049	0.082
0.514*m* 1.448*kg*	1.486	1.484	1.478	1.512
	0.095	0.046	0.071	0.046
	1.484	1.479	1.474	1.513
	0.041	0.036	0.027	0.042

[1]The upper number of each pair is mean period (seconds); the lower number is standard cycle deviation.
*The upper pair reports the values for the right wrist-pendulum systems.
**The lower pair reports the values for the left wrist-pendulum system.

spatial extent—the neural linkages and processes that bridge the distance between the left and right wrist muscle complexes—over which forces are *not* transmitted mechanically; (c) the amplitudes of the composite subsystems are *not* restricted to a single common value; and (d) the natural period of a double wrist-pendulum system is a function of two potentials, an assembled elastic potential and the gravitational potential, whereas the periodicity of an ordinary compound pendulum is a function of only one potential, gravity.

With regard to (a) the same wrist-joint complex can be used to assemble temporarily a compound single pendulum system or a compound double pendulum system. Constraints can be marshaled momentarily or dissolved according to the type of pendulum system required. The functional nature of the field complex identified in (b) identifies the fact that the coordination between the left and right wrist-joint complexes is not enforced by mechanical con-

TABLE 6.8
Angular Excursions of the Double Wrist-Pendulum Systems[1]
Subject 1

Length/Mass Condition of the Left Wrist-Pendulum System	*Length/Mass Condition of the Right Wrist-Pendulum System*			
	0.194*m* 0.568*kg*	0.449*m* 0.644*kg*	0.232*m* 1.468*kg*	0.515*m* 1.544*kg*
0.194*m* 0.568*kg*	68.57*	73.13	56.56	67.31
	4.24	4.69	4.41	3.84
	85.14**	92.82	84.74	94.43
	4.74	5.72	4.31	4.85
0.449*m* 0.644*kg*	77.84	69.05	65.06	67.40
	5.02	4.50	3.17	3.15
	90.65	93.92	89.37	93.69
	4.47	4.20	5.64	3.47
0.232*m* 1.468*kg*	67.39	75.79	63.85	67.41
	5.10	4.43	4.34	4.60
	70.21	82.14	79.03	83.29
	3.24	3.81	3.37	4.16
0.515*m* 1.544*kg*	81.32	74.22	68.70	65.94
	5.46	3.92	4.22	3.85
	77.25	74.98	76.26	76.06
	3.84	3.30	4.21	3.48

[1]The upper number of each pair is mean angular excursion (degrees); The lower number is standard cycle deviation.
*The upper pair reports the values for the right wrist-pendulum system.
**The lower pair reports the values for the left wrist-pendulum system.

straints. The kinetic energy and momentum of a wrist-pendulum system do not transfer across the neural pathways that connect the left and right wrist-joint complexes: The field coupling relating the left and right wrist-joint complexes is *kinematic* rather than *kinetic*. The timing characteristic (temporal topology) of the two wrist systems is preserved invariant (i.e., isochronous) through a soft neural coupling.

By virtue of the soft coupling, the two individual wrist-pendulum systems can be assembled temporarily as a single unit so as to condense out a single natural period. The amplitudes of the individual constitutive wrist-pendulum systems of a compound wrist-pendulum system need not be the same, as noted in (c). The amplitudes of the constitutive wrist-pendulum systems differ (see Tables 6.8 through 6.11) because they are not determined by a global distribution of forces, but rather by the potential energy depots carried in the *local* muscle

TABLE 6.9
Angular Excursions of the Double Wrist-Pendulum Systems[1]
Subject 2

Length/Mass Condition of the Left Wrist-Pendulum System	*Length/Mass Condition of the Right Wrist-Pendulum System*			
	0.192*m* 0.555*kg*	0.454*m* 0.631*kg*	0.232*m* 1.455*kg*	0.518*m* 1.531*kg*
0.192*m* 0.555*kg*	54.97*	56.13	49.11	52.34
	3.45	2.75	2.98	2.67
	47.46**	48.97	50.38	51.88
	2.20	2.85	2.43	3.57
0.454*m* 0.631*kg*	54.12	58.52	48.11	52.70
	3.35	3.05	2.86	3.08
	45.13	52.13	47.41	52.47
	3.24	2.15	2.91	2.73
0.232*m* 1.455*kg*	55.20	57.05	49.69	51.26
	3.19	4.13	3.28	3.35
	43.46	45.89	43.23	43.33
	2.57	2.73	2.62	3.38
0.518*m* 1.531*kg*	56.53	56.57	50.53	53.01
	3.40	3.26	3.59	2.37
	42.92	42.27	39.52	43.88
	2.97	2.76	3.75	1.79

[1]The upper number of each pair is mean angular excursion (degrees); The lower number is standard cycle deviation.

*The upper pair reports the values for the right wrist-pendulum system.

**The lower pair reports the values for the left wrist-pendulum system.

chemistry. Finally, with respect to (d), the role of the assembled elastic potential can be noted by observing the significant contribution of mass over and above its contribution to the equivalent simple pendulum length. The demonstration of this significance must await the detailed analysis presented in Chapters 7 and 8, particularly the latter.

TABLE 6.10
Angular Excursions of the Double Wrist-Pendulum Systems[1]
Subject 3

Length/Mass Condition of the Left Wrist-Pendulum System	*Length/Mass Condition of the Right Wrist-Pendulum System* 0.185*m* 0.581*kg*	0.452*m* 0.657*kg*	0.235*m* 1.481*kg*	0.516*m* 1.557*kg*
0.185*m*	84.41*	106.26	85.77	95.00
0.581*kg*	3.94	3.51	3.01	4.68
	82.67**	88.43	84.74	87.24
	3.00	3.79	3.06	4.09
0.452*m*	85.68	107.37	85.54	93.28
0.657*kg*	4.02	3.27	3.74	3.59
	103.94	101.11	101.98	96.22
	4.22	3.39	4.69	3.11
0.235*m*	80.75	101.17	81.51	96.09
1.481*kg*	4.98	3.06	4.22	4.39
	80.92	82.04	82.21	80.20
	3.69	3.42	3.75	3.12
0.516*m*	79.88	102.36	80.91	92.80
1.557*kg*	4.12	3.52	3.86	3.74
	94.45	93.18	94.42	90.75
	4.01	3.66	4.19	2.34

[1]The upper number of each pair is mean angular excursion (degrees); The lower number is standard cycle deviation.
*The upper pair reports the values for the right wrist-pendulum system.
**The lower pair reports the values for the left wrist-pendulum system.

TABLE 6.11
Angular Excursions of the Double Wrist-Pendulum Systems[1]
Subject 4

Length/Mass Condition of the Left Wrist-Pendulum System	*Length/Mass Condition of the Right Wrist-Pendulum System*			
	0.207*m* 0.472*kg*	0.471*m* 0.548*kg*	0.235*m* 1.372*kg*	0.514*m* 1.448*kg*
0.207*m* 0.472*kg*	66.99*	76.21	55.58	59.76
	4.66	4.41	3.84	5.28
	72.95**	59.45	70.04	59.27
	5.63	5.16	6.79	4.43
0.471*m* 0.548*kg*	56.78	63.57	44.00	53.68
	4.98	3.87	4.13	4.15
	82.07	80.67	85.98	71.78
	4.49	3.91	4.18	4.42
0.235*m* 1.372*kg*	63.49	69.79	53.08	55.98
	4.41	3.31	3.52	3.43
	63.72	51.48	55.37	46.49
	3.95	3.12	4.08	4.63
0.514*m* 1.448*kg*	56.47	63.29	46.41	50.39
	4.82	4.00	4.15	3.87
	67.76	66.61	68,72	56.62
	4.79	3.95	3.52	3.40

[1]The upper number of each pair is mean angular excursion (degrees); The lower number is standard cycle deviation.
*The upper pair reports the values for the right wrist-pendulum system.
**The lower pair reports the values for the left wrist-pendulum system.

TABLE 6.12a
Virtual Lengths and Masses (L_v/M_v) of the Double Wrist-Pendulum Systems
Subject 1

Length/Mass Condition of the Left Wrist-Pendulum System	*Length/Mass Condition of the Right Wrist-Pendulum System*			
	0.194*m*[a] 0.568*kg*[b]	0.449*m* 0.644*kg*	0.232*m* 1.468*kg*	0.515*m* 1.544*kg*
0.194*m* 0.568*kg*	0.191[a] 1.136[b]	0.378 1.212	0.222 2.036	0.476 2.112
0.449*m* 0.644*kg*	0.378 1.212	0.450 1.288	0.332 2.112	0.498 2.188
0.232*m* 1.468*kg*	0.222 2.036	0.332 2.112	0.232 2.936	0.431 3.012
0.515*m* 1.544*kg*	0.476 2.112	0.498 2.188	0.431 3.012	0.516 3.088

[a]length (*m*)
[b]mass (*kg*)

TABLE 6.12b
Virtual Lengths and Masses (L_v/M_v) of the Double Wrist-Pendulum Systems
Subject 2

Length/Mass Condition of the Left Wrist-Pendulum System	*Length/Mass Condition of the Right Wrist-Pendulum System*			
	0.192*m*[a] 0.555*kg*[b]	0.454*m* 0.631*kg*	0.232*m* 1.455*kg*	0.518*m* 1.531*kg*
0.192*m* 0.555*kg*	0.196[a] 1.110[b]	0.383 1.186	0.223 2.010	0.480 2.086
0.454*m* 0.631*kg*	0.383 1.186	0.453 1.262	0.333 2.086	0.501 2.162
0.232*m* 1.455*kg*	0.223 2.010	0.333 2.086	0.231 2.191	0.432 2.986
0.518*m* 1.531*kg*	0.480 2.086	0.501 2.162	0.432 2.986	0.518 3.062

[a]length (*m*)
[b]mass (*kg*)

TABLE 6.12c
Virtual Lengths and Masses (L_v/M_v) of the Double Wrist-Pendulum Systems
Subject 3

Length/Mass Condition of the Left Wrist-Pendulum System	*Length/Mass Condition of the Right Wrist-Pendulum System*			
	0.185*m*[a] 0.581*kg*[b]	0.452*m* 0.657*kg*	0.235*m* 1.481*kg*	0.516*m* 1.557*kg*
0.185*m* 0.581*kg*	0.185[a] 1.162[b]	0.380 1.238	0.224 2.062	0.478 2.138
0.452*m* 0.657*kg*	0.380 1.238	0.451 1.314	0.335 2.138	0.499 2.214
0.235*m* 1.481*kg*	0.224 2.062	0.335 2.138	0.236 2.962	0.432 3.038
0.516*m* 1.557*kg*	0.478 2.138	0.499 2.214	0.432 3.038	0.517 3.114

[a]length (*m*)
[b]mass (*kg*)

TABLE 6.12d
Virtual Lengths and Masses (L_v/M_v) of the Double Wrist-Pendulum Systems
Subject 4

Length/Mass Condition of the Left Wrist-Pendulum System	*Length/Mass Condition of the Right Wrist-Pendulum System*			
	0.207*m*[a] 0.472*kg*[b]	0.447*m* 0.548*kg*	0.235*m* 1.372*kg*	0.514*m* 1.448*kg*
0.207*m* 0.472*kg*	0.204[a] 0.944[b]	0.399 1.020	0.229 1.844	0.479 1.920
0.471*m* 0.584*kg*	0.399 1.020	0.473 1.096	0.341 1.920	0.504 1.996
0.235*m* 1.372*kg*	0.229 1.844	0.341 1.920	0.236 2.744	0.431 2.820
0.514*m* 1.448*kg*	0.479 1.920	0.504 1.996	0.431 2.820	0.515 2.896

[a]length (*m*)
[b]mass (*kg*)

TABLE 6.13
Condition Numbers Identifying Wrist Pendulum Systems

Length/Mass Condition of the Left Wrist-Pendulum System	*Double Wrist-Pendulum Systems* — *Length/Mass Condition of the Right Wrist-Pendulum System* A	B	C	D
A	(1)*	(5)	(9)	(13)
B	(2)	(6)	(10)	(14)
C	(3)	(7)	(11)	(15)
D	(4)	(8)	(12)	(16)

*These numbers (1–16) will be used to identify their respective double wrist-pendulum systems.

REFERENCES

Bell, A. E. (1950). *Christian Huygens and development of science in the seventeenth century*. London: Arnold.

Bernstein, N. A. (1967). *The control and regulation of movements*. London: Pergamon Press.

Dempster, W. (1955). Space requirements of the seated operator. WADC *Technical Report*, 55–159. Report released to the Office of Technical Services, U. S. Dept. of Commerce, Washington, 25, D.C.

Duncan, W. J. (1953). *Physical similarity and dimensional analysis*. London: Edwald.

Feynman, R. P., Leighton, R. B., & Sands, M. (1964). *The Feynman lectures on physics*. New York: Addison-Wesley.

Fowler, C. A., & Turvey, M. T. (1978). Skill acquisition: An event approach with special reference to searching for the optimum of a function of several variables. In G. E. Stelmach (Ed.), *Information processing in motor control and learning*. New York: Academic Press.

Gel'fand, I. M., Gurfinkel, V. S., Tsetlin, M. L., & Shik, M. L. (1971). Some problems in the analysis of movements. In Gel'fand et al. (Eds.), *Models of the structural-functional organization of certain biological systems*. Cambridge, MA: MIT Press.

Graham, D. (1972). A behavioral analysis of the temporal organization of walking movements in the first instar and adult stick insect (Caransis morosus). *Journal of Comparative Physiology*, *81*, 23–52.

Greene, P. H. (1972). Problems of organization of motor systems. In R. Rosen and F. Snell (Eds.), *Progress in theoretical biology* (Vol. 2). New York: Academic Press.

Holst, E. von (1935/1973). *The behavioral physiology of animal and man*. Coral Gables, FL: University of Miami Press.

Hoyle, G. (1983). *Muscles and their neural control*. New York: Wiley.

Iberall, A. S. (1972). *Towards a general science of viable systems*. New York: McGraw-Hill.

Johnstone, R. E., & Thring, M. W. (1957). *Pilot plants, models, and scale-up methods in chemical engineering*. New York: McGraw-Hill.

Poincaré, H. (1914). *Science and method*. Paris: Flammarion.

Shik, M. L., & Orlovskii, G. N. (1965). Co-ordination of the limbs during running of the dog. *Biophysics, 10*, 1148–1159.

Stahl, W. R. (1962). Dimensional analysis in mathematical biology. *Bulletin of Mathematical Physics, 24*, 81–108.

Stein, P. S. G. (1973). The relationship of interlimb phase to oscillator activity gradients in crayfish. In R. B. Stein, K. G. Pearson, R. S. Smith and J. B. Redford (Eds.), *Control of posture and locomotion*. New York: Plenum Press.

Stein, P. S. G. (1977). Application of the mathematics of coupled oscillator systems to the analysis of the neural control of locomotion. *Federation Proceedings, 336, 7*, 2056–2059, June.

Turvey, M. T. (1977). Preliminaries to a theory of action with reference to vision. In R. Shaw and J. Bransford (Eds.), *Perceiving, acting and knowing: Towards an ecological psychology*. Hillsdale, NJ: Lawrence Erlbaum Associates.

Turvey, M. T. (1980). Clues from the organization of motor systems. In U. Bellugi and M. Studdert-Kennedy (Eds.), *Signed and spoken languages: Biological constraints on linguistic form*. Dahlem Konferenzen. Weinheim: Verlar Chemie.

Turvey, M. T., Shaw, R., & Mace, W. (1978). Issues in the theory of action: Degrees of freedom, coordinative structures and coalitions. In J. Requin (Ed.), *Attention and performance VII*. Hillsdale, NJ: Lawrence Erlbaum Associates.

Wesley, J. P. (1974). *Ecophysics: The application of physics to ecology*. Springfield, Illinois: Thomas.

Yates, F. E. (1982). Outline of a physical theory of physiological systems. *Canadian Journal of Physiology and Pharmacology, 60*, 217–248.

7 HAPTIC INFORMATION AS THE ABSTRACTED MORPHOLOGY OF MOVEMENT DYNAMICS

ABSTRACT

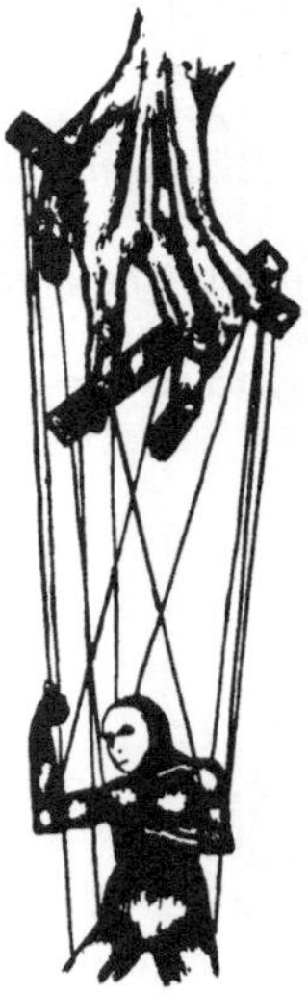

In the course of the main experiment described in Chapter 6, each subject selected a "preferred" period and amplitude for each wrist-pendulum system on eight separate occasions, divided evenly over two days. How were the cyclic parameters arrived at? Wrist-pendulum systems are not mechanical pendulum systems; their cyclicities are not defined by **closed, determinate**, *bookkeeping strategies defined over the conservations. Rather, wrist-pendulum systems are* **open, nondeterminate** *systems whose behavior is based on, but not caused by, information.*

Wrist-pendulum systems are guided primarily by haptic perception and moved by muscular and nonmuscular forces. In concert with the analysis of complex atomisms advanced in Chapter 4, it is suggested that information for the haptic system can be defined as kinematic (or geometric, or spectral) abstractions of the underlying mix of forces. These abstractions are morphological and macroscopic, and describable in a state space of relatively few variables (they are low dimensional). A quasi-formal analysis is provided of the kinetic attractor state (a limit cycle) of a wrist-pendulum system with implications for the apposite abstract description—the kinematic flow field property—by which it is specified.

The generating of these kinematic flow-field properties by kinetic processes, and their playing back as constraints on those kinetic processes, yields the circularly causal process of force→flow→force→flow . . . that may be the hallmark of self-assembling information systems. By the arguments of Chapter 4, the consistencies of such systems rest largely with the lawfulness of the generated information that guides them. It is supposed, therefore, that the periodic behavior of wrist-pendulum systems can be reliable and reproducible because information—in the kinematic, specificational sense of Gibson's—is law based and determinate. There are, as it were, information laws, roughly of the form:

$$\text{kinetic properties} \quad \underset{\text{specify}}{\overset{\text{lawfully generate}}{\rightleftarrows}} \quad \text{kinematic flow field properties.}$$

Consonant with standard law forms, information laws should meet the criteria of structural stability. The data of the main experiment are presented in a way that identifies structurally stable relations among period, amplitude, mass, and length: Over subjects, over hands, over trials, and over days the periodic times and amplitudes of the various wrist-pendulum systems were shown to order invariantly.

7.1. The Self-Reading and Self-Writing Nature of Wrist-Pendulum Systems

Ideally, an oscillating wrist-pendulum ought to be understood as a system that is both auto-oscillatory, in that the *energy flux* by which the system is sustained is determined by properties intrinsic to the system, and self-organizing, in that the *information flux* by which the system is controlled and coordinated is determined by the system's dynamics and, at the same time, modulates them (Kugler, Kelso, & Turvey, 1980). The sought-after understanding refers to a system that, like the termite-nest system, writes and reads information about itself (see Section 4.3). What is at issue here is the precise nature of the self-description exploited by a wrist-pendulum system: What are the properties by which it describes itself? The steps toward an understanding begin with an appreciation of the physically extraordinary nature of a wrist-pendulum system.

7.2. Physically Ordinary Systems and Wrist-Pendulum Systems

In some sense a wrist-pendulum system in oscillation is just a physically ordinary system. It consists of a pendular mass moving relative to the gravitational potential and another elastic potential. A description of its behavior in terms of a play of mechanical forces—originating muscularly and nonmuscularly—seems most reasonable, as does the intuition that there is a particular balance of these forces that defines a steady-state for a wrist-pendulum system. The steady-state of interest is not the stable static one given by the single system or the pair of systems oriented perpendicular to the ground plane. It is, of course, the steady-state of a stable periodic motion that alternates symmetrically kinetic and potential energy forms: The forces in question balance over the cycle.

Now, given that a wrist-pendulum system is not at this dynamic steady state to start with, how does it arrive at this state? If a wrist-pendulum system were truly a physically ordinary system, then the answer would be straightforward. Recalling Section 3.3, any displacement of a physically ordinary system from its equilibrium state gives rise to restorative forces proportional to that displacement. If unopposed, these forces suffice to draw or guide the system to the equilibrium state. The ordinary, simple mechanical pendulum provides an appropriate illustration. It is set into motion by applying a force that displaces it a definite distance from static equilibrium. Then, given negligible friction at the pivot, the pendulum oscillates with an amplitude proportional to the original applied force and with a period proportional to its length. The natural period—the simple pendulum's time-dependent equilibrium state—is a function of the balance between potential energy and kinetic energy (or, synonymously, few between the gravitational and inertial forces) defined over the cycle. Ignoring a details, the energy balance is expressed as:

$$mgl\ \theta^2/2 = \pi m l^2 \theta^2/\tau_0 \qquad (7.1)$$

or

$$\text{Maximum Potential Energy} = \text{Maximum Kinetic Energy} \qquad (7.2)$$

A motion of the pendulum at any period other than the one that satisfies this equation would be in violation of the conservation of energy. In this regard energy conservation restricts the motion of the system to a single periodic time. The mass, the length, the gravitational, and/or elastic potentials define boundary conditions that scale this single periodic time.

Differences between a wrist-pendulum system and the simple mechanical pendulum were alluded to in Chapter 6. First, a wrist-pendulum system is characterized by a lot of friction (i.e., entropy-producing processes), so much so that if an appropriately directed and measured muscular force is not generated each cycle, then the oscillation ends abruptly. Second, because of the presence of entropy-producing processes, a *steady-state*, rather than a conservative equilibrium state, defines the thermodynamically stable condition. Third, the amplitude of a wrist-pendulum system is integral to the steady-state because, unlike the simple mechanical pendulum, only a particular amplitude (possibly from a small set of amplitudes) can be sustained at a given period. Wrapped up in the amplitude variable are the mechanical and thermodynamical conditions for stable periodic movements. Fourth, the energy to start up and to sustain the time-dependent steady-state originates *within* the wrist-pendulum system. This is in notable contrast to the simple mechanical pendulum, where the energy to initiate and sustain the time-dependent equilibrium state is supplied from external sources.

Among the implications of the third contrast are the following: The energy per cycle is prescribed at the outset (by the initial conditions) for the ordinary simple pendulum, whereas this quantity must be arrived at in due course by the wrist-pendulum system; the trade-off per phase angle between competing forces (conservative and nonconservative) to ensure a balance over the cycle is prescribed for the ordinary simple pendulum, whereas it must evolve in the case of a wrist-pendulum system. The wrist-pendulum system must compensate continually for the energy loss incurred by nonconservative forces. This third contrast also distinguishes a wrist-pendulum system from an artifactual clock, despite their greater similarity. The quantity of energy per cycle to be divided among the dissipative (entropy producing) and the two mechanical forms of energy is prescribed fully from the outset, as is the partitioning. This means that the clock's steady-state—defined by the balance of the conservative and nonconservative forces and revealed as a limit cycle—is prescribed fully. In a clock the amount of force injected into the motion each cycle is fixed determinately at the outset by the design of the escapement (see Section 5.2.4).

Putting the matter very simply, the steady-state at which a wrist-pendulum system settles is both *defined* and *discovered* in the course of rhythmic activity. Unlike physically ordinary systems that involve a compliance or stiffness term, the compliance of a wrist-pendulum system is a variable. It has to be assembled macroscopically from a microscopic mixture of elastic and viscous processes, and its assembly extracts an energetic cost. There are, therefore, two ways in which self-assembled compliance states or stiffnesses can vary: They can vary in absorptive/restorative capability and they can vary in the amount of energy degradation required to create and sustain them. From the perspective of a wrist-pendulum system, the fact that its compliance or stiffness is a variable means that the system has potentially many steady-states. For each setting of a compliance value, a balance of forces, muscular and nonmuscular, conservative and nonconservative, is attainable in principle. There is a basis for selecting among the various steady-states, however. Roughly, it is the absorptive/restorative capability of the compliance state relative to its dissipative cost. Thus, although there may be many potential steady-states of a wrist-pendulum system, some states are preferable to others, and one state may be preferable to all others.

Rephrasing the question that has provided the focus of this section: How, then, does a wrist-pendulum system arrive at this *preferred steady-state*? The relatively straightforward answer that applies to physically ordinary systems clearly does not apply to a wrist-pendulum system. It is not drawn to this steady-state by forces. If, to the contrary, it is said to *discover* this steady-state, then how is that discovery to be understood? It might be hypothesized that a wrist-pendulum system computes the minimum energy that can be associated with the rhythmic movement—the person as wrist-pendulum system performs a

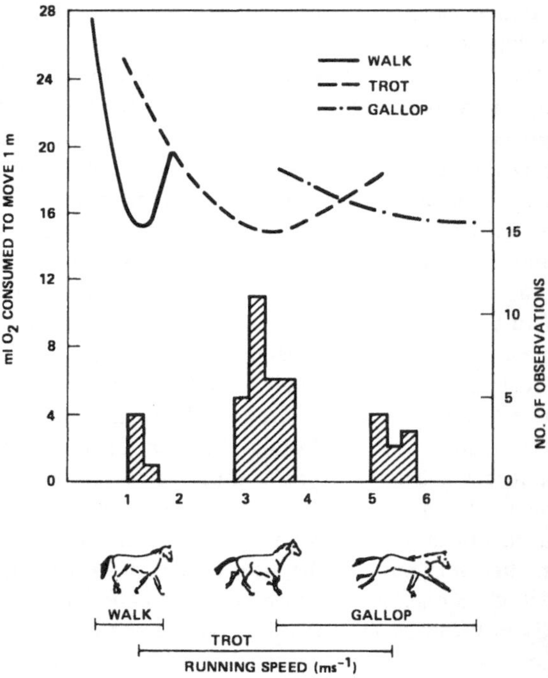

FIGURE 7.1. The metabolic profiles of the major gaits of horse locomotion as measured in the laboratory were matched by the speeds at which horses produce the gaits in the wild. The most frequently occurring speed is that for which the metabolic cost is least (After Hoyt & Taylor, 1981).

kind of formal, optimality analysis. Alternatively it might be hypothesized that information specific to the preferred steady-state is provided by an abstracted description of the dynamics of a wrist-pendulum system. This latter hypothesis is the one pursued here, consistent with the analysis of complex atomisms developed in detail in Chapter 4.

7.2.1. Energetic Gradients and Gait Transitions in Horse Locomotion. Before proceeding to a discussion of the informational support for wrist-pendulum systems, a brief consideration of certain aspects of horse locomotion will illustrate further the notions expressed in the preceding paragraphs and, importantly, help to anticipate the analysis that follows. Figure 7.1 depicts the speeds (observed in the open) and metabolic costs (determined in the laboratory) of horses walking, trotting, and galloping (after Hoyt & Taylor, 1981). At each gait the metabolic cost is roughly a U-shaped function of speed. For the walk and the trot there is a well-defined minimum; for the gallop the metabolic cost varies less sharply with variations in speed. As can be seen, a freely mobile horse will locomote most often at the speed associated with the metabolic minimum of the gait, but it also locomotes at more costly speeds that are both slower and faster. The U-shaped metabolic function identifies the potential well defined by the motions of the limbs under muscular and nonmuscular forces. While the behavior of the horse *qua* oscillatory system must adhere to the surface of the potential well, it can adhere to virtually any point on the surface. It is not confined to the minimum. By providing muscular forces that complement the nonmuscular forces (cf. Bernstein, 1967) a horse can maintain any steady speed that the gait allows. In sum, a horse locomoting in a given gait is not limited to a single steady-state; it is, however, the case that for any given gait there is one steady-state that distinguishes itself as being least costly for the horse, and this state is more or less well marked by the topology of the gait's macroscopic energy-flow characteristics.

7.3. The Haptic Perceptual System

The perceptual system involved in the guidance of the cyclic behavior of wrist-pendulum systems is the haptic perceptual system. Gibson (1966) identifies haptic system with the sensibility of a person both to the properties of his/her body and to the properties of the environment adjacent to his/her body, by the use of his/her body. In contrast to other perceptual systems, the haptic system includes the whole body, almost all of its parts, and all of its surfaces. The extremities—arms/hands and legs/feet—are especially important. They are at one and the same time exploratory sense organs and performatory motor organs. Anatomically, the equipment for feeling adjacent properties of the environment and the equipment for acting with respect to these properties are identical.

To deform the tissues of the body is to stimulate the microscopic mechanoreceptors that interpenetrate them. With respect to a wrist-pendulum system, mechanoreceptors are distributed in and below the skin of the hand(s) and forearm(s), in the wrist-joint(s) and connecting ligaments, and in the muscles and tendons of the forearm(s). They are also wrapped around the dilating and contracting blood vessels that perfuse the muscles.

The muscle/tendon haptic subsystem of a wrist-pendulum system is probably the most important, for two reasons: First, the muscle/tendon subsystem conflates perceiving and acting more than the other haptic subsystems (for example, the skin, or the distensible blood vessels). Second, the muscle/tendon subsystem is, generally speaking, the haptic subsystem principally involved in the perception of inertial properties (Gibson, 1966).

7.3.1. The Question of Haptic Analytic Units: The General, Extrinsic View. Do the distortions of microregions of muscle/tendon tissue appropriately define the mechanical stimulation of (and, hence, the appropriate kind of information for) the haptic system? Do responses of the microscopic mechanoreceptors to these micro-distortions define the process of haptic detection? The ensembles of microscopic distortions and microscopic mechanoreceptor firings are of high dimensionality (in the sense of requiring many state variables for their description). As descriptions of the states of the tissues of a wrist-pendulum system they imply that the means of guiding a wrist-pendulum system involves a very large number of degrees of freedom. Further, they may be taken to imply that the variables of relevance to the haptic functioning of a wrist-pendulum system are those quantities customarily most relevant to the physicist and most directly measurable—quantities such as force per unit area and the magnitude and frequency of impulses (see Stein, 1982, for a review of such variables).

The observable of a wrist-pendulum system that is of concern here is the system's preferred steady-state. Positive answers to the questions that began the preceding paragraph would imply that this observable is actually a composite of simpler observables, with the understanding that (a) the measuring devices or meters that evaluate these simpler observables can be explicated, and (b) from explicating them, an understanding of the measurement of the complex original observable can be constructed (cf. Rosen, 1978). The orthodox reductionistic or extrinsic view of measurement expressed in the foregoing supposes tacitly that (b) is implied by (a) and it supposes explicitly that there is one family of observables, or analytic units, that is valid universally for all behaviors of all systems. The "extrinsic" label derives from the fact that the analytic units are defined independently of the systems and behaviors to which they are applied (see Chapter 2).

7.3.2. The Question of Haptic Analytic Units: The Particular, Intrinsic View. Consideration is to be given to the possibility that the preferred steady-state can be measured more directly than is implied by the reductionistic view of measurement. The question raised is whether there is a macroscopic observable corresponding to the steady-state, and a macroscopic meter by which it can be evaluated. This question is prompted by the attitude toward measurement expressed by Gibson (1966, 1979) and Rosen (1978). In their view, there is not a single family of analytic units; rather, there are many. Which family suits a given situation depends entirely on the nature of the system under investigation and the particular behavior of interest of that system. Given the discussion in Chapter 2, this attitude defines an intrinsic view of measurement.

The foregoing is also prompted by the attitude toward the control and coordination of activity expressed by Bernstein (1967) and Gibson (1979). As intimated several times in preceding chapters, Bernstein and Gibson hold the view that the microscopic, high-dimensional characterization of action problems can be compressed, in a principled way, into a macroscopic, low-dimensional characterization. This joint assertion is that the control of activity is low-dimensional in spite of the high-dimensional complexity to be found in the biokinematic chains and their neural substrate on the one hand, and in the microscopic detail of the stimulation of the sense organs on the other. The contrast being drawn is between a large number of microscopic observables coupled with an evaluative process that is logically deep (that is, involving a large number of formal, computational steps) and a single macroscopic observable coupled with an evaluative process that is logically shallow.

What is required of the sought-after macroscopic observable? To begin with, not only must it yield a basis for the direct measurement of the preferred steady-state of a single wrist-pendulum system but it must also yield a basis for the direct measurement of the preferred steady-state of a pair of wrist-pendulum systems coupled isochronously. Further, as underscored in Section 7.2, the most notable feature of the preferred steady-states of wrist-pendulum systems, single or paired, is that they are created by, and discovered in the course of, the person's rhythmic wrist activity. The sought-after macroscopic observable, therefore, must be created similarly in the course of rhythmic activity; and if it is the basis for discovering the preferred steady state, as supposed, then it must exist continuously during rhythmic movements, even during those that do not coincide with the preferred steady state. There is nothing odd about this requirement. It is a fact of any physical system that its current state implicates its equilibrium or steady states, insofar as the space and time parameters of these states (which may not be actualized under the system's current boundary conditions) scale intrinsically whatever nonequilibrium or nonsteady states the system may assume (see Section 3.13).

Here, then, are the desiderata for the sought-after macroscopic observable. First, it must be definable equally for one wrist-pendulum system and for two isochronously coupled or interacting wrist-pendulum systems. Second, it must be available when the kinetic state that it specifies is not yet actualized. Couched in these terms it is apparent that the preceding query (about the possibility of a haptic observable that corresponds to a preferred steady-state) should be examined in light of the kinematic treatment of information developed in Chapter 4.

It is argued in the following that the sought-after macroscopic observables that guide the organization of wrist-pendulum trajectories are similar to the abstract, relational variables identified for optical flow fields and pheromone flow fields. In the case of wrist-pendulum systems these relational variables are properties of a haptic or biomechanical flow field. It is surmised that the haptic flow field, like the optic and pheromone flow fields, is of low energy (relative to the energy states associated with the various skeletomuscular movement subsystems) and self-organizing (in terms of the relational properties relevant to the guidance of activity).

7.3.3. Haptic Flow Fields: The Informational Support for Wrist-Pendulum Systems. The topology of the microregions of the body's tissues is fixed, permitting an idealization of the haptic system as a slab of matter—say, a cube for simplicity—with the constituent elements (individual microregions of tissue, each served by a mechanoreceptor) standing in an invariant spatial relation.

Because the body is ordinarily immersed in a field of forces, most of the body's tissues are distorted to a greater or lesser degree relative to a ''layout'' of gravitational and other elastic potentials. In terms of the simplification, the ''haptic cube'' is always warped by the presence of potentials.

The force impressed upon any given constituent element depends on both the location of the element and its orientation to the cube's axes. To characterize completely the force felt by an individual element requires the stress (σ) tensor; and to characterize completely the resultant deformation of the element requires the strain (e) tensor. It is because the stress, and in particular the strain, have different values at different points that the ordinarily distorted ''haptic cube'' is a (stress and strain) field in the orthodox sense. With a transformation of posture, the ''haptic cube'' will undergo time-varying distortions. For each microregion a velocity vector de/dt representing the rate of change of strain is defined at each point in time. Further, for each microregion a zero vector representing the microregion's steady-state is defined. There is, therefore, a vector field in a one-dimensional state (strain) space defined for each microregion. And defined across the microregions is an arrangement, order, or array of these undimensional velocity vector fields. For a very simple distortion the partial derivative gives the change in strain per unit distance in any specified direction, for example, $\partial e/\partial x$ for the x-direction of the ''haptic cube'' when the y and z directions and time are fixed.

Suppose that the ''haptic cube'' is stressed locally, as would occur if a wrist-pendulum system were oscillating. If the ''haptic cube'' were a piece of solid matter, then the locally defined stress would propagate via the transport of momentum. In the terms promoted above, a high-energy field would be said to connect the regions of the cube. The ''haptic cube,'' however, is not solid matter and does not equipartition energy and momentum globally (see Figure 7.2, upper panel). Any strain induced locally generates a pattern of neural activity. It is the nervous system, a low-energy field (comparatively speaking), that connects the parts of the ''haptic cube'' (see Figure 7.2, lower panel). Local muscle/tendon stress/strain distortions are propogated globally via the nervous system. These neural propagations function as an irreversible transport mode (analogous in their role to the diffusive transport mode in thermodynamic fields). As a result of the neural transport, the mechanoreceptor consequences of a single wrist-pendulum system in isolation interact with the mechanoreceptor consequences of the person's current postural state.

How should this interaction be characterized? And what does it fashion? The proposed answer to the first question is that it should be characterized in kinematic terms. The proposed answer to the second question is that it fashions morphological properties which are generated by and specific to the kinetics and which serve, in turn, to constrain the kinetics. Recalling the central theme of Chapter 4, for a complex atomism there are orthodox interactions in which the dimension of mass figures prominently, and there are heterodox interactions in

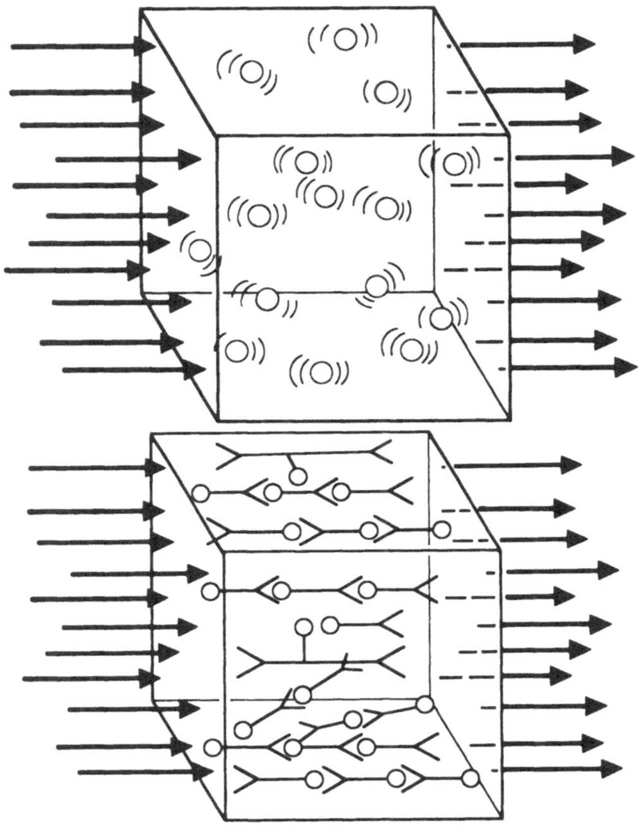

FIGURE 7.2.

Upper panel: Lattice of simple atomisms that transmit perturbations through exchanges of energy and momentum via irreversible diffusion processes.

Lower panel: Lattice of complex atomisms that transmit force perturbations through the exchange of kinematic flow patterns via irreversible neural (radiational) processes.

which the dimension of mass is suppressed. Interactions of the latter type are of importance here. The descriptors for the interaction between mechanoreceptor patterning induced by wrist/forearm activity and mechanoreceptor patterning induced posturally are the morphological properties of the resultant haptic flow field. It is presumed that haptic flow fields are not terribly different from pheromone flow fields and optical flow fields in terms of the kinds of distinct kinematic patterning they can exhibit. Thus the kinds of descriptors promoted in Chapter 4 (with regard to the abstracted, kinematic description of the underlying kinetics during nest building by insects) can be promoted here. When a single wrist-pendulum system oscillates in the context of a given body posture, this particular kinetic state of affairs should submit, in principle, to an alternative description of flow gradients and their related attractors.

If this proposed morphology of the haptic flow field is to meet the second desideratum above then it should specify kinetic states of affairs that are preferred but not yet actualized. The point is that properties of the kinematic field should specify, ideally speaking, states toward which the kinetic field tends as well as states that it is in. That kinematic flow fields in general might provide forward-looking information, that is, information about kinetic states of affairs that are possible under the current conditions but not yet actual, is suggested by the time-to-contact variable, τ (see Section 4.6.3). This variable isolated in optical flow fields specifies the imminence of and time to collision with a surface if the present kinetic conditions (e.g., forces being generated) persist.

An intuitive appreciation of the proposed character of haptic flow fields (as it bears on both of the precedingly identified desiderata) can be enhanced by considering the condition in which the "haptic cube" is stressed locally in two regions at the same time, as would be the case if both wrist pendulum systems are oscillating isochronously (Figure 7.3). By the foregoing, and by the arguments of Chapter 4, each of the two systems generates locally a force description (fundamental dimensions M, L, T) but they interact globally in terms of a flow description (fundamental dimensions L, T), sustained and transported by a low-energy neural medium (see Figure 7.4).

7.3.4. Morphological Invariance and the Problem of Motor Equivalence. The haptic flow field wrought by the activity of the two kinetic sources (against a postural backdrop) is characterizable in terms of attractors and flow gradients specific in their layout to the individual sources and to their interactions. The mesh of haptic flows originating locally can be thought of as giving rise to properties that are specific to the kinetic tendencies, or virtual states, of the two wrist-pendulum systems as a collective or unit (see Figure 7.5). That is, just as the haptic flow field can be said to carry macroscopic observables (properties) specific to the preferred steady-state of a single wrist-pendulum system, so it can be said to carry macroscopic observables specific to the preferred steady-states of wrist-pendulum systems coupled isochronously. In this view, the difference between a system of one biomechanical degree of freedom and a system of two biomechanical degrees of freedom is invisible at the level of the kinematic abstraction. Conventionally it would be said that the two kinds of wrist-pendulum systems—the single and the isochronous pair—comprise an instance of motor equivalence. Alternatively, though relatedly, it might be said that both kinds of wrist-pendulum systems can be regarded as single functional units, or coordinative structures.

This proposal of a common morphological basis for guiding the preferred

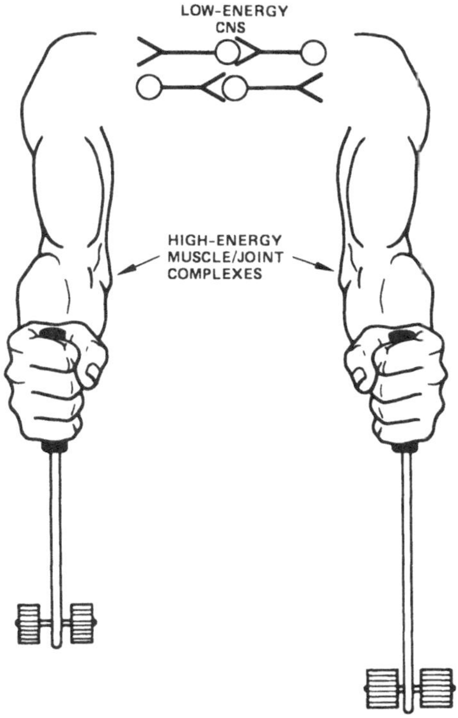

FIGURE 7.3. COOPERATIVITY DEFINED OVER A LATTICE OF COMPLEX ATOMISMS. Oscillatory motions about the wrist joints generate stress/strain patterns that are specific to the local "layouts" of elastic and gravitational potentials. These locally generated patterns spread centrally via the irreversible flow processes associated with neural radiation.

steady-state trajectories of single and double systems is patently a close cousin of the proposal advanced in the preceding chapter to apply Huygens' law for compound pendulums to "convert" two isochronous wrist-pendulum systems into a (virtual) single system. Recall that Huygens' law presupposes rigid, hard-molded connections among the component pendulums and that this presumption is obviously invalid for paired wrist-pendulum systems. But what if wrist-pendulum systems coupled isochronously indeed do abide by Huygens' law? Then it might be useful to consider the argument that the two wrist-pendulum systems approximate (or simulate) a rigid connection by virtue of the fact that their distributed kinetic processes give rise to a single kinematic field property that plays back causally, in the sense of a constraint that guides (with common purpose) both kinetic sources. Put in these terms, the emergence of a single macroscopic attractor that constrains the behavior of the individual (kinetic) subsystems is suggestive of the order parameter prominent in the

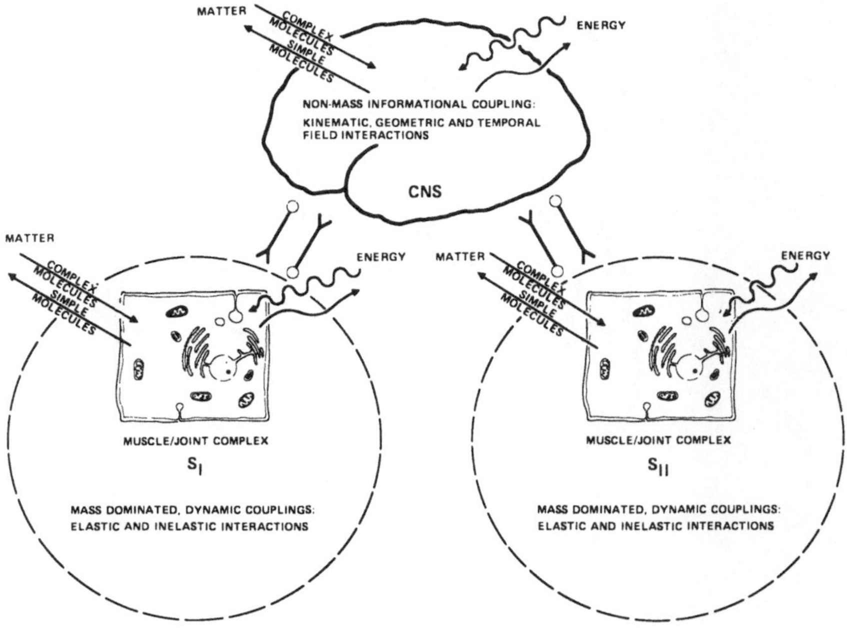

FIGURE 7.4. The cooperative states defined over the lattice of muscle/joint complexes define a hybrid kinetic-informational system where elastic and inelastic interactions dominate the local transport processes and information interactions dominate the global transport processes.

analysis of physical cooperativities (Haken, 1977). When a system is decomposable into many subsystems, the subsystems themselves give rise to a virtual property that serves to organize their behavior[1].

What are the implications of the preceding discussion on the haptic flow field for a person participating in the main experiment described in Chapter 6? The problem facing a participant on each trial is to swing the given pendulum or given pair of pendulums at the most comfortable frequency, that is, at the preferred steady-state. If, as suggested, the solution to the problem is to be found in the morphological properties of the haptic flow field that specify the preferred kinetic tendencies, then a participant's charge is to detect these properties. The aforementioned properties, especially as they are defined for the situation involving two wrist-pendulums, are deserving of Gibson's (e.g., 1966) label "higher-order invariants." Generally speaking, the detection of higher-order invariants is not a *fait accompli*. Differentiating them from the structure at large

[1] The popular example is the mean field theory of the ferromagnet. The different spins of the micro elements generate a macroscopic property, magnetization, which then acts on each individual spin to tell each how to behave.

benefits from practice. As Gibson (1966, 1979) would phrase it, attention to higher-order invariant properties needs to be educated.

7.3.5. Biomechanical Motion as an Example of an "Execution-Driven" Self-Organizing Information System. Chapter 5 concludes with the intimation that the nest-building activities of an ensemble of termites provides an exemplary case of a macroscopic work cycle wrought by synchronizing very many micro work cycles through a few macroscopic flow-field properties. The singularities and gradients evident in the large-scale nonkinetic description of the pheromone diffusion field both originate in and enslave the activities of the micro domain. They are parameters whose appearance and dissolution over the time course of nest building functionally define an escapement mechanism. The macro structure of the flow field regulates the local metabolic energetics of the

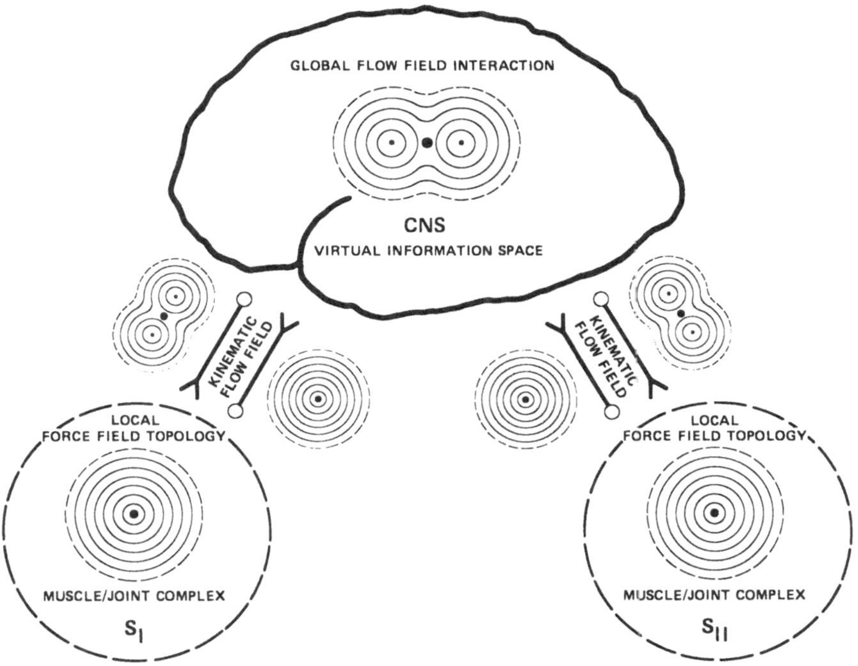

FIGURE 7.5. AN EXAMPLE OF A LOW DIMENSIONAL QUALITATIVE MORPHOLOGY THAT EMERGES DURING A FIELD INTERACTION. There are a number of types of flow dynamics that can result in pattern formation: (i) local flow morphologies that are characteristic of the potential (source-sink) layout; (ii) summation properties that result from field interactions—global cooperative interactions; (iii) subtractive properties that result from field interactions—global competitive interactions. The example presented identifies local flow morphologies that converge centrally, interact, and form a new flow pattern as a result of competitive and cooperative processes.

insect's musculature (insect locomotion). The continuously available and widely distributed energy supply (the metabolic fuels of the very many insects) is marshaled into a succession of macroscopic work cycles, namely, the arches→roof→arches→roof . . . construction sequence (see Figure 4.20). Insofar as an escapement is a mechanism for quantizing or discretizing or, more exactly, classifying the accessible macrostates of a system, then a moral of the insect story is that a basis for quantization can be provided by the macro properties of the flow fields (for a detailed discussion of this point see Chapter 11). It is suggested that the same moral applies to the relationship between the micro-metabolic states (of the functional units of the motor system, see Bloch & Iberall, 1982) and the macroscopic work cycle properties exhibited by the pendular clocking mode of organization.

A number of points made in Chapter 4 can now be usefully rephrased. There are systems such as insect-nest systems and, perhaps, biomechanical movement systems, whose future configurational states depend mostly on their current configurational states rather than on records of prior configurations or on stored prescriptions (see Figure 7.6). These systems are *self-organizing, execution-driven, information systems*. Only crude initial conditions are required for the start-up of their characteristic processes. Once initiated, the characteristic processes become parasitic upon and driven by the flow-field structures that emerge from the current state configurations (at all of the systems' scales). An execution-driven system, therefore, requires only a "crude" (imprecise) knowledge or memory of the conditions required to start its processes. The detail of these processes and the precision with which they are executed exceeds the detail and precision with which the initial conditions are identified. The detail and the precision of the execution-driven system's resultant organization is a function of the *intrinsic structural stability of the dynamic configuration itself* and of the system's self-sensitivity to this dynamic configuration (see Kugler, 1986).

7.4. Wrist-Pendulum Oscillations Constrained by Low-Dimensional, Macroscopic Limit-Cycle Morphologies

Summarizing the main thesis of the chapter to this point, the muscular and nonmuscular forces involved in the oscillation of a wrist-pendulum system generate a complex, spatiotemporal patterning in the haptic system. The adjacent and successive ordering of neural values may be termed a haptic flow field; its relevant properties are macroscopic and qualitative. By hypothesis, this haptic flow field (like the optical flow field, and the pheromone field governing termite nest building) can function as information about and a source of constraint for the force fields that produce it. The field pattern of strain-cum-neural activity is a kinematic abstraction of the field pattern of stress. In the sections that follow, an alternative, low-dimensional macroscopic description of the temporal play of

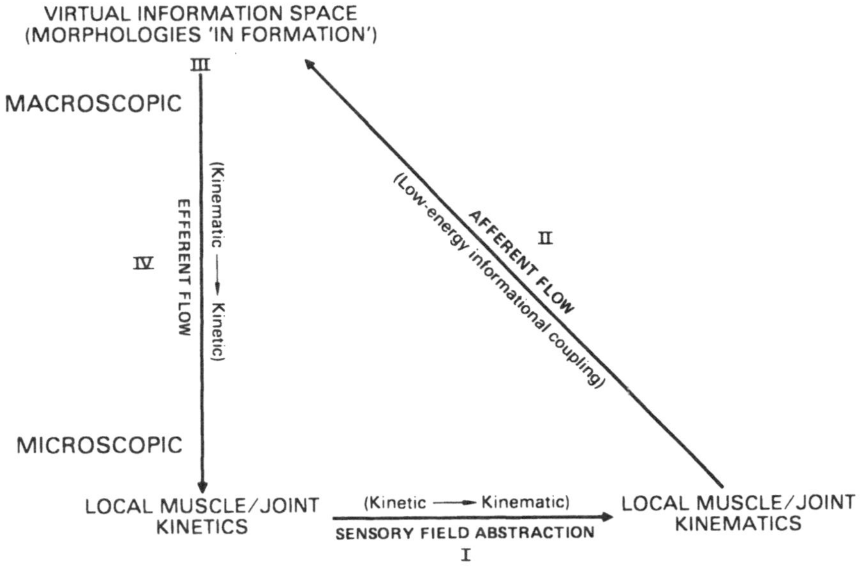

FIGURE 7.6. A SELF-ORGANIZING, EXECUTION DRIVEN, BIOMECHANICAL SYSTEM DOMINATED BY LOW-ENERGY INFORMATIONAL COUPLINGS.

(i): Flow field descriptions are abstracted from the kinetics of the local muscle/joint complex via a (reasonably compact) field of sensory endings;

(ii): The abstracted flow field descriptions radiate centrally via an irreversible (neural propagative) transport mode. During radiational spreading, the field description remains compact;

(iii): Muscle/joint flow field descriptions from different regions of the biomechanical system converge centrally on a localized region evolving new qualitative properties (morphologies) in the interference patterns. The qualitative properties that evolve from the interaction "specify" (in the informational sense) organizational constraints that can be used to coordinate those muscle/joint complexes that are currently contributing to the interference patterns;

(iv): Perceptual sensitivity to these evolving constraints allows them to be mapped back into the local muscle/joint complexes via efferently radiating neural propagation setting up new boundary conditions for the kinetics of the muscle/joint complexes.

forces on the muscle/tendon haptic subsystem is developed. It will be presumed that this abstracted description of the force field is equivalently (a) a description of the haptic field that informs a person, qua wrist-pendulum system, of the preferred steady-state and (b) a limit-cycle morphology.

7.4.1. Fundamental Properties of Flow-Field Abstractions: Attractors, Repellors, Basins, and Separatrices. The mathematical notions of a velocity vector field and its characteristic properties were introduced and casually developed in Chapter 4. The present subsection recapitulates these notions in a slightly more formal way. A flow field consists of a field of tangent vectors (*vector field*, *X*) on a state space (*smooth manifold*, *S*). Analytically, the vector field can be viewed as a "flow" (or group of motions of the state space upon itself). Viewed qualitatively, the flow field can be decomposed into a number of open sets defining *basins*, each of which contains a single state termed an *attractor* or *repellor*. The region that separates one basin from another defines an (ideally thin) set called a *separatrix*. A description in terms of these sets is referred to as an *attractor-basin portrait* of the flow field. (An introduction to these constructs can be found in Abraham & Shaw, 1982, 1986.)

The qualitative "forms" associated with these sets mathematically identify equivalence classes of solutions to differential equations. The individual members of a particular solution space—or equivalence class—are referred to as having *qualitative similarity*. The solution space can be very large or very small, the smallest set containing a single point and identifying a *singular state*. All neighboring states differ qualitatively from the singular states (in solution space). The singular states identify *attractors* and *repellors*. They are low-dimensional subsets of the state space on to which the motion of a system settles rapidly or from which it departs rapidly; they serve, roughly, as the mathematical definition of form.

Complex flow-field patterns or morphologies can be assembled by arranging a *layout* of attractors, repellors, and their connecting basins and separatrices (as seen in the insect nest-building example, Figures 4.1–4.16). The qualitative properties of the patterns will vary as a function of the number of singular states in the pattern and the nature of the qualitative continuation assumed by the solution space in entering and exiting the singular state. For example, a solution space can enter a singular state in qualitative form *X* and take on the qualitative form *Y* defined by the singular state. Upon exiting the singular state, the qualitative form of the solution space can either return to the state *X* associated with the approach state or enter a new state *Z* associated with a new solution space.

Thom's (1975) theory of catastrophes addresses the issue of how many qualitatively different ways solution spaces can approach and exit singular states and what forms they assume. The theorems developed by Thom were based on two assumptions. First, a system of interest is structurally stable, that is, its "form" (defined topologically) remains essentially unchanged during small perturbations of the underlying vector field that describes the system's motion.

Second, the system's motion is governed only by *scalar potentials*. A scalar potential is a function that relates the underlying state space to the (one-dimensional) real numbers; that is, it is a function that assigns a single real number to each point in the state space.

A system is said to be driven by a scalar potential if there is a scalar potential *V* that has the property that at every point in time the system's evolution is in the direction of steepest increase of *V*. Put simply, the time evolution of a system driven by a scalar potential always follows a trajectory that maximizes the increase of *V*. Increases and decreases in the values defined by the potential function are referred to as *gradients* of the potential. Singular states in the potential function identify *equilibrium states* (if no entropy-producing processes are present) or *steady-states* (if entropy-producing processes are present). (The stability of these states can be classified according to the nature of the surrounding gradients. See Appendix D for a discussion of the difference between the methods of classification associated with conventional stability theory versus structural stability theory.)

7.4.2. Flow Field Morphologies of a Damped, Force-Driven Harmonic Oscillator. The analysis of the flow-field properties of relevance begins with the behavior of a damped, force-driven harmonic oscillator that is described by the following relation:

$$F = \frac{\theta L M}{\left[\dfrac{1}{(\omega_r^2 - \omega^2)^2 + 4(B/2M)^2\omega^2}\right]^{1/2}}. \tag{7.3}$$

F = driving force (newtons)
θ = amplitude (degrees)
L = length of the pendulum (meters)
M = mass (kilograms)

ω_r = resonant frequency (hertz)
ω = actual frequency (hertz)
B = damping coefficient (kilograms/seconds)
g = gravity (meters/seconds^2)

$$\text{and: } \omega_r = \frac{1}{2\pi\sqrt{L/g}} \tag{7.4}$$

If only the frequency of oscillation ω is free to vary, then a one-dimensional flow field of force vectors (see Abraham & Shaw, 1982, 1986) can be generated. Figures 7.7 and 7.8 are plots of unidimensional velocity vector fields generated

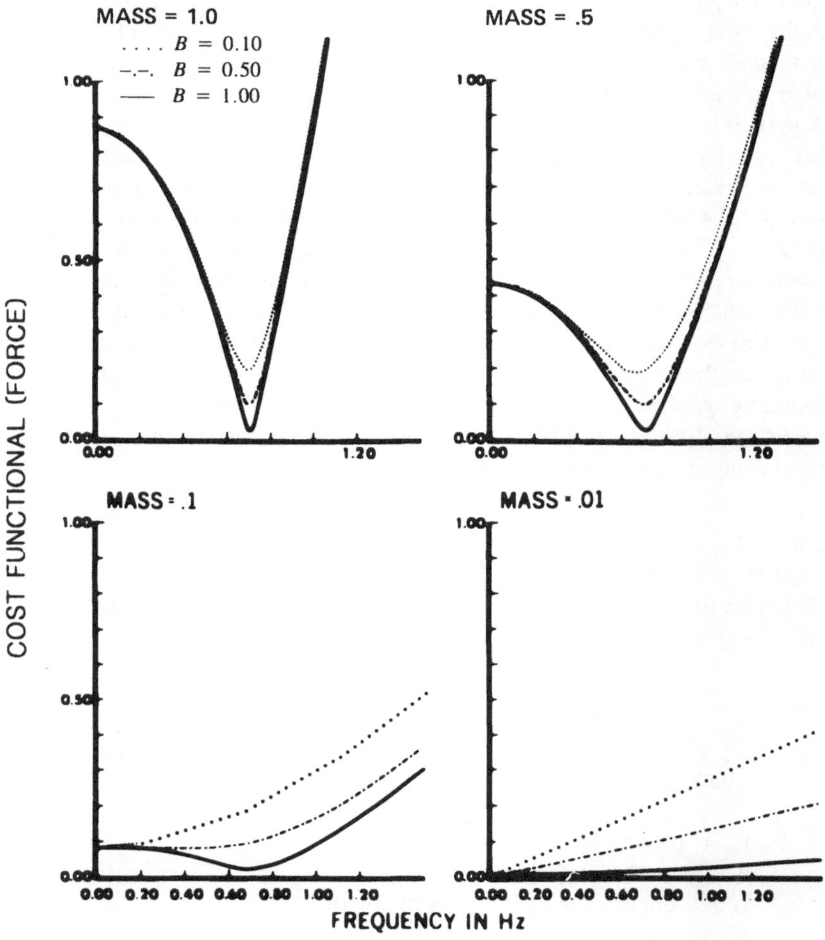

FIGURE 7.7. Force versus frequency response characteristics of a damped, force-driven, harmonic oscillator. Damping (B) is varied within each panel and mass (M) is varied across panels.

in this manner (from Kugler, Kelso, & Turvey, 1980). In Figure 7.7, each curve represents different parameter settings of the mass M and damping coefficient B. Amplitude θ, length L and gravity g are held constant over each set of graphs. Since these graphs are plotted in a two-dimensional space, a third variable is manipulated over the three curves in each graph and a fourth variable is manipulated over the four graphs in each series. In Figure 7.8 the force curves vary as a function of variations in length L and mass M.

Consider first the qualitative properties associated with the "form" of the force field generated by variations in frequency ω. Figure 7.7 identifies three force curves generated by varying the frequency of the oscillator while holding mass, length, amplitude, gravity, and damping constant. All three force curves reveal minima in the force function that are surrounded by ascending force gradients on either side. The minima identify singular states that refer to the oscillator's "natural frequency" ω_0. The natural frequency ω_0 is not to be confused with the resonant frequency ω_r. The resonant frequency ω_r is a property

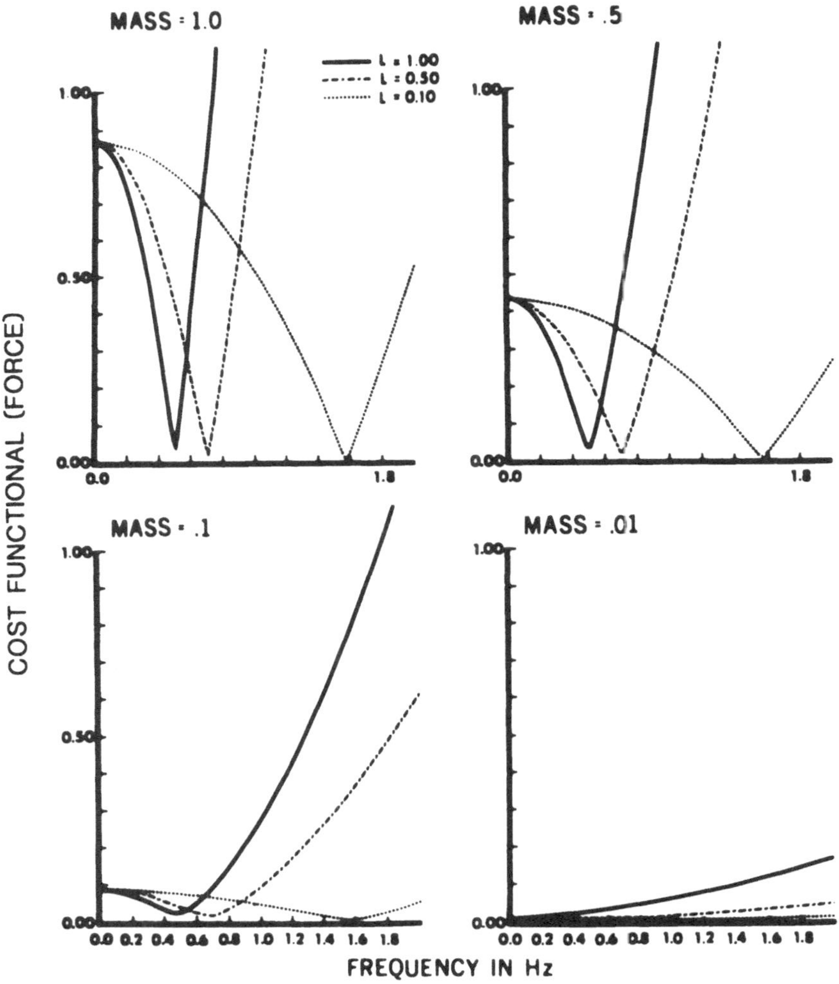

FIGURE 7.8. Force versus frequency response characteristics of a damped, force-driven, harmonic oscillator. Length (L) is varied within each panel and mass (M) is varied across panels.

of the oscillator defined relative to conservative forces (e.g., gravity forces); in contrast, the natural frequency is a property of the oscillator defined relative to the conservative and nonconservative forces (e.g., gravity and friction). It is not an explicit variable in the equation set per se but, rather, it is an emergent property defined relative to the conservative and nonconservative force structures.

Figures 7.7 and 7.8 identify examples of force functions that are associated with variations in mass, damping, and length. Most notable are the changes in location of the steady-state defining natural frequency ω_O with changes in length (see Figure 7.8). Changes in the location of the natural frequency as a function of mass and damping variations are more subtle and in some cases beyond the resolution of the graphs. Variations in natural frequency as a function of mass and damping variations are more apparent in the lower right graphs in Figures 7.7 and 7.8. In these graphs sudden qualitative transitions in the form of the differential solution sets have taken place. Variations of mass, length, and damping have all resulted in translocations of the oscillator's natural frequency towards the origin *(0,0)*. The singular state defining ω_O is annihilated at the origin.

7.4.3. Flow-Field Morphologies of a Wrist-Pendulum Oscillator. The strategy to be taken replaces the wrist-pendulum system, which is a mix of conservative and nonconservative forces at macro and micro scales, by a macroscopic mechanical abstraction. Ideally, the abstraction retains just those interactive capacities that underwrite the real system's periodic timing behavior (see Rosen, 1978). The only forces at work in the abstraction are the gravitational force (F_g) and an elastic force (F_K). It is hypothesized that a single macroscopic elastic potential K is assembled from the multiple elastic and viscous processes in the micro domain. The details of the abstraction are to be found in Chapter 8 (Section 8.3) and the actual micro processes are discussed in Chapter 11. At issue in the present section is the significance of the abstraction. As a purely conservative mechanical system, a wrist-pendulum's behavior would be governed by a position-dependent potential well. That is to say, a single scalar potential $grad\ V = F_{K.g}$ can be entertained with a limit point attractor. By this abstraction the velocity vector field—the flow-field morphology—that supports the selection of the timing behavior of a wrist-pendulum system is simply the gradient vector field of a potential function. A wrist-pendulum system is portrayed, in short, as a gradient system. Variations in periodic timing are reflected as variations in the system's position on the gradient. The abstraction would appear to fly in the face of the customary claim that gradient systems cannot sustain limit cycle behavior. It should be underscored, however, that it is the approach to the eventual limit cycle orbit that is gradient supported, not the production of the limit cycle orbit itself (see Section 7.4.5). The nature and production of the (limit cycle) orbital states is a major concern of Chapter 11.

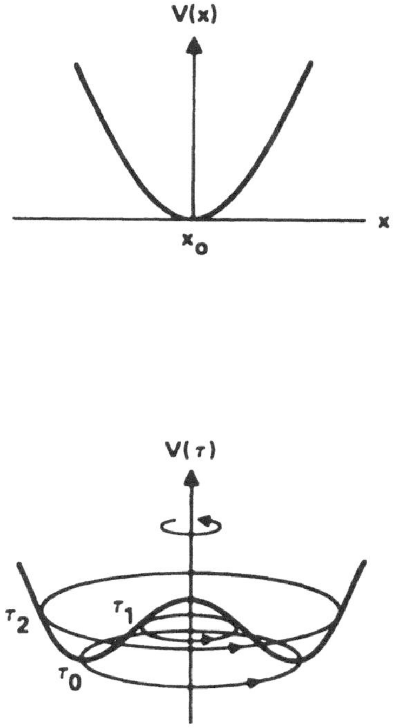

FIGURE 7.9.

Upper panel: Static position-dependent potential function with minimum defined as X_0.

Lower panel: Rotating time-dependent potential function with minimum defined as τ_0.

Figure 7.9 (upper panel) identifies a position-dependent potential well V. This function has a minimum at the point x_o. The minimum defines the stable static resting state of the oscillator. Departures from x_o in any direction result in increases in the potential energy function: x_o is a position-dependent attractor state.

Figure 7.9 (lower panel) identifies the potential energy function associated with oscillation. A rotation around the potential function identifies a period τ_i of oscillation. The potential is measured relative to the period of oscillation $V(\tau)$; it is the elevation of the center of mass averaged over the entire period. The minimum of the potential function defines the dynamic steady-state of the oscillating wrist-pendulum—the natural period $\tau_0 = (2\pi/\omega_0)$. Deviation in the period τ from τ_0 in either direction (that is, $\tau_1 < \tau_0$ or $\tau_2 > \tau_0$) results in an increase in the average elevation of the center of mass. Consequently, any

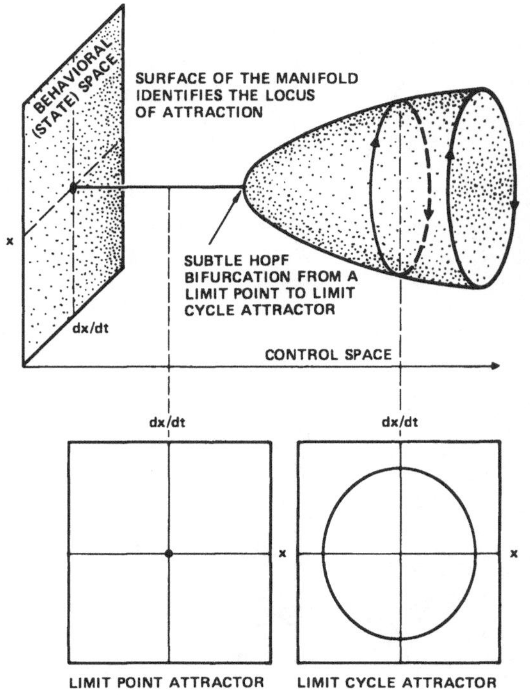

FIGURE 7.10. HOPF BIFURCATION.

deviations of τ from τ_0 will result in an increase in the system's potential energy defined over the cycle, giving rise to gradients on both sides of the minimum that τ_0 defines: τ_0 is a steady-state attractor, with gradients defining a scale for dissipation. A continuous family of potential energy functions can be generated if the length and mass of the pendulum are continuously varied $V(L_i, M_j)$. Each length and mass will define a unique steady-state attractor τ_0.

Figure 7.10 identifies an attractor portrait in which a point attractor similar to the one depicted in Figure 7.9 (upper panel) bifurcates into a limit cycle attractor similar to the one depicted in Figure 7.9 (lower panel). The bifurcation occurs as a function of change in some control parameter. The transition from the point attractor to a limit cycle attractor is referred to as a Hopf bifurcation (named after its discoverer). Fundamentally, the Hopf bifurcation marks the transition from a point attractor to a limit cycle attractor, and vice versa. The coordinates of motion associated with the limit point and limit cycle attractor can be transformed continuously by variations in the mechanical parameters of mass and length. The region in Figure 7.10 labeled as "behavioral (state space)" identifies the macroscopically observable coordinates of motion of displacement (vertical axis) and velocity (horizontal axis). This plotting technique is referred to as a phase plane plot. Consider the dark circle on the surface of the attractor manifold depicted in Figure 7.10 that indicates a limit cycle attractor set. The particular set of points defining the attractor set is itself defined by the parameter setting of the control space. In the case of the wrist-pendulum system the variables defining the control space are mass and length and the variables defining the phase plane are amplitude and frequency. For any given setting of mass and length in the control space, a particular attractor set of points will be defined in the behavior space. Minima in the hypothesized potential function $F_{K,g}(x) = grad\ V(x,t;\ M,L)$ identify the attractor states.

7.4.4. Approaching the Limit Cycle Attractor. As noted several times, the two major contrasts between a wrist-pendulum system and a physically ordinary system are the possession of a deployable on-board source of potential energy and a low-energy perceptual coupling between the system and its field of force vectors. In consequence, the two kinds of systems are not related to their respective attractor state(s) in the same fashion. A wrist-pendulum system can sense the stable tendencies in its dynamics and it can produce forces that are of sufficient magnitude either to resist or to concur with these tendencies. It is this capability that, in part, leads to classifying wrist-pendulum systems as non-determinate. A physically ordinary system, by comparison, must concur with its own dynamical tendencies, it must settle at the attractor, and it must follow the surrounding gradients to the attractor.

Consider again the man-made clock. The clock rides a limit cycle. Because of the continuous dissipation of energy occurring at the pivot (most likely through friction of the static Coulomb type), energy must be replenished to sustain the oscillation. For an accurate timepiece, the amplitude of oscillation will be just that amplitude that fully exhausts the quantity of energy put into the oscillator per cycle. Patently, the period of oscillation is that associated lawfully with the pendulum's length. At all events, if the pendulum's amplitude starts out less than that dictated by the magnitude of the energy squirt, it will grow gradually until it assumes that prescribed amplitude. Conversely, if the initial amplitude is greater than that dictated by the escapement pulse, then it will diminish gradually until the dictated amplitude is attained. This *monotonic* approach to the stable orbit in the phase plane, from a starting point either inside or outside the orbit, is depicted in Figure 7.11 (left panel). This monotonic approach to the limit cycle, consequent to varying initial conditions or perturbations, is guided by the conservation of energy. For larger amplitudes, more energy is dissipated per cycle than is replenished; for smaller amplitudes, the energy dissipated per cycle is less than that received per cycle.

Figure 7.11 (right panel) depicts the *nonmonotonic* approach to the attractor that would be possible, and likely, for a wrist-pendulum system. Recalling the discussion of Section 7.2, the kinetic attractor of a wrist-pendulum system is both defined and discovered in the course of rhythmic activity. The approach of

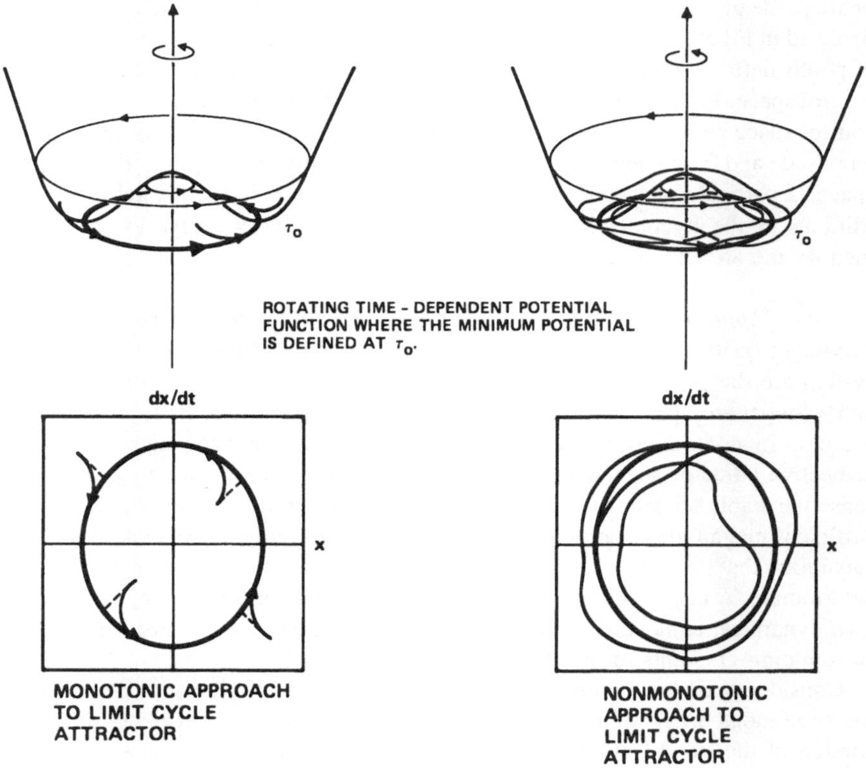

FIGURE 7.11.

Left panel: Determinate energy-and-momentum-dominated limit cycle oscillation.

Right panel: Nondeterminate perception-(information)-dominated limit cycle oscillation.

a wrist-pendulum system to its kinetic attractor is made possible by the information contained in the haptic field about the system's stable dynamical tendencies. The invariant specification of the kinetic attractor, however, is revealed over transformations in the periodic time. Hence, a wrist-pendulum system may jump from an orbit inside the stable limit cycle trajectory to an outside orbit, and vice versa, as it homes in on the attractor state (see Figure 7.11, right panel).

Despite the fact that they approach their kinetic attractors differently, physically ordinary systems and wrist-pendulum systems are alike in that both exhibit stable and reproducible behavior. For both determinate systems—for example, mechanical pendulum systems—guided to their kinetic attractor states by forces, and nondeterminate systems—for example, wrist-pendulum systems—guided to their kinetic attractor states by information, the stability and reproducibility of the pattern of natural periodic times is law based; in each case the law defines a layout of kinetic attractor states.

Given the design of the main experiment it was possible to test the perceptual "sensitivity" of subjects to these attractor states under variations in mass and length of the wrist-pendulum oscillators. Moreover, a manifold of attractor states could be "mapped out" for the purpose of determining the law or laws by which the layouts of attractor states are fashioned. To rephrase the physical theme guiding the present research: The stability and reproducibility of physical events are to be explained in terms of natural laws. The events being investigated in the present monograph are oscillatory movements, and the methodological strategy entails the identification of natural laws to account for the stability and reproducibility of the behavioral patterns over trials, blocks (of trials), days, and subjects.

7.4.5. A Gradient Solution to a Limit Cycle Problem. The notion of a gradient plays a central role in the present analysis of wrist-pendulum, limit-cycle oscillations in two ways. First, as must be the case in producing any work cycle, there is a gradient of energy flow from a source to a sink (associated with entropy production) in accordance with the second law of thermodynamics. Second, the particular work cycle elected is based on the information contained in the haptic flow field, where this information is defined as a flow gradient circumscribing an attractor. This flow gradient is not connected with a potential in the conventional sense; it is, after all, "merely" kinematic. To reiterate, a haptic flow gradient is a kinematic abstraction of an underlying force field. In sum, a wrist-pendulum oscillator has been pictured in the foregoing discussion as a limit-cycle oscillator sustained locally by a continuous metabolic gradient (the degradation of energy from high-order chemical fuel to low-energy heat) and coordinated globally by a kinematic (nonpotential) flow-field gradient.

The limit cycle depicted in Figure 7.9 (bottom panel) has the distinguishing characteristic that its period of oscillation defined at the limit cycle is a function of a single scalar potential. Locating the period of the limit cycle can be achieved through a sensitivity to a single changing scalar quantity. Changes (transformations) in this scalar quantity will generate a one-dimensional flow-field structure (an attractor-basin portrait). This low-dimensional morphology or "attractor layout" is being advanced as the informational basis by which subjects parse the phase space for "comfort" solutions to the oscillation task (see Chapter 11 for a detailed discussion of a thermodynamic basis for comfort states).

7.5. Structural Stability of Period and Amplitude

The subject's task was to discover for each mass/length pendulum condition the tempo that felt most comfortable. For any given mass/length condition the requisite tempo had to be discovered by the subject on a number of separate occasions distinguished principally by time (trial, day) and by hand. Assuming that "most comfortable" refers to a preferred steady-state (characterized by a particular period and amplitude), then consistency should be seen in the ordering of the rhythms elected on the separate occasions by a subject if (as the argument already developed suggests): (a) the preferred steady-states (and their surrounding gradients) are lawfully defined; (b) these kinetic conditions lawfully engender haptic flow-field properties so that information about the steady-states is available; and (c) the subjects can differentiate and detect the information and thereby perceive the preferred steady states. Moreover, and for the very same reasons, the kind of consistent ordering by any one subject should be exhibited by all subjects.

The foregoing is a reference to an important feature of a natural law, namely, its qualitative invariance under quantitative transformations of scale and reference frame. With respect to the experiment, the qualitative invariance relates to the *ordering* of the natural periods and amplitudes as a function of changes in mass and length of the pendulums. Physiological factors, such as fatigue, and general context effects (Helson, 1964) contribute to the drift of natural periods and amplitudes and militate against a subject reproducing exactly the same quantities for a given mass/length condition from one occasion to the next. According to the invariance postulate of natural law, variances such as those observed in period and amplitude from hand to hand, trial to trial, day to day, and subject to subject need not be ascribed necessarily to variability in the natural law; they could be ascribed to variations in the coordinate spaces in which the law is embedded. Subjects in their capacity as coordinate spaces are not exactly identical. Moreover, a subject in his or her capacity as a coordinate space need not be constant from day to day, nor from block of trials to block of trials. The quantitative values of the periods and amplitudes need not be reproduced precisely from one coordinate space to another. What should remain invariant are the *relations* among the quantities—how they order across the mass/length conditions. A reproducible qualitative property of the latter kind identifies a dynamical system that is *structurally stable*.

Roughly speaking, a dynamical system is called structurally stable if a "small change" in the equation of motion leaves the trajectories of the system qualitatively the same; this in turn means that there is a continuous deformation, or transformation, of the state space that carries the unperturbed trajectory onto the perturbed one. Or, stated another way, structural stability means that a modification in the equation of motion can be annihilated by a corresponding coordinate change in the state space, so that the perturbed system in the transformed coordinate space is identical with the unperturbed system in the original coordinates (see Rosen, 1978).

The experiment identified in Chapter 6 investigated conditional structural stabilities in which the independent (control) parameters of mass (M) and length

(L) are manipulated $M \rightarrow M'$, $L \rightarrow L'$, respectively. This introduces perturbations onto the equations of motion defined in state space by the dependent (behavioral) parameters of natural period (τ_O) and natural amplitude (θ_O) such that corresponding transformations of $\tau_O \rightarrow \tau_O'$ and $\theta_O \rightarrow \theta_O'$ preserve invariant the structural stability of the law(s). The invariance defining the structural stability relating the control parameters of length and mass and the behavioral parameters of period and amplitude constitutes the measurement structure of the natural laws governing the pendular clocking mode of movement organization (see Appendix D). This is equivalent to the formal statement (after Luce, 1978) that qualitative relations (structural stabilities) are invariant under automorphisms (identity preserving transformations) of the measurement structure (natural law). The invariance of the natural laws governing the pendular, clocking mode, under relevant transformations, can be investigated using the preceding methodology.

To summarize: If a natural period is construed as an attractor to which a wrist-pendulum system orients, then the ordering of natural periods for, say, a set of wrist-pendulum systems of constant mass and varying length or constant length and varying mass constitutes a layout of attractors or an attractor portrait. Structural stability in the present context is the persistence of this attractor portrait in the face of the "perturbations" introduced into the design of the experiment—viz., hands, subjects, trials, days.

7.5.1. Qualitative and Quantitative Methods of Analysis. The methodology for analyzing a law involves two stages: a qualitative analysis and a quantitative analysis. The qualitative analysis entails the identification of the structurally stable relations associated with the variables implicated in the law. The quantitative analysis requires the identification of the metrics (roughly, the distance functions relating measured quantities) that *locally* provide the scales for the reference frames (and their associated coordinate spaces) in which the law is embedded. (As noted in Chapter 2, the qualitative form of a law can appear to undergo a qualitative transition if the local symmetries that define and distinguish reference frames are not identified.) Once the local metrics are identified, the reference frames can be renormalized and the law transformed into its generic form. The analysis of the present data will proceed according to this protocol. First, a qualitative analysis will be performed on the period and amplitude data to identify structural stabilities associated with period, amplitude, mass, and length. The analysis will focus on identifying intrinsic (qualitative) stabilities that are reproduced from hand to hand (where applicable), trial to trial, day to day, and subject to subject. (The first step consumes the remainder of this chapter.) Second, the form of the scaling laws or the physical laws that relate mass, length, period, and amplitude will be derived theoretically. Third, the derived form of the laws will be used to determine the local metrics that distinguish individual subjects (in their capacity as coordinate spaces). And, fourth, these local metrics will be used to renormalize the coordinate spaces so as to test the invariance of the natural laws in accounting for the stability and reproducibility of the behavioral events. The second, third and fourth steps define the contents of Chapters 8 (for the periodic timing law) and 10, 11, and 12 (for the amplitude law). (The construct of "law" is meant to provide a basis for analyzing the properties of physical events in terms of qualitative and quantitative relations. If, additionally, it is meant to provide a causal account in terms of a theory of mechanism, then an additional methodological consideration must be addressed. A theory of mechanism requires an *interpretive act* and an interpretive act requires a logic. The relation between natural laws and interpretive logics is discussed in Appendix E.)

7.5.2. A Qualitative Analysis of the Period and Amplitude Data. In the experiment described in Chapter 6, two masses (0.1*kg*, 1.0*kg*) and two lengths (from the top of the rod to the added mass, 0.254*m* and 0.559*m*) were manipulated independently by the experimenter to scale the two dependent parameters of preferred period (τ) and preferred amplitude (θ). The other masses and lengths that went into the calculation of the simple pendulum equivalents of the compound single and compound double wrist-pendulum systems were not explicitly manipulated (e.g., mass of hand, distance from the rotation point, mass of the rod, etc.). It will prove helpful for present purposes to focus on the quantities that were manipulated explicitly and to ignore those that were manipulated implicitly. To begin with, this strategy permits a simpler partitioning of the various systems and allows a more parsimonious (masses and lengths are the same for all four subjects) and direct portrayal of the structurally stable nature of the data. Second, and importantly, this strategy highlights the "partial derivative" character of the experiment and the data it generated. Representative quantities of mass and length in a biological system are coordinated through the constancy of density (see Chapter 8). They are not independently manipulable. By contrast, in the experiment reported in Chapter 6 the representative quantities of wrist-pendulum mass and length are largely separate. The mass added to the rod could be changed (e.g., from 0.1*kg* to 1.0*kg*) without a change in the rod length (e.g., 0.254*m*) and vice versa. Table 7.1 gives the mass and length values of the four single and 16 virtual single systems calculated *as if* the only masses and lengths were those manipulated explicitly by the experimenter. Under these conditions Huygens' law is not used to determine the length of a single wrist-pendulum system (it is simply the length of the rod to the site of the added mass) but the law is used to determine the length of a virtual single system. Inspection of Table 7.1 shows that a number of the systems now have the same (explicitly manipulated) mass and a number of them have the same (explicitly manipulated) length. In the observations that follow immediately in this chapter, all reference to "mass" and "length" are to the masses and lengths reported in Table 7.1.

TABLE 7.1
Virtual Single System Magnitudes Computed from the Explicitly Manipulated Lengths and Masses

Explicit Length (m)/Mass (kg) Condition of the Left Wrist-Pendulum System	*Explicit Length/Mass Condition of the Right Wrist-Pendulum System*			
	.559/1.0	.559/0.1	.254/1.0	.254/0.1
.559/1.0	.559/2.0	.559/1.1	.407/2.0	.531/1.1
.559/0.1	.559/1.1	.559/0.2	.282/1.1	.407/0.2
.254/1.0	.407/2.0	.282/1.1	.254/2.0	.254/1.1
.254/0.1	.531/1.1	.407/0.2	.254/1.1	.254/0.2

The experiment investigated the possibility of a number of conditional structural stabilities relating initial conditions defined by the variables of mass and length, and consequent conditions defined by the variables of period and amplitude. Stated in the adjunctive form (see Appendix E), the set of possible conditionals relating the independent variables to the dependent variables is (where the subscripts *l* and *m* denote the constant independent variable):

CONDITIONAL STABILITIES FOR PERIOD (τ)

SINCE: $\{L = \text{constant} \quad \text{and} \quad M_1 < M_2\}$
THEN: $\{T_{1_l} < T_{2_l} \quad \text{or} \quad T_{1_l} > T_{2_l} \quad \text{or} \quad T_{1_l} = T_{2_l}\}$.

SINCE: $\{M = \text{constant} \quad \text{and} \quad L_1 > L_2\}$
THEN: $\{T_{1_m} < T_{2_m} \quad \text{or} \quad T_{1_m} > T_{2_m} \text{ or } T_{1_m} = T_{2_m}\}$.

CONDITIONAL STABILITIES FOR AMPLITUDE (θ)

SINCE: $\{L = \text{constant} \quad \text{and} \quad M_1 < M_2\}$
THEN: $\{\theta_{1_l} < \theta_{2_l} \quad \text{or} \quad \theta_{1_l} > \theta_{2_l} \quad \text{or} \quad \theta_{1_l} = \theta_{1_l}\}$.

SINCE: $\{M = \text{constant} \quad \text{and} \quad L_1 < L_2\}$
THEN: $\{\theta_{1_m} < \theta_{2_m} \quad \text{or} \quad \theta_{1_m} > \theta_{2_m} \quad \text{or} \quad \theta_{1_m} = \theta_{2_m}\}$.

7.5.3. Conditional Stabilities Analyzed Across Trials, Days, Hands, and Subjects. By the design of the experiment each single and double condition was evaluated once in each of eight blocks, four blocks per each of two days. Each block contained a complete and independent set of the relations. It is possible, therefore, to examine each of these blocks with respect to the conditional structural stabilities that have been identified. Each block can be thought of as providing a *local* field (precisely, a context) for evaluating the qualitative reproducibility of the intrinsic relations. While the overall field can fluctuate in quantitative values, this fluctuation need not affect the qualitative, conditional stabilities (whatever they might be) defined internal to the field if, indeed, those stabilities are law governed. An overview of the data is provided by Figures 7.12–7.14. A data point represents the period or amplitude for a wrist-pendulum system in a given block of trials averaged over the four subjects. The plots in each panel are of single or virtual single systems that have the same mass with varying lengths or the same length with varying masses, thereby permitting an identification of the conditional stabilities. Tables reporting the individual subject data are provided to the right of their respective figures.

The eight panels of Figure 7.12 identify the mean periodic times as a function of trial for the single left and single right wrist-pendulum systems. Inspection of each panel reveals that from trial to trial and from day to day there is an invariant ordering of the periodic times associated with wrist-pendulum systems that differ only in length or only in mass. Looking at the tables adjacent to the graphs and tables of the right and left hands it is evident that the periodic times order in the same fashion for the two hands.

Inspection of Figures 7.13 and 7.14 reveals that the periodic times of the virtual single systems also order invariantly over trials and days. For purposes of presentation in Figures 7.13 and 7.14, where there are two instances of a virtual single system—see Table 6.11—periods are averaged. The highly detailed tables of virtual single system periodic times as a function of trials for the individual subjects are not presented to save space. It should be remarked, however, that trial to trial consistency of individual subjects evident in the single systems' data was also evident in the virtual single systems' data.

The mean angular excursion data for the single systems are reported in Figure 7.15. There is a strong tendency for the angular excursion quantities associated with the different mass/length conditions to preserve the same order from trial to trial and from hand to hand. It should be noted that the structural stability of angular excursion may be more difficult to pinpoint and more difficult to rationalize than the structural stability of periodic time. This is suggested by the individual subject data (adjacent tables), which show a substantial number of reversals. These suspected difficulties with angular excursion are confronted in Chapters 10 and 12.

7.5.4. A Summary of Conditional Stabilities Analyzed Across Subjects. The analysis of the structural stabilities associated with the single wrist-pendulum data presented in Figures 7.12 and 7.15 revealed the following conditional stabilities:

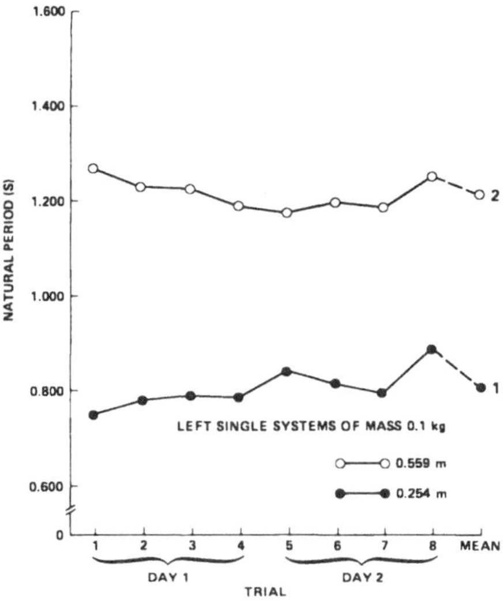

PERIOD (*S*) LEFT SINGLES
TRIAL NUMBER

	1	2	3	4	5	6	7	8	X̄
Subject 1									
2	1.317	1.231	1.279	0.991	1.200	1.094	1.074	1.096	1.160
1	0.816	0.793	0.789	0.751	0.895	0.742	0.718	0.735	0.780
Subject 2									
2	1.818	1.240	1.039	1.221	1.134	1.175	1.269	1.198	1.261
1	0.815	0.795	0.736	0.735	0.766	0.874	0.818	0.762	0.787
Subject 3									
2	1.209	1.114	1.191	1.121	1.171	1.121	1.171	1.266	1.171
1	0.688	0.693	0.721	0.672	0.735	0.699	0.703	0.950	0.733
Subject 4									
2	1.384	1.347	1.405	1.431	1.198	1.398	1.229	1.356	1.344
1	0.711	0.938	0.895	0.948	0.956	0.943	0.906	0.987	0.911

FIGURE 7.12a. Periods (cycle time in seconds) of left singles for condition 1 (mass 0.1kg and length 0.254m) and condition 2 (mass 0.1kg and length 0.559m) as a function of trial.

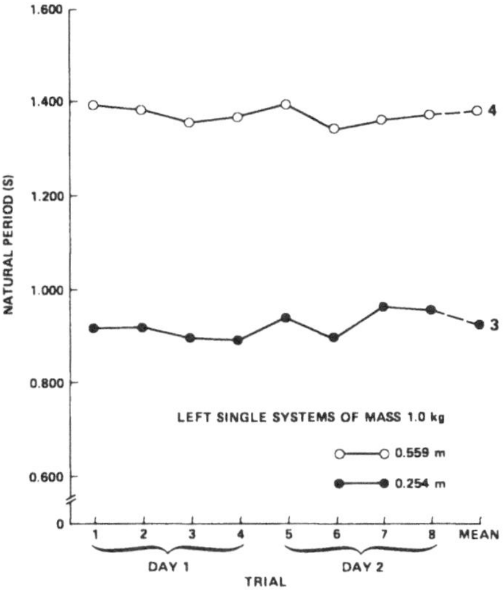

PERIOD (*S*) LEFT SINGLES
TR AL NUMBER

	1	2	3	4	5	6	7	8	X̄
Subject 1									
4	1.425	1.378	1.300	1.311	1.375	1.336	1.319	1.250	1.337
3	0.944	0.905	0.887	0.824	0.918	0.818	0.832	0.780	0.864
Subject 2									
4	1.350	1.373	1.358	1.367	1.414	1.358	1.400	1.345	1.370
3	0.862	0.951	0.850	0.795	0.880	0.930	0.944	0.923	0.892
Subject 3									
4	1.337	1.340	1.317	1.331	1.369	1.342	1.486	1.514	1.397
3	0.868	0.855	0.927*	0.902	0.884	0.833	1.078	1.069	0.927
Subject 4									
4	1.489	1.463	1.456	1.472	1.433	1.453*	1.450	1.406	1.453
3	0.991	0.966	0.948	1.026	1.061	1.000	0.989	1.050	1.004

*indicates where a missing datum (due to a failure in the collection apparatus) was replaced by the subject's mean.

FIGURE 7.12b. Periods (cycle time in seconds) of left singles for condition 3 (mass 1.0kg and length 0.254m) and condition 4(mass 1.0kg and length 0.559m) as a function of trial.

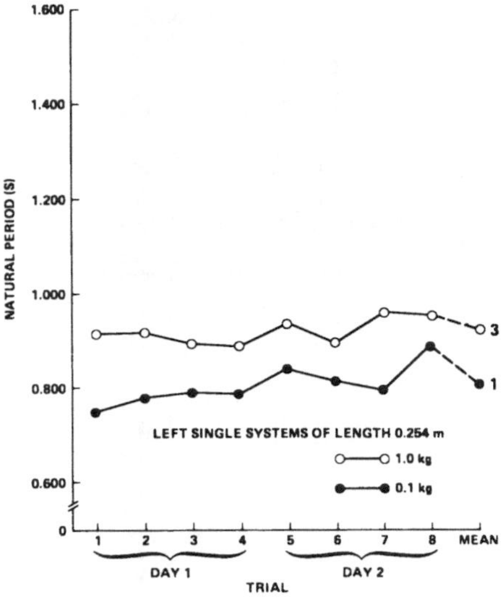

FIGURE 7.12c. Periods (cycle time in seconds) of left singles for condition 1 (mass 0.1kg and length 0.254m) and conditions 3 (mass 1.0kg and length 0.254m) as a function of trial.

PERIOD (*S*) LEFT SINGLES
TRIAL NUMBER

	1	2	3	4	5	6	7	8	
Subject 1									
3	0.944	0.905	0.887	0.824	0.918	0.818	0.832	0.780	0.
1	0.816	0.793	0.789	0.751	0.895	0.742	0.718	0.735	0.
Subject 2									
3	0.862	0.951	0.850	0.795	0.880	0.930	0.944	0.923	0
1	0.815	0.795	0.736	0.735	0.766	0.874	0.818	0.762	0.
Subject 3									
3	0.868	0.855	0.927*	0.902	0.884	0.833	1.078	1.069	0
1	0.688	0.693	0.721	0.672	0.735	0.699	0.703	0.950	0.
Subject 4									
3	0.991	0.966	0.948	1.026	1.061	1.000	0.989	1.050	1.
1	0.711	0.938	0.895	0.948	0.956	0.943	0.906	0.987	0.

*indicates where a missing datum (due to a failure in the collection apparatus was replaced by the subject's mean.

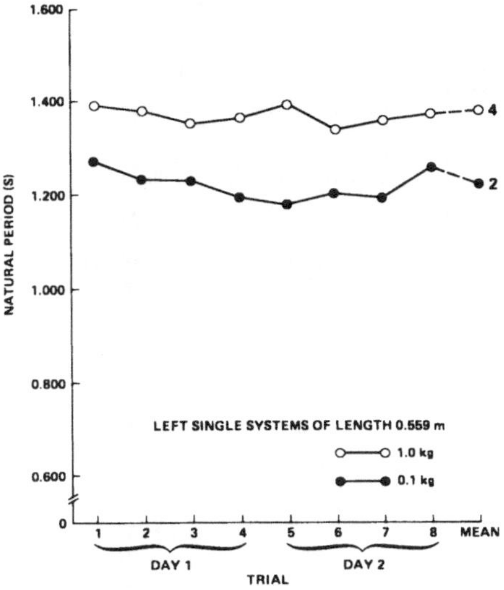

FIGURE 7.12d. Periods (cycle time in seconds) of left singles for condition 2 (mass 0.1kg and length 0.559m) and condition 4 (mass 1.0kg and length 0.559m) as a function of trial.

PERIOD (*S*) LEFT SINGLES
TRIAL NUMBER

	1	2	3	4	5	6	7	8	
Subject 1									
4	1.425	1.378	1.300	1.311	1.375	1.336	1.319	1.250	1.
2	1.317	1.231	1.279	0.991	1.200	1.094	1.074	1.096	1.
Subject 2									
4	1.350	1.373	1.358	1.367	1.414	1.358	1.400	1.345	1.
2	1.818	1.240	1.039	1.221	1.134	1.175	1.269	1.198	1
Subject 3									
4	1.337	1.340	1.317	1.331	1.369	1.342	1.486	1.514	1.
2	1.209	1.114	1.191	1.121	1.171	1.121	1.171	1.266	1
Subject 4									
4	1.489	1.463	1.456	1.472	1.433	1.453*	1.450	1.406	1.
2	1.384	1.347	1.405	1.431	1.198	1.398	1.229	1.356	1.

*indicates where a missing datum (due to a failure in the collection apparatus was replaced by the subject's mean.

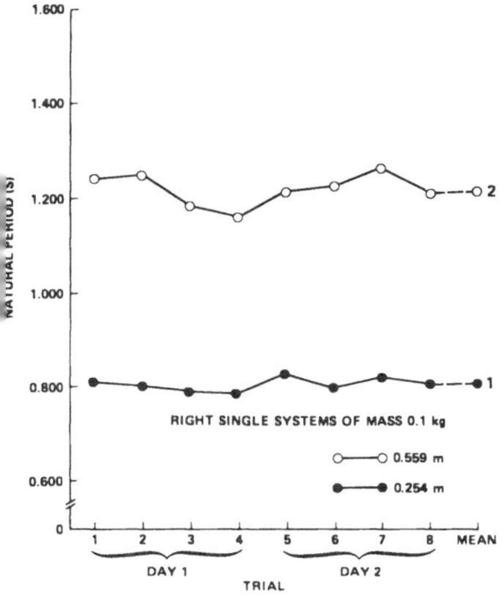

IGURE 7.12e. Periods (cycle time in seconds) f right singles for condition 1 (mass 0.1kig and ength 0.254m) and condition 2 (mass 0.1kg and ength 0.559m) as a function of trial.

PERIOD (S) LEFT SINGLES
TRIAL NUMBER

	1	2	3	4	5	6	7	8	X̄
Subject 1									
2	1.322	1.166*	1.181	1.052	1.264	1.125	1.135	1.075	1.166
1	0.871	0.892	0.811	0.789	0.959	0.772	0.732	0.634	0.808
Subject 2									
2	1.196	1.204	0.931	1.048	1.169	1.326	1.293	1.067	1.155
1	0.815	0.795	0.736	0.735	0.766	0.874	0.818	0.788	0.791
Subject 3									
2	1.102	1.141	1.229	1.116	1.086	1.087	1.212	1.329	1.163
1	0.685	0.635	0.721	0.687	0.714	0.670	0.849	0.920	0.735
Subject 4									
2	1.352	1.413	1.408	1.431	1.350	1.378	1.431	1.375	1.393
1	0.860	0.880	0.882	0.928	0.867	0.877	0.871	0.898	0.833

*indicates where a missing datum (due to a failure in the collection apparatus) was replaced by the subject's mean.

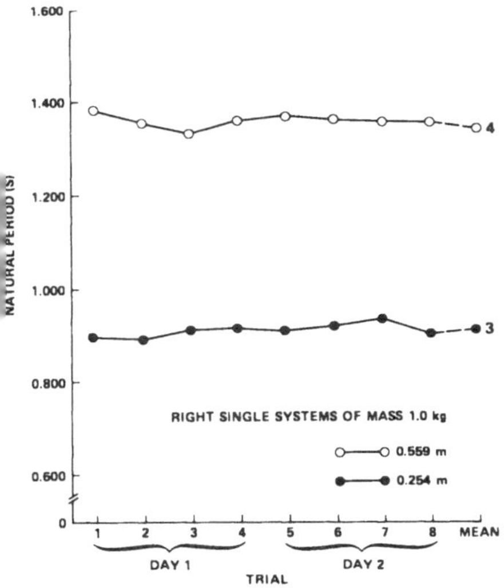

IGURE 7.12f. Periods (cycle time in seconds) of ght singles for condition 3 (mass 1.0kg and length .254m) and condition 4 (mass 1.0kg and length .559m) as a function of trial.

PERIOD (S) LEFT SINGLES
TRIAL NUMBER

	1	2	3	4	5	6	7	8	X̄
Subject 1									
4	1.433	1.350	1.336	1.229	1.367	1.300	1.307	1.248	1.321
3	0.956	0.904	0.880	0.903	0.924	0.836	0.835	0.758	0.874
Subject 2									
4	1.325	1.311	1.293	1.414	1.381	1.379	1.417	1.381	1.362
3	0.850	0.855	0.842	0.925	0.852	1.087	0.937	0.939	0.911
Subject 3									
4	1.351	1.319	1.281	1.327	1.359*	1.319	1.403	1.507	1.359
3	0.808	0.801	0.911	0.853	0.906	0.845	0.991	0.961	0.884
Subject 4									
4	1.447	1.461	1.439	1.489	1.379	1.467	1.425	1.417	1.453
3	0.986	1.008	1.017	0.982	0.961	0.915	0.987	0.961	0.978

*indicates where a missing datum (due to a failure in the collection apparatus) was replaced by the subject's mean.

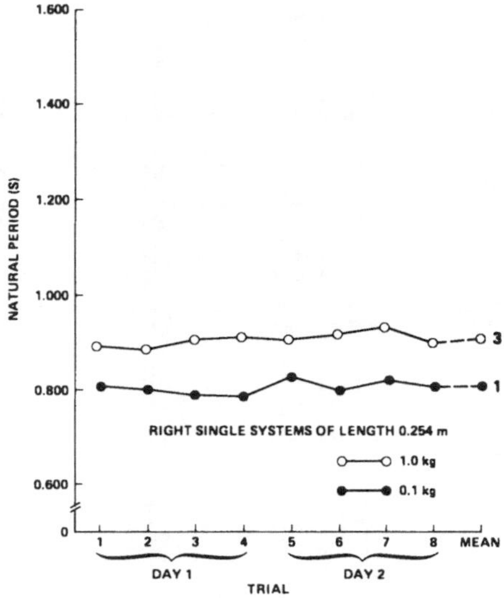

FIGURE 7.12g. Periods (cycle time in seconds) of right singles for condition 1 (mass 0.1kg and length 0.254m) and condition 3 (mass 1.0kg and length 0.254 m) as a function of trial.

PERIOD (*S*) LEFT SINGLES
TRIAL NUMBER

	1	2	3	4	5	6	7	8	X̄
Subject 1									
3	0.956	0.904	0.880	0.903	0.924	0.836	0.835	0.758	0.8
1	0.871	0.892	0.811	0.789	0.959	0.772	0.732	0.634	0.8
Subject 2									
3	0.850	0.855	0.842	0.925	0.852	1.087	0.937	0.939	0.9
1	0.815	0.795	0.736	0.735	0.766	0.874	0.818	0.788	0.7
Subject 3									
3	0.808	0.801	0.911	0.853	0.906	0.845	0.991	0.961	0.8
1	0.685	0.635	0.721	0.687	0.714	0.670	0.849	0.920	0.7
Subject 4									
3	0.986	1.008	1.017	0.982	0.961	0.915	0.987	0.961	0.9
1	0.860	0.880	0.882	0.928	0.867	0.877	0.871	0.898	0.8

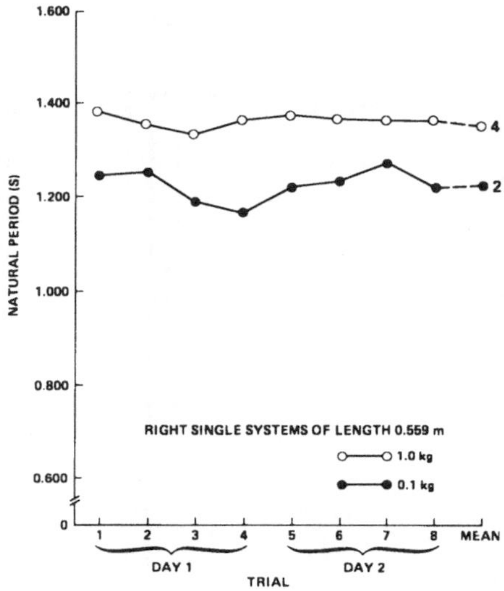

FIGURE 7.12h. Periods (cycle time in seconds) of right singles for condition 2 (mass 0.1kg and length 0.559m) and condition 4 (mass 1.0kg and length 0.559m) as a function of trial.

PERIOD (*S*) LEFT SINGLES
TRIAL NUMBER

	1	2	3	4	5	6	7	8	X̄
Subject 1									
4	1.433	1.350	1.336	1.229	1.367	1.300	1.307	1.248	1.3
2	1.322	1.166*	1.181	1.052	1.264	1.125	1.135	1.075	1.1
Subject 2									
4	1.325	1.311	1.293	1.414	1.381	1.379	1.417	1.381	1.3
2	1.196	1.204	0.931	1.048	1.169	1.326	1.293	1.067	1.1
Subject 3									
4	1.351	1.319	1.281	1.327	1.359*	1.319	1.403	1.507	1.3
2	1.102	1.141	1.229	1.116	1.086	1.087	1.212	1.329	1.1
Subject 4									
4	1.447	1.461	1.439	1.489	1.379	1.467	1.425	1.417	1.4
2	1.352	1.413	1.408	1.431	1.350	1.378	1.431	1.375	1.3

*indicates where missing datum (due to failure in the collection apparatus) was replaced by the subject's mean.

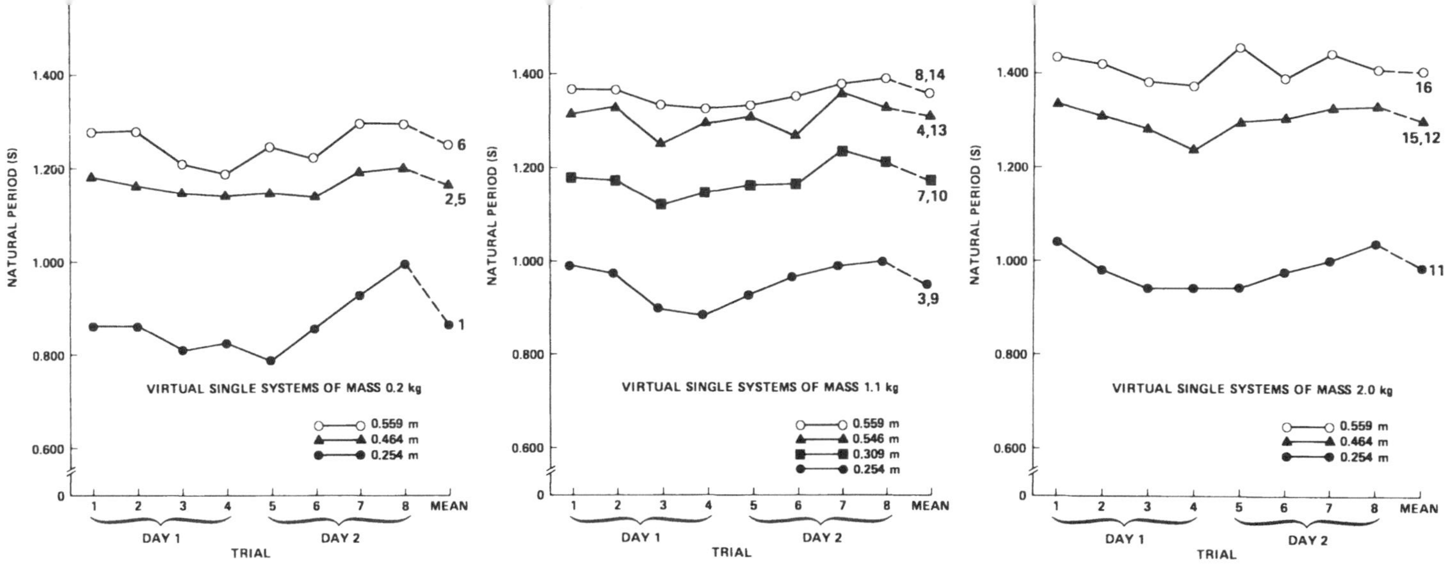

FIGURE 7.13.

Left panel: Period (cycle time in seconds) of virtual single systems for condition 1 (virtual mass 0.2kg and virtual length 0.254m), conditions 2 and 5 (virtual mass 0.2kg and virtual length 0.464m), and condition 6 (virtual mass 0.2kg and virtual length 0.559m).

Middle panel: Period (cycle time in seconds) of virtual single systems for conditions 3 and 9 (virtual mass 1.1kg and virtual length 0.254m), conditions 7 and 10 (virtual mass 1.1kg and virtual length 0.309m), conditions 4 and 13 (virtual mass 1.1kg and virtual length 0.546m), and conditions 8 and 14 (virtual mass 1.1kg and virtual length 0.559m).

Right panel: Period (cycle time in seconds) of virtual single systems for condition 11 (virtual mass 2.0kg and virtual length 0.254m), conditions 12 and 15 (virtual mass 2.0kg and virtual length 0.464m), and condition 16 (virtual mass 2.0kg and virtual length 0.559m).

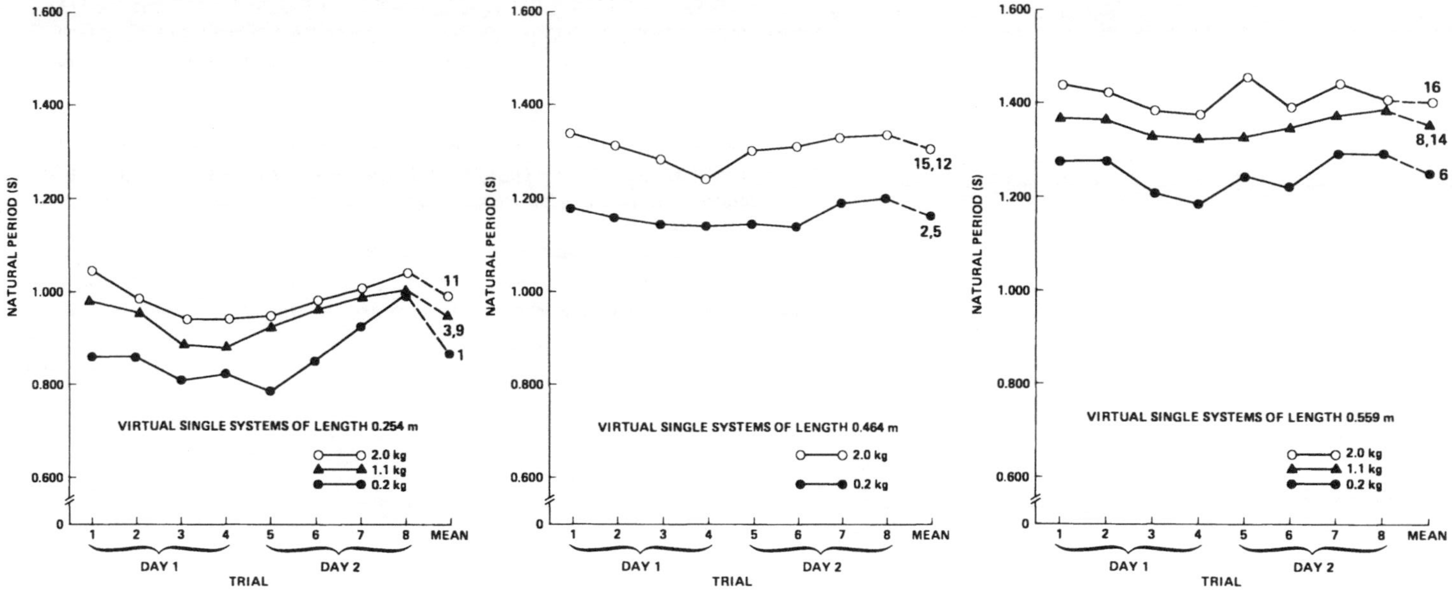

FIGURE 7.14.

Left panel: Period (cycle time in seconds) of virtual single systems for condition 1 (virtual mass 0.2kg and virtual length 0.254m), conditions 3 and 9 (virtual mass 1.1kg and virtual length 0.254m), and condition 11 (virtual mass 2.0kg and virtual length 0.254m).

Middle panel: Period (cycle time in seconds) of virtual single systems for condition 2 and 5 (virtual mass 0.2kg and virtual length 0.464m), conditions 12 and 15 (virtual mass 2.0kg and virtual length 0.464m).

Right panel: Period (cycle time in seconds) of virtual single systems for condition 6 (virtual mass 0.2kg and virtual length 0.559m), conditions 8 and 14 (virtual mass 1.1kg and virtual length 0.559m), and condition 16 (virtual mass 2.0kg and virtual length 0.559m).

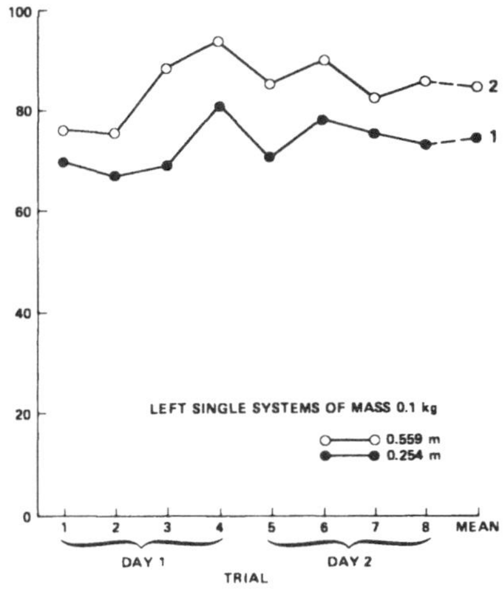

ANGULAR EXCURSION (cycle angle in degrees) LEFT SINGLES
TRIAL NUMBER

	1	2	3	4	5	6	7	8	X̄
Subject 1									
2	84.675	88.454	112.793	103.842	89.799	79.784	79.019	90.190	91.04
1	69.169	76.289	74.095	79.402	75.289	71.461	59.982	74.476	75.54
Subject 2									
2	44.339	41.232	59.762	62.775	57.428	63.384	57.211	59.697	55.75
1	34.845	25.194	41.374	50.579	46.879	52.133	54.070	51.420	44.58
Subject 3									
2	108.594	96.459	103.618	121.113	118.808	126.159	113.397	100.102	111.04
1	82.132	83.263	81.374	91.096	89.640	91.774	96.520	79.196	86.86
Subject 4									
2	67.396	76.771	77.949	88.747	76.204	90.647	81.212	94.489	81.70
1	93.443	84.027	79.522	103.311	71.199	98.873	93.422	88.622	89.04

GURE 7.15a. Angular excursion (cycle an-e in degrees) of left singles for condition 1 nass 0.1kg and length 0.254m) and condition 2 nass 0.1kg and length 0.559m) as a function of al.

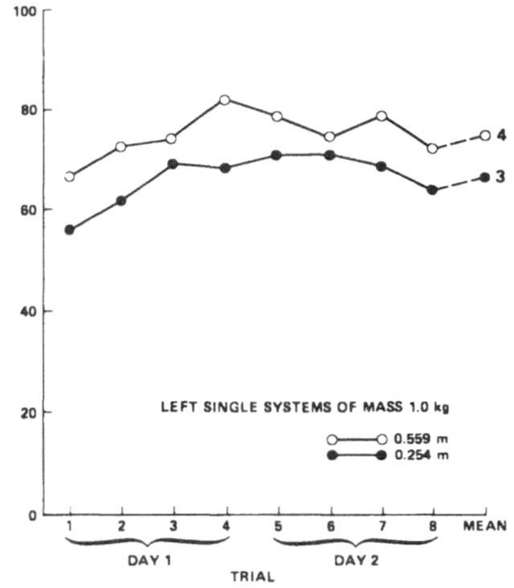

ANGULAR EXCURSION (cycle angle in degrees) LEFT SINGLES
TRIAL NUMBER

	1	2	3	4	5	6	7	8	X̄
Subject 1									
4	68.519	92.359	84.287	91.079	83.694	75.403	67.982	69.301	79.02
3	64.462	81.091	74.586	75.605	73.740	67.841	62.214	71.887	71.45
Subject 2									
4	42.242	40.376	55.042	50.278	52.056	53.876	56.814	52.766	50.42
3	30.202	38.074	48.906	44.239	44.220	53.402	47.649	46.901	44.18
Subject 3									
4	92.813	85.326	95.799	114.030	113.664	107.538	112.346	94.892	102.04
3	80.067	70.496	99.993	89.442	99.860	90.117	94.501	64.754	86.17
Subject 4									
4	63.730	73.580	62.517	74.274	65.603	52.732	79.133	71.026	67.84
3	50.211	58.888	53.248	63.878	66.638	74.658	72.900	71.636	64.00

GURE 7.15b. Angular excursion (cycle an-e in degrees) of left singles for condition 3 nass 1.0kg and length 0.254m) and condition 4 nass 1.0kg and length 0.559m) as a function of al.

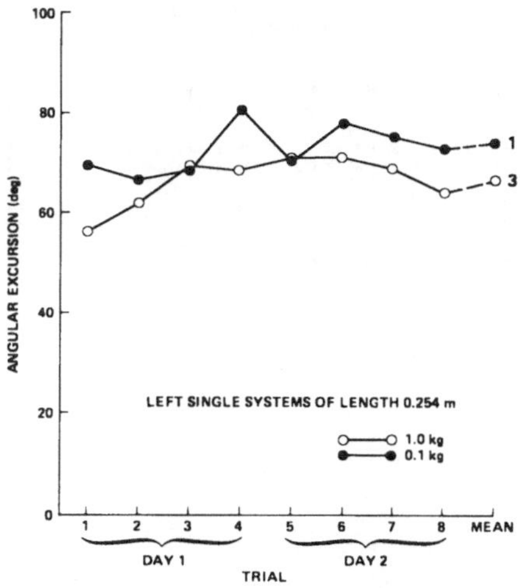

ANGULAR EXCURSION (cycle angle in degrees) LEFT SINGLES

TRIAL NUMBER

	1	2	3	4	5	6	7	8	$\bar{X}$
Subject 1									
1	69.169	76.289	74.095	79.402	75.289	71.461	59.982	74.476	75.
3	64.462	81.091	74.586	75.605	73.740	67.841	62.214	71.887	71
Subject 2									
1	34.845	25.194	41.374	50.579	46.879	52.133	54.070	51.420	44
3	30.202	38.074	48.906	44.239	44.220	53.402	47.649	46.901	44.
Subject 3									
1	82.132	83.263	81.374	91.096	89.640	91.774	96.520	79.196	86.
3	80.067	70.496	99.993	89.442	99.860	90.117	94.501	64.754	86.
Subject 4									
1	93.443	84.027	79.522	103.311	71.199	98.873	93.422	88.622	89.
3	50.211	58.888	53.248	63.878	66.638	74.658	72.900	71.636	64.

FIGURE 7.15c. Angular excursion (cycle angle in degrees) of left singles for condition 1 (mass 0.1kg and length 0.254m) and condition 3 (mass 1.0kg and length 0.254m) as a function of trial.

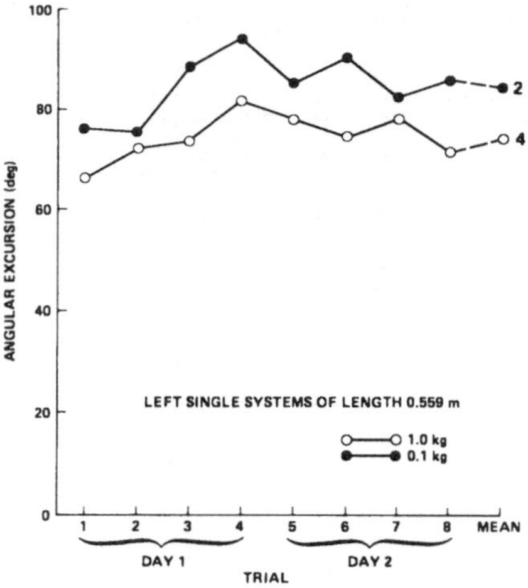

ANGULAR EXCURSION (cycle angle in degrees) LEFT SINGLES

TRIAL NUMBER

	1	2	3	4	5	6	7	8	$\bar{X}$
Subject 1									
2	84.675	88.454	112.793	103.842	89.799	79.784	79.019	90.190	91.
4	68.519	92.359	84.287	91.079	83.694	75.403	67.982	69.301	79.
Subject 2									
2	44.339	41.232	59.762	62.775	57.428	63.384	57.211	59.697	55.
4	42.242	40.376	55.042	50.278	52.056	53.876	56.814	52.766	50.
Subject 3									
2	108.594	96.459	103.618	121.113	118.808	126.159	113.397	100.102	111.
4	92.813	85.326	95.799	114.030	113.664	107.538	112.346	94.892	102.
Subject 4									
2	67.396	76.771	77.949	88.747	76.204	90.647	81.212	94.489	81.
4	63.730	73.580	62.517	74.274	65.603	52.732	79.133	71.026	67.

FIGURE 7.15d. Angular excursion (cycle angle in degrees) of left singles for condition 2 (mass 0.1kgm and length 0.559m) and condition 4 (mass 1.0kg and length 0.559m) as a function of trial.

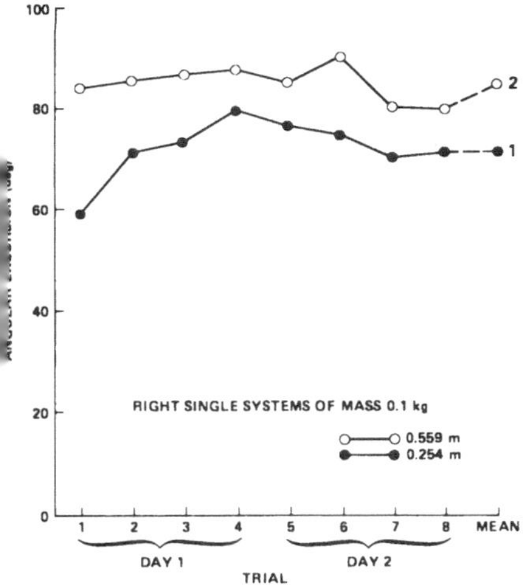

FIGURE 7.15e. Angular excursion (cycle angle in degrees) of right singles for condition 1 (mass 0.1kg and length 0.254m) and condition 2 (mass 0.1kg and length 0.559m) as a function of trial.

ANGULAR EXCURSION (cycle angle in degrees) RIGHT SINGLES
TRIAL NUMBER

	1	2	3	4	5	6	7	8	X̄
Subject 1									
2	94.021	83.14*	92.123	88.804	78.663	87.688	60.615	79.918	83.14
1	59.710	73.847	72.306	77.293	76.506	75.316	59.734	73.040	70.99
Subject 2									
2	62.769	48.681	58.811	51.452	62.833	64.278	60.941	61.828	58.96
1	42.269	48.241	57.444	55.218	69.081	53.805	51.221	57.771	54.37
Subject 3									
2	109.660	108.267	118.296	119.188	129.702	126.242	111.456	105.679	116.08
1	84.693	75.232	77.269	86.199	97.519	90.731	85.395	75.328	84.05
Subject 4									
2	69.965	100.068	77.724	92.367	68.955	83.233	82.249	71.354	80.73
1	50.838	89.862	85.884	99.534	62.267	78.577	83.952	79.206	78.78

*indicates where missing datum (due to a failure in the collection apparatus) was replaced by the subject's mean.

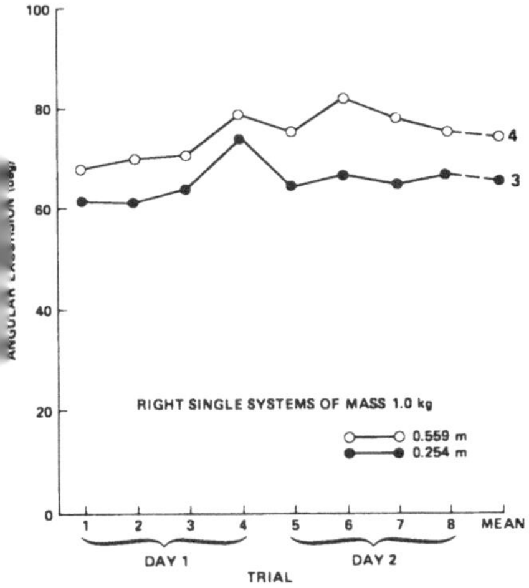

FIGURE 7.15f. Angular excursion (cycle angle in degrees) of right singles for condition 3 (mass 1.0kg and length 0.254m) and condition 4 (mass 1.0kg and length 0.559m) as a function of trial.

ANGULAR EXCURSION (cycle angle in degrees) RIGHT SINGLES
TRIAL NUMBER

	1	2	3	4	5	6	7	8	X̄
Subject 1									
4	77.598	69.833	75.051	82.179	73.580	77.734	70.258	75.366	75.17
3	63.663	69.933	63.543	74.586	64.498	66.100	58.183	69.826	66.29
Subject 2									
4	52.649	46.092	57.867	51.599	53.646	61.519	61.843	62.269	55.92
3	46.279	38.897	53.629	52.583	49.723	52.265	52.296	59.093	50.59
Subject 3									
4	99.788	90.907	102.93*	113.225	102.93*	111.292	109.537	92.802	102.93
3	87.107	73.613	73.717	91.501	88.017	86.202	80.102	71.706	81.47
Subject 4									
4	44.177	75.273	52.203	70.77	58.116	77.698	71.926	72.051	65.26
3	51.942	64.380	67.782	78.325	56.322	64.894	71.340	68.658	65.43

*indicates where a missing datum (due to a failure in the collection apparatus) was replaced by the subject's mean.

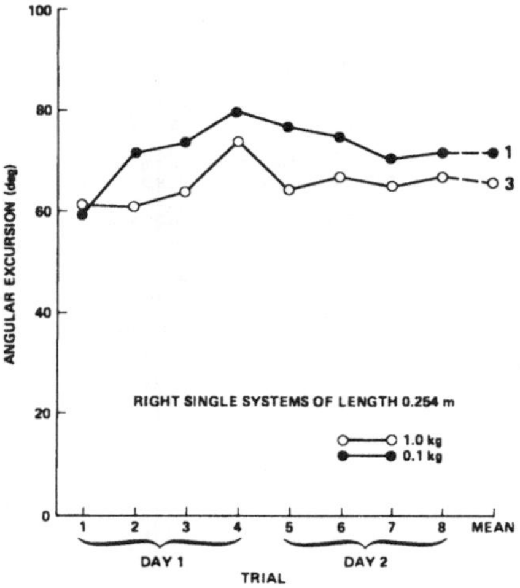

FIGURE 7.15g. Angular excursions (cycle angle in degrees) of right singles for condition 1 (mass 0.1kg and length 0.254m) and condition 3 (mass 1.0kg and length 0.254m) as a function of trial.

ANGULAR EXCURSION (cycle angle in degrees) RIGHT SINGLES
TRIAL NUMBER

	1	2	3	4	5	6	7	8
Subject 1								
1	59.710	73.847	72.306	77.293	76.506	75.316	59.734	73.040
3	63.663	69.933	63.543	74.586	64.498	66.100	58.183	69.826
Subject 2								
1	42.269	48.241	57.444	55.218	69.081	53.805	51.221	57.771
3	46.279	38.897	53.629	52.683	49.723	52.265	52.296	59.093
Subject 3								
1	84.693	75.232	77.269	86.199	97.519	90.731	85.395	75.328
3	87.107	73.613	73.717	91.501	88.017	86.202	80.102	71.706
Subject 4								
1	50.838	89.862	85.884	99.534	62.267	78.577	83.952	79.206
3	51.942	64.380	67.782	78.325	56.322	64.894	71.340	68.658

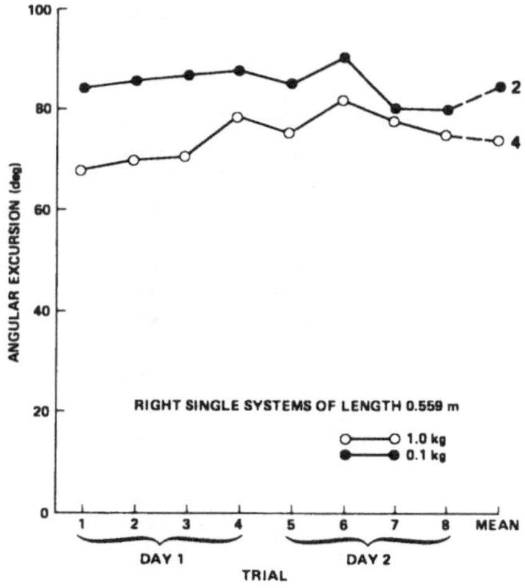

FIGURE 7.15h. Angular excursion (cycle angle in degrees) of right singles for condition 2 (mass 0.1kg and length 0.559m) and condition 4 (mass 1.0kg and length 0.559m) as a function of trial.

ANGULAR EXCURSION (cycle angle in degrees) RIGHT SINGLES
TRIAL NUMBER

	1	2	3	4	5	6	7	8
Subject 1								
2	94.021	83.14*	92.123	88.804	78.663	87.688	60.615	79.918
4	77.598	69.833	75.051	82.179	73.580	77.734	70.258	75.366
Subject 2								
2	62.769	48.681	58.811	51.452	62.833	64.278	60.941	61.828
4	52.649	46.092	57.867	51.699	53.646	61.519	61.843	62.269
Subject 3								
2	109.660	108.267	118.296	119.188	129.702	126.242	111.456	105.679
4	99.788	90.907	102.93*	113.225	102.93*	111.292	109.537	92.802
Subject 4								
2	69.965	100.068	77.724	92.367	68.955	83.233	82.249	71.354
4	44.177	75.273	52.203	70.777	58.116	77.698	71.926	72.051

*indicates where a missing datum (due to failure in the collection apparatus) was re- placed by the subject's mean.

CONDITIONAL STABILITIES FOR SINGLE WRIST-PENDULUM SYSTEMS RELATING PERIOD (τ) TO MASS AND LENGTH VARIATIONS

SINCE: $\{M = 0.1\text{kg} \quad \text{and} \quad L_{0.245m} < L_{0.559m}\}$
THEN: $\{T_{0.245m} < T_{0.559m}\}$.

SINCE: $\{M = 1.0\text{kg} \quad \text{and} \quad L_{0.254m} < L_{0.559m}\}$
THEN: $\{T_{0.254m} < T_{0.559m}\}$.

SINCE: $\{L = 0.254\text{m} \quad \text{and} \quad M_{0.1kg} < M_{1.0kg}\}$
THEN: $\{T_{0.1kg} < T_{1.0kg}\}$.

SINCE: $\{L = 0.559\text{m} \quad \text{and} \quad M_{0.1kg} < M_{1.0kg}\}$
THEN: $\{T_{0.1kg} < T_{1.0kg}\}$.

CONDITIONAL STABILITIES FOR SINGLE WRIST-PENDULUM SYSTEMS RELATING AMPLITUDE (θ) TO MASS AND LENGTH VARIATIONS

SINCE: $\{M = 0.1\text{kg} \quad \text{and} \quad L_{0.254m} < L_{0.559m}\}$
THEN: $\{\theta_{0.254m} < \theta_{0.559m}\}$.

SINCE: $\{M = 1.0\text{kg} \quad \text{and} \quad L_{0.245m} < L_{0.559m}\}$
THEN: $\{\theta_{0.254m} < \theta_{0.559m}\}$.

SINCE: $\{L = 0.254\text{m} \quad \text{and} \quad M_{0.1kg} < M_{1.0kg}\}$
THEN: $\{\theta_{0.1kg} > \theta_{1.0kg}\}$.

SINCE: $\{L = 0.559\text{m} \quad \text{and} \quad M_{0.1kg} < M_{1.0kg}\}$
THEN: $\{\theta_{0.1kg} > \theta_{1.0kg}\}$.

The left wrist-pendulum complex and right wrist-pendulum complex do not differ in terms of these conditional stabilities. A similar set of structural stabilities can be constructed for the virtual single wrist-pendulum systems identified in Figures 7.13 and 7.14:

CONDITIONAL STABILITIES FOR THE VIRTUAL SINGLE WRIST-PENDULUM SYSTEMS RELATING PERIOD (τ) TO MASS AND LENGTH VARIATIONS

SINCE: $\{M = 0.2\text{kg} \quad \text{and}$
$L_{0.254m} < L_{0.464m} < L_{0.559m}\}$
THEN: $\{T_{0.254m} < T_{0.407m} < T_{0.559m}\}$.

SINCE: $\{M = 1.1\text{kg} \quad \text{and}$
$L_{0.254m} < L_{0.309} < L_{0.546m} < L_{0.559m}\}$
THEN: $\{T_{0.254m} < T_{0.309m} < T_{0.546m} < T_{0.559m}\}$.

SINCE: $\{M = 2.0\text{kg}$ and $L_{0.254\text{m}} < L_{0.464\text{m}} < L_{0.546\text{m}}\}$

THEN: $\{T_{0.254\text{m}} < T_{0.464\text{m}} < T_{0.546\text{m}}\}$.

SINCE: $\{L = 0.254\text{m}$ and $M_{0.2\text{kg}} < M_{1.1\text{kg}} < M_{2.0\text{kg}}\}$

THEN: $\{T_{0.2\text{kg}} < T_{1.1\text{kg}} < T_{2.0\text{kg}}\}$.

SINCE: $\{L = 0.464\text{m}$ and $M_{0.2\text{kg}} < M_{2.0\text{kg}}\}$

THEN: $\{T_{0.1\text{kg}} < T_{1.0\text{kg}}\}$.

SINCE: $\{L = 0.559\text{m}$ and $M_{0.2\text{kg}} < M_{1.1\text{kg}} < M_{2.0\text{kg}}\}$

THEN: $\{T_{0.2\text{kg}} < T_{1.1\text{kg}} < T_{2.0\text{kg}}\}$.

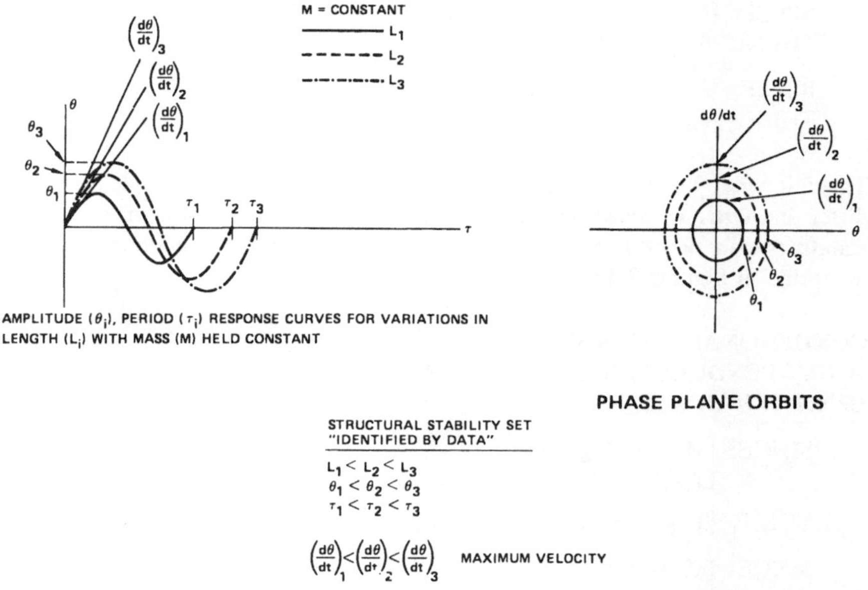

FIGURE 7.16. RESPONSE CURVES (LEFT) AND PHASE PLANE ORBITS (RIGHT) WHEN (EXPLICITLY MANIPULATED) MASS IS HELD CONSTANT.

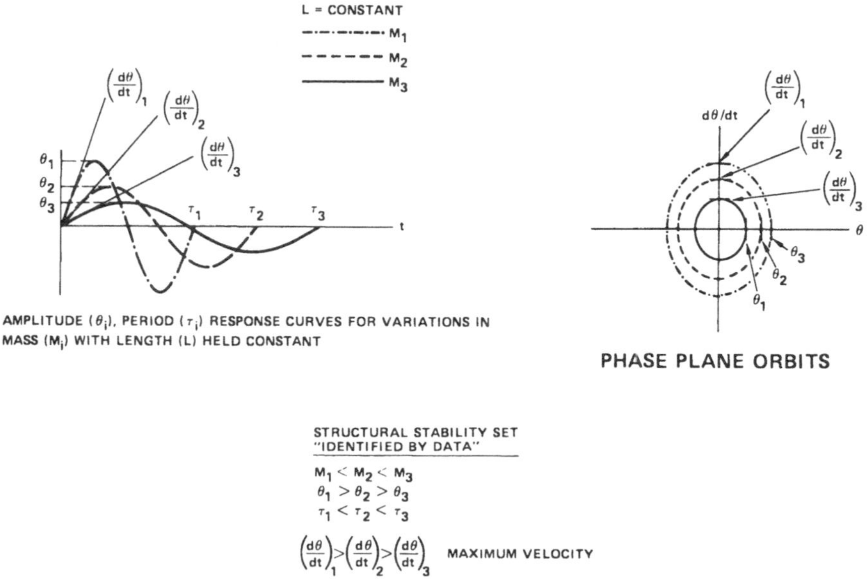

FIGURE 7.17. RESPONSE CURVES (LEFT) AND PHASE PLANE ORBITS (RIGHT) WHEN (EXPLICITLY MANIPULATED) LENGTH IS HELD CONSTANT.

Figure 7.16 summarizes the structural stabilities relating length variation to period and amplitude variations using amplitude/period response curves. In addition, Figure 7.16 depicts the structural stabilities relating velocity $d\theta/dt$ to amplitude θ in terms of phase plane orbits. Figure 7.17 provides a similar summary of the structural stabilities relating mass variations to period, amplitude, and velocity using the amplitude/period response curve and the phase plane.

Figures 7.18 and 7.19 further summarize the structural stabilities identified in the phase planes by plotting them in the form of attractor portraits. The surfaces of the manifolds identify the attractor states. The behavior space is a phase plane and the control space is defined by length (Figure 7.18) and mass (Figure 7.19). These figures highlight the fact that as length increases in the control space (see Figure 7.18), the attractor orbit likewise increases in its maximum amplitude and velocity, but as mass increases in the control space (see Figure 7.19), the attractor orbit *decreases* in its maximum amplitude and velocity. The transition point from a limit cycle to a point attractor identifies the length or mass that would be required to completely eliminate the self-sustained oscillation under its current steady-state regime.

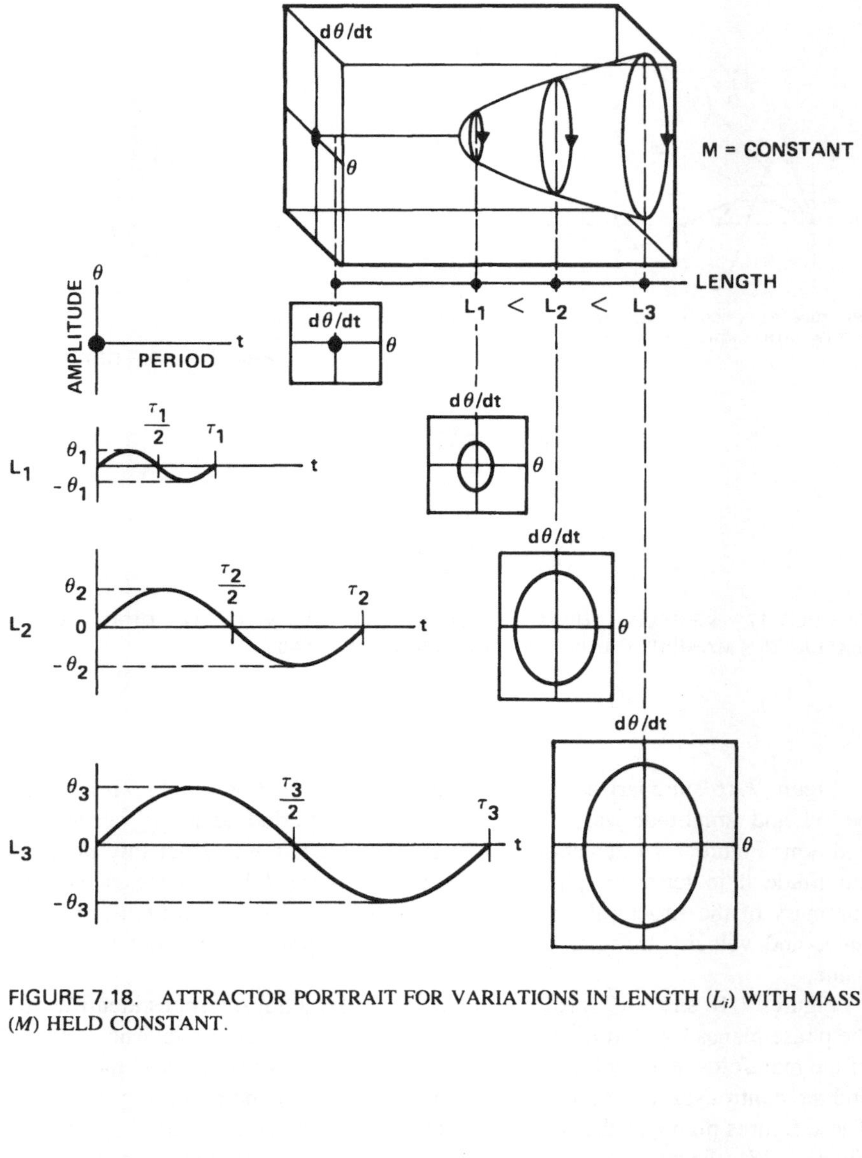

FIGURE 7.18. ATTRACTOR PORTRAIT FOR VARIATIONS IN LENGTH (L_i) WITH MASS (M) HELD CONSTANT.

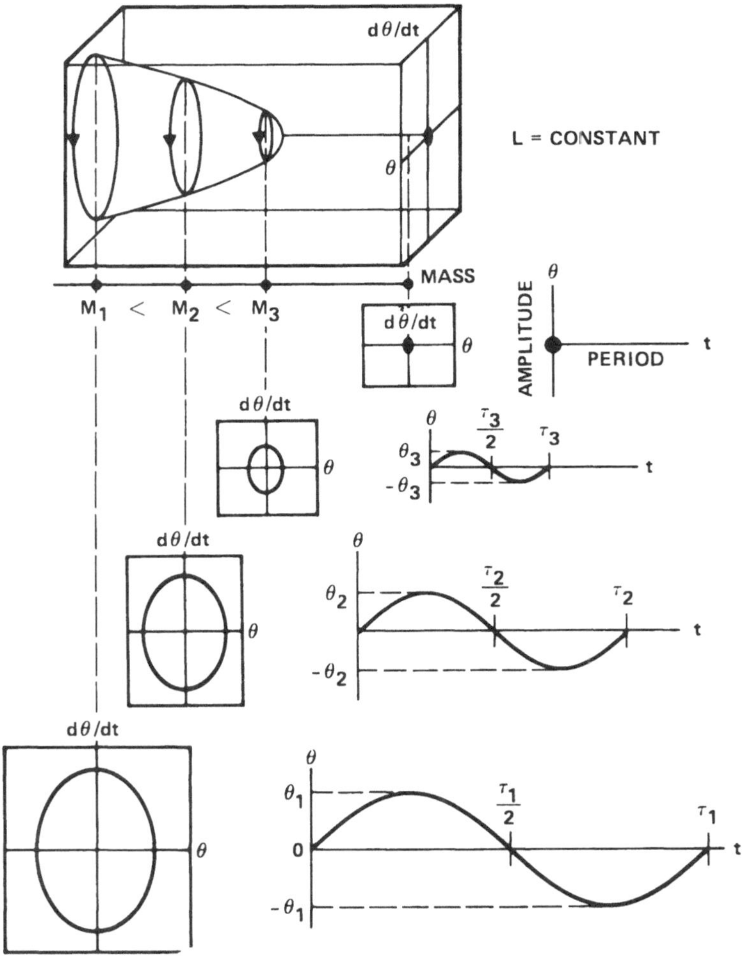

FIGURE 7.19. ATTRACTOR PORTRAIT FOR VARIATIONS IN MASS (M_i) WITH LENGTH (L) HELD CONSTANT.

REFERENCES

Abraham, R. H., & Shaw, C. D. (1982). *Dynamics: The geometry of behavior, Part 1—periodic behavior*. Santa Cruz: Aerial Press.

Abraham, R. H., & Shaw, C. D. (1986). Dynamics—a visual introduction. In F. E. Yates (Ed.), *Self-organizing systems: The emergence of order*. New York: Plenum Press.

Abraham, R. H. (1986). Dynamics and self-organization. In F. E. Yates (Ed.), *Self-organizing systems: The emergence of order*. New York: Plenum Press.

Bernstein, N. A. (1967). *The control and regulation of movements*. London: Pergamon Press.

Bloch, E. H., & Iberall, A. S. (1982). Toward a concept of the functional unit of mammalian skeletal muscle. *American Journal of Physiology: Regulatory, Integrative and Comparative Physiology*, *242*, R411–R420.

Gibson, J. J. (1966). *The senses considered as perceptual systems*. Boston: Houghton Mifflin.

Gibson, J. J. (1979). *The ecological approach to visual perception*. Boston: Houghton Mifflin.

Haken, H. (1977). *Synergetics*. Heidelberg: Springer-Verlag.

Helson, H. (1964). *Adaptation level theory*. New York: Harper and Row.

Hoyt, D. F., & Taylor, C. R. (1981). Gait and the energetics of locomotion in horses. *Nature*, *292*, 239–240.

Kugler, P. N. (1986). A morphological perspective on the origin and evolution of movement patterns. In M. Wade and J. Whiting (Eds.), *Motor skill acquisition in children: Aspects of coordination and control*. The Hague, Netherlands: Martinus Nijhoff.

Kugler, P. N., Kelso, J. A. S., & Turvey, M. T. (1980). On the concept of coordinative structures as dissipative structures: I. Theoretical lines of convergence. In G. E. Stelmach and J. Requin (Eds.), *Tutorials in motor behavior*. New York: North-Holland.

Luce, R. D. (1978). Dimensionally invariant numerical laws correspond to meaningful qualitative relations. *Philosophy of Science*, *45*, 1–16.

Rosen, R. (1978). *Fundamentals of measurement and representation of natural systems*. New York: North-Holland.

Stein, R. B. (1982). What muscle variables does the central nervous system control? *Behavioral and Brain Sciences*, *4*, 535–578.

Thom, R. (1975). *Structural stability and morphogenesis*. Reading, MA: Benjamin.

Turvey, M. T. (1980). Clues from the organization of motor systems. *Signed and spoken languages: Biological Constraints on linguistic form*. U. Bellugi and M. Studdert-Kennedy (Eds.) Dahlem Konferenzen, Weinheim: Verlag Chemie.

8 THE SCALING RELATION OF NATURAL PERIOD TO MASS AND LENGTH

ABSTRACT

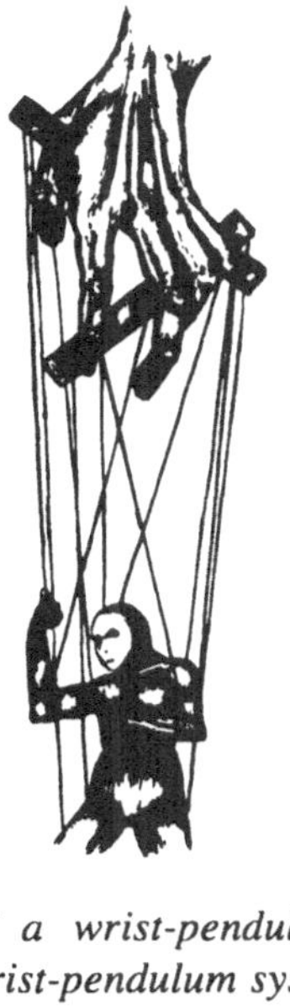

Traditional allometry measures scaling coefficients of laws using invariant (Euclidean) coordinate spaces as 'objective' frames of reference. The method proposed here measures coefficients of the coordinate space using invariant laws as 'objective' frames of reference. This strategy follows directly from the invariance postulate of natural law: Physical laws (or universal constants) constitute invariant standards for measuring objects and events. By postulating a physical law as an 'objective' frame, an allometric plot of the form $X = a_i(\text{physical law})^{c_i}$ *can be used to derive metrics of the coordinate space that embeds the law. The metrical structure of the coordinate space is defined by the deviations of* a_i *and* c_i *from their canonical magnitudes. These indices are physically interpreted in terms of potentials.*

Through the use of similarity analysis, dimensional analysis, and allometry a physical (scaling) law is identified relating periodic time to the (uncorrelated) mass and length of a wrist-pendulum system, namely, $\tau_o = a_i(M^{1/16}L^{1/2})^{c_i}$*. This law of wrist-pendulum systems results from a nesting of transformations—two in scale and one in coordinate space—of a law that relates biological times to biological masses and lengths. Subjects are viewed as nonuniform coordinate spaces for the twice-scaled law; they differ in the strengths of certain scalar and vector* *potentials indexed by* a_i *and* c_i*, respectively. Abiding by the invariance postulate of natural law, departures of* τ_o *from the canonical form,* $(M^{1/16}L^{1/2})^{1.0}$*, are attributed to the properties of the coordinate space. These properties are measurable by casting the paired values* $(\tau_o, M^{1/16}L^{1/2})$ *into double logarithmic coordinates. (For example, the degree of curvature of a coordinate system,* c_i*, is measured by the slope in a log by log plot.) Once measured, the contribution of these properties can be eliminated—through a renormalization of the coordinate space—and the invariance of the timing law thereby revealed.*

The data of the experiment confirm $\tau_o \propto M^{1/16}L^{1/2}$*, that is, the theoretically derived form of the basic scaling law subject to the mechanical constraints wrought by scale transformations of the type of biological system (of which the subject is a token) and the type of biological function (that the subject simulates). The law is also confirmed by two subsidiary experiments, one that directly manipulates one or more potentials of a coordinate space and one that compares in-phase with out-of-phase coordinations. These subsidiary experiments were motivated in part by the issue of whether or not there is a lawful basis for relating locomotory periods to animal size that holds invariant over changes in speed and gait.*

8.1. An Hypothesized Law

It will be hypothesized that for all biological systems there is a single scaling relation or law among the dimensions of time (T), mass (M), and length (L) of the form:

$$T \propto (M^{\rho}L^{\epsilon}), \tag{8.1}$$

where ρ and ϵ are unknown and the monomial $M^{\rho}L^{\epsilon}$ is a member of a class of variables that will be referred to as the *moment class*. The statical moment (dimensionally M^1L^1) and the moment of inertia (dimensionally M^1L^2) are the familiar members of this class.

Scaling Law (8.1) is assumed to relate any representative biological quantities of dimensions T, M, and L. Precisely how the law does relate these quantities will vary with: (a) the type of biological system, (b) the type of biological function, and (c) the token biological system that actually manifests the law in the performance of the function. Factors (a) and (b) scale the law—that is, they make the exponents larger or smaller. Factor (c) determines the coordinate space of the law—that is, the *measurement structure* to which all observations of T, M, and L are referred. The law as stated in (8.1) is in its normal form: It refers to the (topological) connectivity of neighborhoods of points—how time, mass, and length quantities go together regardless of the measurement structures (i.e., coordinate space geometry, clocks, rulers). The law will assume the form of (8.1) only in an ideal, Euclidean measurement structure—i.e., one in which the space and time metrics are uniformly distributed. Factors (a) and (b), akin to mechanical advantages, affect the proportionalities or ratios into which the quantities of the law enter; factor (c) transforms the ideal, Euclidian space of the law into a non-Euclidian space in which the space and time metrics vary as a function of location within the measurement space (see Chapter 2). Simply put, factors (a) and (b) affect the structural stability (internal structure) of the law and factor (c) affects only the geometric setting of the law. Recognizing the invariance of law, in biology as in physics, requires being able to renormalize both scale and coordinate geometry.

8.2. First Scale Transformation: The Type of Biological System

A coarse classification of biological systems suggested by Economos (1982) is of special significance: Animals can be considered similar to one another to the extent that they share similar methods of locomotion and are subject to similar (conservative) reactive forces. The significance of Economos' (1982) taxonomy is that, for terrestrial animals of roughly 20*kg* or more, the transformed Law (8.1) for any representative quantities of *M*, *L*, and *T*, is known (Calder, 1984; Economos, 1982; Gunther, 1975; McMahon, 1975):

$$T \propto (M^{1/4}L^0) \text{ or } T \propto (M^0L^1) \tag{8.2}$$

That is to say, there is a transformation—which might be called "locomotory style *S*"—such that:

$$S: (M^{\rho}L^{\epsilon}) \rightarrow (M^{1/4}L^0) \text{ or } (M^0L^1) \tag{8.3}$$

As long as all species of animals that locomote in style *S* can be considered uniform in their capacity as a (living) coordinate space for embedding the law, then the law in the form of (8.2) holds.

To reiterate, Scaling Relation (8.2) is Law (8.1), relating representative biological quantities of time, mass, and length, as it is manifest within particular mechanical constraints, namely, those defined by a biological system that locomotes in style *S*. How is Scaling Relation (8.2) determined? The manner of its determination follows.

8.2.1. Similarity Analysis and the Scale Equations of Biological Similarity. The goal of a similarity analysis is to compare two things, usually referred to as model and prototype but easily interpreted as two species or two members of the same species or two instances of the same individual organism (say, at two different ages). The comparison is with respect to a property that can be defined in *MLT* units—or any other suitable set of units (Stahl, 1962). Thus, for the property Q exhibited by two systems S_l and S_2

$$Q_1 = M_1^{\alpha} L_1^{\beta} T_1^{\gamma} \tag{8.4}$$

and

$$Q_2 = M_2^{\alpha} L_2^{\beta} T_2^{\gamma} . \tag{8.5}$$

A quantitative relation between the two instances of the property is established by means of Newton's reduction coefficient

$$\Omega = \frac{Q_1}{Q^2} . \tag{8.6}$$

The mass ratio $M_l/M_2 = \mu$, the length ratio $L_l/L_2 = \lambda$, and the time ratio $T_l/T_2 = \tau$ are introduced into Equation (8.6) to give

$$\Omega = \mu^{\alpha} \lambda^{\beta} \tau^{\gamma} . \tag{8.7}$$

The next step is to identify criteria that define a physical similarity. These criteria or constants permit the determination of scale equations which, in turn, specify how one dimension is scaled to another. Economos (1982) has identified one list of constants for biological systems of locomotory style *S*; Gunther and Morgado (1982) have identified another list. The two lists differ minimally: Where Gunther and Morgado emphasize a constancy of auto-oscillatory phenomona, Economos emphasizes the constancy of tensile strength. Two notions are central to both lists of constants. The first is the thesis of elastic similarity suggested by Rashevsky (1960) and developed by McMahon (1973, 1975, 1984). The second is the expansion of the *MLT* system of dimensional analysis in physics to an *MLDT*-system for biology, where *D* is diameter. For several reasons the cylindrical shape of animals mandates that *L* and *D* be treated as dimensionally distinct (McMahon, 1975).

Scaling Law (8.1) in the form (8.2) is here derived by dimensional analysis from the three constants of muscle tissue, namely, density, elasticity, and tensile strength:

(a) constancy of density: $m = c_1 l d^2$; (8.8)

(b) constancy of elasticity: $l = c_2 d^{2/3}$; (8.9)

and (c) constancy of tensile strength: $f = c_3 A = c_3' d^2$, (8.10) where m is muscle mass, l is muscle length, d is muscle diameter, A is cross-sectional area (proportional to d^2), f is force, and the cs are constants. The constancies (a) and (b) give the scaling equations (where $\delta = D_l/D_2$):

$$\mu \cdot \lambda^{-1} \cdot \delta^{-2} = 1 , \tag{8.11}$$

$$\lambda \cdot \delta^{-2/3} = 1 , \tag{8.12}$$

respectively; and together they give the scaling equation:

$$\lambda \cdot \mu^{-1/4} = 1 . \tag{8.13}$$

Writing force in its dimensional form, constant (c) can be expressed:

$$\mathrm{MLT}^{-2}\mathrm{D}^{-2} = c_3' . \tag{8.14}$$

When Scaling Equation (8.11) is brought into (8.14), the following scaling equation is obtained:

$$\tau \cdot \lambda^{-1} = 1 . \tag{8.15}$$

Given (8.13), it follows that

$$\tau \cdot \mu^{-1/4} = 1 . \tag{8.16}$$

The reduction coefficient, Equation (8.7), can now be expressed as a composite power of mass:

$$\Omega = \mu^{p\alpha + q\beta + r\Upsilon} \tag{8.17}$$

where p, q, and r are parameters determined by Scaling Equations (8.13) and (8.15). With respect to any representative temporal quantity (for example, natural period) the dimensional exponents are $M^0L^0T^1$ and the parameters based on (8.13) and (8.15) are $p = 1$, $q = 1/4$, and $r = 1/4$. Substituting these exponent and parameter values into Equation (8.17) gives

$$\Omega = \tau = \mu^{1/4} . \tag{8.18}$$

And completing the above procedure for length gives

$$\Omega = \tau = \lambda . \tag{8.19}$$

Taken together these two expressions for τ give the interdependency of mass and length

$$\mu = \lambda^4 \tag{8.20}$$

The result can be expressed, therefore, as a proportionality written, in simple dimensional terms, as either

$$T \propto (M^{1/4}L^0) \text{ or } T \propto (M^0L^1) \tag{8.21}$$

that is, the form that Scaling Law (8.1) takes when scaled by the mechanical constraints of biological systems of locomotory style *S*.

8.3. Second Scale Transformation: The Appliance Simulated

Imagine a function to be performed by a biological system. The function can be likened to an appliance and performing the function can be likened to simulating the appliance. An appliance implicates mechanical constraints. Therefore, if Scaling Law (8.1) is applied in the simulation of a given appliance by a biological system of locomotory style *S*, then a further scale transformation of the law is to be expected.

In the main experiment reported in Chapter 6, a subject (abstractly, a biological system that locomotes in style *S*) simulated a mass spring/pendulum device with a part of his or her body. Figure 8.1 depicts a hybrid mass-spring/pendulum characterization of a wrist pendulum system. A substantial experimental literature implies that a unit comprised of a group of muscles and the joint they activate behaves qualitatively like a mass attached to a spring with a

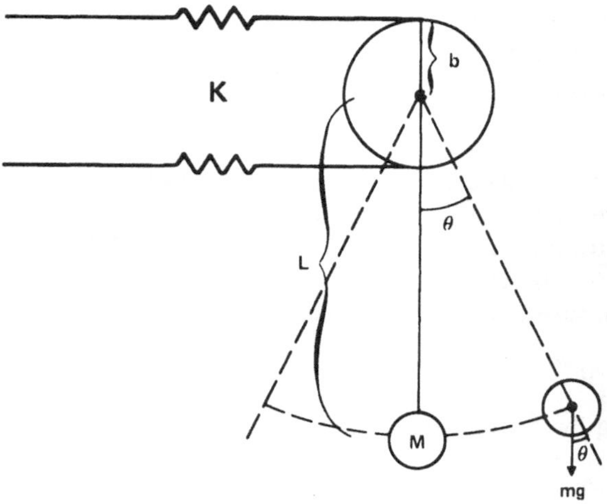

FIGURE 8.1. HYBRID MASS-SPRING CHARACTERIZATION OF A WRIST-PENDULUM SYSTEM.

stiffness that is invariant for the movement, the spring constant K (e.g., Bizzi, Polit, & Morasso 1976; Fel'dman, 1966a, 1966b, 1974; Fel'dman & Latash, 1982; Kelso & Holt, 1980). As hypothesized in Chapter 7, when K is ascribed to a muscle-joint system, it is a macroscopic state variable (in the laboratory frame of reference) that is functionally dependent on muscle elasticity *and* viscosity (Wainwright, Biggs, Currey, & Gosline, 1976). That is to say, at a less macroscopic scale (and in another frame of reference), restorative *and* dissipative forces can be identified that functionally yield the restorative constant K identified at the more macroscopic scale. In short, friction is implicitly included in K and need not be considered explicitly in the characterization of a wrist-pendulum system. With this caveat in mind, the state variables of the system depicted in Figure 8.1 are K, the stiffness; b, the distance from the ulna to the center of the hand/pendulum handle component (a mechanical advantage term); I, the moment of inertia of the pendulum of length l and mass m; g, the gravitational constant; τ_o, the natural period; θ_o, the amplitude at natural period. Accordingly, a wrist-pendulum system oscillating at its natural period τ_o is defined by the following functional relation, which can be likened to an equation of state since it is a relation among a system's state variables that characterizes the system at equilibrium (see Rosen, 1978):

$$\tau_o = \emptyset(K, b, I, g, \Theta)\ . \tag{8.22}$$

Dimensional considerations (e.g., Bridgman, 1922; Krantz, Luce, Suppes, & Tversky, 1971) are sufficient to derive the form of the functional $\emptyset$. Dimensionally, taking mass (M), length (L), and time (T) as the basic dimensions, we have:

$$K \stackrel{d}{=} MT^{-2}\ ,\ b \stackrel{d}{=} L\ ,\ I \stackrel{d}{=} ML^2\ ,\ g \stackrel{d}{=} LT^{-2}\ ,\ \text{and}\ \Theta \stackrel{d}{=} 1, \tag{8.23}$$

(where $\stackrel{d}{=}$ means "dimensionally equal to"). Note that amplitude θ in radians or degrees, is dimensionless. This means that θ does not enter into any interactions with m and l, the independent variables, and that it cannot affect the form of the equation of state that relates τ_o to m and l (which is not to say that θ cannot affect τ_o). By dimensional homogeneity, and by allowing I to be approximated as ml^2 and θ to be neglected, it follows that the simplest form of the function, up to a numerical factor, is

$$\tau_o = \left[\frac{1}{b}\frac{I}{Kg}\right]^{1/4} = \left[\frac{l}{b}\frac{m}{K}\frac{l}{g}\right]^{1/4} \tag{8.24}$$

At a minimum, any vibrational phenomenon requires two properties: One property ensures that the system in question tries to return to its equilibrium state, and one makes the system in question overshoot its equilibrium state (Main, 1978). In Equation (8.24) the returning or restorative properties are gravity and stiffness and the overshoot property is the moment of inertia, approximated by ml^2. Because gravity acts vertically and the motion is rotational, the gravitational restorative force (a torque) is not proportional to the displacement (rather, it is proportional to the sine of the displacement). This means, simply, that the system depicted in Figure 8.1 and modeled by Equation (8.24) cannot be linear, although it might approximate a linear system for small displacements. Further, there is reason to expect that the restorative force due to stiffness contributes to the system's nonlinearity (Asatryan & Feldman, 1965).

The equation of state (8.24) models *any* appliance suitably described by the state variables τ_o, I, b, K, g; it is indifferent to the magnitude of the appliances it models and to the materials from which they are constructed. In the present context, Equation (8.24) is advanced as modeling a ("flesh and blood") biological system but in a different context it could be advanced as modeling a ("nuts and bolts") mechanical system. The form of Equation (8.24) holds no allegiances other than to dimensional homogeneity.

Scaling Equations (8.18) and (8.19) define the proportionalities that must hold among the dimensions T, L, M if a given biological system is to be accepted as a token of the type "locomotory style S." In comparison, the equation of state, Equation (8.24), defines the proportionalities that must hold among the quantities τ_o, I, b, K, g if the simulated appliance is to be at equilibrium.

According to Equation (8.24), the representative quantities of time (natural period) and length (pendulum length) scale ($\propto$) as

$$\tau_o \propto l^{1/2} \tag{8.25}$$

and the representative quantities of time (natural period) and mass (pendulum mass) scale as

$$\tau_o \propto \mathrm{m}^{1/4} \tag{8.26}$$

Embedding the relations among the dimensions given by Equations (8.18) and (8.19) within the relations among the state variables given by (8.25) and (8.26) gives the relations among the quantities that must hold in the simulation of a mass-spring/pendulum appliance by a biological system of locomotory style S; namely,

$$\tau_o \propto (l^1)^{1/2}, \text{ that is, } \tau_o \propto l^{1/2} \tag{8.27}$$

and

$$\tau_o \propto (\mathrm{m}^{1/4})^{1/4}, \text{ that is, } \tau_o \propto \mathrm{m}^{1/16} \tag{8.28}$$

The masses and lengths of the wrist-pendulum systems are quasi independent; an increase in one of these quantities is not strictly proportional to an increase in the other. The appropriate moment variable, therefore, is given simply by the product of $m^{1/16}$ and $l^{1/2}$ (in comparison to the derivation of the moment variable for the type of biological system, Section 8.2). That is to say, for the conditions of the experiment—a biological system simulating mass-spring/pendulum systems with uncorrelated masses and lengths—Scaling Law (8.1) relating representative quantities of T, M, L is transformed in scale to:

$$T \propto (M^{1/16}L^{1/2}) . \qquad (8.29)$$

8.4. A Transformation of the Coordinate Space: The Token Biological System

An individual subject of the present experiment is a *token* of the type of biological system of locomotory style S. It is the individual subject, not the biological type (an abstraction) that simulates the mass-spring/pendulum appliance. Each token system must abide by Law (8.1) in the scaled form of (8.29), but the token systems are not uniform. Each individual token system provides a *unique physical context* for the expression of the scaled law (8.29). The method of allometry is used to identify these individually defined coordinate spaces, to which the scaled law applies.

8.4.1. The Principles of Similitude and Allometry. The concepts of magnitude and direction are prominent in any discussion of the forms and functions of living things (Thompson, 1917/1942; P. S. Stevens, 1974). To discuss the growth of living things entails, in addition, the concept or dimension of time. Imagine a completely spherical animal that grows over some interval of time t. Choosing a linear dimension, for example, the animal's radius r, it is obvious that the linear dimension will increase in magnitude as the animal grows. For successively larger values of r, measurements of the animal's volume and surface area will reveal that the growth rates of those two properties are not identical: Volume grows more rapidly with increases in r than does the surface area. The common expressions for spherical volume and surface area are $4\pi r^3/3$ and $4\pi r^2$, respectively. From these expressions it can be seen that volume (V) and surface area (S) relate to the linear dimensions as $V \propto r^3$ and $S \propto r^2$. Of the forces determining the form of the animal, some will be in proportion to surface area and some (notably gravity) will be in proportion to volume. Clearly, as the absolute scale of the imaginary animal increases, the balance of the form-determining forces will undergo change.

With increments in r, the variables that enter into an equation of state for some particular capability of the animal will remain the same but their relative values will alter. To conserve the equation of state, and *a fortiori* the capability, requires that the state variables maintain a fixed, quantitative relation. The latter statement is the principle of similitude. The principle's primary consequence is that regular distortions in the form of an animal must occur if particular physical quantities or processes are to be invariant. (A most visible tendency in morphology is for surface form to alter so that surface area keeps pace with volume.)

It is worthwhile to note that there is nothing uniquely biological about the principle of similitude. The principle, and the tendencies it expresses, are very general (Buckingham, 1915; Rosen, 1978). In this respect the choice of an imaginary spherical animal, taking note of the different disproportionalities of volume and surface with respect to a linear dimension, is not so arbitrary. It is a choice that encourages an understanding of morphology and function through a coupling of geometry and the dynamics of open systems (Gould, 1966; Iberall, 1974; Thompson, 1917/1942).

Because of the very general physical grounds for the principle of similitude, it is not surprising that equations of similitude, equations of the type

$$Y = aX^b, \tag{8.30}$$

are commonplace expressions of many phenomena of metabolism, biochemistry, physiology, evolution and the like (Gunther, 1975; Huxley, 1931/1972; Schmidt-Nielsen, 1972). These equations—often termed allometric (from *allois*, meaning difference) because they express the deviation from geometric similarity, or isometric similarity, of the relation between two variables (Calder, 1984; Huxley, 1931/1972; McMahon, 1977; McMahon & Bonner, 1983; Peters, 1983; Schmidt-Nielsen, 1977, 1984a,b)—are the simplest expressions of relative growth, that is:

$$\frac{dY}{dt} / Y = b \frac{dX}{dt} / X . \tag{8.31}$$

What might be referred to as the principle of allometry is the fact that the ratio of relative increase of variable *Y* with respect to another variable *X* is a constant (Rosen, 1967; von Bertalanffy, 1960).

The joint implication of the principles of similitude and allometry is that lawful relations defined over the class of biological things must invariably take the form of an allometric equation: If the biological or physiological variables *Y* and *X* are lawfully linked, then they will relate as a straight line in logarithmic coordinates. A careful review of the large number of allometric relations (Gunther, 1975; Huxley, 1931/1972) suggests that allometric equations index a fundamental design feature of open, evolving systems, namely to maintain nonequilibrium steady states in the face of changes in scale. The various systemic processes must be mutually dependent (Adolph, 1949; von Bertalanffy, 1960). As originally pointed out by Huxley (1931/1972), when any two properties of an animal are growing exponentially, they are necessarily related by an allometric equation (see discussion in Gunther, 1975).

8.4.2. Mathematical and Physical Allometry. Typically the tools of allometry are used to measure the scaling relationship between two variables. The method of allometry is depicted in Figure 8.2. Lines (1), (2), and (3) in the left-hand graph depict trajectories of some set of data points relating an independent parameter x and a dependent parameter y. The $[x,y]$ system of coordinates defines a rigid Euclidean measurement space. In a Euclidean space the slope and intercept of a function can be used to help define the relationship that maps the y variable onto the x variable. Slope and intercept are helpful in the foregoing respect, however, if and only if the relation is linear (monotonic) and of the first order. Trajectories (1) and (2) of the left-hand graph of Figure 8.2 meet these criteria. Consequently, for these trajectories the slope identifies a scaling constant of proportionality and the intercept identifies a constant by which x and y differ for all values. Using the slope and intercept, the functional mapping can be put into a mathematical form. For trajectory (1) the mapping function is $y = 1/2(x)$ and for trajectory (2) the mapping function is $y = x$. Given that both trajectories intercept the origin, there are no constant differences associated with the variables (only proportional differences). Trajectory (3), however, identifies a mapping function that is not easily analyzed because the slope of the relationship is continuously changing. If this mapping function is a smooth and continuous power relation, that is, $Y = aX^c$, then it is possible to derive the exact scaling relation by transforming the coordinate space in which the data are plotted from a Euclidean space into a logarithmic coordinate space. The logarithmic coordinate space transforms the power relation $Y = aX^c$ into a linear relation defined by $Y = a + cX$, where $Y = logY$, $a = loga$, and $X = logX$. The plot of Y versus X in logarithmic coordinates is a straight line, with *slope* $= c$ and *Y-intercept* $= a$. The values of a and c can be estimated graphically, or computationally using simple linear regression analysis. The use of the linear regression method provides a least-squares line measure, a numerical estimate of the slope, and a standard error of estimation indicating a goodness-of-fit between the parameter estimate and the data.

The right-hand graph of Figure 8.2 identifies a logarithmically transformed space together with the three trajectories. In the logarithmic space trajectory (3) forms a straight line. The slope of the line identifies the exponential scale factor that relates the x and y variables. Specifically, the slope of a mapping function in logarithmic coordinates identifies the exponential scaling relationship of the two variables. A *slope* $= 1$ identifies a first order scaling relationship, a *slope* $= 2$ identifies a second order scaling relationship, and so on. A logarithmic transformation can be used, therefore, as a methodological tool for deriving the scaling relationships that map variables—with particular utility for scaling relationships of order greater than one.

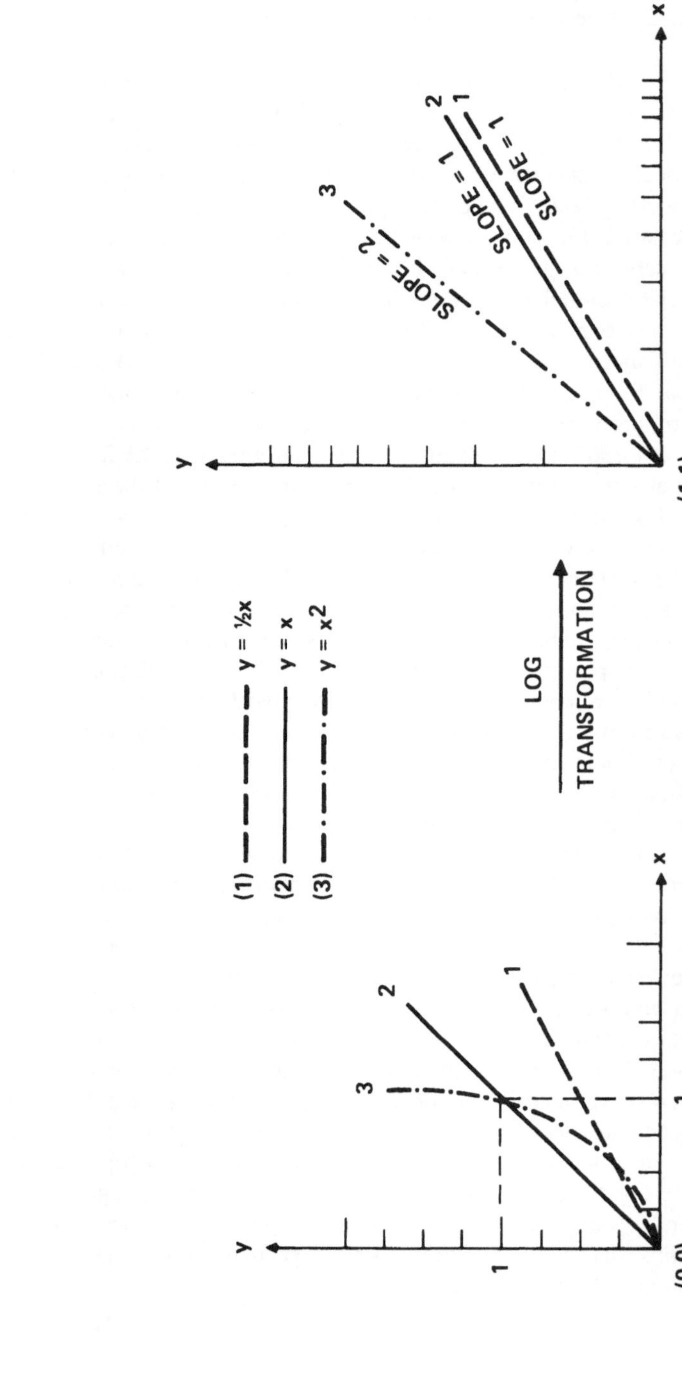

FIGURE 8.2. ALLOMETRIC METHOD OF COORDINATE TRANSFORMATIONS.

The aforementioned methodology constitutes the framework for mathematical allometry. Arrays of data are mapped into logarithmic coordinate spaces for the purpose of deriving the order of the scaling relationship. A second form of allometry, *physical allometry*, uses the same method for deriving the order of the scaling relationship but rationalizes the derived order in terms of physical laws. For example, if the data of length (independent parameter) and natural period (dependent parameter) for a simple pendulum were plotted in logarithmic coordinates, then the slope would be equal to $1/2$. In physical allometry this scaling relationship must be rationalized by a physical law. For the simple pendulum the law is $\tau = 2\pi\sqrt{L/g}$. Whereas mathematical allometry requires only the identification of a scaling relationship, physical allometry requires, in addition, the identification of the law that rationalizes the scaling relationship.

8.4.3. Intrinsic Geometries and Physical Laws as "Objective" Frames of Reference. Measurement involves the identification of relationships between invariant metrics (clock, ruler, etc.) and an object (i.e., "world point"). These invariant metrics define *objective frames of reference* that are used to identify invariant relationships over sets of "world points". Insofar as an object of interest occupies space-time, and *geometry* is used to describe the object's position and motions in space-time, certain assumptions regarding the mathematical descriptions of the structure of space-time must be made before a physical frame of reference can be chosen to anchor observations (measurements). Measurement then is the process of locating (specifying the *coordinates*) of a "world point" in some *frame of reference*. A reference frame can be thought of as a "container" embedded in a geometric universe. It may have more, fewer or as many dimensions as the nominated geometry. It need not have sides at right angles to each other. It can be circular, spherical, cylindrical, skewed,. . . . Events take place inside the container and are specified by distances or angles changing with respect to nominated reference points in or on the container.

As noted in Chapter 2 criteria for choosing a geometric scheme have traditionally emerged from either of two hypotheses. The first holds that the geometry should be independent of events being measured. Within this geometry reference frames can be chosen arbitrarily for convenience. This assumption is fundamental to extrinsic theories of measurement. (For example, the space in which an event occurs is often assumed to be a flat (zero curvature) three-space with the coordinates of the hypersurfaces intersecting orthogonally.) When such an extrinsic view is taken, then physical laws will be invariant from one frame to another, provided that the frames of reference do not accelerate relative to other such rigid frames in which events are occurring and being compared. At subluminal velocities, time and length metrics may differ across frames that are in rectilinear, uniform motion with respect to each other; the intrinsic structure of (conservation) laws, however, remains invariant across frames—as in the case of special relativity (which assumes a flat four-dimensional space-time).

The second hypothesis assumes that the appropriate geometry is not independent of events—instead, the geometric structure of events and the geometric structure of space-time itself are linked. It may then happen that the description of the geometry increases in complexity (compared to Euclidean geometry), but the physical concepts become simpler. This search for natural geometries intrinsic to the physical phenomena being observed constitutes the intrinsic approach to measurement.

The use of intrinsic geometries makes it possible to describe, by invariant laws, events in all (even accelerated) physical frames of reference—the principle of general covariance; and to "geometrize" many physical phenomena—the principle of geometrization of natural phenomena. Einstein's theory of general relativity provided the first successful utilization of the principle of geometrization of natural phenomena. Although the theory of general relativity focuses on the motions of large bodies of matter in gravitational fields, the concept of intrinsic geometries applies to any set of physical observables governed by field potentials.

Events, measurements, constants and coordinates are the building materials for physical theories. There is a restriction, though; laws must be invariant over transformations of coordinates. In the development of a physical theory for biology one can begin by assuming a law as the 'objective' frame of reference and looking for a geometry in which it can be expressed most simply. Or one can begin by positing a geometry as the 'objective' frame of reference and look at world points and lines in search of laws. The latter route to discovery is exhibited in the changing views of space, time and mechanical motions from Galileo and Newton through to Einstein's special theory of relativity. The former route to discovery is expressed by Einstein's general theory. It proposes an interdependence of geometry and events and does not posit a geometry. Newton explicitly posited a geometry; so did Einstein in special relativity. General relativity, in contrast, makes curvature a physical *variable*. General relativity dissolved the distinction between inertial and noninertial frames of reference and the problem of inertial mass disappeared—a major achievement in similitude!

The strategy of intrinsic measurement follows directly from the invariance postulate of natural laws: Natural laws (or universal constants) constitute 'objective' frames of reference with respect to which measurements can be compared and contrasted. This stands in sharp contrast to the extrinsic approach which starts by assuming that the objective frame of reference is an externally nominated geometric structure (see Chapter 2 for an extended discussion). The program of measurement evolves from the extrinsic to the intrinsic strategy. Once the invariant characteristics of laws have been identified (Newton's program), they can be used as the universal metrics against which coordinate space geometries are measured (Einstein's program). The law replaces rigid geometry as the objective frame of reference for all measurements. Characteristics of nonrigid geometries can be measured using rigid (structurally stable) laws.

8.4.4. Allometry as a Method for Measuring Displacement and Curvature Coefficients of Coordinate (Mesh) Systems. Typically, allometry uses the logarithmic transformation of coordinates to measure scaling relations between variables. The structure of the logarithmic coordinate space is used as the invariant metric against which the structure of a law is measured and evaluated. In the spirit of the extrinsic approach to measurement, the structure of the coordinate space is nominated independent of the events being measured and is assumed to hold invariantly across token systems. Less typical in allometry is the use of *relational properties of the law* as the invariant metric against which the structure of the coordinate space is measured. This approach requires an a priori commitment to the validity of the structure of the law. If the structure of the law exhibits theoretical and experimental validity, then allometry can proceed to use the invariant properties (such as characteristic exponents) of the law to measure the geometrical properties (such as characteristic exponents) of the measurement space. This is the intrinsic perspective on measurement; to repeat, it uses the structure of a physical law (or universal constant) as an objective frame of reference for evaluating the geometry of the space.

To sketch this alternative allometric strategy, consider again the preceding case of the simple pendulum in which the variables of length and period are compared to determine the relationship defined between the variables. To measure the variations in the coordinate space, the data arrays must be arranged so that the independent and dependent quantities relate according to the scaling predicted by the law. In the simple pendulum example, since the law predicts that the time scales to length according to $T^{1} \propto L^{1/2}$, then the data arrays must be arranged accordingly. When the data arrays are so arranged, the transformation to a logarithmic coordinate space should yield a slope of unity. Any departure of the slope from unity would not be due to variations in the law, or in the variables themselves, but rather to variations in the coordinate space in which the law is embedded. In sum, the tools of allometry can be applied to define not only the symmetries associated with laws but also the symmetries associated with the coordinate spaces that house the laws.

8.4.5. Allometry and Its Application to Wrist-Pendulum Systems. Returning to a wrist-pendulum system and the law that relates period to mass and length, a useful developmental image is that of a wrist pendulum system as a growing system becoming heavier and/or longer with increments in the mass and/or length of the pendulum. The following allometric problem can be considered: How does the natural period of a wrist-pendulum system change over successive, larger instances of the system? Posing the question in this form leads to the expectation that when plotted logarithmically against $(M^{1/16}L^{1/2})$ the natural periods of oscillation of wrist-pendulum systems should yield for each subject a straight line with a positive slope if each subject can be represented by a coordinate space with uniform metrical distribution. More precisely, for any subject, there is an allometric equation:

$$\tau_o = a_i(M^{1/16}L^{1/2})^{c_i} . \qquad (8.32)$$

The quantities a_i and c_i—the intercept and slope, respectively, in double log coordinates—will both be unity (the antilog of a zero intercept in linear coordinates equals unity) if and only if natural period is determined solely by $(M^{1/16}L^{1/2})$—and if the coordinate (mesh) systems have uniform metrics (Section 2.1). It follows, therefore, that any departures of a_i and c_i from unity must index symmetries in the coordinate (mesh) systems in which the scaling relation $T \propto (M^{1/16}L^{1/2})$ is manifest.

8.4.6. Scalar Potentials, Vector Potentials, and Allometric Relations. Following Rosen (1967), an analogy can be drawn between the line connecting systems in an allometric plot and the trajectory of a single point-mass in momentum and position coordinates. The point-mass is the paradigmatic system of particle mechanics. Consider its motion as governed by Newton's law of inertia. This law can be expressed either as a force (Newton's way) or as a constant difference between kinetic energy and potential energy (Lagrange's way). If the particle's equation of motion is known in the form of a Lagrangian function then the state of the particle at time t_2 can be derived from the knowledge of its state at time t_1 (where $t_2 > t_1$) *and* the path between the two states can be determined: It is that path for which the integral of the Lagrangian is a minimum (Hamilton's principle).

Rosen (1967) proposes that similar optimality considerations apply to an allometric relation between two biological observables. It is simplest to think of this proposal in regard to a single growing animal. The two observables might be the animal's head length and total length (cf. Huxley, 1931/1972). Growing can be likened to a continuous displacement, and the path of the head length and total length coordinates can be likened to a trajectory. The claim, by analogy, is that there is an *equation of motion* defined over these variables for which a kind of *Lagrangian function* can be constructed and its minimum determined. This claim is generalized to comparisons among species—say, metabolism from mice to elephant—where the species variation can be thought of as a continuous transformation in form.

The central point of Rosen's (1967) argument is that *an allometric trajectory indexes forces in equilibrium* (cf. Adolph, 1949; von Bertalanffy, 1973). If the forces are changed then the trajectory will change, but it will remain, as before, optimal—tracking a minimum state. Pursuing the analogy further, suppose that the motion of a particle in a field is examined and its trajectory in [*displacement (X), displacement (Y), potential (V)*] coordinates determined on three separate occasions, designated y_1, y_2, y_3 in Figure 8.3.

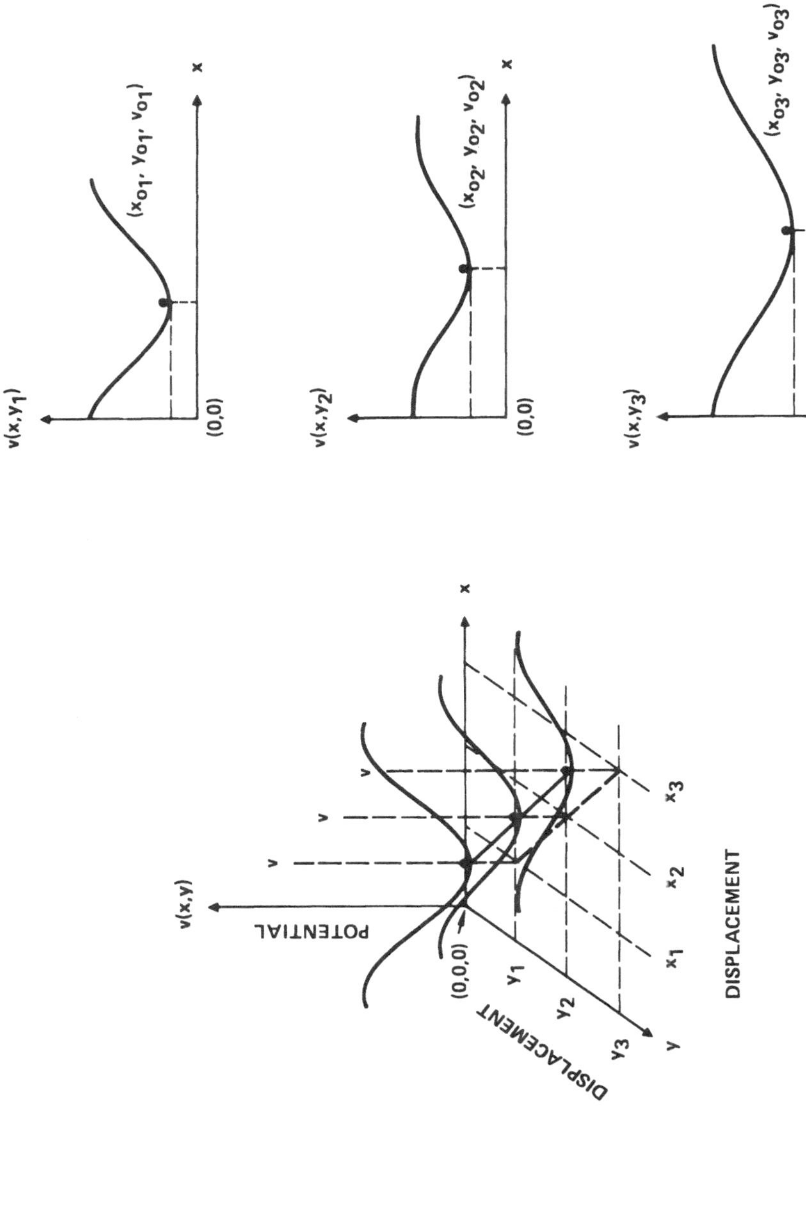

FIGURE 8.3. TRAJECTORY OF CRITICAL STATES DEFINING MINIMA IN THE POTENTIAL FIELD.

In identifying the location of the particle in the field (x_i, y_i, v_i) it will be assumed that each location defines a minimum of a potential well. The resulting trajectory of points defines the layout of potentials in the field. If the potentials were uniformly distributed throughout the field, the particle would exhibit a straight line trajectory such as the particle trajectory shown in Figure 8.3 (the line connecting the minima). Figure 8.4 (left panel) identifies a trajectory of the same particle, under the influence of the same natural law, but in a different field. The new trajectory is to be attributed to the symmetry of the new field, not to a change in the symmetry of the law. The law remains invariant in accordance with the invariance postulate.

The points identifying the trajectory in Figures 8.3 and 8.4 are associated with singularities in the potential function. In this regard the trajectory can be viewed as a method of measuring the potential field. Curvatures in the trajectory identify properties relative to the layout and strength of the potentials. The geometrical properties of slope and intercept in Euclidean coordinates relate to these potential characteristics in the following manner: Slope differences implicate differences in (one or more) scalar potentials (i.e., potentials that only vary in magnitude); curvature differences implicate differences in one or more vector potentials (i.e., potentials that very in magnitude and direction). If the trajectories were compared in logarithmic coordinates, slope differences due to scalar potential differences would show up in the intercepts and curvature differences due to vector potential differences would show up in the slopes (see Figure 8.2). In logarithmic coordinates, the uniform distribution of coordinate triplets, as indexed by their collinearity (see Figure 8.4, right), signifies the uniform curvature of the field into which the law relating the quantities is inserted.

Returning to the allometric equation (8.32), the trajectory of points in the $\log \tau_o \times \log (M^{1/16}L^{1/2})$ plane will exhibit variations in intercept and slope as a function, respectively, of the scalar and vector potentials characterizing the coordinate systems in which the lawful relation between τ_o and $(M^{1/16}L^{1/2})$ is being observed. Only two types of potentials can be operating: gravity and a generalized compliance (an elastic potential). Gravity will appear as a scalar; the generalized compliance, however, can show up in either scalar or vector form. In the present situation there are several possible sources of variation in compliance, most notably due to differences in subjects, hands, and task (swinging one pendulum or a pair of pendulums isochronously). Strength can contribute scalar and vector characteristics that will differ from subject to subject and may be expected to differ between hands. (By the dimensional considerations of Section 8.2., strength—which is proportional to the transverse sectional area of the muscle—should vary with the individual's height to the third power.) The transformation of chemical energy into mechanical energy can also create potential field properties of a scalar and vector nature. The possible variability in mechanical efficiency (external work divided by total energy exchanged) from

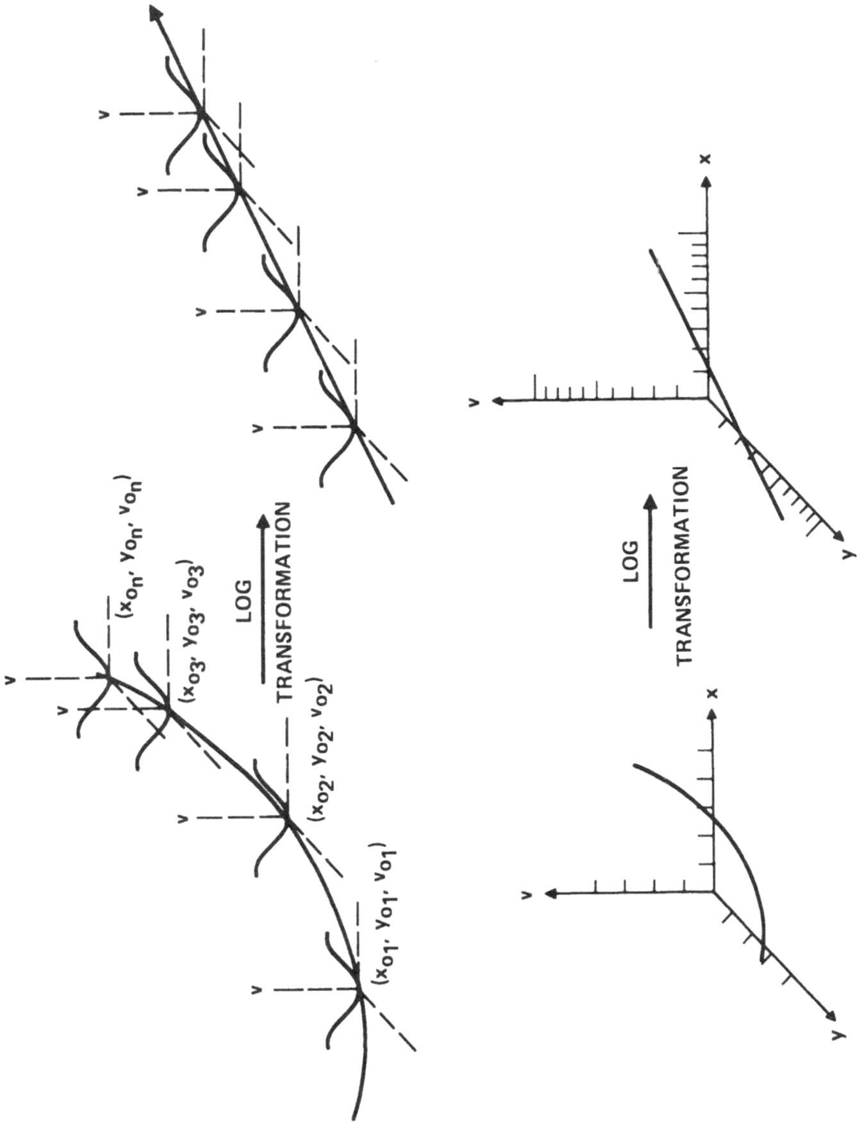

FIGURE 8.4. TRAJECTORY OF MINIMA (x_{0i}, y_{0i}, v_{0i}) IN A NORMAL COORDINATE SPACE AND A LOGARITHMIC COORDINATE SPACE.

subject to subject (Astrand & Rodahl, 1970) implicates an additional source for variability in the scalar and vector properties of the energy field.

To the degree that the scalar and vector potentials are inhomogeneous, coordinate systems will differ and the invariant relation of τ_o to $(M^{1/16}L^{1/2})$ will be obscured. As noted, the law is written for a "normal space," meaning a rigid Euclidean coordinate (mesh) system—i.e, a physical space in which potential is uniformly distributed (Chapter 2). To demonstrate the invariance of the lawful relation, the different coordinate systems in which the law is manifest must be *renormalized*, that is to say, the scalar and vector potentials must be identified and removed. Revealing Scaling Law (8.1) in the scaled form of Equation (8.21) means introducing a transformation on the individual coordinate (mesh) systems that restores the law's canonical form, i.e., Euclid's mesh system.

8.4.7. Mechanical Similarity and Coordinate Transformations. The invariance of natural laws under coordinate transformations is fundamentally linked to the invariance of conservations. For example, the multiplication of a Lagrangian energy equation by any constant does not affect the equations of motion. The constant can be associated with a scale transformation on the coordinate (mesh) system in which the physical events occur. Because of the invariance of the conservations it is possible, in a number of cases, to make a direct evaluation of the intrinsic scaling relationships without the necessity of actually integrating the equations of motion. Such cases include those situations in which the potential energy is a homogeneous function of the coordinates—that is, it satisfies the condition

$$U(\alpha r_1, \alpha r_2, \ldots, \alpha r_n) = \alpha^k U(r_1, r_2, \ldots, r_n) , \qquad (8.33)$$

where α is any constant and k is the degree of homogeneity of the function (Landau & Lifshitz, 1976).

Consider the case where the coordinates are transformed by a factor α and the time by a factor β: $r_a \rightarrow \alpha r_a$, $t \mapsto \beta t$. Then all the velocities $V_a = dr_a/dt$ are changed by a factor α^2/β^2 (because *energy* $\propto$ $(velocity)^2 = L^2/T^2$). The potential energy is multiplied by α^k. If α and β are such that $\alpha^2/\beta^2 = \alpha^k$, that is, $\beta = \alpha^{1-k/2}$ then the result of the transformation is to multiply the Lagrangian by the constant α^k, that is, to leave the equations of motion unaltered.

A change of all the coordinates of the particles by the same factor corresponds to the replacement of the paths of the particles by other paths, geometrically similar but differing in size. The result is that if the potential energy of the system is a homogeneous function of degree k in the (Cartesian) coordinates, the equations of motion permit a series of geometrically similar paths, and the times of the motion between corresponding points are in the ratio

$$(T'/T) = (L'/L)^{1-k/2}, \qquad (8.34)$$

where L'/L is the ratio of linear dimensions of the two paths. Not only the times but also any mechanical quantities at corresponding points at corresponding times are in a ratio that is a power of L'/L. For example, the velocities V, energies E, and action H are such that

$$V'/V = (L'/L)^{k/2},\ E'/E = (L'/L)^{k},\ H'/H = (L'/L)^{1-k/2}. \qquad (8.35)$$

8.5. A Summary of the Hypothesized Scaling Law and Its Transformations

Before proceeding to an evaluation of the law with the data of the experiment, and law and its transformations are reviewed:

Form of the Law	*Transformation*	*Type of transformation*
$T \propto M^{\rho}L^{\epsilon}$		
↓	type of biological system (locomotory style S)	[transformation of scale]
$T \propto M^{1/4}L^{0}$		
↓	type of appliance to be simulated (a mass-spring/pendulum)	[transformation of scale]
$T \propto M^{1/16}L^{1/2}$		
↓	token biological system simulating the appliance (a subject)	[transformation of coordinate space]
$\tau_0 = a_i(M^{1/16}L^{1/2})^{c_i}$		

8.6. Evaluating $T \propto (M^{1/16}L^{1/2})$: Linear Regressions of Data on Model

Figure 8.5 and Table 8.1 give the linear regressions of *log* τ_o on *log* $(M^{1/16}L^{1/2})$ for the 4 right and 4 left single systems and for the 10 virtual single systems (given by ignoring right-left differences in composition). A more detailed partitioning of the various virtual single systems follows (Table 8.2). In the meantime, Table 8.1 and Figure 8.5 suffice to make the opening points, namely, that the moment variable and periodic time are highly correlated for both single and virtual single systems [($r(2) = .990$, $\rho<.01$; $r(8) = .872$, $\rho<.001$)] and that the slopes in all instances closely approximate unity. These two points suggest that the scaling law for timing is structurally stable across both (single and double) categories of systems. While the system categories and subjects do not distinguish themselves with reference to the structure of the law, they do distinguish themselves in terms of the slope and intercept values assumed within the measurement space.

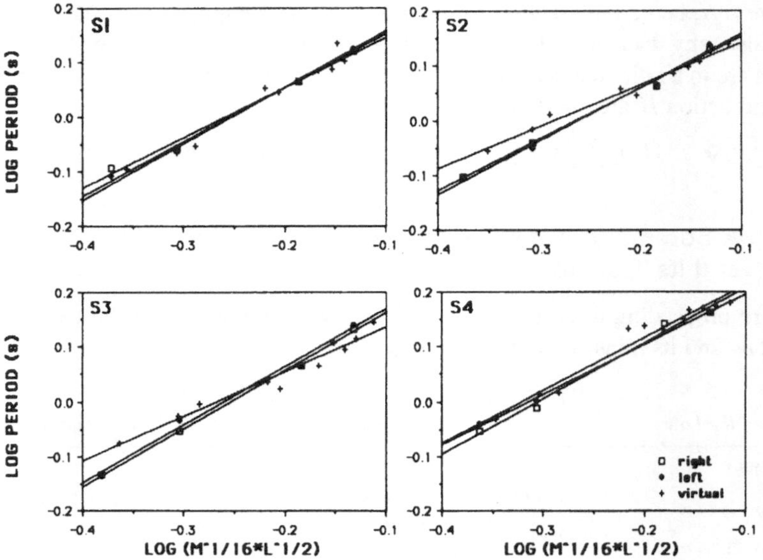

FIGURE 8.5. Regressions of log natural period on log moment variable ($M^{1/16}L^{1/2}$) for the subjects in the main experiment.

TABLE 8.1
Summary of Scaling of Period (τ_o) Against the Moment Variable ($M^{1/16}L^{1/2}$) for Right and Left Single and Double (Virtual Single) Systems

		Subject			
Property	*System Category*	1	2	3	4
Slope (c_i)	Right	.921*	.940	1.056	1.009
	Left	.993	.974	1.057	.913
	Double	1.029	.762	.817	.961
τ_o-intercept (a_i)	Right	.238**	.248	.267	.309
	Left	.252	.255	.275	.287
	Double	.258	.218	.218	.308
Correlation Coefficient	Right	.994	.997	.999	.992
	Left	.996	.996	.996	.999
	Double	.983	.987	.980	.979

*Each value identifies the slope (c_i) of the regression line $\tau_o = a_i[M^{1/16}L^{1/2}]^{c_i}$, where τ_o is in *s*, *M* is in *kg* and *L* is in *m*.

**τ_o-intercept values (a_i) refer to the intercept of the above regression line in log (base 10) coordinates.

8.6.1. The Significance of Mass and a Verification of Its Theoretically Derived Exponent. The derivation of the periodic timing law was somewhat lengthy. A more simple derivation might have ignored the mass of the wrist-pendulum systems and considered these systems solely within the context of a mechanical similitude (Gunther, 1975). The fact that gravity is implicated in the cyclic motion of a wrist-pendulum system points to *1/2* as the exponent of length (from $\tau = 2\pi\sqrt{L/g}$). It is true, nevertheless, that the wrist-pendulum systems under inquiry are not fully accounted for by the criteria of mechanical similitude; they are governed simultaneously by several physical regimes, not one. In short, the mass of a wrist-pendulum system cannot be ignored (as the analysis in the latter part of Chapter 7 underscores). The system's mass relates to an elastic potential in the same manner that the system's length relates to the gravitational potential.

Using all 24 systems, a multiple regression of log τ_0 on log M and log L was conducted for each subject. For Subject 1 ($r^2 = .981$, $p < .0001$), the mass exponent was .037 ($p = .0001$, with lower and upper 95% confidence limits of .013 and .061, respectively) and the length exponent was .497 ($p = .0001$, 95% limits .463, .532). For Subject 2 ($r^2 = .962$, $p < .0001$), the mass exponent was .087 ($p = .0001$, 95% limits .055, .119), and the length exponent was .410 ($p = .0001$, 95% limits .366, .455). For Subject 3 ($r^2 = .966$, $p < .0001$), the mass exponent was .093 ($p = .0001$, 95% limits .06, .125) and the length exponent was .443 ($p = .0001$, 95% limits .398, .488). And for Subject 4 ($r^2 = .951$, $p < .0001$), the mass exponent was .059 ($p = .002$, 95% limits .024, .094) and the length exponent was .485 ($p = .0001$, 95% limits .429, .540).

A further multiple regression was conducted on the data of the four subjects taken together (that is, for a total of 96 systems). For this overall analysis ($r^2 = .873$, $p < .0001$), the mass exponent was .056 ($p = .0002$, 95% limits .027, .084), and the length exponent was .469 ($p = .0001$, 95% limits .428, .510). Two conclusions are to be drawn. First, the predicted (.0625) and the observed (.0560) mass exponents, and the predicted (.50) and the observed (.47) length exponents are closely similar. Second, variations in mass contributed significantly to variations in periodic timing.

There is a further simple verification of the contribution of mass over and above the contribution of length. (Recall that the lengths in question are those of simple pendulum equivalents of compound pendulums. Mass enters into their calculation through Huygens' law.) The lengths of the four virtual single systems that were symmetrical (that is, the right and left wrist-pendulum systems that comprise them are identical in mass and length) and the lengths of the four single systems are approximately equal (see Tables 6.1 and 6.12). The two kinds of systems, however, differ in mass with the mass of a virtual single system being twice that of its single system mate. Consequently, if mass is a significant variable, then the periodic times of a virtual single (τ_v) and a single (τ_s) system should enter into a ratio that exceeds unity: $\tau_v/\tau_s > 1$. More particularly, if the exponent of mass is small and of the order of .0625, then τ_v/τ_s should equal the ratio $M_v^{1/16}/M_s^{1/16}$—which is a constant 1.044 given that $M_{virtual}$ for these particular virtual systems is always twice M_{single}. The mean $\tau_{virtual}/\tau_{single}$ ratios for each of the four subjects in order were: 1.025, 1.079, 1.063, and 1.040. Again it is to be concluded that mass contributed significantly in the preferred steady states of the wrist-pendulum systems, thereby sustaining the general but roughly hewn conclusion of Chapter 7.

8.6.2. The Relation of Slope and Intercept. As remarked, the deviations of a_i and c_i from unity in the twice-scaled version of the scaling law, namely, $\tau_o = a_i(M^{1/16}L^{1/2})^{c_i}$, could be ascribed to differences in the relevant potentials comprising the different coordinate (mesh) systems. To better evaluate these variations in a_i and c_i, Table 8.2 reports the linear regressions in double log coordinates for each of the four subjects for the (1) four symmetric virtual single systems, (2) six asymmetric-right virtual single systems, and (3) six asymmetric-left virtual single systems. Asymmetric virtual single systems are those for which the two wrist-pendulum systems differ in either length or mass or both. The distinction between the virtual single systems' categories (2) and (3) is based on the wrist-pendulum system with the lower value of $(M^{1/16}L^{1/2})$. The use of $(M^{1/16}L^{1/2})$ as a measure for classifying systems is motivated by its relationship

TABLE 8.2
Scaling of Period (τ_o) Against the Moment Variable ($M^{1/16}L^{1/2}$) For Symmetric and Asymmetric Virtual Single Systems

Subject	*Category*	*Correlation Coefficient*	*Slope*	*Intercept**
1	Symmetric	.986	1.091	.279
	Asymmetric Right	.997	1.010	.259
	Asymmetric Left	.988	.962	.233
2	Symmetric	.993	.770	.222
	Asymmetric Right	.965	.713	.194
	Asymmetric Left	.993	.817	.239
3	Symmetric	.988	.824	.223
	Asymmetric Right	.969	.874	.233
	Asymmetric Left	.953	.790	.203
4	Symmetric	.998	.944	.293
	Asymmetric Right	.964	.855	.298
	Asymmetric Left	.961	.996	.319

*in log (base 10).

to the natural period of the wrist pendulum system in isolation. Asymmetric-left and asymmetric-right classify the coupled wrist-pendulum systems according to the qualitative categories of "high" and "low" natural frequencies relative to a left/right symmetry. Asymmetric-left is a class of oscillators with the "higher" frequency oscillator (of the pair) always in the left hand and asymmetric-right is a class of wrist-pendulum systems with the "higher" frequency oscillator (of the pair) always in the right hand.

To the three system categories identified in Table 8.2 can be added the right single and left single systems whose regression parameters were reported in Table 8.1. These five systems can be thought of as five relatively distinct evaluations of the periodic timing law per subject. Each system type identifies a different setting or coordinate (mesh) system for the law. The question can be raised, therefore, of whether the intercepts and slopes change independently from one coordinate (mesh) system to another, reflecting independently changing potentials.

The trajectories of data points represented in Table 8.1 and 8.2 lie for the most part in the quadrant comprised of positive $log\ \tau_o$ values and negative $log\ (M^{1/16}L^{1/2})$ values. If the mass, length, and time magnitudes are expressed in grams, centimeters, and milliseconds, then all pairs of τ_o, $M^{1/16}L^{1/2}$ values lie in the fully positive quadrant of the log by log space. In that quadrant the data are easier to visualize. After recomputing the regression equations for the data in grams, centimeters, and milliseconds the slopes and the logarithms of the intercepts for the five system types were correlated in linear coordinates for each subject. The correlations for all subjects were negative and significant at the .01 level, pointing to a systematic increase in the slope (c_i) as the logarithm of the intercept decreased. The marked linear relation for each subject signifies that the slopes and intercepts were not affected independently by system type and that the five allometric functions formed a convergent family—that is, that the five functions pass, theoretically, through a common point. These conclusions apply equally to the 20 allometric functions provided collectively by the 4 subjects. The regression of slope on the logarithm of intercept for the 20 functions yielded a correlation coefficient of -.96 and a slope of -.97.

It is possible to obtain an objective estimate of the point at which the functions ideally intersect. The argument is due to J. C. Stevens (Stevens & Rubin, 1970; Stevens, 1974). For any pair of intersecting functions $\tau_o = a_1(M^{1/16}L^{1/2})^{c_1}$ and $\tau_o = a_2(M^{1/16}L^{1/2})^{c_2}$ it follows from the equality of τ_o at the intersection point that

$$a_1(M^{1/16}L^{1/2})^{c_1} = \tau_o = a_2(M^{1/16}L^{1/2})^{c_2}. \quad (8.36)$$

On arranging terms the latter equation becomes

$$(\log a_1 - \log a_2)/(c_2 - c_1) = \log(M^{1/16}L^{1/2}). \quad (8.37)$$

Equation (8.37) means that the slope of the regression of c_i on *log* a_i, when inverted and multiplied by *-1*, gives a quantity whose antilogarithm is the magnitude of the value of $M^{1/16}L^{1/2}$ at which the functions intersect. This value ($gm^{1/16}cm^{1/2}$) proved to be 9.775, 9.895, 9.030, and 11.738, for Subjects 1 through 4 respectively and was equal to 10.877 when all 20 functions were considered together.

When Equation (8.37) is applied to psychophysical data, the estimated value of the independent variable at which a family of functions converge can sometimes be rationalized. For example, it is often found to be the value of a very strong stimulus level that approximates the pain threshold and which possibly represents the ceiling level for the modality (Stevens, 1974). Is the convergence value of $M^{1/16}L^{1/2}$ for the present data rationalizable? Because the functions distinguished by hand, symmetry, and subject converge theoretically on a common value (10.877), it could be argued that this value represents a quantity that is indifferent to coordinate (mesh) system variation. That is to say, the convergence value might define a critical quantity whose component quantities are *coordinate independent scalars* (see Section 9.6 for a discussion of such properties and their relationship to intrinsic coordinate space geometries). Tempering this hope is Gould's (1966; White & Gould, 1965) observation that points of convergence in allometry can be changed by arbitrary changes in the unit of measure; he is, therefore, skeptical that slope-intercept correlations can have any biological meaning. Bearing this skepticism in mind, the other coordinate of the intersection point was determined. The regression of $M^{1/16}L^{1/2}$ on τ_o was computed for each of the five system types for each of the four subjects. The slopes of these functions were then correlated with the logarithmic value of their intercepts and the value of τ_o on which the functions converged determined according to Equation (8.37), intersubstituting $M^{1/16}L^{1/2}$ and τ_o. For the 20 functions the correlation coeficient was -.997 and the theoretical convergence value of τ_o was 1.124*s*. For the individual subjects (in order), the functions converged on τ_os of 1.119, 1.168, 1.039, and 1.282*s*. In like fashion the regression parameters of $\tau_o/M^{1/16}$ against *L* for the five system types were cast into a correlational analysis for each subject separately and for all subjects together. The collective correlation coefficient was 0.95 and the theoretical *L* convergence value was 45.5*cm*. For the individual subjects (in order), the convergence *L*s were 42.5*cm*, 37.8*cm*, 33.2*cm*, and 52.4*cm*. Given the theoretical convergence values of $M^{1/16}L^{1/2}$ and *L* it was possible to calculate the theoretical convergence value of *M*. Collectively this value was 2,076*gm* and for the four subjects (in order) the convergence *M*s were 650*gm*, 2,011*gm*, 1,323*gm*, and 2,304*gm*.

The foregoing analysis suggests a general convergence on the triplet 2,076*gm*, 45.5*cm*, 1.124*s*. It is not obvious, however, that these quantities are critical in any significant physical biological sense. What is obvious, however, is the coherence among the independent and dependent quantities across tasks (system types) and subjects. All the observed allometric relations are members of a single family. What they have in common is their lawful form. Where they differ is in terms of an apparently continuous coordinate-space transformation. The transformation and its continuity are revealed in the linearly correlated changes of the coefficients and the exponents. These points are brought to focus in the subsection that follows.

8.6.3. A Renormalization. The a_i and c_i metrics identified in Table 8.1 were used to renormalize the coordinate spaces so as to evaluate the scaling law. As previously noted, renormalization should reveal that a single scaling law, $T \propto (M^{1/16}L^{1/2})$, links natural period to mass and length across subjects, hands, and tasks (single or double system). For each subject the natural periods of the 24 conditions—4 right singles, 4 left singles, and 16 virtual singles—were computed from the allometric equation $\tau_o = a_i(M^{1/16}L^{1/2})^{C_I}$. The a_i, c_i parameters from the right single-systems' data were used for the right singles computations; the a_i, c_i parameters from the left single-systems' data were used for the left singles computations; and the a_i, c_i parameters for the virtual single systems' data were used for the virtual single systems' computations. If the law $T \propto (M^{1/16}L^{1/2})$ holds for all coordinate spaces, then the relation of the $[24 \times 4]$ observed natural periods to the $[24 \times 4]$ predicted natural periods should be a single line of slope one and intercept zero in *linear* coordinates.

This further analysis is required, in large part, because the strength of an individual allometric regression is not always sufficient to determine how lawlike is the bivariate relation (Smith, 1980). The correlations and regressions reported in Table 8.1 and Figure 8.5 need to be complemented by an examination of the residuals. The question is: How well do the allometric regressions *predict*? Following Smith (1980) the percentage prediction error

$$[\text{modulus (observed} - \text{predicted)/predicted}] \times 100$$

was computed for right singles, left singles, and virtual singles. Averaged over the four subjects these prediction errors amounted to 1.58%, 1.61%, and 2.78%, respectively. For right singles the range was 0% to 3.88%; for left singles it was 0.55% to 3.52%; and for virtual singles it was .08% to 8.57% percent.

Figure 8.6 plots the observed periods against the predicted periods in linear coordinates. The intercept and slope of the computed regression very closely approximate zero and one, respectively. There are two conclusions to be drawn. First, there was a single common basis for the periodic timing behavior of the single wrist-pendulum systems and the virtual single wrist-pendulum systems of all four subjects. (The periodic timing law is invariant over coordinate (mesh) systems.) Second, the residuals were not of constant size. The periods of some virtual single systems were predicted less well than others. This conclusion is taken up in the chapter's Postscript (Section 8.10).

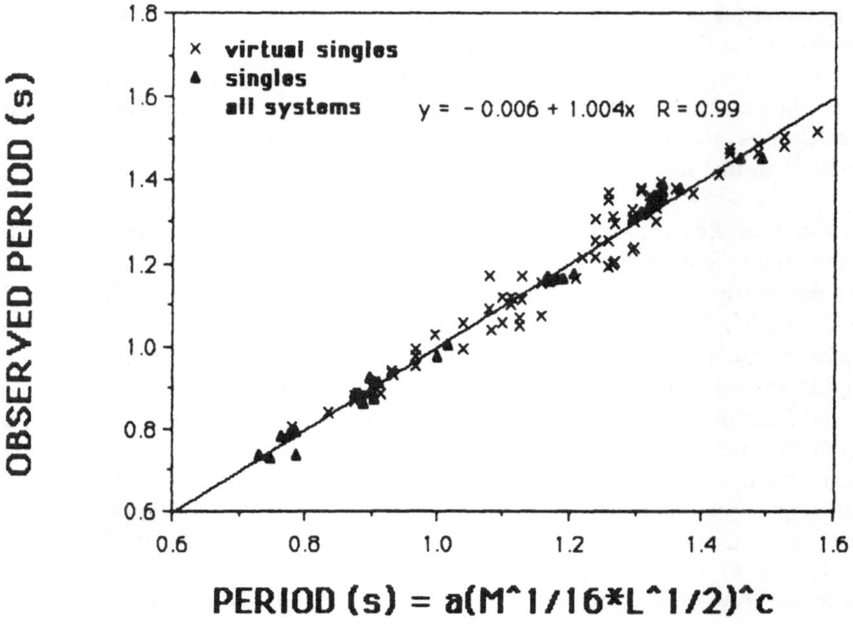

FIGURE 8.6. Renormalized data plotted in linear coordinates of predicted period versus the observed period for the single and virtual single systems of all four subjects.

8.7. The Locomotory Gaits: A Scalar Potential Difference Based on Integer Multiples of Gravity?

It is evident that the subjects in the experiment reported in Chapter 6 followed closely the relation between length and period dictated by pendular motion in a gravitational field. That is to say, the coherence of the periodic timing over conditions and subjects is attributable largely (but not exclusively) to the constancy of acceleration due to gravity. This constancy in the gravitational field intensity at the terrestrial level would also seem to be an important basis for the coherence in locomotory periodic times over gaits and species.

Although there are no universally agreed upon criteria for parsing gaits, those advanced by Pennycuick (1975) have a certain thermodynamic appeal. A gait, for any given animal, is a continuous series of stepping patterns. If, on gaining speed, an animal can move smoothly from one pattern of leg movements to another then the two patterns are viewed as instances of the same gait. Thus, very slow and very fast quadruped walking entail differences in phase and kinematic detail, but both are regarded as examples of a single gait given the continuous transition from one to the other through intermediary patterns. In comparison, where one movement pattern changes to another discontinuously, without intermediates, the two patterns are instances of different gaits. Only three gaits are distinguished in this fashion: the walk, the trot, and the canter. Roughly and approximately, the walk is an asymmetrical gait (limbs of the same girdle out of phase) in which the body is never unsupported (there is always one limb in contact with the ground); the trot is a symmetrical gait in which diagonal pairs of legs move together; and the canter is an asymmetric gait in which the animal passes its weight around from one leg to another, using the four legs in succession. (The gallop in its various forms is reached from the canter by a continuous transformation and so is regarded as a fast canter rather than as a separate gait.) It is worth recalling (from Section 7.2.1) that these three gaits, when exhibited by the horse, are associated with distinct metabolism-by-speed profiles (Figure 7.1).

Pennycuick (1975) measured the stepping frequencies of adult members of 14 different mammal species (and the stepping frequency of the young of one species, the Gnu) walking, trotting, and cantering freely in the Serengeti. Table 8.3 reports the species and their mean periodic times (unparenthesized) as a function of gait. (Note that not all animals use all three gaits.) The length measure reported by Pennycuick (1975) was shoulder height (the height above the ground of the top of the back over the shoulder). In Table 8.4 this measure is replaced by the length of the virtual single system equivalent of the four limb-pendulum systems, under the assumption that the four limbs are of approximately the same mass and length. On the basis of length and mass measures of an adult Gnu (see Pennycuick, 1975) and suggestions for computing the center of mass of a limb (Williams & Lisner, 1977) the equivalent simple pendulum length was calculated (see Chapter 6) to be approximately .62 times the shoulder height. This proportionality factor was applied to the mean shoulder height of each species identified in Table 8.3 to derive for each species a virtual single leg-pendulum system length. The mass of a virtual single leg-pendulum was taken as .35 times the body mass, given that the legs contribute roughly between 30% and 40% of an animal's body mass.

Regression analyses in double log coordinates revealed that for all three gaits periodic times scaled to mass raised approximately to the 1/8 power (.16 for walking, .13 for trotting, and .12 for cantering) and to length raised approximately to the 1/2 power (.49 for walking, .48 for trotting, and .44 for cantering). These exponents are expected of a vibratory system of masses, springs, and levers representing an elastically similar animal (McMahon, 1975, 1977). Because of the interdependency of biological mass and biological length, Equation (8.20), the moment variable governing the gait periodicities is,equivalently, either $(M^{1/8}L^{0})$ or $(M^{0}L^{1/2})$. Figure 8.7 plots the regression of periodic times on this moment variable for each of the three gaits.

As animals grade up from a slower to a faster gait, the quantity of chemical energy irreversibly liberated as mechanical energy increases. It is a scalar transformation, however, that leaves invariant the relation of periodic time to the monomial in mass and length as can be seen from inspection of Figure 8.7. The

TABLE 8.3
Observed and predicted (in parentheses) periodic Time (*s*) for Walking, Trotting, and Cantering as a Function of Species' Mass (*kg*) and Length (*m*)

			Gait		
Species	*Mass**	*Limb Length (Virtual)*	*Walk*	*Trot*	*Canter*
Thompson's gazelle	20	.372	.800 (.865)	.426 (.463)	.385 (.387)
Warthog	85	.391	.784 (.887)	.439 (.478)	
Gnu (calf)	18	.471	.909 (.974)		.450 (.436)
Spotted Hyena	65	.484	.930 (.987)		.476 (.441)
Grant's gazelle	62	.465	.995 (.968)	.588 (.517)	.408 (.433)
Impala	50	.508	1.053 (1.013)		.465 (.453)
Lion	150	.546	1,212 (1.048)	.617 (.561)	
Kongoni	170	.670	1.190 (1.163)		.541 (.520)
Topi	130	.732	1.176 (1.215)		.526 (.543)
Zebra	350	.806	1.111 (1.274)	.645 (.681)	.450 (.570)
Gnu (adult)	300	.856	1.143 (1.314)		.500 (.587)
Buffalo	750	.868		.667 (.707)	.588 (.592)
Black Rhinoceros	1,400	1.748	1.538 (1.877)		
Giraffe	1,000	1.662	2.083 (1.830)		.833 (.818)
Elephant	6,000	2.012	1.905 (2.012)		

*Mass is the average mass of the species as reported in Garland (1983).

question arises, therefore, of what is constraining this quantity such that the proportional change from the walk to the trot to the canter is the same for all of the species in Table 8.3. It would appear that not all periodic times count as preferred steady states (Section 7.2). More to the point, not all periodic times are accessible. The spectrum of preferred (stable) gait periodicities is discontinuous like the spectrum of gaits itself. Are there other *spectral lines* in addition to those depicted in Figure 8.7? Data provided by Alexander and Moloiy (1984) (and by

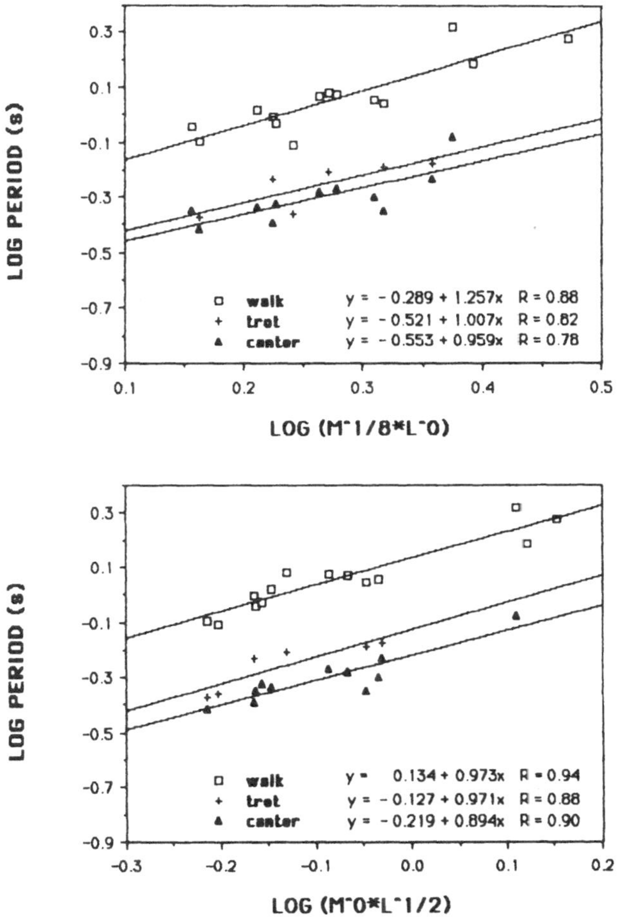

FIGURE 8.7. Regressions of log period on log ($M^{1/8}L^0$) and log ($M^0L^{1/2}$) for the three quadruped gaits.

Alexander, personal communication) suggest that where primates canter on all fours, their cantering does not relate to representative quantities of mass and length in the same way as the cantering of quadrupeds. Figure 8.8 shows this additional regression of τ_o on ($M^{1/8}L^0$) and ($M^0L^{1/2}$) based on three primates (bushbaby, vervet monkey, Papio) who canter regularly and for which all the requisite measures are available.

Talk of spectra evokes the related notions of spectral series (such as those that characterize atomic structures) and multiple normal modes (such as those that characterize vibratory systems of multiple degrees of freedom). It is of interest

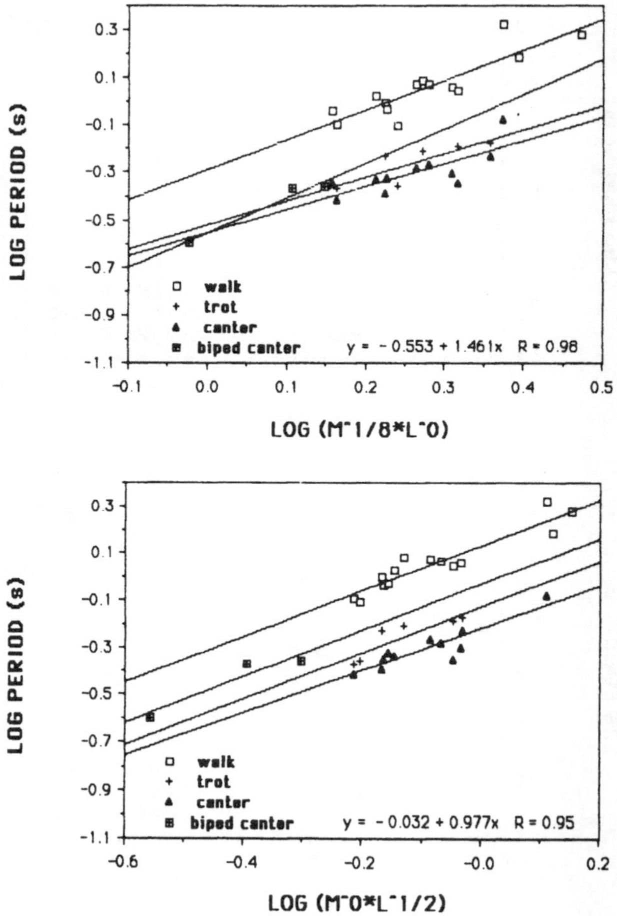

FIGURE 8.8. Regressions of log period on log $(M^{1/8}L^0)$ and log $(M^0L^{1/2})$ for the three quadruped gaits and for bipedal cantering.

to consider whether there is a principle of quantization governing the gait spectrum analogous to the principle of quantization governing the atomic spectra. That is, do the gait periodicities form an integer series? And if they do, what constant is entering into these integer multiples? The most likely candidate is the constant of acceleration due to gravity.

Any length L' and the period τ' that corresponds to it by the gravitational law for simple pendulums give a pair of units in terms of which a "gravitational" coordinate space can be defined. Let $L' = 20cm$ and $\tau' = 898ms$. Then lengths and times of pendulum systems can be expressed as π-*numbers*: $\pi_L = L/20$ and

$\pi_\tau = \tau/898$, respectively. For simple physically ordinary pendulums $\pi_\tau = \sqrt{\pi_L}$. In $log\ \pi_\tau$ by $log\ \pi_L$ coordinates the periodic times of physically ordinary simple pendulums fall on a line passing through the origin ($\pi_L = 1$, $\pi_\tau = 1$) at an angle of 26.57 degrees. Plotting the walking, trotting, and cantering data of Table 8.3 in this $log\ \pi_L$ by $log\ \pi_\tau$ space yields three straight line trajectories at 24.76 degrees, 25.82, degrees and 24.09 degrees, respectively.

In the ideal case the three trajectories would parallel the trajectory through the origin and for each the correlation of π_τ and π_L would be 1.0. In the absence of perfect parallelism and perfect correlation, the point on a trajectory closest to the origin can be taken as the point giving the π_L and π_τ values whose relation is most representative of the relation binding time and length for that gait. The calculated coordinates for the "closest points" were (1.134, .762) for walking, (1.459, .458) for trotting, and (1.517, .394) for cantering. These coordinates would be closely, approximated in any π_L by π_τ space (i.e., whatever the choice of unit length and period). The coordinates of the "closest points" were converted back into lengths and times and each coordinate pair was cast into the following form of the simple pendulum equation:

$$\tau = 2\pi\sqrt{L/ng}\ . \tag{8.38}$$

For the physically ordinary simple pendulum $n=1$. For walking, trotting, and cantering the value of n was found to be 1.95, 6.96, and 9.73, respectively. For purposes of comparison, using another unit length and unit period, namely, 34.286*cm* and 1176*ms* (where period is conveniently the square of length) the n values were 2.13, 7.08, and 10.08. This analysis might be read as suggesting that the gaits are organized around different scalar multiples of the gravitational field intensity at the ecological scale. It might also be read as suggesting that the quadruped gaits are organized around integer multiples (2, 7, and 10) of the gravitational field intensity. With respect to the canter of bipeds, the analysis suggests that this particular gait is organized around gravity multiplied by 5. It would appear, in sum, that the gravitational field intensity is quantized, in some sense, by the mammalian gaits.

The latter reading must be accepted with reservations (included among these reservations is the approximate nature of the data that gave rise to the integer numbers). But with these reservations in mind an integer based equation for gait periods is presented and evaluated. Its inspiration is Balmer's simple formula for atomic spectral series, namely,

$$\lambda = \lambda_o n^2/\ (n^2-4)\ , \tag{8.39}$$

where λ is wavelength, λ_o is a constant and n is a variable integer.

Consider an animal characterizable as a virtual single limb-pendulum of length L_v. A periodic time can be assigned to this length through the simple pendulum equation for lg. This is the periodic time the virtual limb would exhibit if it swung freely from the "shoulder" end solely under the influence of g (multiplied by l). Let this periodic time be designated as τ_l. Then the stable periodic times (or attractor states or normal modes) for a terrestrial mammal's locomotions are given by:

$$\tau = \frac{\tau_l}{\sqrt{n}}, \tag{8.40}$$

where n is one of a restricted number of integers and τ_l is constant for an animal (set by the equivalent simple pendulum length of its limbs) but variable across animals. Written in terms of angular frequency (ω), Equation (8.40) becomes:

$$\omega = \omega_l \sqrt{n} \tag{8.41}$$

To evaluate Equation (8.40) τ_l was determined for each species represented in Figure 8.8 using Equation (8.38) with $n = l$, and the period of each of the gaits exhibited by a species predicted from (8.40) with n taking on the values of 2, 5, 7, and 10 for the quadruped walk, the biped canter, the quadruped trot, and the quadruped canter, respectively. (The predicted periods for the quadruped gaits are the parenthesized quantities in Table 8.3.) A correlation in linear coordinates between observed and predicted periods yielded a correlation coefficient of .975, an intercept of .024, and a slope of .944 (see Figure 8.9). The intercept's small displacement from the origin must reflect, in part, the inaccuracies in estimating the lengths of the virtual single limbs. The very close proximity of the slope to unity and an r^2 of .95, however, point to the reasonableness of the estimates of the integer numbers (the scalar multiples of g) and of the formula for the gait spectral series (i.e., Equation (8.40)). The preferred periods of the gaits represent normal modes. The four limbs oscillate at approximately the same frequency for most of the performances of a given gait (see von Holst, 1935/1973) and pass through their equilibrium points at approximately the same time. The gait frequencies are multiples of the lowest mode frequency. Can anything more be said? For example, is it fair to say that the spectrum of gait periods is an expression of the quantizing of a constant (gravity) at the ecological scale? An affirmative answer poses, in turn, the question of whether or not quantizing is a general physical strategy providing for any system at any scale a series of discrete periodicities at which system stability is ensured.

8.8. Manipulating the Geometry of the Coordinate Space: (Subsidiary Experiment 1)

Section 8.7 noted that changes in gait can be associated with changes in the metrical coeficients of the measurement space, rather than changes in the timing

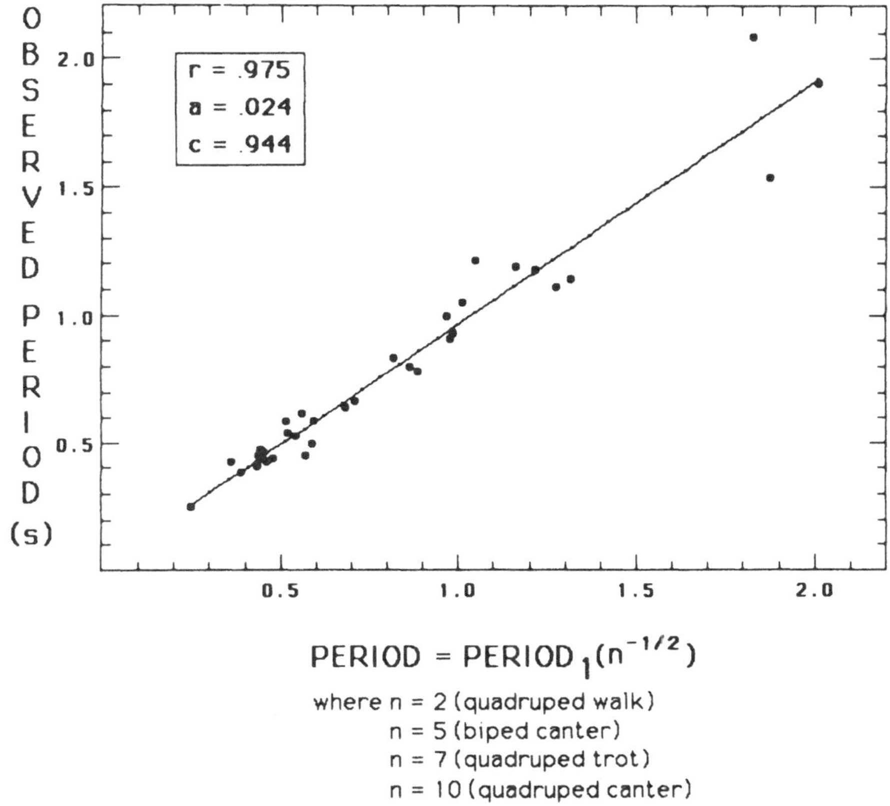

FIGURE 8.9. Fit in linear coordinates between period predicted from Equation (8.40) and observed period for the three quadruped gaits and the biped canter.

law. Qualitative changes associated with gait transitions can be further associated with quantized changes in the scalar metric a_i of the measurement space. While the exact value of the a_i coefficient can vary from system to system, the ratios of a_i's associated with different gaits remain invariant across systems.

Figure 8.10 plots the preferred periods (τ) for each subject in the main experiment against the moment variable ($M^{1/16}L^{1/2}$). Linear regressions are performed on data sets consisting of the 8 single systems and the 16 virtual single systems. While the intercept of Subject 4 differed markedly from the intercepts of the other three subjects, the slopes were closely similar for all four subjects. In the terms of Chapter 2, it is as if the mesh system to which Subject 4's periodic times and ($M^{1/16}L^{1/2}$) quantities were referred was stretched uniformly in the direction of the *y-axis* relative to the mesh-systems of Subjects 1, 2, and 3.

By the preceding, locomoting faster in the comfort mode—the increase in

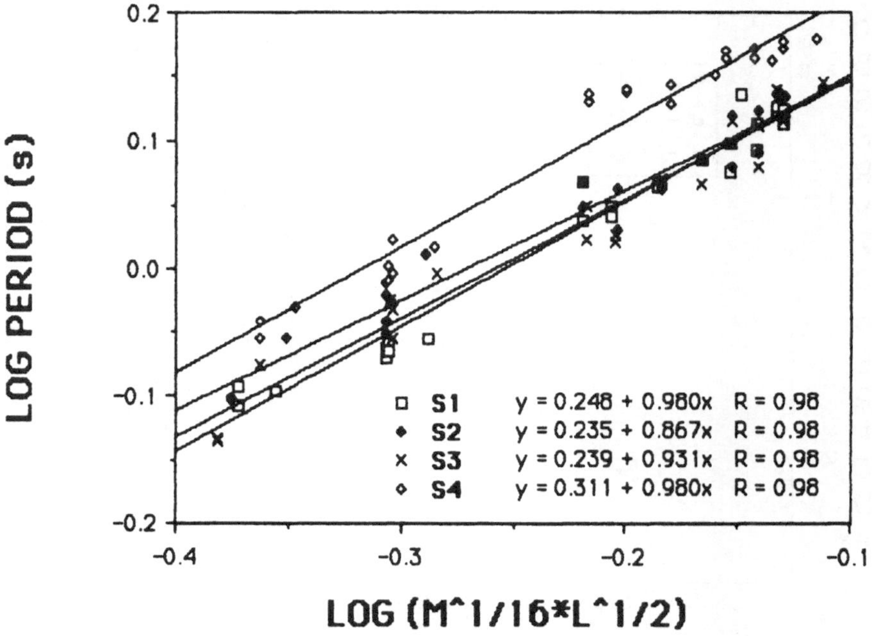

FIGURE 8.10. Regressions of log natural period on log ($M^{1/16}L^{1/2}$) for the twenty four systems of each of the four subjects.

speed accompanying the transition from walk to trot to canter—amounts to a compression in the vertical dimension of the mesh-system to which periodic time and the moment variable of locomotion are referred.

Changes in locomotory speed under conditions of constant gravity, mass and limb length involve changes in internal parameters (such as stiffness, resting length, etc.). A wrist-pendulum situation was created to analyze the timing law ($\tau \propto M^{1/16}L^{1/2}$) under conditions involving variations of internal parameters. The experimental set-up required that subjects squeeze a dynamometer at designated levels with one wrist-joint complex while performing the preferred oscillation task with the other wrist-joint complex (see Figure 8.11). Subjects were required to squeeze the dynamometer at one of three proportions (0, .25, .50) of their maximum strength for 10 seconds while swinging a weighted pendulum with the other hand at the most comfortable period. The external parameters (mass and length of pendulum and gravity) were held constant. The squeezing activities of the nonoscillating hand were used to induce internal parameter transformations. The purpose of this experiment was to examine the effects of internal parameter variations on the structure of the timing law.

The maximum squeeze of the right hand and that of the left hand were determined for each subject in a series of tests conducted prior to the experiment

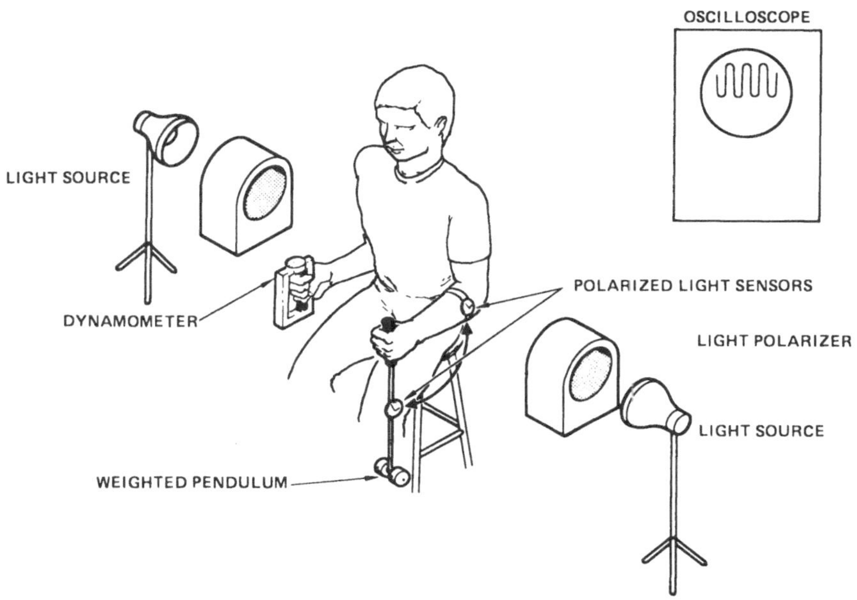

FIGURE 8.11. EXPERIMENTAL SET-UP FOR SQUEEZE EXPERIMENT. The subject was required to maintain a dynamometer at an experimenter designated percent of maximum capacity. While maintaining the squeeze level constant at the prescribed level, the subject was required to swing a pendulum in the other wrist-joint complex at its most "comfortable" frequency of oscillation.

proper. None of the four subjects (all males) had participated in the main experiment. The single wrist-pendulum systems were the same as those described for the main experiments (namely, A, B, C, D, see Section 6.3.1), as were the instructions to the subject. The natural period was determined on eight separate occasions for each wrist-pendulum system for each hand as a function of each of the squeeze conditions. The four right and the four left wrist-pendulum systems were randomized within each block of eight trials as were the three squeeze conditions. There were eight blocks of trials.

Table 8.4 reports for each subject and each hand the periodic time (averaged over the eight trials) associated with each of the four systems (A, B, C, D) for each of the three dynamometer readings corresponding to 0% (no exertion) 25% and 50% of the subject's maximum exertion. Table 8.4 identifies that wrist-pendulum frequency increased (period shortened) as the muscular exertion of the other hand increased. The change in preferred frequency is viewed as a consequence of change in an internal parameter (see Chapters 10, 11, and 12 for

TABLE 8.4
Periodic Times for Single Wrist-Pendulum Systems as a Function of Hand and Proportion of Maximum Force Exerted by the Other Hand

		*Right Hand SYSTEM**				*Left Hand SYSTEM*			
Subject	*Exertion*	*A*	*B*	*C*	*D*	*A*	*B*	*C*	*D*
	0	.850	.887	1.242	1.307	.771	.865	1.149	1.311
1	.25	.871	.932	1.193	1.270	.779	.875	1.139	1.280
	.50	.847	.925	1.141	1.287	.805	.885	1.124	1.289
	0	.481	.716	.974	1.254	.569	.705	1.077	1.335
2	.25	.411	.654	.832	1.067	.544	.759	.938	1.214
	.50	.576	.661	.753	.982	.633	.669	.932	1.162
	0	.646	.737	.957	1.113	.584	.661	.829	1.045
3	.25	.599	.662	.840	.932	.562	.601	.770	.907
	.50	.588	.716	.825	.954	.472	.541	.707	.837
	0	.679	.785	1.065	1.295	.649	.753	.983	1.282
4	.25	.652	.787	.960	1.141	.601	.722	.944	1.138
	.50	.602	.724	.907	1.191	.614	.710	.843	1.134

*From Section 6.3.1, system refers to the particular combination of rod length (*m*) and added mass (*Kg*). A = .294/0.1, B = .599/0.1, C = .294/1.0, D = .599/1.0.

a discussion of the thermodynamical and mechanical aspects of internally induced transformations).

The masses and lengths of the simple pendulum equivalents of each wrist pendulum system for each subject were calculated in the manner detailed in Chapter 6. The periodic times for a given exertion level were averaged over a subject's left and right wrist-pendulum systems. The regression of these periods on $(M^{1/16}L^{1/2})$ are presented in Figure 8.12. For all of the regression lines the two quantities correlated at the .05 level or greater. *The timing law* $\tau \propto M^{1/16}L^{1/2}$ *was preserved structurally invariant under the "squeezing transformation"*. To identify what properties of the measurement space were effected by the transformation a multiple regression analysis was performed individually on each subject's triplet of lines. The analysis proceeded as follows. First, it examined whether there was any difference among the three lines, that is, did period change as a function of exertion? Second, if a difference was found then the three lines were compared pairwise to see whether exertion level and moment variable interacted. Third, if no interaction was found for a given pair of lines then the lines were compared for a difference due to exertion level. The second and third steps determined, respectively, whether the exertion manipulation resulted in a change in the slope or a change in the intercept of the linear relation (in double

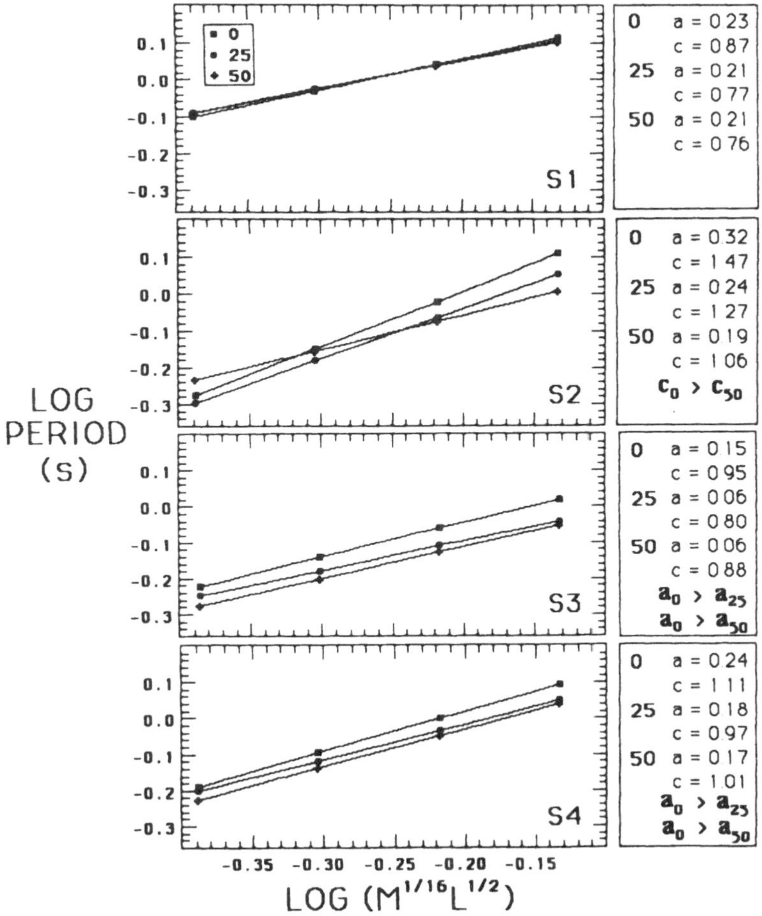

FIGURE 8.12. Regressions of log period on log $(M^{1/16}L^{1/2})$ as a function of proportion of maximum squeeze for each subject. Regressed data are averaged left and right periods. Points on regression lines are period values as predicted by the regression equations.

log coordinates) between peridic time and $(M^{1/16}L^{1/2})$. As noted above, the slope and intercept are physically viewed as indicies of external (such as gravity) or internal potentials (such as elasticity) that populate the biomechanical work space.

No differences among the lines was found for Subject 1, $F(4, 18) = 0.403$. The regression lines representing Subject 2's periodic times under the three levels of exertion were significantly different, $F(4, 18) = 2.961$, $p<.05$. The slope for the 50% exertion line proved to be significantly different from that of the 0% exertion line ($\rho<.02$); and there was a suggestion of a difference between the intercepts of the 25% and the 0% exertion lines ($\rho<.07$). Subject 3's regression lines differed significantly, $F(4, 18) = 3.56$, $\rho<.03$. The difference was in the intercepts. The 0% exertion intercept was higher than both the 25% exertion ($\rho<.004$) and the 50% exertion ($\rho<.006$) intercepts. For Subject 4 it was similarly the case that the regression lines differed $F(4, 18) = 4.351$, $\rho<.01$ with the difference attributable to the 0% intercept being larger than the 25% ($\rho<.01$) and the 50% ($\rho<.003$) intercepts. The slope and intercept values for each subject's regression lines are given in Figure 8.12 together with a summary of the significant slope and intercept differences.

The task of squeezing a dynamometer in one hand while producing pendular clocking behavior in the other hand brought about a decrease in the (comfort mode) periods for three of the four subjects. This result of contralateral muscular exertion, however, did not affect the relative order and position of the coordinate pairs (τ_o, $M^{1/16}L^{1/2}$). The timing law is preserved invariant under the squeezing transformation; the effect of the transformation is revealed only in modifications of the slope and intercept values (see Figure 8.12).

8.9. Comparing "Gaits": An Experimental Evaluation (Subsidiary Experiment 2)

In the quadruped gaits of walking and trotting limbs of the same girdle—shoulder or pelvic—are 180 degrees out of phase. In fast cantering, limbs of the same girdle are in phase. Do the periodic times of these out of phase and in phase gaits obey the same scaling law for timing? Pennycuick's (1975) observations of animals in the wild (Figure 8.7) provide initial evidence in support of this claim. The notion of virtual systems further suggests that the timing law should be invariant over in-phase and out-of-phase constraints. A second subsidiary experiment was designed to examine the effect of phase relation on the timing law. In this experiment subjects were required to swing two pendulums either in phase or 180 degrees out of phase while identifying the most comfortable period of oscillation for the virtual system. The experiment consisted of 12 conditions. Conditions 1-6 involved virtual systems 1, 2, and 3. Conditions 7-12 involved virtual systems 4, 5, and 6 (see Table 8.5). Four subject were tested on the 12 conditions. The subjects swung the pendulums in phase in the odd numbered conditions and out of phase in the even numbered conditions. (None of the subjects—all males—had taken part in either the main experiment or the first subsidiary experiment.) Each subject experienced each condition 8 times. On each trial in each condition the subject's task was to swing the two pendulums rhythmically, at a common comfortable tempo. The 96 trials were divided into 6 blocks of 16 trials each with each block comprised of 4 conditions repeated 4 times. The four conditions were made up of two in phase conditions (e.g., 1 and 2) and their out of phase equivalents (7 and 8). Within a block the trials were

TABLE 8.5
Composition of Symmetric and Asymmetric Systems in the Second Subsidiary Experiment

Virtual System Type		*rod length/added mass* Right Hand	*rod length/added mass* Left Hand
	1	.447m/.25kg	.447m/.25kg
SYMMETRIC	2	.447m/.50kg	.447m/.50kg
	3	.447m/1.0kg	.447m/1.0kg
	4	.294m/1.0kg	.599m/.10kg
ASYMMETRIC	5	.294m/.10kg	.599m/.10kg
	6	.599m/1.0kg	.294m/.10kg

ordered randomly. And across subjects the conditions were so arranged that each in phase/out of phase pair appeared equally often in each block.

The simple pendulum equivalents for the left and right wrist-pendulum systems and the resultant virtual single system magnitudes (in kilograms and meters) were computed in the manner described in Chapter 6. A three way analysis of variance with factors of Symmetry (see Table 8.5), Virtual System and Phase Relation was conducted on the mean periodic times. Symmetry was significant reflecting the fact that the $M^{1/16}L^{1/2}$ quantity was greater, on the average, for the asymmetric than the symmetric systems, $F(1, 3) = 11.94$, $p<.05$. Within the symmetric/asymmetric classification, virtual systems differed significantly, $F(1, 3) = 21.14$, $p<.02$, consistent with the differences in their moment variable magnitudes. There was no effect of Phase Relation, $F(1, 3) = 1.08$, $p>.05$, and there was no interaction between Symmetry and Phase Relation, $F(1, 3) = .53$.

Figure 8.13 presents for each subject the allometric regressions of periodic time on the moment variable for in phase and out of phase coordination. (The bivariate relation was significant at the .01 level in all cases.) For two of the subjects (2 and 3) the exponent of the regression equation was closer to two. Subjects 2 and 3 exhibit steeper slopes than were seen in the data of subjects in the main experiment and the first subsidiary experiment. Their steeper slopes, however, are accompanied by higher intercepts. The implication is that the trajectories of Subjects 2 and 3 are related to those of the other subjects through a coordinate transform, not a transformation in the structure of the timing law.

The preceding experiment (see Figure 8.13) suggests that the periodic timing law, when defined over two wrist-pendulum systems, is preserved invariant under transformations induced by phase relations. The same characteristic is observed for the pendular, clocking mode defined over four limbs (Figure 8.7). In both situations the comfort period is defined in terms of global (virtual single) mass and length properties that are derived from two or more locally distributed masses and lengths. The conclusion to be drawn from the locomotory data and

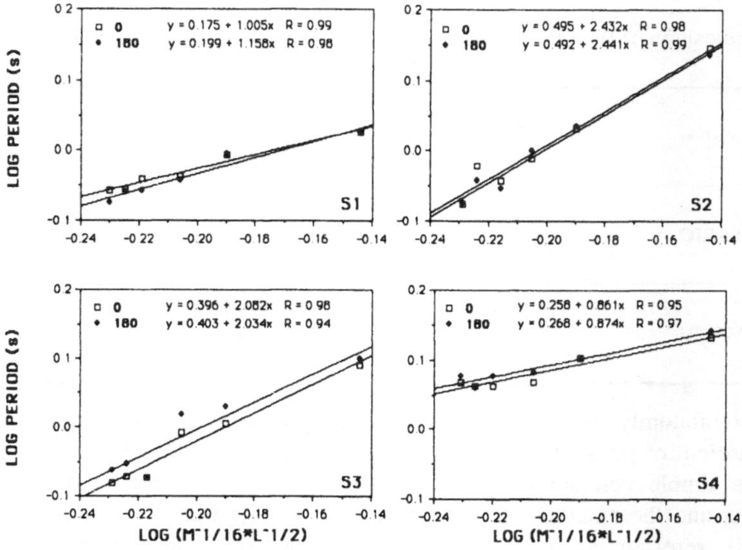

FIGURE 8.13. Regressions of log period on log ($M^{1/16}L^{1/2}$) as a function of phase relation (0 degrees, 180 degrees) for each subject.

the two subsidiary experiments is that steady state (comfort) periodic timing—in wrist-pendular behavior and locomotion—is lawfully linked to single quantities of mass and length and that this linkage is structurally stable under (a) changes in the preferred periodicities of the local components (Experiment 1) and (b) changes in the phase relations defined between local oscillatory components (Experiment 2).

8.10. A Postscript on the Experimental Analysis

It will be recalled from Chapter 6 that the transformation of a compound pendulum to a simple pendulum is anchored in the conservation of energy. This conservation-based transformation has two outstanding features. First, it is an *alternative description* that requires little detail. The specifics of a compound pendulum's volume or shape or girth, for example, are not retained in the simple pendulum description. Second, it is a symmetry operation (see Chapter 2): The transformation of the compound pendulum into a simpler version of itself leaves invariant a characteristic behavioral property of the compound pendulum, namely, periodic timing.

When the transformation is applied to a single left or right wrist pendulum system the law is transparent to most local detail. Global timing is a function of Huygens' law which lawfully relates a single magnitude defining the distance from the axis of rotation in the wrist to a point where the total mass of the rod, added weights, and the hand is concentrated dynamically. This distance is the length of an equivalent simple pendulum. This simple pendulum, if left to oscillate freely under gravity's influence, would oscillate with the period of the wrist-pendulum system, if it, too, were left to oscillate freely, and could oscillate freely, under gravity's influence.

The detail of a double wrist-pendulum system exceeds that of either a left or a right single wrist-pendulum system. There are twice as many masses oscillating about twice as many axes at twice as many distances from these axes. There are (at least) twice as many muscles (at least the radial and ulnar flexors of the left and right wrists). There are (at least) twice as many motoneurons and (at least) twice as many interneurons. The differences in detail could be continued but enough have been identified to make the point. Under the transformation rendered by Huygens' law, the greater detail of a double wrist-pendulum system relative to a single wrist-pendulum system is made invisible. A double wrist-pendulum system is replaced, just as a single wrist pendulum system is replaced, by a simple pendulum system. The analyses of this chapter reveal that periodic timing, of either a single or a double wrist-pendulum system, is lawfully related to a single length quantity, as given by Huygens' law, and a single mass quantity, as given by the sum of the various masses.

This postscript is intended to highlight two aspects of double wrist pendulum systems that are taken up in earnest in Chapter 9. The first is contained in the preceding remarks on the dynamical equivalency of the pendular, clocking mode assembled over double and single wrist pendulum systems. As noted, double wrist-pendulum systems function as virtual single systems. The entire mass distributed over the two hands acts as if it were concentrated at a single point at a specific distance from a (virtual) axis of rotation. The fact that Huygens' law applies successfully to the double wrist-pendulum systems means that there is an equivalency, of sorts, between a compound system in which the component pendulums are linked through neural processes and a compound system in which the component pendulums are linked through mechanical forces. The equivalency can be seen in these terms: In the assembling of the pendular clocking mode over two wrist-pendulum systems neural processes simulate certain aspects of solid matter—the neural processes sustain the kinematic patterns of the local kinetics. With respect to Figure 7.3, the processes depicted in the lower panel behave as if they were the processes depicted in the upper panel.

The second aspect of double wrist-pendulum systems to be highlighted in this postscript contrasts with the first. It is the fact of dissimilarities between double systems and single systems and among double systems. To begin with, in the main experiment the slope and intercept of the trajectory of the double (virtual single) systems were notably different from the slope and intercept of the single systems' trajectory for two of the four subjects (see Subjects 2 and 3 in Figure 8.5 and Table 8.1). In approximate terms: Does this difference point to a

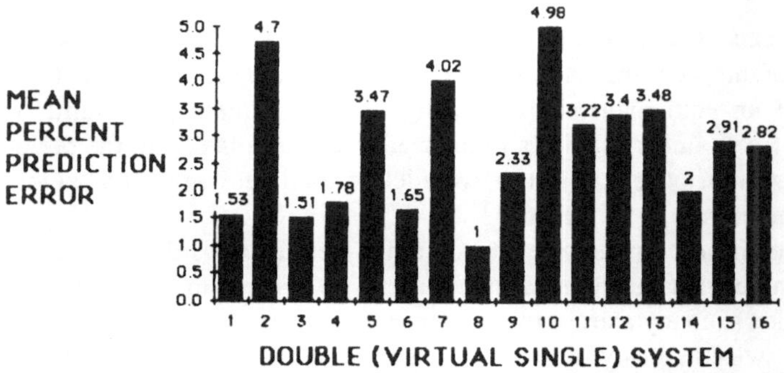

FIGURE 8.14. MEAN PERCENT PREDICTION ERROR AS A FUNCTION OF DOUBLE (VIRTUAL SINGLE) SYSTEM.

tendency for the double systems to induce, with other things being equal, a relatively larger transformation in the coordinate space housing the periodic timing law? A consideration of the first subsidiary experiment (see Figure 8.12) suggests that, with single wrist-pendulum systems, coordinate space changes of similar magnitude and direction can be induced through the presence of an internally generated (tonic) force. This force was in addition to the forces at the wrist producing the pendular, clocking behavior. By analogy, do double (virtual single) wrist-pendulum systems entail an additional force (to a greater or lesser degree depending on the system and on the person)?

The impression that double (virtual single) systems are not dynamically homogeneous as a set is suggested by Figure 8.6. The allometric regressions predict the periodic timing of some systems less well than others. The interpretation of an allometric trajectory as indexing an equilibration of forces, that is, as a succession of local potential minima, suggests that residuals reveal subtle (but possibly important) differences in the underlying force structures of systems that are largely similar. Figure 8.14 presents the mean prediction error for each of the 16 double (virtual single) systems of the main experiment using the formula of Smith (1980) identified in Section 8.6.3. If the differences exhibited in Figure 8.14 are reliable then the question is whether or not they can be tied down to physical properties of the double (virtual single) systems. Can the dynamical inhomogeneity of double (virtual single) systems be characterized in terms of their mass and length magnitudes? Can the basis for the putative force-structure differences be identified and measured?

REFERENCES

Adolph, E. F. (1949). Quantitative relations in the physiological constitution of mammals. *Science*, *109*, 579–585.

Alexander, R. M. N., & Maloiy, G. M. S. (1984). Stride lengths and stride frequencies of primates. *Journal of Zoology, London*, *202*, 577–582.

Asatryan, D. G., & Fel'dman, A. G. (1965). Functional tuning of the nervous system with control of movement or maintenance of a steady posture—I. Mechano-graphic analysis of the work of the joint on execution of a postural task. *Biophysics, 10*, 925–935.

Astrand, P., & Rodahl, K. (1970). *Textbook of work physiology*. New York: McGraw-Hill.

Bertalanffy, L. von. (1960). Principles and theory of growth. In W. W. Nowinski (Ed.), *Fundamental aspects of normal and malignant growth*. Amsterdam: Elsevier.

Bertalanffy, L. von. (1973). *General system theory*. Harmondsworth, England: Penguin.

Bizzi, E., Polit, A., & Morasso, P. (1976). Mechanisms underlying achievement of final head position. *Journal of Neurophysiology, 339*, 435–444.

Bridgman, P. W. (1922). *Dimensional analysis*. New Haven, CT: Yale University Press.

Buckingham, E. (1915). On physically similar systems. *Physical Review, 4*, 345–370.

Calder, W. A. (1984). *Size, function, and life history*. Cambridge, MA: Harvard University Press.

Economos, A. C. (1982). On the origin of biological similarity. *Journal of Theoretical Biology, 94*, 25–60.

Fel'dman, A. G. (1966a). Functional tuning of the nervous system with control of movement or maintenance of a steady posture—II. Controllable parameters of the muscle. *Biophysics, 11*, 565–578.

Fel'dman, A. G. (1966b). Functional tuning of the nervous system with control of movement or maintenance of a steady posture—III. Mechanographic analysis of execution by man of the simplest motor tasks. *Biophysics, 11*, 766–775.

Fel'dman, A. G. (1974). Control of length of the muscle. *Biophysics, 19*, 766–771.

Fel'dman, A. G., & Latash, M. L. (1982). Afferent and efferent components of joint position sense; interpretation of kinaesthetic illusions. *Biological Cybernetics, 42*, 205–214.

Gould, S. J. (1966). Allometry and size in ontogeny and phylogeny. *Biological Review, 41*, 587–640.

Gunther, B. (1975). Dimensional analysis and theory of biological similarity. *Physiology Review, 55*, 659–699.

Gunther, B., & De La Barra, B. Leon (1966). Theory of biological similarities, non-dimensional parameters and invariant numbers. *Bulletin of Mathematical Biophysics, 28*, 91–102.

Gunther, B., & Martinoya, C. (1968). Operational time and theory of biological similarities. *Journal of Theoretical Biology, 20*, 107–111.

Gunther, B., & Morgado, E. (1982). Theory of biological similarities revisited. *Journal of Theoretical Biology, 96*, 543–559.

von Holst, F. (1935/1973). *The behavioral physiology of animal and man*. Coral Gables, FL: University of Miami Press.

Huxley, J. S. (1931/1972). *Problems of relative growth*. New York: Dial Press.

Iberall, A. S. (1974). Discussion paper: Growth, form, and function in mammals. *New York Academy of Science, 231*, 71.

Kelso, J. A. S., & Holt, K. (1980). Exploring a vibratory systems account of human movement production. *Journal of Neurophysiology, 43*, 1183–1196.

Krantz, D. H., Luce, R. D., Suppes, P., & Tversky, A. (1971). *Foundations of measurement* (Vol. I). New York: Academic Press.

Landau, L. D., & Lifshitz, E. M. (1976). *Mechanics*. New York: Pergamon Press.
MacMahon, T. A. (1973). Size and shape in biology. *Science*, *179*, 1201–1204.
MacMahon, T. A. (1975). Using body size to understand the structural design of animals: Quadrupedal locomotion. *Journal of Applied Physiology*, *39(4)*, 619–627.
MacMahon, T. A. (1977). Allometry. In *McGraw-Hill yearbook of science and technology*, 48–57.
MacMahon, T. A. (1984). *Muscles, reflexes and locomotion*. Princeton, NJ: Princeton University Press.
MacMahon, T. A., & Bonner, J. T. (1983). *On size and life*. New York: Scientific American Library.
Main, I. G. (1978). *Vibrations and waves in physics*. New York: Cambridge University Press.
Pennycuick, C. (1975). On the running of the gnu (Connochaetes taurinus) and other animals. *Journal of Experimental Biology*, *63*, 775–799.
Peters, R. H. (1983). *The ecological implications of body size*. New York: Cambridge University Press.
Rashevsky, N. (1960). *Mathematical biophysics* (3rd ed.), 2 Vols. New York: Dover.
Rosen, R. (1967). *Optimality problems in biology*. New York: Plenum Press.
Rosen, R. (1978). *Fundamentals of measurement and representation of natural systems*. New York: North-Holland.
Schmidt-Nielsen, K. (1972). Energetic cost of locomotion: Swimming, running, and flying. *Science*, *177*, 222–227.
Schmidt-Nielsen, K. (1984a). Problems of scaling: Locomotion and physiological correlates. In T. J. Pedley (Ed.), *Scale effects in animal locomotion*. New York: Academic Press.
Schmidt-Nielsen, K. (1984b). *Scaling*. New York: Cambridge University Press.
Schuring, D. J. (1977). *Scale models in engineering*. New York: Pergamon Press.
Smith, R. J. (1980). Rethinking allometry. *Journal of Theoretical Biology*, *87*, 97–111.
Stahl, W. R. (1962). Dimensional analysis in mathematical biology. *Bulletin of Mathematical Physics*, *24*, 81–108.
Stevens, J. C., & Rubin, L. L. (1970). Psychophysical scales of heaviness and the size-weight illusion. *Perception and Psychophysics*, *8*, 225–230.
Stevens, J. C. (1976). Families of converging power functions. In H. R. Moskowitz, et al. (Eds.), *Sensation and measurements*. Dordrecht, Holland: D. Reidel Publishing Company.
Stevens, P. S. (1974). *Patterns of nature*. Boston: Little, Brown and Company.
Thompson, D. W. (1917/1942). *On growth and form*. London: Cambridge University Press.
Wainwright, S. A., Biggs, W. D., Currey, J. D., & Gosline, J. M. (1976). *Mechanical design of organisms*. Princeton, NJ: Princeton University Press.
White, J. F., & Gould, S. J. (1965). Interpretation of the coefficient in the allometric equation. *American Naturalist*, *99*, 5–18.
Williams, D. A., & Lissner, H. R. (1977). *Biomechanics of human motion*. New York: Saunders.

9 MEASURING THE ENERGY FOR COORDINATION

ABSTRACT

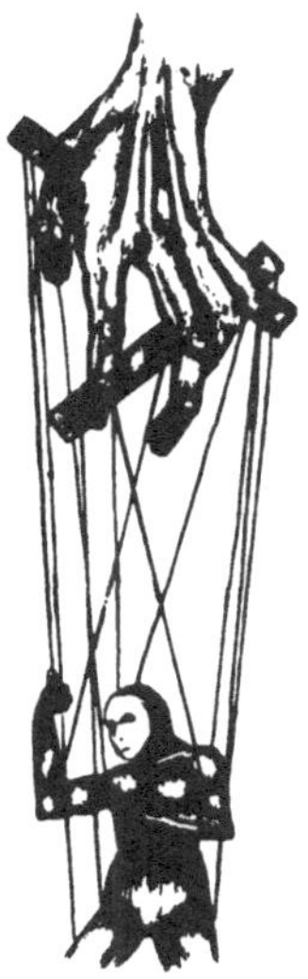

Coordination as instanced by the isochrony of a pair of limbs is brought within the province of the physics of cooperativity. A virtual single wrist-pendulum system is identified as the cooperative level of a minimal ensemble of two entities. The support for this cooperative level comes from the dynamical relation between (a) the two component wrist-pendulum systems and (b) the experimentally required relations between them. The former are viewed as atomisms operating from "below" and the latter are viewed as field boundary conditions or nonholonomic constraints operating from "above." Three aspects of this cooperative level construal of a virtual system are emphasized: It is a unitary process, signified by single quantities of mass and length (see Chapter 6), that is assembled in significant part by the conservations; it behaves like a rigidly connected, conservative field—a local perturbation is responded to by the cooperativity as a whole; and it determines, by its particular steady-state demands, the degrees of local deviations from steady state and, therefore, the interactive forces manifest at the atomistic level. The main line of empirical support for the latter point identifies a candidate dimensionless number that expresses the mismatch between the equilibrium states of the cooperative and atomistic levels. It is proposed that the magnitude of this number indexes the magnitude of restorative forces required to

maintain the virtual single system ***qua*** *cooperativity. The larger coefficients of curvature in the [log T_o x log $(M^{1/16}L^{1/2})$] plots of the virtual single systems reported in Chapter 8 are attributed to this additional potential, namely, the energy required to coordinate the two limbs. The energy for coordination for each virtual system is determined by the classical mechanics procedure for assessing the internal energy of a system of particles. So determined, the magnitude of the proportion of total mechanical energy devoted to coordination is found to correlate positively with the magnitude of the proposed dimensionless number.*

9.1. An Energetic Cost for Soft Synchronization or Coordination

There is no hard coupling of the limbs in locomotion. It has been suggested by Shik and Orlovskii (1965) that the achievement of an approximately constant phase relation between limbs of the same girdle is a matter of a (softly constructed) synchronizing function whereby the phase angle of one limb affects the threshold angle at which the other limb makes the transition from the support phase to the transfer phase. Be that as it may, the 180 degrees out of phase relation between limbs of the same girdle that characterizes the alternate-stepping gaits is preserved (up to a limit) even when the two limbs are forced to cover different distances per cycle, as in turning (Kulagin & Shik, 1970). The experiment described in Chapter 6 exaggerates the latter situation by pairing wrist-pendulum systems of different lengths and masses. For each of the 16 pairings human subjects are able to maintain an approximately constant phase relation at an approximately constant periodic timing for the duration of the task.

It is axiomatic that there is an energy cost to move the individual limbs. What is not as evident, but equally true given the second law of thermodynamics, is that there is an energy cost to achieve a soft synchronization or coordination of the limbs for any duration. To be explored is the hypothesis that it is this source of dissipation—this additional concentration of energy that is commonly termed the energy of interaction, but here termed the energy for coordination—that is magnifying the coefficient of curvature in the plot of τ_o against the moment variable $(M^{1/16}L^{1/2})$ for the virtual single systems (see Figures 8.5, 8.6, and Table 8.2). The questions of focal concern are: (1) What is the physical interpretation of coordination?; (2) What determines the quantity of the energy for coordination?; and (3) How is the energy for coordination to be measured?

9.2. Coordination Physically Interpreted as Cooperativity

"To coordinate" means "to bring parts into proper relation." "To cooperate" means "to concur in producing an effect." An instance of a mutual dependency among relatively independent entities that results in a unitary process and a consequent loss of degrees of freedom is likely to be called "coordination" by the student of movement but "cooperation" by the physicist. The entities that are said to "coordinate" are, for example, muscle-joint complexes, whereas the entities that are said to "cooperate" are atomisms (where the term atomism refers to anything that is atomlike at the scale of the ensemble to which it belongs). The two words do differ significantly: "Coordinate" makes implicit reference to an agent, "cooperate" does not. (The agent implied in the application of "coordination" to movement is usually the brain.) If the term "cooperate" can stand for the term "coordinate" in all those circumstances to which it is applied, however, the physical perspective on cooperative phenomena can be brought to bear on coordinative phenomena.

9.2.1. The Cooperative Level of Organization. The first point to be made is that cooperativity is a state of affairs of an ensemble that is maintained from below by the activity of the atomisms of the ensemble and from above by the field boundary conditions. By the analysis of Iberall and Soodak (1986), this three tiered organization holds for each systematic slice of a natural structure. For example, with regard to a large-scale fluid field, there are atomisms at the lowest level (e.g., the molecules), field boundary conditions at the highest level (e.g., the walls of a conduit), and a cooperative level in between (e.g., vortices of a material flow field). The dynamic relation between the lower and upper levels is the support for the cooperative level.

The three tiered analysis can be applied to a coordinative structure (CS) or muscle linkage, informally defined as a group of muscles, often spanning several joints, that is constrained to act as a single functional unit (Turvey, Shaw, & Mace, 1978; Turvey, Fitch, & Tuller, 1982). A pair of wrist-pendulum systems, coupled to a common period at a fixed phase relation, is an exemplary coordinative structure. The two wrist-pendulum systems are independent. There is no causal link between the two muscle-wrist joint complexes in the sense of activity in one complex giving rise inexorably to activity in the other. Yet, when the subject meets the demands of the experiment, the two independent muscle joint complexes behave as a single unit. This is the essence of the virtual single-system characterization (of an isochronous pair of wrist pendulum systems) derived from Huygens' law.

An isochronous pair of wrist-pendulum systems is divisible into three levels as follows (see Figure 9.1). At the lowest level are the two wrist-pendulum systems' coordinative structures CS_i and CS_j—the atomisms of the virtual system—with inertial magnitudes I_i and I_j. At the highest level are the field boundary conditions in the form of constraints, namely, (a) $\emptyset_i = (180\ degrees + \emptyset_j)$, where $\emptyset_i$ and $\emptyset_j$ are the phase angles of CS_i and CS_j, respectively, and (b) $\tau_v = \tau_{o_v}$, where τ_v is any period, and τ_{o_v} is the natural period, of the pair of systems. At the middle level is the virtual single wrist-pendulum system, CS_v, a

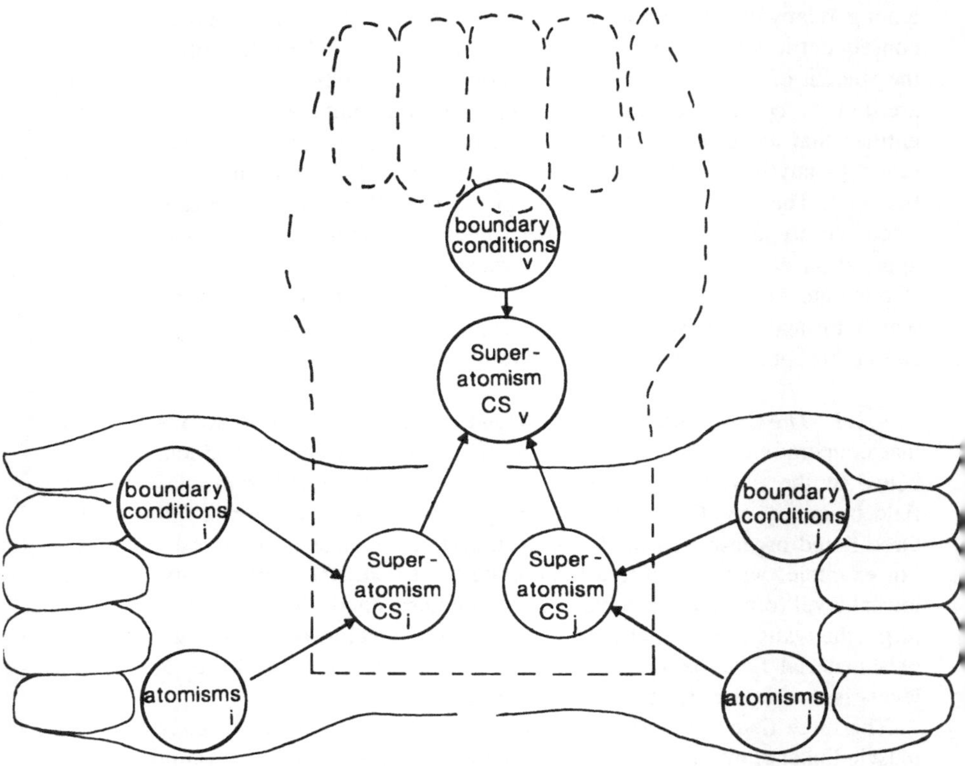

FIGURE 9.1. SCHEMATIC OF A COORDINATIVE STRUCTURE.

superatomism, with inertia I_v. The atomisms from below and the boundary conditions from above relate dynamically to support the existence of the coordinative level or cooperative level (the superatomism) in between. As can be seen from Figure 9.1, this three tiered organization applies to all the involved coordinative structures.

The constraints identified in the preceding are physically heterodox in this important sense: They are the physical interpretations of the subject's intentions. They are, in a sense, nonholonomic constraints that simulate holonomic constraints. Proponents of the ecological approach to perception and action have argued that an important step toward naturalizing intentionality would be taken by analyzing how intentions fit into the nomic (lawful) context (see Turvey, Shaw, Reed & Mace, 1981). The aforementioned three tiered description of a coordinative structure may prove helpful to such an analysis (see Chapter 13).

9.2.2. The Steady State of the Cooperative Level. The second point to be made is that the distributional characteristics of the conserved entities (that define the field) vary as a function of the scale at which the cooperativity is defined. At the cooperative scale the stress-strain characteristic of the field dynamic is approximately uniformly distributed throughout the field. That is, no portion of the field sustains a disproportionately large amount of stress-strain (relative to the scale of the organizational atomistic "units" defining the cooperativity). In an ideal equilibrium state, a uniform distribution of potential is prescribed for all atomistic scales by the equipartition theorem. In contrast, during a cooperativity, the uniform distribution prescribed by the equipartition theorem is only satisfied macroscopically. Satisfying the equipartion theorem macroscopically is achieved at the expense of elaborate microscopic spatial and temporal flow patterns defined on the microscopic potential distributions. A cooperativity, therefore, is an organization on the level between atomisms and boundary conditions that is achieved at the expense of departures from steady state in the lower atomistic level (cf. Prigogine, 1980).

The cooperativity is stable by virtue of the irreversible processes or work cycles operating at the atomistic level. The stability of the cooperativity is evident in its resistence to perturbation. The cooperativity preserves the macroscopic steady state potential distribution through systematic adjustments at the atomistic level. This cooperative resistence of the macroscopic field dynamic to perturbations is one of the hallmarks of a physical cooperativity and it has been advanced as a fundamental property of a coordinative structure (Fowler, Rubin, Remez, & Turvey, 1980; Kelso, Holt, Kugler, & Turvey, 1980; Kelso, Tuller, & Harris, 1983; Kugler, Kelso, & Turvey, 1980; Schmidt, 1980; Turvey, Shaw, & Mace, 1978, for reviews). The coordinative dynamics of many movement patterns, defined over ensembles of muscles and muscle/joint complexes, have been shown experimentally to exhibit this resistence to perturbation (see Asatryan & Fel'dman, 1965; Fel'dman, 1966a,b; Bizzi, Polit, & Marasso, 1976; Kelso & Holt, 1980; Kelso, Tuller, Bateson, & Fowler, 1984) and it is the overarching feature of quadruped locomotion (Arsharvskii, Kots, Orlovskii, Rodionov, & Shik 1965; Boylls, 1975). Bernstein (1967) summarized his analysis of locomotion this way:

> Movements react to changes in one single detail with changes in a whole series of others which are sometimes very far removed from the former both in space and in time, and leave untouched such elements as are closely adjacent to the first detail, almost merged with it. In this way *movements are not chains of details but structures which are differentiated into details*; they are structurally whole, simultaneously exhibiting a high degree of differentiation of elements and differing in the particular forms of the relationships between these elements. (p. 69)

9.2.3. The Steady State of the Atomistic Level. The third point to be made is that the steady state of the atomisms of the lower level is *defined at that level.*

This steady state is continuously defined and it continuously scales the forces of interaction among the atomisms even though the steady state configuration may never actually materialize (see Chapter 3). At the steady state the interactive forces are minimal. They increase with departures from the steady state. Such departures arise with the formation of the cooperative level. The atomistic level must assume the state that is consonant with the steady state of the cooperative level. In this sense, therefore, the degree of "disequilibrium" of the atomistic level and, therefore, the magnitude of the interactive forces, is set by the steady state of the cooperative level relative to the steady state of the atomistic level. Prigogine (1980) and his colleagues (Glansdorff & Prigogine, 1971; Nicolis & Prigogine, 1977) have chosen to emphasize the far-from-equilibrium characteristics of the atomistic level during the formation of a cooperative level, whereas Iberall, Soodak, and Yates (Iberall, 1977-78; Iberall & Soodak, 1986; Soodak & Iberall, 1986; Yates 1982, 1986) have chosen to emphasize the near-equilibrium state of the cooperative level.

Two concepts from von Holst (1935/1973) can be characterized in this context. The maintenance tendency refers to the disposition of an oscillating limb to be characterized continuously by, and to gravitate toward, its preferred tempo, that is, its steady state, when coupled to another oscillating limb; the magnet effect refers to an oscillating limb's tendency to draw its coupled partner toward that tempo. The maintenance tendency described by von Holst can be interpreted more physically as referring to the changing dynamics of the force structure that locally binds the atomisms. If an atomism is displaced from its steady state, a local force structure will be developed that is directly proportional to the magnitude of the displacement. At a critical distance from the local steady state, instabilities (i.e., nonlinearities) can arise in the force structure. For example, in the case of locomotion, the instabilities may appear in the form of gait transitions (see Kugler et al., 1980; Kugler, Kelso, & Turvey, 1982). In the context of the present experiment the virtual single system defines a global steady state (natural period at the virtual level) that displaces the atomistic constituents (local wrist-pendulum systems) away from their locally defined steady states (natural periods in isolation) onto the globally defined steady state.

Importantly, a virtual system's periodicity cannot be understood as resulting from the tendency of the coupled systems (the atomisms) to attract each other, like magnets, to their respective natural periods. The periodic timing of the virtual system is specified by the mechanical law $T = a(M^{1/16}L^{1/2})^c$. The individual wrist-pendulum systems (the atomisms) oscillate as best they can with the natural period (that is, at the steady state) defined by the moment variable of the virtual single system (the cooperative level). This understanding is further developed below.

9.2.4. The Invariance of Conservations from Atomistic to Cooperative Level. The fourth point is that in the creation of the cooperative level the only properties of the atomistic level that are preserved invariantly are the conservations. As previously noted, the steady state parameters of the atomistic level are not preserved; they play no role in defining the steady state of the cooperative level. Because the cooperative level is macroscopic, the distributed conserved quantities of the atomistic level must take on, at the cooperative level, *single* values. At the cooperative level the conservations are not viewed as distributed properties of an ensemble of atomisms but rather as point properties of a super-atomism. It is the case, however, that those field boundary conditions that were external to the atomistic level and that kept the ensemble from dissolving now become the field boundary conditions that keep the super-atomism from dissolving.

Reiterating a central claim and observation, the natural period of a virtual single wrist-pendulum system is determined by a law that relates any representative biological quantity of time to the appropriately scaled moment variable $M^{\rho}L^{\epsilon}$. The law is applicable to pairs of isochronous wrist-pendulum systems in addition to single wrist-pendulum systems because each pair, like each single system, is defined by a *single* mass quantity and a *single* length quantity. Where do these *single* quantities for an isochronous pair of wrist-pendulum systems come from?

Huygens' law specifies the length L_v of a compound system whose center of gravity in oscillation would be raised and lowered that distance that would produce a work cycle equal to the summed work cycles of the multiple component systems. In the transformation of an ensemble of individual simple systems of length L_i and mass M_i to a single simple system of length L_v and mass M_v, mass, angular momentum, and energy are conserved. The parallel should be noted between Huygens' strategy (to derive his law) and the strategy of statistical mechanics. In both, a single state parameter is sought to represent an ensemble of atomisms at equilibrium or steady state; in both, the search is conducted with the aid of the conservations (implicitly in the case of Huygens; explicitly in the case of statistical mechanics).

It is the conservations, therefore, that define a single quantity of mass and a single quantity of length for an isochronous pair of wrist pendulum systems. This observation is consistent with the general physical principle that *a unitary process* (a single system) *is assembled minimally by the conservations* (Iberall & Soodak, 1986). Given this observation, the fact that a subject swings two pendulums together at the period lawfully associated with $(M_v^{1/16}L_v^{1/2})$ means that *the nervous system is transparent to properties specifying the conserved quantities*.

9.2.5. Statistical Mechanics. The fifth point is that cooperativity (the reduction of degrees of freedom in an ensemble, the assembling of a unitary process via the conservations) is a phenomenon within the province of statistical mechanics. An alternative claim might be that, where the cooperativity of oscillatory processes is at issue, the phenomenon comes under the province of the mathematical theory of coupled oscillators (e.g., Abraham & Shaw, 1982; Minorsky, 1962; Stoker, 1950; Winfree, 1980). By this theory, a unitary oscillatory process would be constructed out of the variables of period, amplitude, and phase—the quantities associated with the individual oscillatory elements—related by continuous mathematical functions. The connectivity of the elements is construed mathematically rather than physically; the transitions (in the phase plane) of the unitary process's behavior are seen in terms of breaks in the topology rather than in terms of preserving the conservations. The mathematical theory of coupled oscillators is in the tradition of Kepler and Galileo rather than in the tradition of Newton: It provides a rich medium for describing and predicting but it does not explain in the sense of identifying the causal basis. The absence of a firm physical foundation reduces the appeal of the mathematical theory of coupled oscillations. The appeal is reduced further by other related considerations. The sources of energy for coupled oscillators in the case of animal movement are multiple, distributed, and (relatively) independent. (For example, in the present experiment the sources are the two wrist-muscle complexes.) In comparison, the mathematical theory of coupled oscillators assumes, for the general case—after Rayleigh, Duffing, and Van der Pol—a single source of energy that is driving all the oscillators. The image is of a number of oscillatory elements mechanically linked in some fashion inside a box that is being shaken rhythmically by an external, time-varying force (Abraham & Shaw, 1982). Further, the paradigmatic case to which the mathematics refers is of a physical system that is markedly different from the system that is of interest here, namely, an animal. In the paradigmatic case of the mathematical theory, the oscillatory elements transport, and are linked by, the source-determined energy and momentum. In the case of animals, the energy and momentum originating in a source are *not transported globally*. The linkage of the oscillatory elements is through very low-energy "signals." These "signals" are sustained by neural conductions that preserve field-like (near continuous) kinematic properties of the respective oscillator's kinetics (Section 6.5.1, and the following). These comparisons are captured in Figure 9.2 with reference to the coupling of two wrist-pendulum systems.

Bernstein's (1967) problem of degrees of freedom is a variant of the many-bodies problem, as is Sherrington's problem of multi-synaptic reflexes. While Sherrington (1906) considered an integrative account of the single reflex arc—fundamentally, a monosynaptic structure of two interacting cells—to be imminent, an integrative account of multiple reflex arcs seemed hopelessly elusive. Consonant with the quandary faced by classical particle mechanics, the interaction of two bodies could be understood, but the interaction of more than two bodies could not. Poincaré showed that there is no formal quantitative solution to the many-bodies problem. The solution, he underscored, is to be found in *qualitative* dynamics. His celebrated solution of the many-bodies

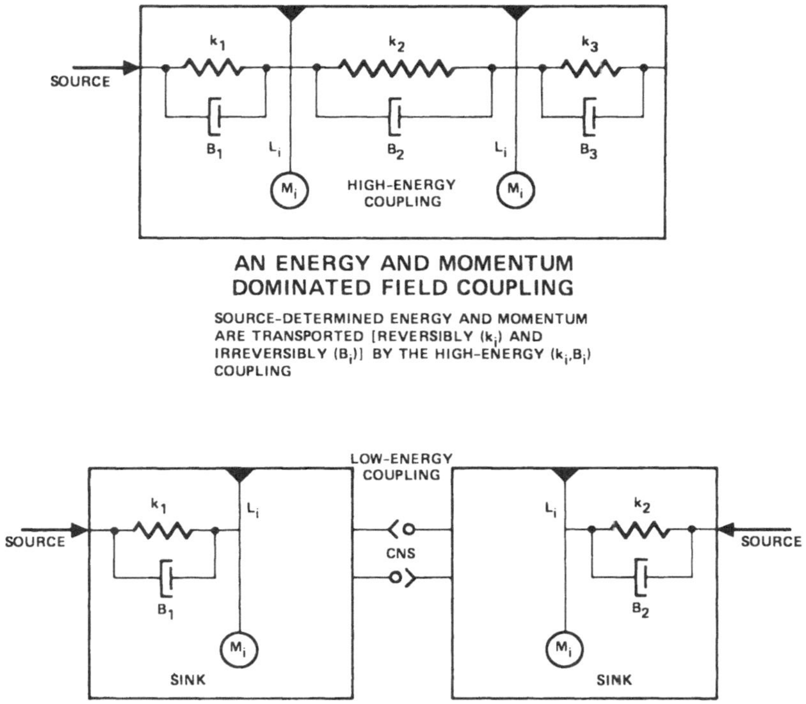

AN ENERGY AND MOMENTUM DOMINATED FIELD COUPLING

SOURCE-DETERMINED ENERGY AND MOMENTUM ARE TRANSPORTED [REVERSIBLY (k_i) AND IRREVERSIBLY (B_i)] BY THE HIGH-ENERGY (k_i,B_i) COUPLING

AN INFORMATION OR PERCEPTUALLY DOMINATED FIELD COUPLING

SOURCE-DETERMINED ENERGY AND MOMENTUM ARE NOT TRANSPORTED BY THE LOW-ENERGY (CNS) COUPLING

FIGURE 9.2. CONTRASTING PERSPECTIVES ON COUPLED OSCILLATORS.

problem is at the core of the mathematical theory of coupled oscillators. Less celebrated, and often unrecognized, is the statistical mechanical solution proffered by Helmholtz, Boltzmann and Maxwell. Statistical mechanics solves the many-bodies problem by asking: What holds invariant over any number of bodies in any number of interactions? The answer—the cooperativity of many bodies in a physical field is enforced by the mandatory sharing and conservation of number, mass, momentum, and energy—provides a foundation on which to build a lawful account of the behavior of an ensemble of many degrees of freedom.

9.2.6. A Kinematic Conservation: Temporal Topology? It must be emphasized that, though significant, the orthodox conservations are not sufficient to account for the cooperativity instanced by locomotion, or by an isochronous pair of wrist-pendulum systems. First, the quantities of mass, momentum, and energy are not transported between the limbs, as noted. This means that these conservations cannot be the basis for the isochronous linkage. In the more orthodox physical field, it is the transport of the aforementioned quantities that keeps one region of the field informed, as it were, about what is occurring at another region. Second, a field boundary constraint must refer to a conservation. But the constraints noted previously, with respect to the three tiered analysis of a coordinative structure (Section 9.2.1), do not appear to refer, even obliquely, to the orthodox conservations. Where known conservations are inadequate to the task, the hegemony and success of the conservational approach to cooperativity is license to hunt for a further conservation. It was intimated in Section 6.5.1 that a conservation of topological properties associated with kinematic flow fields ('critical regions' associated with the initiation, termination, convergence, or divergence of flows) may have to be introduced to provide a physical account of limb coordination. It has been argued elsewhere that this conservation does seem to be required by the ecological approach to perception and action (e.g., Kugler, Turvey, Carello, & Shaw, 1985). If a parallel is needed, it might be found in the hypothesis advanced in fluid mechanics that the vorticity of fluid particles remains constant during the turbulent mixing of the fluid. This hypothesized conservation of vorticity does not have the generality of the conservations of energy and momentum, for example, but it is expected to hold with equanimity in all situations to which it applies.

9.3. Specifying the Quantity of Energy for Coordination

9.3.1. A Steady State Metric for Locomotion. From the physical interpretation of coordination it follows that the force "to bring parts into the proper relation" (alternatively, the energy for coordination) is determined by the degree to which the atomistic level is made to depart from steady state by the conservative requirements of the cooperative level. In the act of terrestrial locomotion two, four, or more limbs cooperate in a single pendular clocking mode of organization. Where each limb is a local instance of the pendular, clocking mode of organization, collectively the two, four, or more limbs constitute a global instance of the pendular, clocking mode. It is important that the limbs in locomotion meet the compound pendulum criterion of isochrony. By the physical interpretation of coordination, therefore, there will be a cooperative level of locomotion defined by single quantities of mass, length, and time in comparison to an atomistic level defined by two, four, or more mass quantities, length quantities, and time quantities. (This interpretation quided the analysis in Section 8.7.) Although limbs of the same girdle are closely alike, limbs of different girdles are not. Small but significant differences are to be expected, therefore, in the aforementioned quantities. The amount of energy to be put into coordination will be defined for the individual limbs by the degree to which the state of the locomotory act in full is not the steady state of the individual limbs. The situation is captured by the following *π-number*:

$$\pi_{eq} = \left[\frac{\text{steady state of cooperative level}}{\text{steady state of atomistic level}}\right] \tag{9.1}$$

9.3.2. Center of Percussion. The concept of center of percussion will be introduced to help identify some of the dynamic stabilities and instabilities of relevance to Equation (9.1), namely, those that relate the center of mass (gravity) to the center of oscillation. The center of percussion coincides with the center of oscillation and refers to the situation in which the pendulum is struck by another object.

If a rigid compound pendulum, free to move in a plane, is struck with an impulsive blow J at a point P, and the line of force is perpendicular to the line from P to the center of mass cm, then the initial motion of the body is rotation about the center of rotation R relative to P (see Figure 9.3). The center of percussion is defined as that point P that can be struck without causing an acceleration at another designated point R, defined as the center of rotation. Point P (the center of percussion) identifies the same point as the center of oscillation. The concept of center of percussion has a functional application for objects such as bats (baseball and cricket), hammers, tennis rackets, swords, etc. Figure 9.3 identifies a baseball bat that is struck at its center of percussion with a baseball. L_{cm} identifies the distance from the center of rotation R to the center of mass cm, and L_P identifies the distance from the center of rotation R to the center of percussion P.

The center of percussion is that point on the bat at which the ball must be struck to secure the maximum effect with the least effort. If the point of contact is not at the center of percussion, then a "sting" will be felt that is associated with a shock wave resulting from a sudden transfer of *linear momentum* to the pendulum axis. There is no such momentum transfer, however, if the center of percussion is the point of contact. The sting is a function of the linear transformation of the pendulum axis. In the case where the ball strikes the bat at the center of percussion, no "sting" is felt since the position of the hands constitutes the center of rotation; there is no tendency for the point of rotation R to move (transfer momentum linearly) forward or backward while the blow is struck. When the ball is struck at the center of percussion the bat will smoothly rotate about point R. The same principle applies in the design of a hammer. The shape of a hammer is constructed with the center of percussion located in the hammer head, the point of rotation is located at the hand-grip portion of the

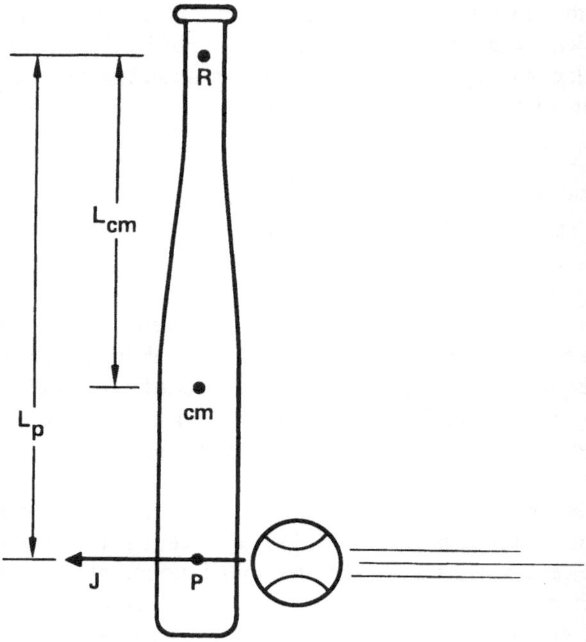

FIGURE 9.3. CENTER OF PERCUSSION (see text for details).

handle. Blows struck on the hammer head result in a smooth rotational motion about the hand-grip portion of the handle; in a well designed hammer, only angular momentum is transferred to the handle during striking.

The linear momentum acquired by the bat or hammer can be derived from the following expression:

$$p = m\bar{v} \tag{9.2}$$

and the angular momentum J about the axis of rotation R is equal to the linear momentum p times the radial distance L_P:

$$J = pL_P\ . \tag{9.3}$$

Or, expressed in terms of inertia, $I = ml^2$ (l is measured from the center of mass to the center of rotation),

$$J = I\omega \tag{9.4}$$

where $\bar{v} = \omega l$. Solving the aforementioned equations for L_P leads to

$$L_p = \frac{I}{ml_{cm}} = \left[\frac{\text{inertial moment}}{\text{statical moment}}\right] \tag{9.5}$$

Comparing Equation (9.5) with Equation (6.5) shows that, as noted previously, the center of percussion coincides with the center of oscillation (cf. Sears, 1950).

Figure (9.4) identifies a series of multi-flashed photographs illustrating the motion of a pendulous body suspended by a cord when the body is struck a horizontal blow perpendicular to the pendulum axis. The center

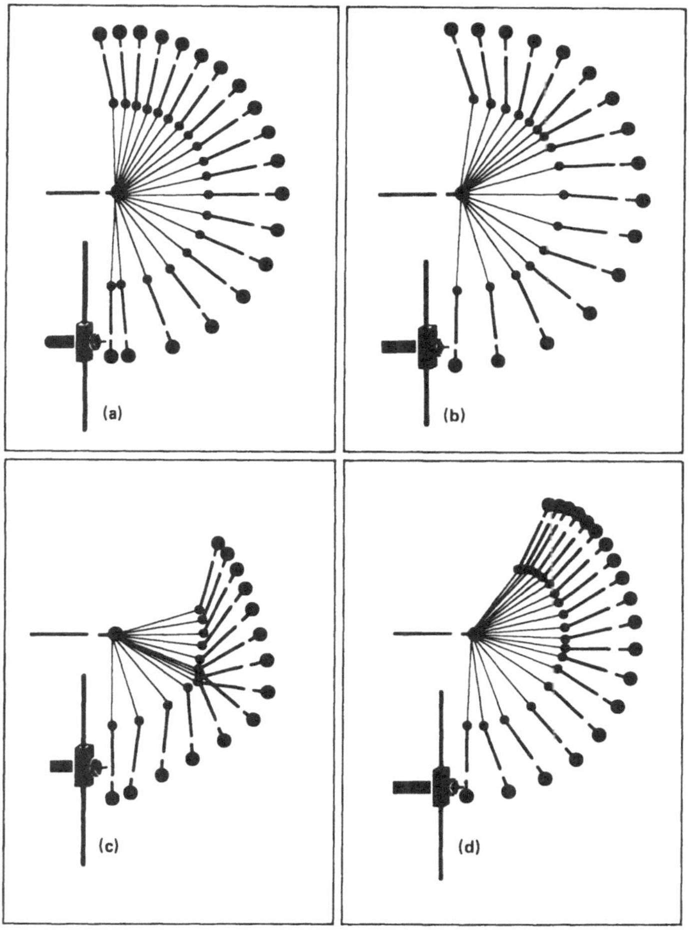

FIGURE 9.4. MOTION OF A SUSPENDED PENDULOUS BODY (Adapted from Sears, F. W. & M. Zemansky, 1951; see text for details).

of mass is marked by a white band. In (a), the pendulous body is struck at its center of percussion relative to the fixed point at the upper end of the cord, and it starts to swing smoothly about the point of rotation. The axis of rotation does not undergo any transfer of linear momentum from the strike. The system would rotate smoothly about the same axis of rotation even in the absence of a rigidly defined fixed point constraint. In (b) the body is struck at its center of mass, which is (always) above the center of oscillation. Note that it does not start to rotate about the pivot, but that its initial motion is one of pure *translation*. In this case rotation is a function of the fixed point constraint and *not* the geodesic of the angular momentum. It is evident from this demonstration that the center of percussion does not coincide with the center of mass. In (c), the body is struck above, and in (d) below its center of percussion. In these cases the axis of rotation will move in the direction opposite to the direction of the strike.

9.3.3. The Cooperative Level's Center of Gravity vs. the Atomistic Level's Center of Gravity. It will be argued that a π_{eq} can be constructed from estimates of the vertical coordinates of the centers of gravity of the cooperative and atomistic levels and that the π_{eq} can be used to identify instabilities arising as a function of atomistic departures from local steady states. It will be argued further that this π_{eq} can be used to rationalize gait transitions in locomotory systems involving multiple oscillatory components.

There is an equivalent simple pendulum system for two, four, or more limbs of individual mass and length M_i and L_i, respectively, whose center of gravity is at the length L_g given by the law of statics:

$$L_g = \frac{\sum M_i L_i}{\sum M_i} \tag{9.6}$$

This length, L_g, defines the static equilibrium distribution of masses. The lengths L_i in Equation (9.6) can be expressed as natural periods on the basis of the law of simple pendulums:

$$\tau_{o_i} = 2\pi\sqrt{L_i / g} \tag{9.7}$$

that is,

$$L_i = (g\tau_{o_i}^2) / 4\pi^2 \tag{9.8}$$

Substituting for each L_i its corresponding τ_{o_i} will bring in the local elastic (e) potentials in addition to the local effects of the gravitational (g) potential:

$$L_{(g,e)} = \frac{g\sum M_i \tau_{o_i}^2}{4\pi^2 \sum M_i} \tag{9.9}$$

Equation 9.9 identifies the length $L_{(g,e)}$ of an equivalent simple atomistic-level pendulum with an elastic component, and is a measure of the atomistic level's steady state.

The cooperative level is defined by a virtual system with a center of oscillation given by Huygens' law. Because it is assumed that the cooperative level is a simulation of a *simple* pendulum system, the center of oscillation of the virtual system is *coincident* with its center of gravity. The distance of the virtual center of oscillation/gravity from the axis of suspension is expressible in terms of the virtual system's natural period after the law of simple pendulums:

$$L_v = (g\tau_{o_v}^2) / 4\pi^2 \tag{9.10}$$

The foregoing quantity is a measure of the cooperative level's steady state. A dimensionless ratio can now be assembled as follows:

$$\pi_{eq} = \left[\frac{\text{steady state of the cooperative level}}{\text{steady state of the atomistic level}}\right] = \frac{(\sum M_i)\tau_{o_v}^2}{\sum M_i \tau_{o_i}^2} \tag{9.11}$$

The larger the departure from unity, as measured by the modulus $|1-\pi_{eq}|$, the greater is the displacement of the atomistic level from steady state and the greater the restorative forces or, equivalently, the energy of interaction. (As will be apparent in the next section, π_{eq} in the main wrist-pendulum experiment correlates positively with the mechanical energy directed toward keeping a pair of wrist-pendulum systems isochronously coordinated.) In the more general case of locomotion, the differences between or among limbs are constant. What does vary, however, is the velocity of locomotion, that is, the periodic timing of the locomotory act in full. It is to be expected that the strain on the atomistic level will increase to the degree that the cooperative level departs from its steady state. This expectation is captured in the following scalar modification of Equation (9.11)

$$\pi_c = (\tau_v/\tau_{o_v})\pi_{eq}\, , \tag{9.12}$$

where τ_v is any periodic timing. The significance of the dimensionless quantity π_c is that it can be regarded as a generalized Reynolds number for locomotion. Critical values of π_c, or a function similar to Equation (9.12), might be expected to identify gait transitions. At one or more values of π_c, the nonuniformity of the conservations at the cooperative level exceeds the dissipative limit of the atomistic level, under its current organization. The nonholonomic field boundary condition defining the topology of the gait can no longer be sustained. The old cooperativity must give way to a new cooperativity.

9.3.4. Calculating $|1-\pi_{eq}|$ for the Virtual Single Wrist-Pendulum Systems. Using Equation (9.11), $|1-\pi_{eq}|$ was calculated for each of the virtual systems

TABLE 9.1
Virtual Single Systems' $|1-\pi_{eq}|$ Values and Mean Normalized Internal Energy Values

System	Subject 1	Subject 2	Subject 3	Subject 4	Average $\lvert 1-\pi_{eq}\rvert$	Average Normalized Internal Energy ($\times 10^{-3}$)
1	.019	.249	.310	.021	.150	1.976
2	.136	.138	.152	.345	.193	7.352
3	.037	.220	.145	.125	.132	2.947
4	.138	.020	.089	.148	.099	1.552
5	.200	.329	.130	.327	.247	4.033
6	.095	.106	.033	.065	.075	4.167
7	.274	.428	.099	.426	.307	11.337
8	.061	.009	.021	.064	.039	4.598
9	.007	.222	.253	.069	.138	7.150
10	.447	.255	.299	.563	.391	34.427
11	.033	.303	.198	.104	.160	4.550
12	.218	.052	.259	.407	.234	13.271
13	.060	.154	.030	.189	.108	1.149
14	.032	.032	.008	.112	.046	7.164
15	.122	.315	.152	.358	.237	5.561
16	.053	.056	.040	.083	.058	4.611

for each subject. For any given virtual system of Table 6.11, the τ_{o_v} value was computed, as before, as the mean of the two periods given in the corresponding cell of the subject's table of periods for the double wrist-pendulum conditions (see Chapter 6). The τ_{o_i} values of the left and right wrist-pendulum systems composing a virtual single system were the natural periods of those systems as reported in Tables 6.3-6.6.

The 16 virtual single system $|1-\pi_{eq}|$ values are collected in Table 9.1. The values for the individual subjects are reported together with the values averaged over subjects. Inspection of this table reveals that higher modulus values tend to be associated with virtual single systems composed of subsystems that differ in length and mass. According to the immediately preceding arguments, the modulus values in Table 9.1 should predict the quantity of energy required in each case to coordinate the wrist-pendulum systems comprising the virtual single system. Table 9.1 also includes, therefore, estimates of this energy quantity, averaged over the subjects. How these estimates were arrived at is the topic of the next section.

9.4. A Mechanical Analysis of a Virtual Single System in the Laboratory and Center of Inertia Frames of Reference

9.4.1. An Analogy. The total energy E_T of an isochronous pair of wrist-pendulum systems is:

$$E_T = \sum_{n=2} (I_i \omega_i^2) / 2 \quad (9.13)$$

where I_i and ω_i (where $\omega_i = 2\pi\theta_i/\tau_i$ and θ_i = *maximum amplitude*) are the inertia and maximum angular velocity, respectively, of a component wrist pendulum system. The intuition under elaboration is that some part of this total energy is wrapped up in the bonding of the two subsystems—keeping them on-line, as it were. The cooperative level perspective on an isochronous pair of wrist-pendulum systems is a perspective of the two subsystems *as coordinated*, that is, as a virtual single wrist pendulum system. In comparison, the atomistic level perspective is of the two subsystems doing what must be done to sustain the coordinated state.

For an analogy, consider two ballroom dancers or figure skaters performing a routine. They can be looked at as a single "unit"—the "team"—moving systematically across the dance floor or the rink. (Indeed, they are judged as a unit.) Alternatively, they can be looked at less abstractly as two individuals, that is, as two units. While most of the variously directed motions of the two dancers or skaters contribute to the translations and rotations of the team across and on the surface, some of their motions are for the purpose of keeping them together *as a team*. They continuously adjust to each other's inevitable departures from the ideal state. In consequence, an energy bookkeeping should reveal that the total energy expended by the two dancers or skaters will exceed the translational and rotational energy of the unit they form. This analogy is pursued with the tools of classical particle mechanics.

9.4.2. An Analysis from the Laboratory Reference Frame (the L-frame). The virtual single system of inertial magnitude I_v, a system of two atomisms, is oscillating with an angular velocity ω_v. The kinetic energy of the virtual single system or superatomism in the L-frame is:

$$E_{v_L} = (I_v \omega_v^2) / 2. \quad (9.14)$$

The two component atomisms, of inertial magnitudes I_1 and I_2, oscillate in the L-frame with angular velocities ω_1 and ω_2; the sum of their kinetic energies gives the *total mechanical energy* of the virtual single system, namely:

$$E_{i_L} = \sum I_i \omega_i^2 / 2 = (I_1 \omega_1^2 + I_2 \omega_2^2) / 2. \quad (9.15)$$

The difference between (9.15) and (9.14) is a measure of how much of the total mechanical energy available to the virtual single system is not showing up in the oscillation (or "rotation") of the virtual single system.

9.4.3. An Analysis from the Center of Inertia Reference Frame the (C frame). In addition to the L-frame of reference, the classical mechanical analysis of a system of atomisms identifies a frame of reference that is rigidly fixed to the center of inertia (or mass or gravity) of the system and which translates with respect to inertial frames such as the L-frame. This other frame of reference is referred to as the frame of the center of inertia, or C-frame, and is distinguished by the fact that within it the total momentum is zero (Irodov, 1978; Landau & Lifschitz, 1976).

The maximum kinetic energy in the C-frame is the internal energy of a system of atomisms. To calculate the internal energy requires that the angular velocities of the individual atomisms in the L-frame be transformed to their respective values in the C-frame. This is achieved by the classical law of velocity transformations (Irodov, 1978):

$$\omega_1' = \omega_1 - \omega_v \tag{9.16}$$

$$\omega_2' = \omega_2 - \omega_v \,. \tag{9.17}$$

It follows, therefore, that the kinetic energy of a virtual single system in the C-frame, the system's internal energy $\overline{E}_{v_c}$, is:

$$\overline{E}_{v_c} = \frac{[I_1(\omega_1 - \omega_v)^2 + I_2(\omega_2 - \omega_v)^2]}{2} \,. \tag{9.18}$$

This simplifies to:

$$\overline{E}_{v_c} = \frac{[I_1 I_2]\,[\omega_1 - \omega_2]^2}{2[I_1 + I_2]} \,. \tag{9.19}$$

Figure 9.5 summarizes the energy analysis of a virtual single system in the L-and C-frames of reference.

9.4.4. Calculating the Internal Energy of the Virtual Single Systems. Because it is an abstraction, a virtual single system's angular velocity is not measurable

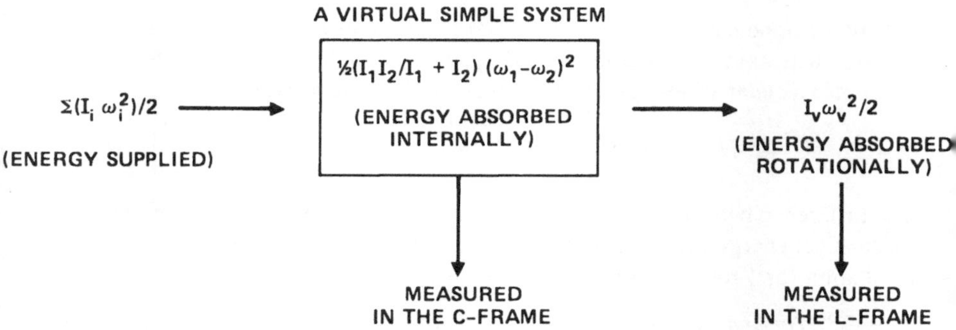

FIGURE 9.5. ENERGY MEASURED FROM THE C-FRAME OF REFERENCE AND THE L-FRAME OF REFERENCE.

directly but it is calculable on the basis of the conservation of angular momentum:

$$\omega = \frac{(I_1\omega_1 + I_2\omega_2)}{(I_1 + I_2)} \tag{9.20}$$

The internal energy of a system can, therefore, be calculated either by Equation (9.19) or by the difference between the quantities given by Equations (9.15) and (9.14). Table 9.1 reports for each of the 16 virtual systems of the main experiment the internal energy values—using Equation (9.19)—expressed as proportions of the total mechanical energy—given by Equation (9.15)—averaged over the four subjects.

9.5. The Periodic Timing Law, Energy for Coordination, and the π_{eq}-*Number*

The analyses reported in Chapter 8 suggested that not all virtual single systems were alike. Some systems seemed to be less tightly bound to the law-based trajectory relating periodic time to the moment variable $M^{1/16}L^{1/2}$ than others. The thesis of this chapter is that virtual systems differ in the degree to which the inclinations of the cooperative and atomistic levels of organization are compatible. In brief, some systems are internally less dynamically compatible than others. The point of developing the quantity $|1\text{-}\pi_{eq}|$ was to provide a metric for this internal compatability. And the point of pursuing the internal energy equations of classical particle mechanics was to provide a measure of the energy demand incurred by a system's degree of internal compatability. It would be a mistake to assume that this latter measure does full justice to the energy demands of coordination. For example, it fails to accommodate the phase differences that must go hand in hand with trying to couple, in an isochronously harmonic fashion, two physically different pendulum systems. Nevertheless, there are reasons to suspect that the derived, renormalized internal energy roughly tracks the internal compatability metric $|1\text{-}\pi_{eq}|$. These suspicions were evaluated by correlating the two measures. This was done in two ways. First, the correlations were with respect to the 16 pairs of mean values reported in Table 9.1 (see Figure 9.6) and second, they were with respect to the 64 pairs of individual values (the total number of pairings given by 4 subjects and 16 virtual single systems). Both correlations were significant, $r=0.727$ ($p<.01$) and $r=0.503$ ($p<.001$), respectively. The implication is that a system's internal compatability and its energy for coordination are related. At the same time the relation falls far short of the exactitude that one ordinarily expects from a physical analysis. As noted, one of the two measures is less than ideal and both may be open to improvement. A further correlation, between the mean $|1\text{-}\pi_{eq}|$ values and the mean percent prediction errors reported in Figure 8.14, was computed. This bivariate relation of .754 is depicted in Figure 9.6.

The impression that not all virtual single systems are alike can be confirmed in another way. In Chapter 8 the various systems were divided into three groups

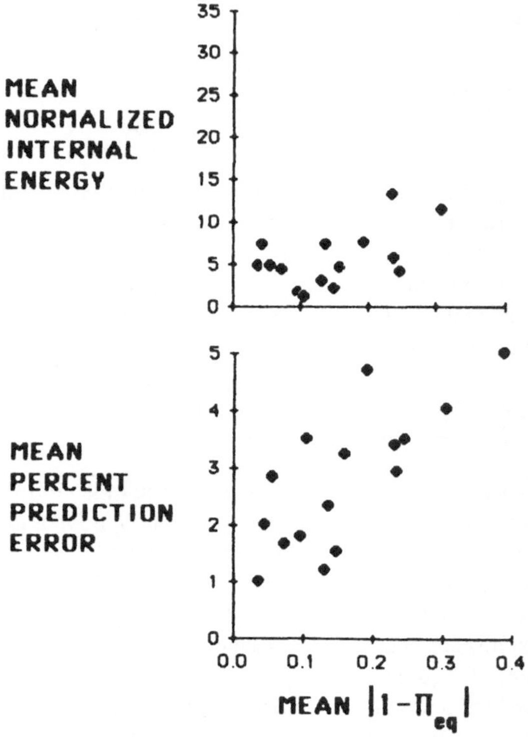

FIGURE 9.6. Mean normalized internal energy and mean percent prediction error expressed as a function of mean $|1\text{-}\pi_{eq}|$.

for the purpose of evaluating the periodic timing law (see Table 8.2). Instead of parsing the various systems according to the criteria advanced in Chapter 8 they can be parsed according to the internal compatability metric. For each subject the five systems with the highest $|1\text{-}\pi_{eq}|$ values and the five systems with the lowest $|1\text{-}\pi_{eq}|$ values can be compared with respect to their trajectories in the $log\ \tau_o$ by $log\ M^{1/16}L^{1/2}$ space. The comparison is shown in Figure 9.7. Roughly interpreted, the outcome agrees with the analyses of Chapter 8. It suggests that differences among systems—here, differences in the degree of competition between cooperative and atomistic levels—are largely realized as differences among the mesh systems housing the periodic timing law. The law itself is invariant.

9.6. A Methodological Postscript: Frame-Independent Dimensional Quantities and Criterial Nondimensional Quantities

The preceding chapter (particularly Sections 8.6.2 and 8.7) and the present chapter make tentative stabs at identifying, respectively, constant dimensional quantities and criterial dimensionless quantities underlying the soft assembling of the pendular, clocking mode of organization. The tentative nature of the effort should not detract from the worth of the enterprise—identifying the dimensional constancies that ultimately define natural units and the relationships among them and identifying the pure (dimensionless) quantities that ultimately define criteria for parsing rhythmic movement into its natural categories. The focus of the final section of this chapter is the primacy of scalar quantities and their potentially decisive role in a strategic physics.

9.6.1. Vector and Scalar Perspectives on Conditional Periodicities. An oscillation is (strictly) periodic if the values of physical quantities, varying in the

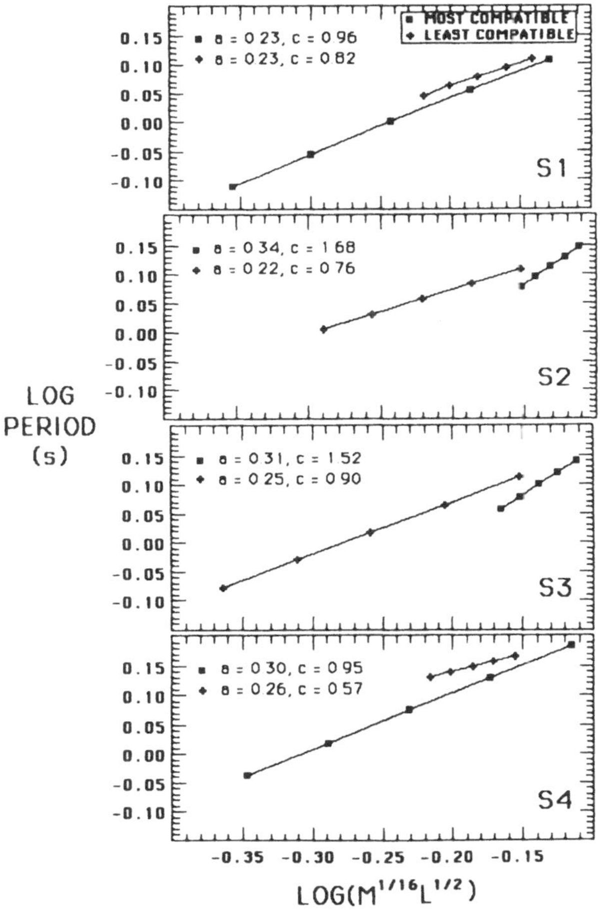

FIGURE 9.7. Regression log period on log $(M^{1/16}L^{1/2})$ for systems of high internal compatibility and systems of low internal compatibility. Points are values predicted from regression equations.

process of oscillation, are repeated in successive equal intervals of time. The period τ is the minimum time interval after which the magnitudes of *all* the physical quantities characterizing the oscillatory motion are repeated. As a consequence, any periodically varying physical quantity S is dependent on time t. That is,

$$S = S_0 + x(t) , \tag{9.21}$$

where S_o is a constant and $x(t)$ is a periodic function of time,

$$x(t + \tau) = x(t) . \tag{9.22}$$

A periodic system's oscillatory behavior is described through a differential equation ($dx/dt - w^2x = 0$, for simple harmonic motion) thereby underscoring time as the independent variable governing the system's behavior. Time is a "proper" independent variable for this reason: It is plainly not affected by the other variables in the equation and these other variables, which form the system's state, vary in time. It is illuminating to contrast mechanical equations (of periodic systems) with the equations of thermodynamics. Suppose that the pressure, volume, and temperature of a gas are given. Then it is possible to find from the thermodynamical equation of state the value that any one of these variables will assume at some temporal delay after a change in the other two. Thermodynamical equations do not discriminate among the quantities describing a state in the sense of identifying one of them as invariably "independent." As a consequence thermodynamical equations are much less elegant than mechanical equations. They are, moreover, less strongly oriented to temporal predictions than are mechanical equations and more strongly oriented to the prediction of unknown variables on the basis of a set of known ones, with all the variables having reference to the same state (Margeneau, 1950).

An oscillatory or vibrational motion that is not strictly periodic is termed *conditionally periodic*. Real, oscillatory motions, as contrasted with ideal oscillatory motions, are of this kind. In a purely descriptive language, a conditionally periodic system is one in which each of the coordinates undergoes periodic motion but the associated frequencies are not all rational fractions of each other. As a result the motion of the system as a whole is not periodic; that is, it is not totally degenerate. Expressed causally, a conditionally periodic system is one whose physical quantities do not change solely as a function of time. Frequently such a system is causally conditioned by thermal and mechanical properties whose magnitudes need not be time dependent. (A prosaic example of a conditionally periodic system is the pendulum with dry friction.) It is customary to treat conditionally periodic systems as if they were simply periodic. Differential equations are used to model their motions with time appearing explicitly as an argument and/or implicitly through the derivative (see Sections 5.2.2 and 5.2.3). It will be recalled that the derivative or velocity vector defines a system's tendency to move with an habitual velocity at a given point in its state space. It will be recalled further that for a system governed by a differential equation (or a set of differential equations), a velocity vector field can be constructed that gives the instantaneous habitual velocities at every point in the system's state space. This construction has proven to be a most useful analytic tool (see Abraham & Shaw, 1982, and Chapter 7). The construction is so useful, in fact, that it encourages equating a conditionally periodic system with its velocity vector field.

Despite its appeal, the vector perspective on conditionally periodic systems has serious shortcomings. In the vector perspective, as noted, time is characterized purely mathematically (as the ultimate independent variable or as the parameter of a one parameter continuous group of transformations) but assigned a physical, causal role. Realist ontology (Bunge, 1977; Gibson, 1979; Turvey, Shaw, Reed, & Mace, 1981) denies time a causal role for the reason that it is not a self-existing object but rather a network of relations among factual items. An important lesson of relativity theory is that there is no absolute or thing-free time (Bunge, 1977). In consequence, mathematical properties are not truly assigned to time but to changing things. Paraphrasing Gibson (1979), time is an abstraction from events (changing surface layouts). The construction of a velocity vector field proceeds according to sentences of the kind: "The state of system X at time t." But sentences of this kind really mean, "The state of system X when the reference frame is in state t." A reference frame, it will be recalled, is a concrete thing of a certain kind; it is not merely conceptual, like a coordinate system (see Chapters 2 and 3, and the following). The problem, in a nutshell, is that in embracing the vector perspective on conditionally periodic systems, science puts itself in the potentially dangerous position of relinquishing causal explanation to description (however powerful). Metaphorically speaking, the danger is that of mistaking the deputy for his or her constituency.

In contrast to the vector perspective, a *scalar* perspective on conditionally periodic systems focuses on the physical rather than the mathematical, on the mechanical and thermodynamical propertied things that are causally efficacious rather than the conceptual abstractions that are causally impotent. Promoting a scalar perspective over a vector perspective is continuous with themes aired in Chapter 2 on the improper uses of arithmetic and logic. More positively, the significance of the scalar perspective is that it directs research to the quantities that shape the characteristic behavior of conditionally periodic systems. These quantities are in the form of: (a) conservations defining what must be shared among component processes, (b) dimensional constants defining limits on the rates and graininess of interactions, (c) nondimensional numbers indexing how the governing forces configure, thereby providing a basis for expressing qualitative similarities among systems that differ in scale, and (d) criterial values of these nondimensional numbers defining natural category boundaries—locations in the state space at which a system's qualitative character changes.

9.6.2. Conserved Quantities. The invariance of the quantities of energy, mass, and momentum over all conservative mesh systems is the basis of their utility in constructing and verifying dynamical equations. The invariance of these quantities under the transformations from macro to micro frames, and vice versa (see Figure 3.5) is the basis for relating processes at different scales (see Chapters 3 and 4). The conservations or summational invariants have been in the limelight for much of this text. The strategic feature of the summational invariants brought into focus in the present chapter is that of looking for one or more conserved quantities whenever an interaction between systems entails a loss of degrees of freedom. At the cost of being redundant, the rationale for the strategy is given a further brief exposition.

Where two systems interact, a dynamics is induced in one or both systems. The interactions customarily investigated (the interactions between two particles) have proven, at bottom, to involve an exchange of conserved quantities—for example, momentum and/or energy. Given that it is conserved quantities that are exchanged, the interacting particles exhibit a kind of coordination or cooperativity. The ''equation of constraint'' is of the form: Total energy before collision minus the energy of one particle after collision equals the energy of the other particle after collision. That is to say, because of the summational invariant of energy, the behaviors of the two particles are not independent. Though the example is hackneyed, the elastic collision portrays succinctly the relation between the concepts of conserved quantity and cooperativity. The traditional success of the relation merits the following general strategy: When the known conservations are unable to account for the coupling (and loss of degrees of freedom) of two or more systems, then another conservation must be sought. (For a recent example in elementary particle physics, the quantity of *strangeness*—a property configured from electric charge, baryon charge, and the projection of the isotropic spin—has been advanced as a summational invariant for strong and electromagnetic interactions.) In response to the incompleteness of an orthodox conservational account of the coordination defined over two or more biokinematic units, the advocated strategy is to inquire (Section 9.2.6): Is there an additional scalar at the ecological scale whose conservation, when considered in conjunction with the familiar conservations, fully rationalizes the coupling of body parts and the perceiving-based coupling of the body as a unit to the persisting and changing properties of surface layout?

9.6.3. Dimensional (Universal) Constants. The states of a system are frame (of reference) dependent. Units of duration and units of distance are always local. This is the larger moral of relativity theory and the essence of intrinsic measurement (see Chapter 2). As was just noted, a reference frame is a concrete thing of a certain kind. The fact that each and every reference frame can be represented by a mesh system does not render them conceptual. Frames are substantial. They are assigned substantial properties. The paradigm is the mass qua inertial frame moving at uniform velocity.

Although it is traditional to think of reference frames as perfectly rigid bodies, this conception is limiting intellectually, and probably misguided. Because a frame must function as a clock (again, there is no thing-free time) it must have movable parts. What qualifies as a clock, fundamentally, is any thing whose states exhibit strict partial order—that is, any thing whose states are asymmetric and transitive, so that simultaneity of two distinct states is prohibited (Bunge, 1977). The mobility/clocking requirement implies that frames will be more generally field-like than rigid-body-like.

Where analysis focuses on a single complex system (say, a single person) that stratifies into multiple reference frames according to scale, the conservations provide, in principle, a basis by which the frames can be stitched together (see Chapter 3). But what of the problem of relating states defined relative to frames of reference attached to *different* complex systems? The pendular clocking mode of organization—a simple systemic state—is characterized by spatial and temporal properties. These properties can be defined extrinsically with respect to "absolute" standards of angular excursion and duration. From the perspective of the system assembling the pendular clocking mode, however, the spatial and temporal properties are defined relative to its own chemical, thermal, mechanical (field) states (see Chapter 11). If the assembling is law based, as argued in this book, then the laws should be the same in passing from any one complex system (e.g., a cockroach) to any other complex system (e.g., a human being). That is, the laws should be the same from any one frame to any other frame. To establish this as a fact (beyond the kind of evidence presented in Chapters 8 and 12) would require minimally a means of connecting the frames defined by the indefinitely many instances of individual biological systems. The conservations are of little help in this problem. Constant dimensional numbers, on the other hand, promise resolution.

An ultimate physical theory, it has been suggested (Lieber, 1968), is one that is directly based upon and explicitly invokes at least one universal constant of nature. Among the best-known dimensional universal constants are the dimensional number that identifies the upper limit on any possible material velocity (the constant speed of light, c) and the dimensional number that identifies the lower limit of physical interactions (Planck's constant of action, h). Two others of great note are the Boltzmann gas constant (k) and Einstein's constant elastic curvature (G) associated with gravitation. Dimensional universal constants are the most fundamental examples and manifestations of nature's invariance. Arguably, they are the ultimate basis and expression of natural law. If such is the case, then *all natural phenomena, including biological, should be grounded ultimately in dimensional universal constants* (Lieber, 1968). Insofar as this statement suggests the existence of a set of ultimate rational units of which all phenomena are constituted, it would appear to counter the notion of states as frame dependent and, thereby, the thesis of intrinsic measurement. Special relativity theory clarifies the situation. Lengths and times are dependent on the frame of reference. They are not absolute. But a length x and a time t in any one inertial frame can be converted into a length x' and a time t' in any other inertial frame (traveling with respect to the first at velocity v) by transformations that explicitly invoke c:

$$x' = (x - vt) / 1 - v^2 / c^2)^{1/2} , \tag{9.23}$$

$$t' = (t - vx / c^2) / (1 - v^2 / c^2)^{1/2} . \tag{9.24}$$

These are, of course, the well-known Lorentz transformations of the special theory of relativity. (The explicit invocation of c makes this theory an ultimate physical theory, by the aforementioned convention.) In the present context Equations (9.23) and (9.24) should be viewed as illustrating how different frames of reference are connectible through a dimensional universal constant. The illustration speaks to the problem posed previously of building a relativistic physical biology of the pendular clocking mode of organization valid for all frames to which this mode of organization can be referred. The physical biology would be relativistic because of a frame-dependency of the characteristic quantities.

The central claim—that the dimensional universal constants are never incidental in any natural phenomenon—is not in accord with the correspondence principle. This principle, advanced originally by Bohr, addresses the interface between phenomena successfully accommodated by an older and less complete physical theory (e.g., classical mechanics) and phenomena that can be accommodated only by a younger and more complete theory (e.g., relativistic mechanics). The division is formally based on one or more universal constants (e.g., speed of light). On one side of the division (the "classical" side), the dimensional constant is irrelevant; on the other side of the division (the "nonclassical" side), the constant is mandatory. Lieber (1968) has referred to this as an asymptotic correspondence principle because at the interface certain terms become vanishingly small. Although it has proved to be methodologically useful and formally convenient, Bohr's (asymptotic) correspondence principle is ontologically irksome. It allows that there are domains of natural phenomena to which no dimensional universal constants apply. But if dimensional constants constitute the ultimate underpinnings of nature, then each and every domain of phenomena that nature gives rise to must be founded on at least one dimensional constant. What is implied is a correspondence principle that is universal rather than asymptotic (Lieber, 1968).

By the preceding, classical mechanics cannot be a correct theory, even for the macroscopic domain to which it has been traditionally applied with success.

Classical mechanics lacks universal constants in the formulation of its propositions. It will probably take a nonclassical variant of the macroscopic domain, however, to reveal this inadequacy of the classical portrayal. The chemical-thermal-mechanical phenomena (see Chapter 11) at the heart of life processes may provide such a test. These phenomena are close to the classical macroscopic domain and they pose a most basic question: What are the laws fashioning classical mechanical forces from chemistry? The intimation of the universal correspondence principle is that the discernment of these laws rests ultimately on the uncovering of a dimensional constant (or constants) that fetters chemical-thermal-mechanical processes.

It is noteworthy that the discovery of each of the four dimensional universal constants identified in the foregoing resulted in a radical transformation in fundamental axioms, and that each—c, h, k, and G—ushered in a branch of physical science—relativistic mechanics, quantum mechanics, statistical mechanics, and general relativity, respectively. In sum, the history of science and the universal correspondence principle encourage a search for frame independent dimensional biological constants or, more precisely, frame independent dimensional constants at the ecological scale on which are founded the phenomena of (mutually defined) animals and their surroundings (Gibson, 1979). The discovery of such constants would be expected to strongly constrain the axiomatic foundations of the sciences of living things (e.g., biology, physiology, psychology). One consequence of advocating a *set* of ultimate rational units is that quantization emerges as a potentially very general property of nature. The analysis presented in Section 8.7 would be consistent with such an interpretation.

9.6.3. Criterial Dimensionless Quantities. Two wrist-pendulum systems that differ in mass and length are, self evidently, systems that differ in scale. The same truth holds for two different quadrupeds such as a Thompson's gazelle and a giraffe. These differences in scale, however, do not propagate without limit; not *all* aspects are different. This is also self-evident. Thus, both wrist-pendulum systems oscillate; both quadrupeds walk. Despite the difference in scale, the two wrist-pendulum systems act similarly and the two quadrupeds act similarly. But once these similarities have been identified, the question becomes: How deep do the similarities go? For example, do the two wrist-pendulum systems manufacture their stable oscillations in precisely the same fashion? Are the processes governing walking in the gazelle the same as those governing walking in the giraffe? These questions can be interpreted in more than one way. They could be asking whether or not the two systems are mechanically similar. Or they could be asking whether the similarity is in terms of thermodynamic processes or in terms of the associated chemical kinetics, and so on. Whatever the target of the question, however, the answer requires a metric—a graduated scalar quantity that can be assigned to the systems in question and on the basis of which they can be compared reliably. If the two systems are similar on the targeted dimension, then the quantity in question should be identical, or nearly so, for the two systems. Similarity amounts to an invariant under a transformation of scale, that is, a property that is unchanged by a change in dimensional quantitites. This observation suggests that the sought-after metric must be comprised of pure or nondimensional numbers (π-*numbers*) indexing qualitative states of affairs. Previously, qualitatively identical behavior in the face of (small) perturbations in a system's magnitudes was identified as structural stability (see Chapter 7). On examination, similitude and structural stability prove to be closely related concepts (Rosen, 1978).

The major nondimensional quantities or similitude criteria at work in physical theory and engineering practice are assembled from laws. The laws in question are those governing the phenomenon of interest. Only occasionally does the number of governing laws exceed two. In fact it has been contended that principal (that is, law based) π *numbers* derived from more than two laws are rarely useful (Schuring, 1977). Partly because of the way they are constructed, principal π-*numbers* admit of a characterization in terms of two "competing" force structures (with the caution that the characterization not be applied too liberally). Thus the Reynolds number $Re = ve/u$ (where v is the characteristic velocity of the fluid for the given problem, e is the characteristic linear dimension, and u is the kinematic viscosity of the fluid), which applies to streaming viscous fluids, can be characterized as the ratio of inertial to viscous forces. The Froude number $Fr = v^2/gl$ (where v is the velocity of the fluid at a point far from the body it is flowing around, l is the characteristic linear dimension of the body, and g is the acceleration due to gravity), which applies generally to the motions of accelerated masses in the gravitational field, can be characterized as the ratio of inertial to gravitational forces. And the Mach number $M = v/c$ (where v is the fluid velocity at the point being considered and c is the local velocity of sound in the fluid at the same point), which applies to the steady motion of a compressible fluid at high velocity, can be characterized as the ratio of elastic to inertial forces.

This ratio-of-forces perspective on principal π-*numbers* is apparent in the development of π_{eq} in the pages of the present chapter. The perspective is helpful in thinking about similitude. For example, the systems governed by the same laws are similar if, and only if, the corresponding π-*numbers* are numerically equal. In the ratio-of-forces perspective, similarity of systems means that the patterning of forces (a qualitative property) is the same in the two systems even though the actual magnitudes of the forces in the two systems are different (see Section 8.4.1). The perspective is helpful in another, related way—with respect to thinking about the degenerate behavior associated with the principle of similitude. As noted in Section 8.4.1, at some configuration of forces the phenomenon in question can no longer be sustained. (For example, the walking gazelle at some velocity becomes a trotting gazelle.) At the transition point it is to be supposed that the patterning of forces changes dramatically. The lesson of the critical Reynolds number (at which laminar flow becomes turbulence) is that there comes a point with change in scale (in the case of *Re* this is ordinarily a change in velocity) at which one force can no longer successfully "balance" the other. At the critical point the two force structures (inertial, viscous) reduce to virtually one force structure (inertial). It is roughly apparent that in a complex biological system such degeneracies might abound, given the potentially rich play of laws that the complexity allows. The degeneracies (the transition points) are landmark properties. They can serve, therefore, as a natural basis for parsing the system's states. If the degeneracies do function in the aforesaid manner—if the shifting patterns of forces function as natural boundary-making devices—then it becomes important to explore and evaluate principal π-*number* characterizations of complex biological systems. There is a sense in which the present chapter engaged in such an exploration/evaluation, albeit on a small scale.

It is essential to note, however, that because complex biological systems are governed more by information than by forces, the degeneracies indexed by critical π-*numbers* must be properties of kinematic flow fields more so than properties of kinetic configurations (see Chapters 4 and 7). An important question, therefore, is whether or not the principle of similitude and its applied analytic strategies, which are oriented naturally to kinetic states of affairs, translate to nonmaterial flows? Of some significance in this regard is the discovery of dimensionless quantities that parse the optical flow field into distinct specificational states. The first derivative of $\tau(t)$ (see Section 4.5.3) has a critical value of $-.5$ (Lee, 1980). Below this value the upcoming collision between point of observation and substantial surface will involve an exchange of momentum (if the current kinetic conditions persist). Equal to and above this value, the upcoming collision will involve no exchange of momentum (the point of observation's velocity will go to zero at the moment of contact if the current kinetic conditions persist). A ratio of $\tau(t)$s—roughly, the time to contact the horizontal plane and the time to contact the orthogonal vertical plane passing through the point of observation—also proves to be a dimensionless quantity distinguishing the specificational states of local symmetries of the optical flow (Todd, 1981). Its relevance is to matters such as where a projectile (on an ordinary parabolic trajectory) will land with respect to a point of observation. When the ratio is unity, the landing will be at the point of observation. When the ratio exceeds unity, the landing will be behind the point, and when it is less than unity, the landing will be ahead of the point.

These nondimensional similitude criteria for kinematic flow fields have yet to be rationalized through governing laws. Nevertheless, it is supposed that they share with the major principal π-*numbers* a common basis in law. Insofar as the optical π-*numbers* just identified mark off distinct specificational states with regard to the control of action, and insofar as they are lawfully based, it follows that there is an objective foundation for the most fundamental "psychological" categories—those having to do with the regulation of everyday locomotion.

REFERENCES

Abraham, R. H., & Shaw, C. D. (1982). *Dynamics: The geometry of behavior, Part one—Periodic behavior*. Santa Cruz: Aerial Press.

Arsharvskii, Yu. I., Kots, Ya. M., Orlovskii, G. N., Rodionov, I. M., & Shik, M. L. (1965). Investigations of biomechanics of running by the dog. *Biophysics, 10*, 665-672.

Bernstein, N. A. (1967). *The coordination and regulation of movements*. London: Pergamon Press.

Asatryan, D. S., & Fel'dman, A. G. (1965). Functional tuning of the nervous system with control of movement or maintenance of a steady state posture—I. Mechano-graphic analysis of the work of the joint on execution of a postural task. *Biophysics, 10*, 925-935.

Bizzi, E., Polit, A., & Morasso, P. (1976). Mechanisms underlying achievement of final head position. *Journal of Neurophysiology, 339*, 435-444.

Boylls, C. C. (1975). A theory of cerebellar function with application to locomotion. II. The relation of anterior lobe climbing fiber function to locomotion behavior in cat. *COIN Technical Report* 76-1. Deptartment of Computer and Information Science, University of Massachusetts.

Bunge, M. (1977). *Treatise on basic philosophy. Ontology I: The furniture of the world.* Boston, MA: D. Reidel.

Fel'dman, A. G. (1966a). Functional tuning of the nervous system with control of movement or maintenance of a steady posture—II. Controllable parameters of the muscle. *Biophysics, 11*, 565 578.

Fel'dman, A. G. (1966b). Functional tuning of the nervous system with control of movement or maintenance of a steady posture—III. Mechanographic analysis of execution by man of the simplest motor tasks. *Biophysics, 11*, 766-775.

Fowler, C. A., Rubin, P., Remez, R. E., & Turvey, M. T. (1980). Implications for speech production of a general theory of action. In B. Butterworth (Ed.), *Language production*, New York: Academic.

Gibson, J. J. (1979). *The ecological approach to visual perception*. Boston: Houghton Mifflin.

Glansdorff, P., & Prigogine, I. (1971). *The thermodynamic theory of structure, stability and fluctuations*. New York: John Wiley and Sons.

von Holst, F. (1935/1973). *The behavioral physiology of animal and man*. Coral Gables, FL: University of Miami Press.

Iberall, A. S. A field and circuit thermodynamics for integrative physiology. I, II and III. *American Journal of Physiology: Regulatory Integrative and Comparative Physiology*, *2*, R171 R180, 1977; *3*, R3-R19, 1978; *3*, R85-R97, 1978.

Iberall, A. S., & Soodak, H. (1986). A physics for complex systems. In F. E. Yates (Ed.), *Self-organizing systems: The emergence of order*. New York: Plenum Press.

Irodov, I. E. (1978). *Fundamental laws of mechanics*. Moscow: Mir Publishers.

Kelso, J. A. S., & Holt, K. (1980). Exploring a vibratory system account of human movement production. *Journal of Neurophysiology, 43*, 1183-1196.

Kelso, J. A. S., Tuller, B., Bateson, E., & Fowler, C. A. (1984). Functionally specific articulatory cooperation following jaw perturbations during speech: Evidence for coordinative structures. *Journal of Experimental Psychology: Human Perception and Performance*, *10*, 812-832.

Kelso, J. A. S., Tuller, B., & Harris, K. S. A. (1983). A "dynamic pattern" perspective on the control and coordination of movement. In P. Macneilage (Ed.), *The production of speech*. New York: Springer Verlag.

Kelso, J. A. S., Holt, K., Kugler, P. N., & Turvey, M. T. (1980). On the concept of coordinative structures as dissipative structures: II. Experimental lines of convergence. In G. E. Stelmach & J. Requin (Eds.), *Tutorials in motor behavior*. New York: North-Holland.

Kugler, P. N. (1985). Information as the morphology of dynamics. In W. H. Warren & R. E. Shaw (Eds.), *Persistence and change: Proceeding of the first international conference on event perception*. Hillsdale, NJ: Lawrence Erlbaum Associates.

Kugler, P. N., Turvey, M. T., Carello, C., & Shaw, R. E. (1985). The physics of controlled collisions: A reverie about locomotion. In W. H. Warren, Jr. & R. E. Shaw (Eds.), *Persistence and change*. Hillsdale, NJ: Lawrence Erlbaum Associates.

Kugler, P. N., Kelso, J. A. S. K., & Turvey, M. T. (1982). On the control and coordination of naturally developing systems. In J. A. S. Kelso & J. E. Clark (Eds.), *The development of movement control and coordination*. New York: John Wiley and Sons.

Kugler, P. N., Kelso, J. A. S., & Turvey, M. T. (1980). On the concept of coordinative structures as dissipative structures: I. Theoretical lines of convergence. In G. E. Stelmach & J. Requin (Eds.), *Tutorials in motor behavior*. New York: North-Holland.

Kulagin, A. S., & Shik, M. L. (1970). Interaction of symmetrical limbs during controlled locomotion. *Biophysics*, *15*, 171-178.

Landau, L. D., & Lifshitz, E. M. (1976). *Mechanics*. New York: Pergamon Press.

Lee, D. N. (1980). Visuo-motor coordination in space-time. In G. E. Stelmach & J. Requin (Eds.), *Tutorials in motor behavior*. New York: North-Holland.

Leiber, P. (1968). Constants of nature: Biological theory and natural law. In C. H. Waddington (Ed.), *Towards a theoretical biology*, Volume I. Chicago: Aldine.

Margeneau, H. (1950). *The nature of physical reality*. New York: McGraw-Hill.

Minorsky, N. (1962). *Nonlinear oscillations*. Princeton, NJ: D. van Nostrand.

Nicolis, G., & Prigogine, I. (1977). *Self-organization in nonequilibrium systems: From dissipative structures to order through fluctuations*. New York: Wiley-Interscience.

Prigogine, I. (1980). *From being to becoming*. San Francisco: Freeman and Co.

Rosen, R. (1978). *Fundamentals of measurement and representation of natural systems*. New York: North-Holland.

Schmidt, R. A. (1980). On the theoretical status of time in motor program representations. In G. E. Stelmach & J. Requin (Eds.), *Tutorials in motor behavior*. New York: North-Holland.

Sears, F. W. (1950). *Mechanics*. Cambridge, MA: Addison Wesley.

Sears, F. W., & Zemansky, M. (1951). *Mechanics, heat and sound*. Cambridge, MA: Addison-Wesley.

Sherrington, C. (1906). *The integrative action of the nervous system*. New Haven: Charles Scribner's Sons.

Shik, M. L., & Orlovskii, G. N. (1965). Co-ordination of the limbs during running of the dog. *Biophysics*, *10*, 1148-1159.

Shuring, D. J. (1977). *Scale models in engineering*. New York: Pergamon Press.

Soodak, H., & Iberall, A. S. (1986). Thermodynamics and complex systems. In F. E. Yates (Ed.), *Self-organizing systems: The emergence of order*. New York: Plenum Press.

Stoker, J. J. (1950). *Nonlinear vibrations*. New York: Wiley Interscience.

Todd, J. T. (1981). Visual information about moving objects. *Journal of Experimental Psychology: Human Perception and Performance*, *7*, 795-810.

Turvey, M. T., Fitch, H. L., & Tuller, B. (1982). The Bernstein perspective: I. The problem of degrees of freedom and context conditioned variability. In J. A. S. Kelso (Ed.), *Human motor behavior: An overview*. Hillsdale, NJ: Lawrence Erlbaum Associates.

Turvey, M. T., Shaw, R., & Mace, W. (1978). Issues in a theory of action: Degrees of freedom, coordinative structures and coalitions. In J. Requin (Ed.), *Attention and performance*. Hillsdale, NJ: Lawrence Erlbaum Associates.

Turvey, M. T., Shaw, R. E., Reed, E. S., & Mace, W. M. (1981). Ecological laws of perceiving and acting: In reply to Fodor and Pylyshyn (1981). *Cognition*, *9*, 237-304.

Winfree, A. T. (1980). *The geometry of biological time*. New York: Springer-Verlag (Series in Biomathematics, Vol. 8).

Yates, F. E. (1982). Outline of a physical theory of physiological systems. *Canadian Journal of Physiology and Pharmacology*, *60*, *3*, 217-248.

Yates, F. E. (Ed.) (1986). *Self-organizing systems: The emergence of order*. New York: Plenum Press.

10 SPACING IN THE PENDULAR CLOCKING MODE: THE THEORY OF CONDITIONAL PERIODICITY AND ADIABATIC INVARIANTS IN OPEN CONSERVATIVE AND NONCONSERVATIVE SYSTEMS

ABSTRACT

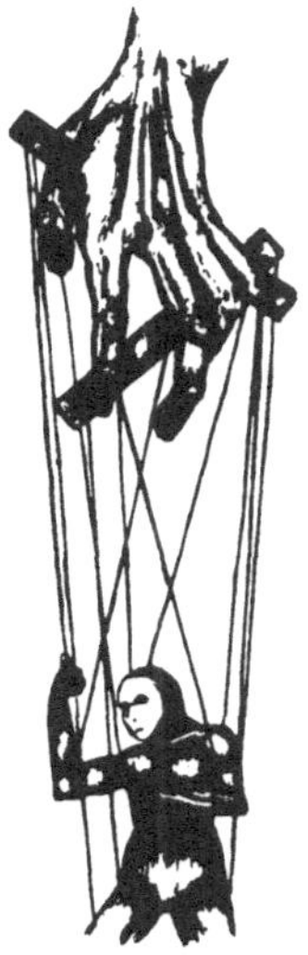

As Chapter 5 highlights, the step-cycles of locomotion are not understandable in the terms of the conventional theory of auto-oscillatory phenomena for two major reasons. First, although the limbs are of fixed mass and length, the step-cycles can operate at indefinitely many near-stable periodicities. Second, this fact of indefinitely many near-stable periodicities implicates a flexible but lawful energy flow process for assembling and sustaining the locomotory step-cycles. With regard to the data of the main experiment, there are two outstanding questions whose resolution would appear to require a principled understanding of the energy flow processes responsible for the aforementioned facts of locomotion. Both questions have to do with the spacing of events: (1) What is the nature of the dependency in the steady state of a single left or right wrist-pendulum system's amplitude on its mass and length? and (2) what law governs the amplitudes assumed by a single left or right wrist-pendulum system of fixed dimensions when it is caused (by isochronous couplings) to assume different, stable periods? With respect to the first question, the data indicate that amplitude tends to vary inversely with mass and directly with length; they also suggest, however, that the signs of the exponents may be changeable from subject to subject. With respect to the second question, the data show that as the periodic time of a wrist-pendulum system

shortens, its amplitude sometimes increases, sometimes decreases, and sometimes remains constant within and across subjects. The present chapter and the next two (Chapters 11 and 12) are devoted largely to a theory that accommodates these data. If the dynamics behind the spacing-timing linkage (question two) can be understood, then the link between amplitude and mass and length might become apparent (question one).

For a system's timing to vary, some physical parameter(s) must change. Systems whose timing varies as a function of variations in physical dimensions of the operational components are referred to as **conditionally periodic** *systems. For the wrist-pendulum paradigm, the mechanical parameters of mass and length are prime candidates. When a wrist-pendulum system of fixed mass and length exhibits variations in timing, however, the "internal" candidate parameters are not so obvious. Of particular interest are the class of internal transformations defined as isothermal and adiabatic. These transformations require that the system's action and entropy production remain time-invariant throughout the transformation. The time-independence of action defines a constraint that scales the energy flowing into the system to the frequency of oscillation. The constraint operates as a* **soft** *conservational constraint. Ehrenfest's theorem defines the adiabatic linkage between a system's period of oscillation, energy carried within the oscillation and the mechanical parameter that is conditionally responsible for (a) bringing new energy into the system and (b) changing the system's operating period.*

The chapter divides into two parts. The first part addresses conditional periodicity in open and closed conservative systems. The second part addresses the problem in the context of open nonconservative systems. Within each of these parts, equation sets are developed identifying adiabatic relationships between energy changes and frequency changes and between amplitude changes and frequency changes.

With respect to the energy by frequency relation, both conservative and nonconservative systems exhibit a positive first order linear relationship whose constant of proportionality identifies the action of the system. One notable feature of the nonconservative case is a nonzero, positive frequency intercept. Another is that the geometry of the nonconservative ***[energy X frequency]*** *space exhibits symmetry with respect to the energy and frequency values in an adiabatic trajectory of states. The positive energy value coincident with twice the intercept frequency is equal to the value of the negative energy intercept. Moreover, whereas the conservative equation for action constancy links amplitude to period changes through a monotonic, nonlinear function, the resultant nonconservative equation links amplitude to period changes through a nonmonotonic, nonlinear function.*

The physical concepts underlying these mathematical descriptions are discussed in detail. The focus is on understanding the physical basis for changes in coordinates of motion as a function of changes in mechanical parameters.

The bottom line is that solutions to conditionally periodic problems do **not** *necessarily require explicit formulation of the equations of motion if the transformations resulting in different periods satisfy adiabatic assumptions. Under these conditions the energy of oscillation varies strictly as a function of period and a constant of proportionality (an adiabatic invariant). Ehrenfest's theorem offers the most concise statement of the relationship between adiabatic invariants and mechanical transformability in conditionally periodic systems. To anticipate, the experimental evidence presented in Chapter 12 suggests that the coordinates of motion defining "comfort" states for wrist-pendulum motion are those that satisfy adiabatic assumptions. The goal of the next two chapters will be to unravel slowly the physical principles underlying adiabatic transformations as they bear on comfort states in conditionally periodic biomechanical organizations.*

10.1. Spacing in the Pendular, Clocking Mode: Identifying the Problem

The pendular, clocking mode of movement organization involves quantities of both time and space, metrics of both duration and distance. In wrist-pendular behavior, for example, the initiation of radial flexion and the initiation of ulnar flexion are separated by an amount of time and by an amount of space. These initiations are events that are observable in the laboratory frame of reference. The timing of one event can be measured relative to the other in seconds and the spacing of one event can be measured relative to the other in degrees of angular excursion. These extrinsic measures of timing and spacing in the wrist-pendular, clocking mode rely on the disclosure of the system's intrinsic metrics (Chapter 2) for their rationalization. One important lesson of general relativity, emphasized several times previously, is that these metrics are not a priori, or strictly geometric, but depend on conservational distributions of a momentary nature.

The primary focus of the preceding chapters has been the question of timing—in particular the question of what determines periodic timing in the steady (or comfort) state. In steady-state quadruped locomotion a single duration metric is common to the four limbs. A single duration metric similarly underwrites the steady state behavior of two coupled wrist-pendulum systems. In previous chapters the single duration metric, the common periodic timing, was seen to scale lawfully to a monomial in mass and length. In turn, the single mass and length quantities composing the monomial were seen to follow lawfully from conservational requirements. The conclusion drawn was that periodic timing in the steady state reflects a unitary process assembled largely through the conservations. It was noted in Chapter 9 that a duration metric, a "clock", is given by dynamical states that order asymmetrically and transitively, so that no two states are simultaneous. The unitary process assembled through the

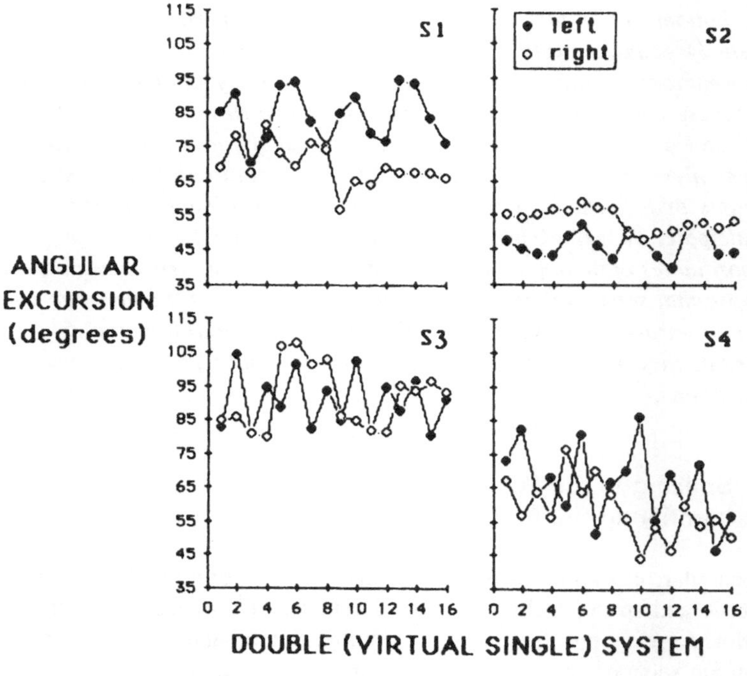

FIGURE 10.1. Angular excursions of left and right components of the virtual single systems of each subject in the main experiment.

conservations can, therefore, be given a more precise identification: It is a single, strict partial ordering of dynamical states.

The focus is now turned to the distance metric, to the task of providing a principled account of the spacing of events that characterizes steady state pendular, clocking behavior. The data from the main experiment reported in Chapter 6 identify the challenge. To begin with, where two coupled limbs clock with a common timing they need not clock with a common spacing. This fact is made clear in Figure 10.1 which presents the left and right wrist pendulum angular excursions for each double (virtual single) system for each subject (see Tables 6.8 to 6.11). As is evident from the inspection of Figure 10.1, the spacings of the left and right wrist-pendular motions are not identical. If the one tempo implicates a single set of dynamical conditions governing the timings of both systems then the two angular excursions implicate a double set of dynamical conditions, one set governing the spacing of the left system and one set governing the spacing of the right system. In other words, the data presented in Figure 10.1 imply that the spacing of events in the pendular, clocking mode assembled over two wrist-pendulum systems is determined primarily (but probably not exclusively) by local conditions, those operating at the sites of the individual wrist-pendulum systems. Spacing contrasts, therefore, with timing.

The latter sentences contain the idea that the lawful bases of spacing and timing in steady state pendular, clocking behavior may be distinct. This idea is suggested further by a consideration of the angular excursions of the single wrist-pendulum systems reported in the main experiment. These are presented in Figure 10.2. The four single systems are ordered by the magnitude of $M^{1/16}L^{1/2}$, from lowest (System A) to highest (System D). (The labeling convention is that of Table 6.1.) The ordering, therefore, is in terms of increasing natural period. It is evident from inspection of Figure 10.2 that angular excursion is not ordered by the moment variable governing periodic timing. In contrast to the A, C, B, D order of natural periods, angular excursions order as B, D, A, C for the left and right systems of Subjects 1, 2, and 3. For Subject 4 the ordering of angular excursions is A, B, D, C for the left systems and B, A, C, D for the right systems.

There are a number of systematicities in the single system angular excursion data reported in Figure 10.2 (and Table 6.3). The difference between System A and System C and the difference between System B and System D is mainly one of mass. From the "partial derivative" perspective of Section 7.5, that is,

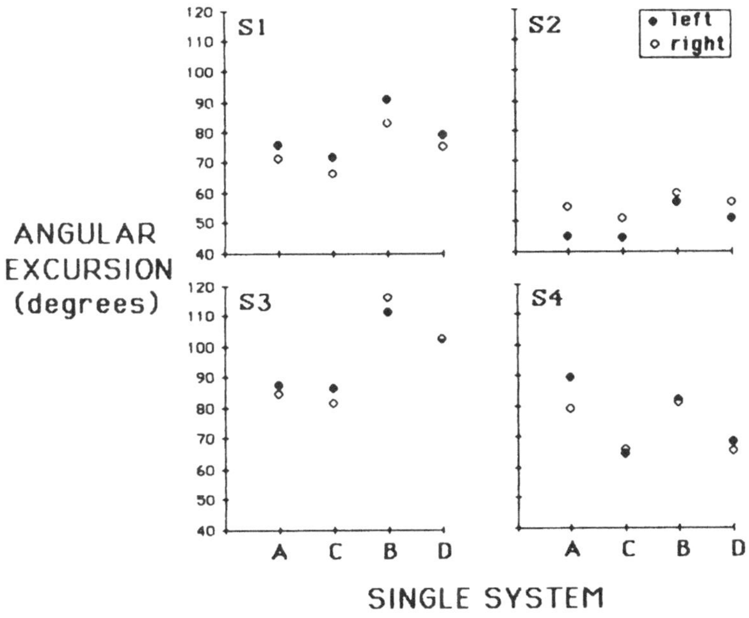

FIGURE 10.2. Angular excursions of single systems in the main experiment. Letter labels are interpreted in Table 6.1.

assuming a change in mass with length held constant, the conclusion would be that angular excursion is inversely related to mass (Figures 7.16 and 7.18 make just that point). For all four subjects, and for both left and right systems, the angular excursions of A and B were greater than the angular excursions of C and D, respectively. Similarly, from the "partial derivative" perspective (assuming now a change in length without a change in mass) the conclusion would be that angular excursion is directly related to length (a point that is made by Figures 7.15 and 7.17). Length more than mass distinguishes System B from System A (B is longer) and System D from System C (D is longer). Inspection of Figure 10.2 and Table 6.3 reveals that, in most instances, the angular excursions of B and D exceed, respectively, those of A and C. Instances where they do not are to be found in Subject 4's data. The angular excursion of left System A (89.04 degrees) was greater than that of left System B (81.70 degrees). A difference in this direction was evident on 6 of the 8 trials (see Figure 7.15a). Further, the angular excursion of right System C (65.43 degrees) was very slightly greater than that of right System D (65.26 degrees). A large difference in this direction was present on three of the eight trials (see Figure 7.15f).

Together, the systematic features and the anomalies of the single systems' angular excursion data suggest that the scaling relation governing spacing may not be of the same *type* as that governing timing. Where the scaling relation governing timing appears to be "absolute," the scaling relation governing spacing may be "relative." Where the direction of the relation between representative biological quantities of *time*, mass and length (increase mass, increase time, increase length, increase time) was invariable from frame to frame the direction of the relation between representative biological quantities of *space*, mass, and length was variable (for example, increase length, increase space, but, in another frame, increase length, decrease space). A difference between the timing and spacing scaling relations is suggested by other considerations. Quadruped locomotion exhibits a spacing/mass relation similar to that seen in the single wrist-pendulum data. Maximum angular excursion at the hip diminishes as body size increases. This amplitude reduction, in proportion to $M^{-1/8}$, is said by McMahon (1975; 1977) to be consistent with elastic similarity criteria (see Chapter 8). It is not clear, however, that these criteria can give rise simultaneously to a positive scaling of angular excursion to length, such as in the single wrist-pendulum behavior of Subjects 1, 2, and 3. Nor is it clear how these criteria alone could give rise to both positive and negative exponents, as in the data of Subject 4. The spacing of events in the pendular, clocking mode is scaled to a monomial in mass and length with, perhaps, the following characteristics: The exponents derive, in part, from lawful relations among quantities that are not essentially spatial and, in part, from the properties of the coordinate spaces in which these lawfully related quantities are embedded.

Figure 10.2 and Figure 8.5 show, respectively, that differences in angular excursion and differences in periodic timing are tied to differences in the

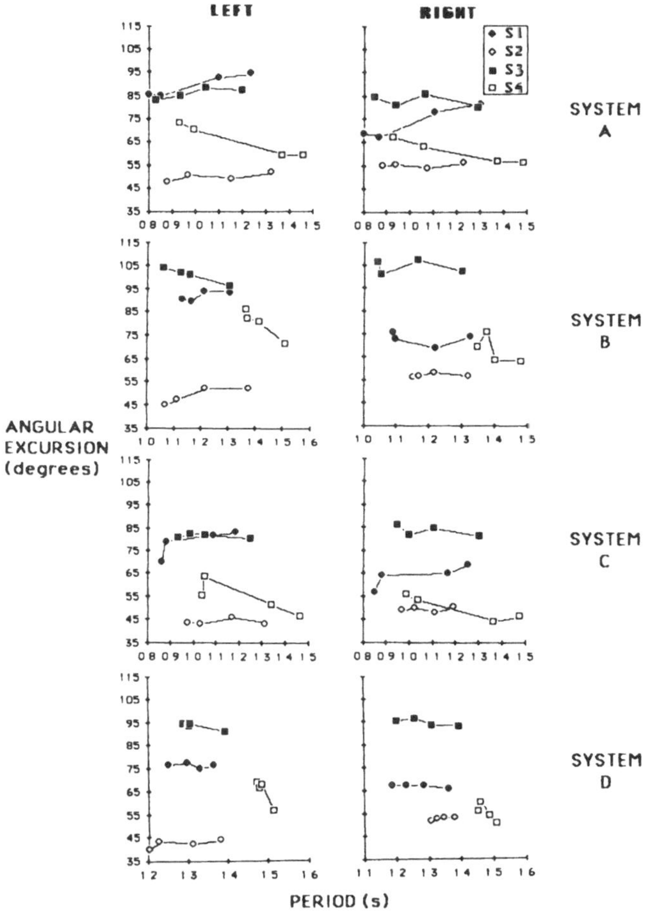

FIGURE 10.3. Angular excursions as a function of period for each left and right component system in the 16 virtual single systems of the main experiment. See text for details.

mass/length magnitudes of the wrist-pendulum systems. Renewing the developmental image of Section 8.4.4, as a wrist-pendulum system becomes longer and/or heavier its spacing and timing change. The "growing" is a mechanical transformation. Changing the mass/length magnitudes changes the pattern of forces which changes, in turn, the mechanical properties of angular excursion (twice amplitude) and period. Angular excursion and period are shown to change jointly in Figure 10.3. The joint changes depicted in this figure do not follow, however, from changes in the mass/length magnitudes. Each panel of the figure presents the oscillatory parameters of a single wrist-pendulum system of *fixed* mass and length. The trajectories of (angular excursion, period) coordinates depicted in a panel follow from changing the wrist-pendulum system to which the depicted fixed wrist-pendulum system is coupled. By the design of the main experiment, any particular wrist-pendulum system (e.g., left A) was coupled with each of the four wrist-pendulum systems of the other wrist (right A, B, C, D, as in the top left panel of Figure 10.3). As a result it clocked at four different periods. That is to say, its periodic timing was conditional on the couplings.

From the perspective of the fixed wrist-pendulum system, the succession of couplings is a sort of mechanical transformation. Although the system's mass and length are not altered by the couplings, the forces at play are altered (Chapter 9). The result is a change in the systems's oscillatory parameters. Inspection of the various curves in the 8 panels of Figure 10.3 reveals that as the timing changes so does the spacing, more or less. The relation between timing and spacing is not fixed, however. Sometimes angular excursion increases as period increases. Sometimes angular excursion decreases as period increases. And sometimes angular excursion appears to change very little, if at all, as period increases. The increases in period follow lawfully from the increases in the $M^{1/16}L^{1/2}$ magnitudes of the four virtual single systems in which the fixed single system functions as an atomism. That is, the changes in the fixed system's duration metric are understandable through the periodic timing law. But how are the corresponding changes and nonchanges in its distance metric to be understood? What relates the changes in the spacing of a wrist-pendulum's radial and ulnar flexions to the changes in their timing?

In Chapter 5 it was noted that the step cycles of locomotion make very particular demands on a theory of auto-oscillatory phenomena. The limbs are of fixed mass and length. Nevertheless, the step cycles can and do operate at indefinitely many near-stable periodicities. A limb is mechanically transformable, with the transformation effected "softly" through internal tissue properties. The multiple near-stable periodicities of locomotion that result from this particular style of mechanical transformation implicate a flexible but lawful energy flow process for assembling and sustaining step cycles. In both steady state wrist-pendulum behavior and locomotion there is, therefore, an issue of how the mechanical transformability of the biomechanical oscillator is to be interpreted. The interpretation given for both instances of rhythmic movement must carry with it an account of the spacing that reflects the energy demands of the mechanically transformed system. Can there be a very general solution to this problem of relating steady state spacing and timing in rhythmic biological systems? By the scalar perspective detailed in the preceding chapter, any such general solution must be tied to one or more constant quantities.

A constant quantity linking spacing to timing will be advanced (in the present chapter and the next two) that is a hybrid of two different quantities, a conserved quantity (energy) and a quantity of motion (frequency). As is true of the conservations, this hybrid quantity is invariant only with respect to transformations (symmetry operations) of a particular kind. In general terms the hybrid quantity is referred to as an adiabatic invariant and the relevant automorphisms are referred to as adiabatic transformations. That such a quantity and its automorphisms might be in evidence at all of nature's scales and able, thereby, to play a strategic role in all of her processes, is suggested by the work of Ehrenfest (1913, 1914, 1916, 1917, 1959). Where the insights of Huygens guided the efforts in preceding chapters to understand the principled basis of timing in the steady-state pendular, clocking mode it is the insights of Ehrenfest that guide the efforts in this chapter and the next two to understand the principled basis of spacing.

The present chapter is especially technical and lengthy. It is so for two reasons. First, the extension of the concept of adiabatic invariant to the nonconservative province of living things and their movements breaks new ground. New equation sets are required. Second, the concept of adiabatic invariant does not receive much treatment in standard physics textbooks. Consequently, much of the groundwork needs to be laid out in detail. The development of the concept of adiabatic invariant starts with the basic concepts of rotational inertia, torque, and the conservation of angular momentum.

10.2. Mechanical Transformability in "Rotating" Conservative Systems

10.2.1. Rotational Inertia. The tendency for an object in rotation to continue to rotate is referred to as *rotational inertia* and constitutes the rotational equivalent of Newton's first law, the law of inertia. The law of rotational inertia can be stated as follows: "A rotating body continues to rotate uniformly unless forces act to change this motion." Rotational inertia is associated with the mass of an object and its distributional characteristics.

10.2.2. Angular Momentum. The rotational equivalent of linear momentum is *angular momentum*. For a particle rotating about a fixed point O, as in Figure 10.4, the angular momentum is defined as the *cross product* (x) of the radius vector r (from the particle's center of mass to O) and the particle's linear momentum p (Landau and Lifshitz, 1976). In symbols, the angular momentum J is defined as:

$$J = r \times p = r \times mv \,. \qquad (10.1)$$

Angular momentum is a vector, its direction being given by the right-hand rule for cross products. The vector J points in the direction perpendicular to the plane containing the radius vector and the linear momentum (or velocity) vector (see Figure 10.5). Angular momentum has the dimensions of linear momentum multiplied by length, (ML^2/T). The units of angular momentum are $gm\text{-}cm^2/sec$ (or *erg-sec*) or $kg\text{-}m^2/sec$ (or *joule-sec*). The component of J along any line (or

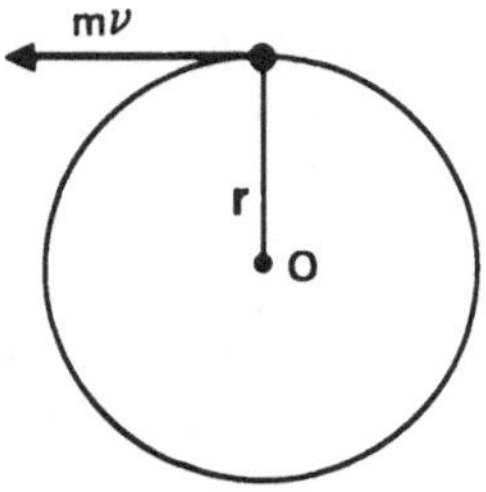

FIGURE 10.4. ANGULAR MOMENTUM ($mv \times r$) OF A ROTATING OBJECT.

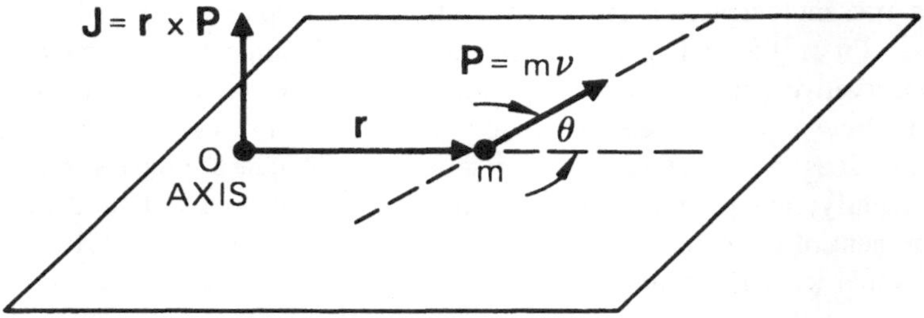

FIGURE 10.5. ANGULAR MOMENTUM ($J = r \times P = r \times mv$) OF A PARTICLE OF MASS m, VELOCITY v.

axis) passing through the fixed reference point O is the angular momentum of the particle about the axis r.

The total angular momentum of a collection of particles (or an extended body) is the vector sum of the individual angular momenta of the particles (or composite parts). If the mass of the ith particle is m_i, and the radius of its path is r_i, and the speed is v_i (the particle's velocity vector being perpendicular to its radius vector), then the magnitude of the angular momentum of the ith particle relative to the axis of rotation is

$$J_i = m_i v_i r_i \ . \tag{10.2}$$

The direction of the angular momentum is the same as that of its angular velocity vector ω (see Figure 10.6). The particle's speed v_i is related to the angular speed ω by $v_i = r_i\omega$, and can be stated in terms of angular momentum as

$$J_i = m_i r_i^2 \omega \ . \tag{10.3}$$

The total angular momentum of the rigid ensemble is derived by summing the contributions to J from all composite particles, recognizing that all particles have the same angular velocity ω:

$$J = \sum_i m_i r_i^2 \omega \tag{10.4}$$

where the angular momentum J and the angular velocity ω are all taken with reference to the same center of rotation. For bodies that are symmetrical with respect to the axis of rotation, the moment of inertia is a scalar multiple of the angular velocity; for nonsymmetrical bodies, the moment of inertia is a tensor and the direction of the angular momentum is not necessarily along the direction of ω.

10.2.3. Torque and the Constancy of Angular Momentum. The torque (or turning moment) N about the fixed point O is defined by:

$$N = r \times F \tag{10.5}$$

where F is the force acting on the particle. The relationship between torque and angular momentum can be derived by differentiating Equation (10.1) with respect to time

$$\frac{dJ}{dt} = \frac{d}{dt}(r \times p) = \left(\frac{dr}{dt} \times p\right) + \left(r \times \frac{dp}{dt}\right) . \tag{10.6}$$

Because the vector product of any two parallel vectors is zero, the first term in (Equation 10.6) can be eliminated.

$$\frac{dr}{dt} \times p = v \times mv = 0 . \tag{10.7}$$

The second term is a statement of Newton's second law in an inertial frame of reference,

$$r \times \frac{dp}{dt} = r \times F = N . \tag{10.8}$$

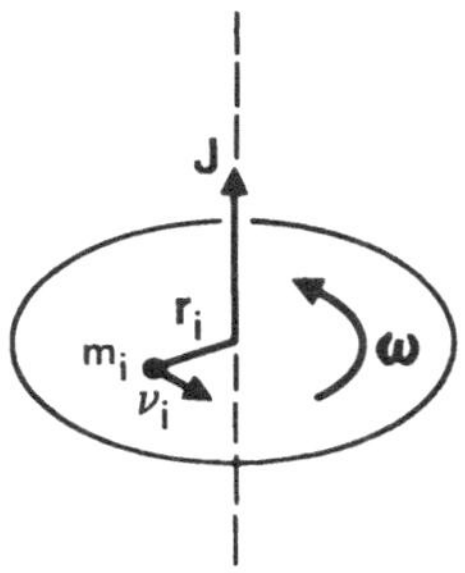

FIGURE 10.6. ANGULAR MOMENTUM ($J = m_i r_i v_i$) OF A PARTICLE i OF A RIGID BODY ROTATING AT ANGULAR VELOCITY ω.

From Equations (10.6), (10.7), and (10.8), the following relationship can be defined between torque *N* and angular momentum *J*,

$$\frac{dJ}{dt} = N; \tag{10.9}$$

the time rate of change of angular momentum is equal to the torque. If a system is isolated from a resultant external torque N_{ext}, the rate of change of the total angular momentum is zero, and the total angular momentum remains constant. An external torque is the resultant of an externally generated force applied to a body. In contrast, an internal torque is the resultant of an internally generated force exerted by one part of a body onto another part of a body. Internal torques are *independent* of external torque. Variations in external torque are associated with variation in angular momentum. In contrast, angular momentum is conserved over variations in internal torque. The conservation of angular momentum applies to closed orbits, open orbits, and all collision processes.

$$\text{If: } \sum N_{ext} = 0; \text{ Then: } \frac{d}{dt}\sum J = 0; \text{ And: } \sum J = \text{constant.} \tag{10.10}$$

If $N = 0$, then J is constant in time. From Equation (10.2), the angular momentum is conserved if either $F = 0$ or F (radial force) is proportional to r.

10.2.4. Constancy of Angular Momentum and Mechanical Transformability in an Open Conservative System. The angular momentum of a system is invariant ($J = constant$) in the absence of any external torques ($N_{ext} = 0$). The constancy of angular momentum, however, does not imply the constancy of (maximum kinetic) energy. For example, consider the rotational situation depicted in Figure 10.7, where a particle of mass *m* is moving in a central force field derived from a potential $V(r)$. The radial axis r can be varied by pulling the string that passes through a tube at the axis of rotation. The only restrictions on this transformation are that the variations in the mechanical parameter must cause neither a degeneracy in the coordinates of the motion nor a generation of external torque. For conservative systems these requirements can be satisfied if the mechanical transformation is performed very slowly relative to the period τ_o of the motion (cf. Landau and Lifshitz, 1976). Extremely slow changes of this kind are named adiabatic. In the case of the transformation in r_o, the adiabatic criterion can be satisfied if

$$\tau_o \frac{\Delta r_o}{dt} \ll r_o\,. \tag{10.11}$$

The mass is rotated at a velocity v_o with a string of radial length r_o. The angular momentum of the rotating particle is defined by Equation (10.2) as

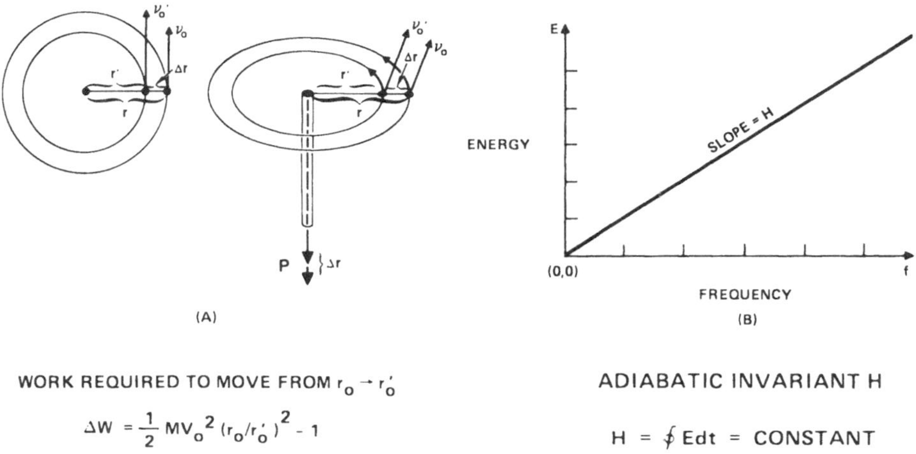

FIGURE 10.7. ROTATING MASS WITH A VARIABLE RADIUS OF ROTATION: AN OPEN CONSERVATIVE SYSTEM.

$$J_o = mv_o r_o \,. \tag{10.12}$$

The kinetic energy of the particle is defined as

$$E_o = \frac{1}{2} mv_o^2 \,. \tag{10.13}$$

Let the radial distance r_o be shortened to r_o' by pulling very slowly on the string at P. If the force F_i on the particle resulting from the shortening of the string is radial and proportional to r_i, so that $F_o/r_o = F_o'/r_o'$, the resulting external torque is zero. The angular momentum is invariant under the mechanical transformation of $r_o \rightarrow r_o'$:

$$mv_o r_o = mr_o' v_o' = \text{constant} \,. \tag{10.14}$$

The velocity v_o' associated with the new length r_o' can be derived from Equation (10.14):

$$v_o' = v_o[r_o / r_o'] \,. \tag{10.15}$$

The kinetic energy associated with the motion at r_o' can be calculated in terms of v_o'

$$E_o' = \frac{1}{2} mv_o'^2 = \frac{1}{2} m_o v_o^2 [r_o / r_o']^2 \,. \tag{10.16}$$

The change in energy from E_o to E_o' is associated with the work W introduced from outside the system in the shortening of the string from r_o to r_o':

$$W = \frac{1}{2} mv_o^2[(r_o / r_o')^2 - 1] . \tag{10.17}$$

The angular momentum acts on the radial motion as an effective repulsive potential energy: Extra work W must be put into the system in order to move the particle from distance r_o to distance r_o' against the potential gradient, $\nabla V(r)$. Put simply, the rotating system must be *open* to the flow of new energy into the system if angular momentum is to be invariant under changes in the radius of rotation. This fact is highlighted in the next section.

10.2.5. Constancy of Energy and Mechanical Transformability in a Closed Conservative System. In the preceding example angular momentum is invariant under rotational transformations that change periodic timing. The invariance of angular momentum under rotational transformation is the principle of angular momentum conservation. Angular momentum, however, is not invariant under all rotational transformations in conditionally periodic systems. Consider a particle that rotates freely on a string attached to a fixed peg of diameter d (see Figure 10.8). As the particle rotates, the string winds up on the surface of the peg, resulting in a systematic decrease in the radius of the trajectory. The radius decreases by $2\pi d$ with each complete rotation. In contrast to the preceding situation (Section 10.2.4), the changing radius r_o is not associated with a corresponding change in total energy E_o. The kinetic energy of motion remains constant, indifferent to changes in the radial length. Under these conditions, the change in radius, $r_o \rightarrow r_o'$, is coupled to a change in angular momentum, $J_o \rightarrow J_o'$. The rapid change in the radial length gives rise to an external torque that decreases the angular momentum. Recall: $J = 0$ and $N = 0$, if $r_o = constant$ or $r_o/r_o' = F_o/F_o'$. In the situation in which the kinetic energy is constant, the central force F_o no longer relates proportionally to variation in the radial axis r_o.

10.3. Constancy of Action and Mechanical Transformability in an Open Conservative (Pendulum) System

The phenomenon of the mechanical transformability of periodic timing (i.e., conditional periodicity) is now taken up in the context of a simple (string) pendulum system (see Born, 1945; ter Haar, 1982 for additional details). The system is composed of a point mass attached to a string that passes vertically through a hole in a horizontal board, such that the length of the pendulum can be decreased by slowly pulling the string up through the hole at the axis of rotation (see Figure 10.9). The rate of decrease of length must be extremely slow compared with the frequency of swing, so that one can still speak of a

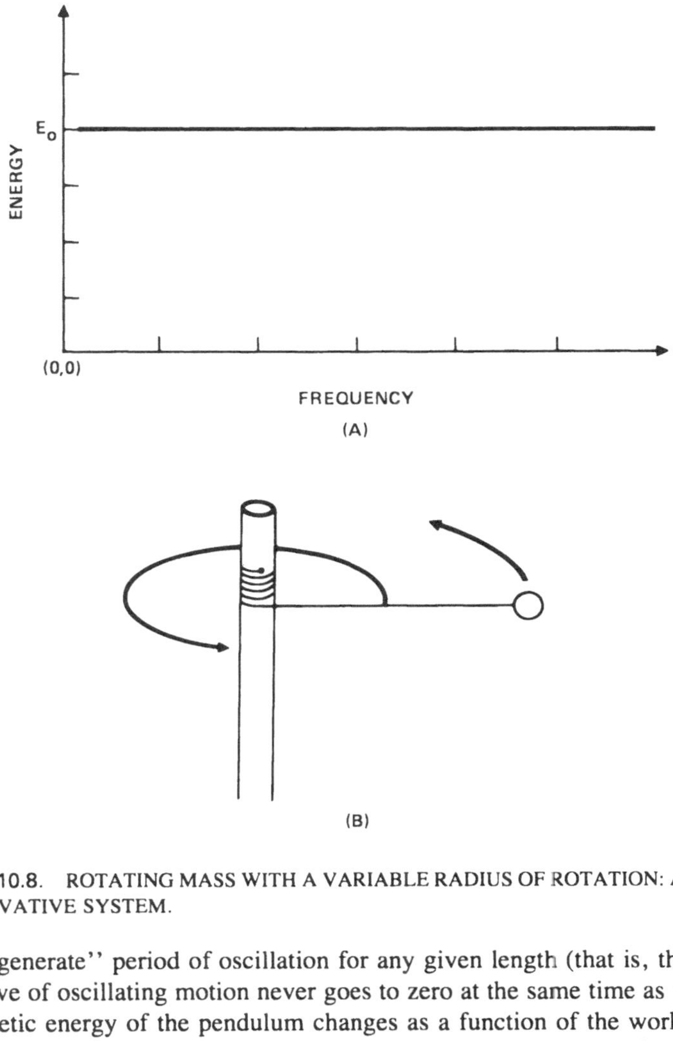

FIGURE 10.8. ROTATING MASS WITH A VARIABLE RADIUS OF ROTATION: A CLOSED CONSERVATIVE SYSTEM.

"nondegenerate" period of oscillation for any given length (that is, the second derivative of oscillating motion never goes to zero at the same time as the first). The kinetic energy of the pendulum changes as a function of the work done in pulling against the tension of the string. The frequency of the pendulum changes as a function of the length change. If the externally induced change in the length of the string pendulum is at a rate that is slow compared to the intrinsic frequency (that is, if the change is adiabatic) then the *[energy/frequency]* ratio will be constant throughout the process. The proof follows.

When the string is shortened, work is done, first against gravity, and second against the centrifugal force of the oscillating pendulum. If λ is the length of the pendulum, m the mass (concentrated at a point), and f the frequency, then

$$2\pi f = \sqrt{g/\lambda} \tag{10.18}$$

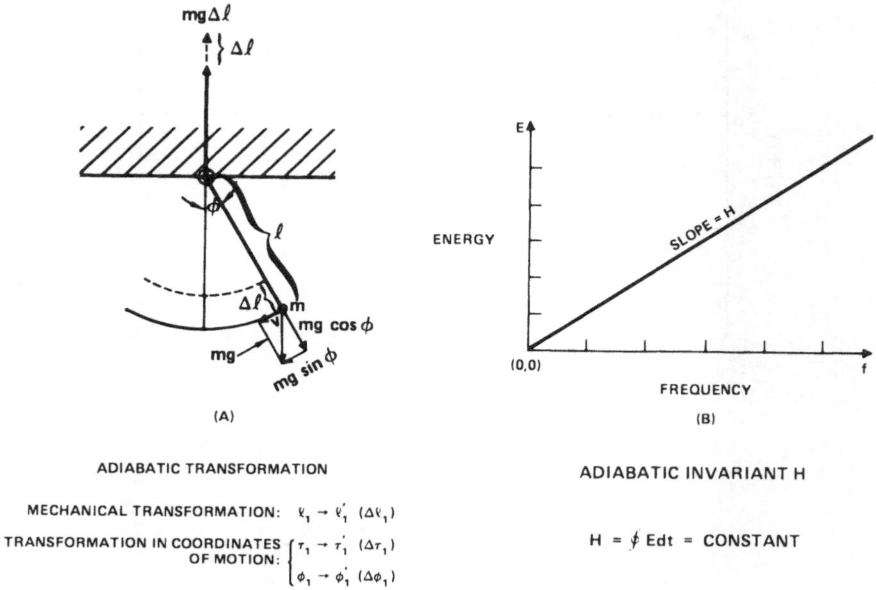

FIGURE 10.9. SIMPLE PENDULUM OF VARIABLE LENGTH: AN OPEN CONSERVATIVE SYSTEM.

Allow the length of the pendulum to be altered from λ to $\lambda + \Delta\lambda$; in the present example let $\Delta\lambda$ be a negative number, so that the pendulum is shortened (instead of elongated). The component of the weight that stretches the string is $mg\cos\phi$ (ϕ = the angle of the instantaneous deflection), and the centrifugal force is $m\lambda[d\phi/dt]^2$. The work W done against gravity and centrifugal force is

$$W = \int_{\lambda}^{\lambda+\Delta\lambda} [mg\cos\emptyset + m\lambda(d\emptyset/dt)^2][-\Delta\lambda] \; . \qquad (10.19)$$

Assume that the shortening of the string takes place very slowly, so that during the process the pendulum oscillates a great many times. The next step is to integrate over the motion so as to obtain

$$W = -[mg\overline{\cos\emptyset} + m\lambda\left(\frac{d\emptyset}{dt}\right)^2]\Delta\lambda \; , \qquad (10.20)$$

where the bar indicates averaging over the undisturbed motion. If the amplitude is small then $\cos\emptyset$ can be replaced by $[1-(\emptyset^2/2)]$ yielding

$$W = -mg\Delta\lambda + [mg\overline{(\varnothing^2/2)} - m\lambda\overline{\left(\frac{d\varnothing}{dt}\right)^2}]\Delta\lambda$$
$$= -mg\Delta\lambda + \Delta E\,. \tag{10.21}$$

The first term corresponds to the elevation of the position of equilibrium. The second term, that is, the product of $\Delta\lambda$ and the expression in brackets, represents the increase ΔE in the energy of the pendulum motion. The energy of the undisturbed motion is

$$E = (m/2)\lambda^2\left(\frac{d\varnothing}{dt}\right)^2 + mg\lambda(1 - \cos\varnothing)\,, \tag{10.22}$$

where the first term represents the kinetic energy, and the second represents potential energy relative to the resting position. Replacing $(1 - \cos\varnothing)$ by the approximation $(\varnothing^2/2)$, reduces the preceding expression to

$$E = (m/2)\lambda^2\left(\frac{d\varnothing}{dt}\right)^2 + mg\lambda(\varnothing^2/2)\,. \tag{10.23}$$

This is the energy function for a linear oscillator with the linear amplitude ($q = \lambda\varnothing$, see Section 10.4.1). The motion is therefore simple harmonic oscillation $\varnothing = \varnothing_o \cos\omega\tau$, and

$$\bar{\varnothing}^2 = (\varnothing_o^2/2)\,, \quad \overline{\left(\frac{d\varnothing}{dt}\right)^2} = (\varnothing_o^2\omega^2/2)\,, \tag{10.24}$$

from which (since $\omega = 2\pi f = \sqrt{g/\lambda}$) it follows that

$$E = (m\lambda^2\varnothing_o^2\omega^2)/2 = mg\lambda(\varnothing_o^2/2)\,, \tag{10.25}$$

while, from Equation 10.21,

$$\Delta E = -\frac{1}{4}(m\lambda\omega^2\varnothing_o^2)\Delta\lambda = -\frac{E}{2\lambda}\Delta\lambda\,. \tag{10.26}$$

Rearranging terms yields,

$$\frac{\Delta E}{E} = -1/2\frac{\Delta\lambda}{\lambda}\,. \tag{10.27}$$

And, since f varies as a function of $1/\sqrt{\lambda}$ (from $2\pi f = \sqrt{g/\lambda}$), it follows that

$$\frac{\Delta f}{f} = -1/2\frac{\Delta\lambda}{\lambda}. \tag{10.28}$$

Substituting from above,

$$\frac{\Delta E}{E} = \frac{\Delta f}{f}. \tag{10.29}$$

This is a differential equation for E as a function of the frequency f, and its solution is

$$\frac{E}{f} = \text{constant} = H. \tag{10.30}$$

Therefore, during the extremely slow (adiabatic) shortening of the pendulum, the quantity of action H remains constant while the period and maximum amplitude of oscillation co-vary. Figures 10.10a,b identify a coordinate space defined by the trajectory of equilibrium states associated with the adiabatic transformation. It is an "intrinsic coordinate space" because the coordinates are linked to internal states of the system. The following relations relate the equilibrium state space (δ, mean free path) and time (τ, mean relaxation time) metrics to the mechanical length (λ) parameter:

$$E = f(\tau,\delta;\lambda); \tag{10.31}$$

$$E \propto [\lambda\delta^2/\tau^2]; \tag{10.32}$$

$$H = g(E,f); \tag{10.33}$$

$$f = (1/\tau); \tag{10.34}$$

and finally

$$H = [E/f] \propto [(\lambda\delta^2/\tau^2)/(1/\tau)], \tag{10.35}$$

which reduces to

$$H \propto [\lambda\delta^2/\tau] = \text{constant}. \tag{10.36}$$

To repeat, an adiabatic transformation of the length of the pendulum results in the *constancy of action*, H (see Rayleigh, 1902, and Ehrenfest, 1913, 1916, 1917). Action, a scalar quantity, is the cycle integral of kinetic energy with respect to period. It should not be confused with the vector quantity of angular momentum, J. (Recall that angular momentum is the cross product of the radius

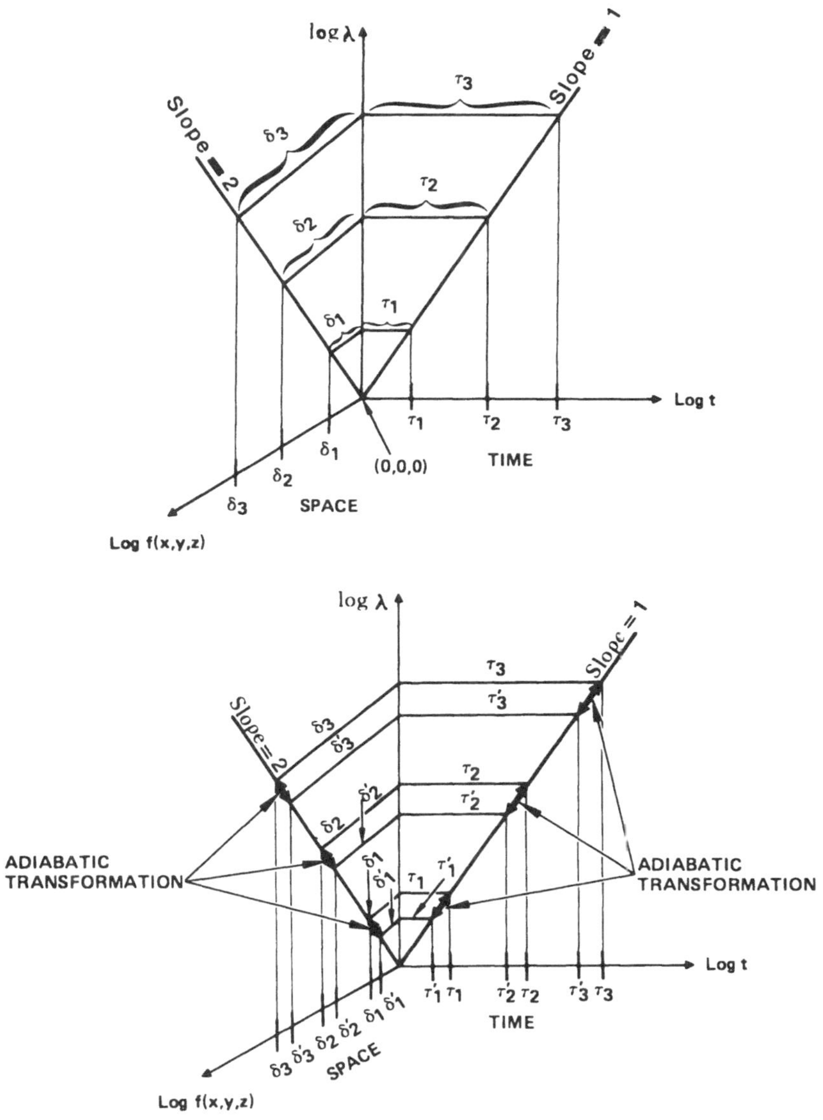

FIGURE 10.10. SCALING RELATIONSHIP BETWEEN PERIOD AND MAXIMUM AMPLITUDE OF OSCILLATION DEFINED BY CONSTANT ACTION: THE BUILDING MATERIALS FOR AN INTRINSIC COORDINATE SPACE. (a) Three periods (τ_1, τ_2, τ_3)) and their respective amplitudes (δ_1, δ_2, δ_3) that are generated by three different lengths (λ_1, λ_2, λ_3) of the string pendulum identified in Figure 10.6. If the initial energy conditions of the string oscillator are scaled in accordance with the same action constant then the log slope of $(\lambda/\tau) = 1$ (actually, -1, but the sign is ignored for convenience) and the log slope of $(\lambda/\delta) = 2$. These identify the two scaling coefficients relating period and amplitude of oscillation, respectively, to the length of the string pendulum. (b) Covariations in observed periods and amplitudes under slow (adiabatic) transformation in the lengths of the oscillators ($\lambda_1 \rightarrow \lambda'_1$, $\lambda_2 \rightarrow \lambda'_2$, $\lambda_3 \rightarrow \lambda'_3$).

vector and linear momentum. Both action and angular momentum have the same dimensions, ML^2/T.)

By way of concluding, it is valuable to note that Equation (10.30) is an instance of a general formulation: the *action integral of the kinetic energy E_o taken over a period, is constant*:

$$H = 2\int_0^{\tau} E_o dt = 2\tau E = \frac{2E}{f}. \tag{10.37}$$

10.3.1. Adiabatic Invariants, Quantization of Action and Bohr's Original Quantum Mechanical Formulation. The quantization of invariant quantities (action) in systems undergoing changes in periodic states was central to Bohr's original formulation of the quantum theory (Kramers, 1957). Note the similarity between Bohr's original quantum equation

$$nh = 2\int_0^{\tau} E_{kin} dt \qquad (h = \text{Planck's constant}), \tag{10.38}$$

and Equation (10.37) where the quantum of action, H, in Equation (10.37) is replaced by the principal quantum number n and Planck's constant. (Recall that in the foregoing examples constancy of action was distilled from very slow changes that guaranteed no degenerate states. Intermediate values between the n_i's would give rise to such degeneracies in the coordinate motion.) At first the Bohr formulation was interpreted as entailing a "nonclassical" constraint. Through Ehrenfest's efforts (see Klein, 1970) it became evident, however, that these periodic mechanical systems, characterized by magnitudes whose values do not change when the systems are subjected to slowly varying forces, could be understood without the aid of nonclassical "quantum" assumptions. Ehrenfest termed these "semi-permanent magnitudes" *adiabatic invariants* (see also, d'Abro, 1952; Dugas, 1948, for additional historical accounts). Their recognition led to a dramatic revision in the interpretation of conservational constraints.

Classical mechanics views conservational constraints as rigidly fixed in either space (angular and linear momentum) or time (energy and matter). In comparison, an adiabatic invariant (such as action) is a "hybrid" constraint built out of a product of conserved entities (energy, E; momentum, p) and coordinates of motion (frequency, f; amplitude, q): *action* $= \Delta E/\Delta f$ and *action* $= \Delta p/\Delta q$. These hybrid conservation-like, soft constraints form a bridgework that links the classical equations of motion to the equilibrium states of statistical mechanics and thermodynamics (and, ultimately, the nonequilibrium states associated with the second law of thermodynamics). Ehrenfest's conceptual bridgework linking classical mechanics to statistical thermodynamics is the topic of the next section; it identifies Ehrenfest's generalization of adiabatic invariants from the special case of simple harmonic motion to the general case of ensembles of periodic systems at equilibrium). Using adiabatic invariants, Ehrenfest was able to generalize the statistical foundation of the second law of thermodynamics while still satisfying the potential "undermining" effects of the quantum theory.

10.4. Ehrenfest's Generalization of Adiabatic Invariants and Mechanical Transformability to Ensemble Conservative Systems

Ehrenfest's theoretical generalization relating thermodynamic properties to statistical mechanical properties via adiabatic invariants is summarized. The summary draws on Klein's account from Ehrenfest's personal notes (1970, pp. 279-282):

Consider the mechanics of periodic motions defined over an ensemble of particles and the corresponding conditions on mechanical transformability that result in the constancy of action. The motion of a system is described completely by expressing the energy as a function of the coordinates q_k and the momenta p_k, the so called Hamiltonian function $H(q_1, q_2, \ldots, p_1, p_2, \ldots)$ (see Goldstein, 1950, and Whittaker, 1937, for a review of the Hamiltonian and its relation to the action integral.) For ordinary rectangular coordinates the momentum p_k becomes $m_k\dot{q}_k$ and the equations of motion can be written as (the dot denotes differentiation with respect to time):

$$\dot{q}_k = \frac{\partial H}{\partial p_k}, \text{ and } \dot{p}_k = -\frac{\partial H}{\partial q_k}, \tag{10.39}$$

where H in these equations refers to the Hamiltonian function (not to action).

In ensemble systems, composed of many degrees of freedom, the energy of each particle is replaced by a generalized energy measure of an average molecule (see Tolman, 1938, for a review). The average measure is derived from a distribution function that assigns to each particle a mean kinetic energy. This energy state is identified with particular coordinates of motion (*displacement*-q_k; *velocity*-p_k). Figure 10.11 identifies an example of an ensemble system undergoing an adiabatic transformation. An ideal gas is enclosed in a chamber with a piston that is manipulated externally so as to vary conditionally the periodic motion of the particles.

A generalized force corresponding to changes in λ, is given by the negative derivative, $-\partial\epsilon/\partial\lambda$, where ϵ is the energy, defined as

$$\epsilon = (q_1, \ldots, q_k, p_1, \ldots, p_k; \lambda_i), \tag{10.40}$$

or, in a more concise notation,

$$\epsilon = \epsilon(q, p; \lambda). \tag{10.41}$$

Let $f(q, p; E, \lambda)$ identify a stationary (time-independent) distribution function in

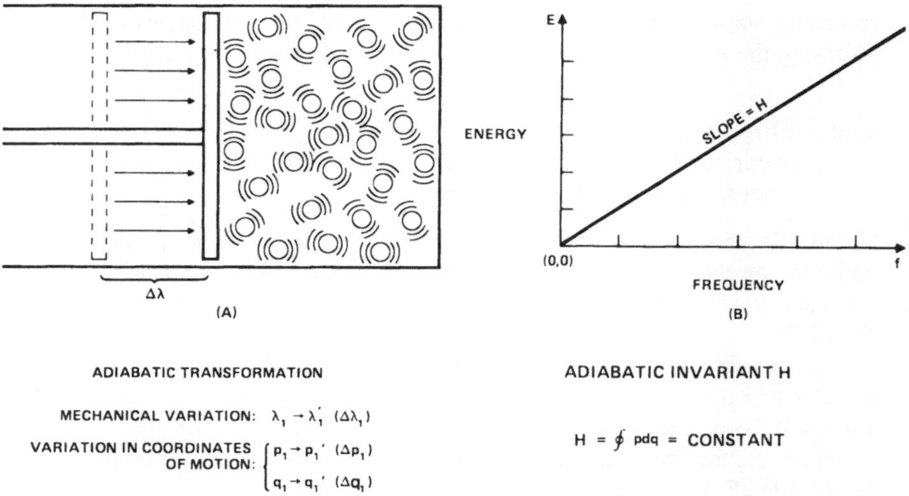

FIGURE 10.11. A CONDITIONALLY PERIODIC ENSEMBLE SYSTEM OPEN TO THE FLOW OF ENERGY. The atoms in the enclosed chamber are bounced back and forth between perfectly reflecting walls. If the piston is pushed very slowly (i.e., adiabatically—so that the work done moving the piston against the pressure of the gas all goes into the kinetic energy of the atoms), then the ratio of the kinetic energy of the atoms to the frequency of their collisions with the interior walls is constant. The ensemble particle system is undergoing an adiabatic transformation when $H = E/f = constant$ as conditioned by changes in λ.

phase space, where E is the total energy of the system of N molecules. The force F_λ exerted by the system is given by the equation

$$F_{\lambda_i} = -\int d\tau f \frac{\partial \epsilon}{\partial \lambda_i}, \tag{10.42}$$

where $d\tau$ stands for $dp_1 \ldots dp_r$, and the integral is over all of the molecular phase space. The work, W, done by the system in any process in which the parameter λ and the energy E are changed is defined as

$$W = \sum_i F_{\lambda_i} \delta\lambda_i = -\int d\tau f \delta\epsilon , \tag{10.43}$$

where $\delta\epsilon = \sum_i (\partial\epsilon/\partial\lambda_i)\delta\lambda_i$. The total energy E of the system is derived by the equation

$$E = \int d\tau f \epsilon . \tag{10.44}$$

The energy change δE in this process can be written as

$$\delta E = \int d\tau \epsilon \delta f + \int d\tau f \delta\epsilon . \tag{10.45}$$

The heat δQ supplied to the system in such a process is defined as the sum of the system's energy change and the work done by the system,

$$\delta Q = \int d\tau \epsilon \delta f \, . \tag{10.46}$$

A statistical weight function *G(q, p; λ)* is introduced (Ehrenfest, 1914), where $Gd\tau$ is the weight attached to the phase space volume $d\tau$ at the point *(q, p)* where the external parameter has the value λ. And the probability *P* of a given distribution *f* is identified as

$$P = \Pi(Gd\tau)^{fd\tau}\left[\frac{N!}{\Pi(fd\tau)!}\right], \tag{10.47}$$

where the products are taken over all volume elements in phase space (and Π means products). The most probable distribution is the one that maximizes *P* for a given total energy *E*. The most probable distribution is identified (Ehrenfest, 1914) as

$$f(q, p; \lambda) = (N/Z)G(q, p; \lambda)e^{\{-\mu\epsilon(q, p; \lambda)\}} \, . \tag{10.48}$$

Here $Z = \int d\tau G e^{-\mu\epsilon}$, and μ are determined by the energy constraint (10.44). With this explicit form for *f*, the expression (10.46) for the heat δQ can be restated in the form

$$\mu\delta Q = \delta(\mu E + N\ell nZ) - (N/Z)\int d\tau e^{(-\mu\epsilon)}\delta G \, . \tag{10.49}$$

The quantity δG is the change in the weight function *G* due to the changes in the λ parameter(s);

$$\delta G = \sum_i (\partial G/\partial\lambda_i)\delta\lambda_i \, . \tag{10.50}$$

Consider the quantity *lnP*. It can be calculated for the most probable distribution using (10.46) and (10.47), with the following result

$$\delta\ell nP = \delta(\mu E + N\ell nZ) \, . \tag{10.51}$$

Equations (10.49) and (10.51) differ only by the extra term on the right-hand side of (10.49), the term involving the change in the weight function. According to Boltzmann's theorem, thermodynamics relates to statistical mechanics such that $\mu\delta Q$ and δlnP are equal. Accordingly, μ is an integrating factor for the differential heat, and so, up to a scale factor, μ is the reciprocal of the absolute temperature and *lnP* is the entropy. The validity of this theorem (proved by Boltzmann for the special case that *G(q, p; λ)* is everywhere constant) requires that the extra term involving δG must vanish. Ehrenfest (1914) showed that, because this condition must hold for all μ and all λ, the necessary and sufficient condition for the validity of Boltzmann's theorem can be expressed in the form

$$\int_{\epsilon=\epsilon}^{\epsilon=\epsilon} d\tau \delta G = 0 . \tag{10.52}$$

This condition is identified as the "*δG-condition*" (Ehrenfest, 1914). It expresses the fact that the weight function must be adiabatically invariant if the statistical foundations of the second law of thermodynamics are to remain valid. The ratio of energy to frequency *(1/τ)* is invariant under slow parameter variation. What is required, therefore, to preserve the statistical interpretation of the second law is that the weight function in phase space be a function of the adiabatic invariant.

Returning to the previous equations of motion (Equation 10.39), the theorem of the conservation of energy follows immediately from the total differential coefficient of the Hamiltonian function H with respect to time:

$$\frac{dH}{dt} = \sum_k \left\{ \frac{\partial H}{\partial q_k} \dot{q}_k + \frac{\partial H}{\partial p_k} \dot{p}_k \right\} = 0 . \tag{10.53}$$

In words, the energy of the system, $H(q_k, p_k) = E$, is constant. The motion of the system is described completely by stating the energy as a function of the coordinates q_k and p_k, $H(q_1, q_2, \ldots, p_1, p_2, \ldots)$. In general it is possible to replace the pair of variables (q_k, p_k) by an arbitrary pair of canonically conjugate variables, where canonical variables are defined by the fact that they satisfy equations of motion of the type given above. The investigation of such canonical transformations leads to a mathematical problem (the so-called Hamilton-Jacobi differential equation) that in many cases can be solved without explicit solutions to the equations of motion. If it can be assumed that the system's heat production is constant throughout the transformation process, then the problem of periodic timing reduces to a simple scaling problem relating energy to frequency via a constant of proportionality. For an equilibrium state transformation, the heat production equals zero throughout the transformation process. For a steady-state transformation, however, the heat production equals some positive value which similarly remains constant through the transformation (see Chapter 11).

The derivation of the energy-periodic timing relationship follows from stating the equations of motion in terms of new canonical variables (J_k, W_k), where J_k is a momentum coordinate and W_k is a conjugate position coordinate, in which the energy depends only on the quantities J_k, and not on the quantities W_k. Then by the equations of motion and using the Hamiltonian function, H

$$\frac{J_k}{dt} = - \frac{\partial H}{\partial W_k} = 0, \tag{10.54}$$

so that J_k, called an *action variable*, is constant throughout the motion. In addition, $[dW_k/dt] = V_k = [\partial H/\partial J_k]$ is *constant* over time, (where the symbol V_k indicates a constant *vibration number* for the cyclic coordinate W_k), then the constancy of the quantities J_k and W_k increases linearly with time: $W_k = V_k T$. Thus the integration problem is solved in the new coordinates, and it only remains to make the reverse transformations.

The canonical variables J_k and W_k can be determined conveniently in the case where the system is separable, that is, once the equations of motion have been solved, each p_k depends only on the corresponding q_k (see Goldstein, 1950; ter Haar, 1971):

$$p_k = p_k(q_k) \ . \tag{10.55}$$

Here it may be shown that the action variable J_k corresponding to the kth period is given by

$$J_k = \oint p_k \, dq_k \ , \tag{10.56}$$

where the definite integral, indicated by the symbol $\oint$, is to be taken over the whole period. These integrals, which are called *action integrals*, can be calculated only for conditionally periodic systems; that is, for systems for which coordinates can be found such that each goes through a cycle as a function of time, independently of the others (see Section 9.6.1). Sometimes the coordinates can be chosen in several different ways, in which case the shapes of the quantized orbits depend on the choice of coordinate systems, but the energy values do not. Ehrenfest (1914) proved that the action variables J_k are adiabatic invariants, that is, that the quantities J_k can be quantized.

The adiabatic invariants for other systems can be determined, theoretically, in a similar manner. It is the case, however, that the direct method using the Hamiltonian is, in general, extremely troublesome, and other methods for discovering invariants may be required. The next section describes how it is possible to do so by means of a geometrical interpretation of the invariant quantity $J = w/f$.

10.4.1. Phase-Plane and the Action Integral. The energy of a pendulum system of length λ and mass m can be expressed using the following notation $q = \lambda \emptyset$, $p = mq$, $k = mg/\lambda$ as

$$E = \left[\frac{1}{2m}\right]p^2 + \left[\frac{k}{2}\right]q^2 \ . \tag{10.57}$$

In a *pq-plane* this equation represents an ellipse with the semi-axes

$$a = \sqrt{2mE}\ ,\ b = \sqrt{2E / k}\ , \tag{10.58}$$

where a and b follow from expressing Equation (10.57) in the form

$$\frac{p^2}{2mE} + \frac{q^2}{2E/k} = 1\ . \tag{10.59}$$

The area of an ellipse is

$$\oint pdq = ab\pi\ , \tag{10.60}$$

such that in the preceding case

$$\oint pdq = 2\pi E\sqrt{m/k}\ . \tag{10.61}$$

(The symbol $\oint$ means that we have to integrate over a whole period. Here it means integrating over the whole circumference of the ellipse.) Given that $2\pi f = \sqrt{k/m}$, then

$$\oint pdq = \frac{E}{f} = J\ . \tag{10.62}$$

The adiabatic invariant is, therefore, simply the area of the ellipse. This relation can be applied not only to systems with one degree of freedom, but also in complicated systems with many degrees of freedom. The extension to more than one degree of freedom depends on the fact that in many cases coordinates q_1, q_2, . . . can be introduced such that the momenta associated with them have the property that p_1 depends only on q_1, p_2 on q_2, and so on (that is, the system is assumed to be *separable*; see Goldstein, 1950; Sommerfeld, 1950). In general, these systems are multiply periodic. In this case the motion can be resolved into the superposition of simple periodic vibrations and their harmonics (so-called Lissajous figures).

As an example, consider the case of motion in a plane, where the rectangular coordinates x and y oscillate with two frequencies f_1 and f_2. If f_1 were equal to f_2, the path would be a circle, an ellipse, or a straight line, according to the relation of the phases. If the ratio of f_1 to f_2 is a rational number, the paths are again closed orbits. If f_1 and f_2 are incommensurable, that is, if the ratio is irrational, the orbit does not close, but gradually fills the whole rectangle within which the variable ranges. For multiply periodic motions in general, the orbits are of this type. If, however, an orbital curve closes after a finite number of revolutions, there is in reality only one period, and there will be only one quantized condition of the type

$$\oint p dq = J . \tag{10.63}$$

If the curve does not close, that is, if two or more incommensurable periods are present, then there will be as many quantized conditions as there are periods:

$$\oint p_1 dq_1 = J_1 \tag{10.64}$$

and

$$\oint p_2 dq_2 = J_2 . \tag{10.65}$$

This is the general case, also referred to as *nondegenerate*, whereas the case of coincident or commensurable periods is called a case of *degeneracy*. If u is the number of periods and v the number of degrees of freedom, then $w = (v - u)$ is called the *degree of degeneracy*.

10.4.2. Ehrenfest's Hypothesis. The concept of "adiabatic invariant" originated in thermodynamics. It was advanced with successive theorems by Clausius, Szily, Boltzmann, Rayleigh, and finally generalized by Ehrenfest (1913, 1914, 1916) to any systems satisfying adiabatic assumptions (see Ehrenfest, 1914 for references identifying the sources for the aforementioned theorems). This theoretical extension, known as *Ehrenfest's hypothesis*, provided the backdrop for the preceding sections. It is brought to the fore in the present section.

In an adiabatic change of state in thermodynamics, the coordinates that determine the heat motion are not affected directly, but only indirectly, if no external heat is added (isothermal) and the conditions of the system are mechanically transformed (for example, volumetric changes, positional changes with respect to the gravitational field, and so forth). In generalizing the concepts of adiabatic transformability and invariance, the only requirement is that the external manipulations are brought to bear on one or more parameters of the system, not directly on the coordinates of the motion.

In thermodynamics, an adiabatic change of state is regarded as a chain of reversible states in thermal equilibrium; in the more general case, the adiabatic transformation from the original to the final state has to occur through a sequence of intermediate reversible states that are very near, or at, equilibrium. The heat production must be invariant with respect to time. Again, the critical restriction is that the coordinates of motion must *not* go through a state of degeneracy. Any quantities that remain unaltered during a transformation restricted in this fashion are adiabatic invariants.

There are, therefore, three necessary and sufficient conditions for an adiabatic transformation in open conservative systems:

1. The transformation occurs through reversible, equilibrium states.

2. The effect is not on the coordinates of the motion but on parameters of the system that enter into relations that remain (invariant) throughout the transformation.

3. The transformation is not systematically coupled to the phases of the ensemble's oscillatory motion (in which case the external transformations would linearly drive the coordinates of the oscillatory pattern).

The relating of thermodynamic properties and statistical mechanical properties exemplified by the behavior of the system depicted in Figure 10.11 can now be drawn to a close. A very slow, essentially adiabatic, statistical mechanical transformation is the mechanical analogue of a reversible, adiabatic, thermodynamic transform. In the statistical mechanical case, the representative ensemble is readjusted to a new canonical energy distribution after change in an external coordinate $\Delta\lambda$. It is assumed that under the transformation $\Delta\lambda$ a thermodynamic ensemble exhibits no net heat exchange with the surroundings and the canonical energy distribution remains uniform. The requirement of the initial stage of a finite change in the coordinate $\Delta\lambda$ would be an infinitesimal readjustment of energy ΔE that is inserted into the ensemble to secure a canonical redistribution of the total energy.

The standard move is to correlate the definite values assigned to mechanical quantities on the basis of thermodynamics with the mean values that they have in a suitable representative ensemble. The energy E of a thermodynamic system is correlated with the mean energy E in the corresponding ensemble. This correlation makes it possible to consider a thermodynamic, adiabatic transformation as a transformation in which this *mean* energy is not affected by thermal flows. (This is to be preferred to the overly strong claim that the transformation brings about no readjustments at all in local energy distributions.) Moreover, having initially introduced canonical distribution as the suitable representation for the condition of thermodynamic equilibrium, it is possible to regard the same kind of distribution as representing the successive conditions of thermodynamic equilibrium through which a system would pass in a finite, reversible, adiabatic change.

10.4.3. Ehrenfest's Theorem. The various discussions to this point on transformable conservative systems can be summarized explicitly as a theorem. Given a conservative system of n degrees of freedom, the Lagrangian of the system may contain, in addition to the n coordinates (q, dq/dt) a set of parameters $\lambda_1, \lambda_2, \ldots, \lambda_r$. (This includes, for example, the coordinates of certain fixed centers of force, or quantities on which the moments of inertia of certain rotating masses can depend, such as radial length.)

The motion of the system is periodic for all values of the parameters λ_1, λ_2, . . . , λ_r and for initial values of the *q's* and *dq/dt's*. (The period τ_o may depend in any way on these initial values q_o, dq_o/dt and on the parameters λ_1, λ_2, . . . , λ_r.) The following adiabatic transformation is defined on the system: The system is sustaining an oscillatory motion, with period τ_o, for fixed values of the parameters λ_1, λ_2, . . . , λ_r. The parameters λ_1, λ_2, . . . , λ_r are varied infinitely slowly until they have assumed the values λ_1', λ_2', . . . , λ_r' and then the parameters remain constant. The system will now exhibit another periodic motion, of period τ_o'.

If the "frequencies" $(1/\tau_o)$ and $(1/\tau_o')$ are denoted by f_o and f_o', respectively, and the average kinetic energies of the first oscillation and the second oscillation are denoted by $\bar{E}_o$ and $\bar{E}_o'$ respectively, then Ehrenfest's theorem states: *The average kinetic energy of the system increases in the same proportion as the frequency under an adiabatic transformation*;

$$H = \frac{\bar{E}_o}{f_o} = \frac{\bar{E}_o'}{f_o'} \tag{10.66}$$

This is a general theorem that applies to all classes of conditionally periodic systems subject to changes that satisfy the criteria of adiabatic transformations.

Rayleigh (1902) came to the same conclusion while investigating the conditions for oscillating chambers in musical instruments. He stated the conclusion in the form of a theorem. Rayleigh's theorem refers to the behavior of the energy of a normal mode under a sufficiently slow variation of the dimensions of the enclosure: If the reflecting walls of the enclosure are made to approach each other infinitely slowly, then the energies of all proper vibrations change in exactly the same proportion, and are directly proportional to their frequencies (see Figure 10.11). In other words, the ratio $[E_\lambda/f]$ is invariant in such a process, during which f varies inversely as the dimension of the enclosure. The motion of the walls must be slow enough for the normal mode to retain its identity (i.e., no degeneracies occur), so that many oscillations occur during any appreciable displacement of the walls. If the walls are made to approach each other very slowly (i.e., "adiabatically"—so that all the work done moving the walls against the pressure of the gas goes into the kinetic energy of the particles), then the ratio of the kinetic energy of the molecules to the frequency of their collisions with the walls is constant.

Finally, it is of some importance to note that Equation (10.66) can be transformed as follows:

$$H = \frac{\bar{E}_o}{f_o} = \tau_o \bar{E}_o = \tau_o \frac{\int_o^\tau E_o dt}{\tau_o} \int_o^\tau E_o dt \,. \tag{10.67}$$

The root of Equation (10.67) is

$$\delta' \int_0^T E_0 dt = 0 , \tag{10.68}$$

where δ' means an adiabatic variation. Equation (10.68) identifies the close relation between Ehrenfest's theorem and the principle of least action.

10.5. Toward Adiabatic Transformability and Invariance in Biological Systems

The analysis of adiabatic transformability to this point has been dominated by the themes of infinitely slow change and zero heat production. Neither theme can be a reality for biological systems. Changes in the frequency of a rhythmic movement, for example, ordinarily occur in finite time and not necessarily at a fixed rate. Moreover, such changes ordinarily incur heat production. It will be shown, however, that neither of the two themes constitutes conditions sine qua non for adiabatic transformability. That is, Ehrenfest's hypothesis and theorem can be extended to non-rate-limited transformations and nonconservative systems.

Consider once again the open conservative system undergoing an adiabatic transformation that is depicted in Figure 10.7. The "openness" of the system is with respect to the energy brought in to modify the radius; it is energy derived from an energy reservoir that is external to the rotating body. The "adiabatic transformability" of the system is due to the fact that the shortening of the radius can be conducted (and is only conducted) infinitely slowly. To conduct the shortening at some definite rate would incur an external torque N and a consequent change in the body's angular momentum (Section 10.2). The ultimate result would be a change in the body's coordinates of motion.

An external torque applied to a body can be generated only by an interaction of the body with one or more external frames of reference. (An external torque is a measure of the influence of one frame on another.) Interactions between parts of the body that take place in the frame of reference centered in the body cannot produce an external torque and, therefore, cannot produce a change in angular momentum (Section 10.2). If the changes in the radius of the rotating body depicted in Figure 10.7 originated in an energy reservoir localized in the frame centered in the rotating body, then those changes, whatever their rate, would take place without incurring an external torque on the body. Angular momentum changes and degeneracies in the coordinates of motion could not arise. The result would be a potentially adiabatically transformable system—changes in rotational energy proportionate to changes in rotational frequency—where the transformation (in radius) would not have to occur infinitely slowly. Two non-rate-limited

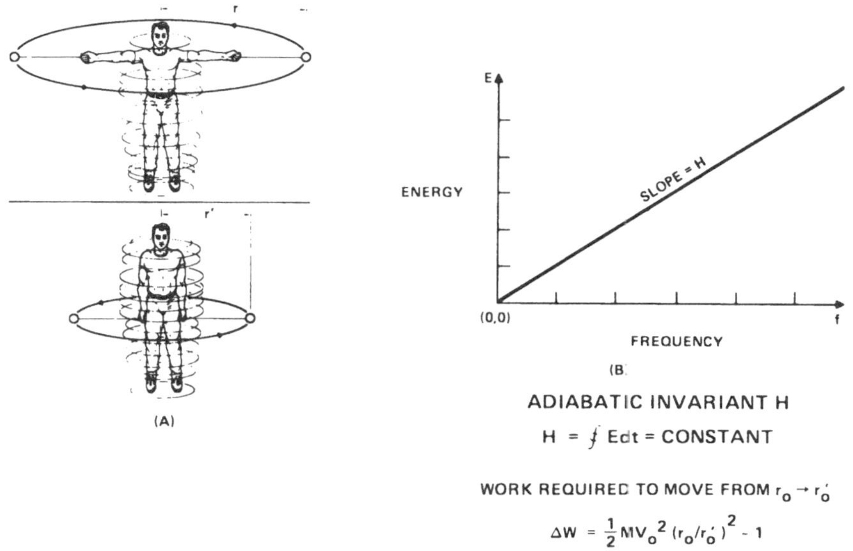

FIGURE 10.12. ADIABATIC TRANSFORMATION IN AN OPEN NONCONSERVATIVE SYSTEM THAT IS NONRATE LIMITED.

examples follow that bring the notions of adiabatic transformability and invariance closer to the biological domain.

10.5.1. Conservative Adiabatic Transformations that are not Rate Limited. Figure 10.12 identifies a rotating system and an energy reservoir that are both defined within the same gravity-free space (such as an orbiting space lab). The man and the rotating hand-held objects define a single frame of reference. By pulling his arms down the man increases the frequency of rotation of his body and the hand-held masses. The amount of energy the man puts into the rotating system, through the lowering of his arms, and the frequency of rotation are directly proportional. The transformation is not an infinitely slow change in the radius of rotation but it is, nevertheless, an adiabatic transformation. Only internal torque is generated in the act of injecting energy into, and taking energy out of, the rotating component. The example suggests, therefore, that an adiabatic transformation is not necessarily rate limited.

Consider a second example, one in which the transformation occurs within a gravitational field. Figure 10.13 identifies the following situation: A cat is raised a short distance above the ground surface and rotated such that its feet are facing upward and then it is released. The force of gravity powers the cat's descent to the ground surface. During the descent, however, the cat can use its on-board energy reservoirs to generate internal torque for the purpose of reorientation. The

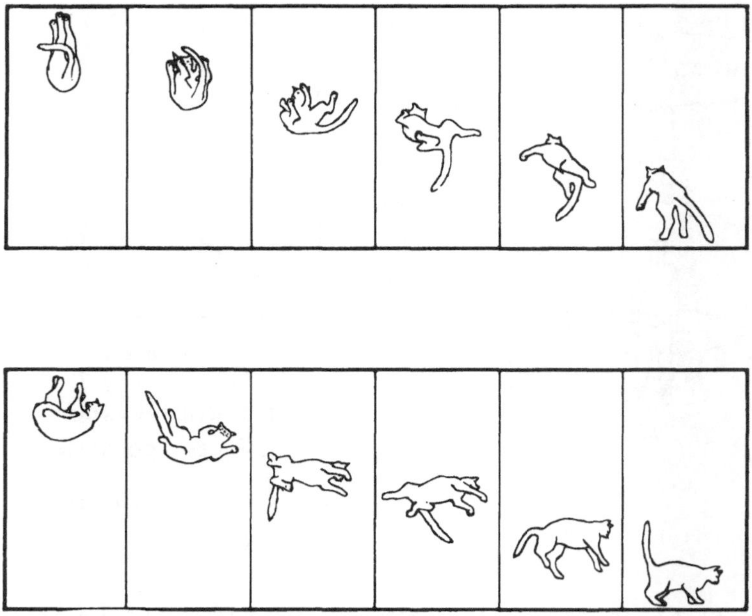

FIGURE 10.13. Example of a nonrate-limited adiabatic transformation in a gravitational field (artist's sketches from a set of high speed photographs that appeared in *Comptes Rendus of the Paris Academy*, 1874, p. 714).

cat changes its orientation through the generation of mechanical variations in the centers of oscillation distributed throughout the cat's limb segments. These changes can involve the addition of new energy into the rotational motion during shortening of the rotational radii or deletions of energy during elongations.

For example, the cat can rearrange its limbs and tail, thereby changing its moment of inertia I and its angular speed ω so that its feet and tail are properly oriented for a soft landing. The cat achieves this goal by suitable rotation of the anterior extremities coupled with opposite rotation of the posterior ones. While the moment of inertia and angular speed change continuously throughout the descent, the angular momentum remains constant. Internal forces are unable to impart motion to the center of mass, even in a system with flexible joints and on-board energy sources, like the cat. In order to move the center of mass, the internal forces must be applied to an external support, thereby generating an external force. Angular position, however, can be varied *ad libitum* with the use of internal forces alone—that is, without an external support to push against. The motions of the cat during postural realignments do not create any new external torques as long as the cat is free of external surface contact resulting in the constancy of angular momentum. While the angular momentum of the cat's motion is constant, the energy of oscillation is free to vary as the cat changes its limbs' radii of oscillation. The cat's total energy of rotation (work) divided by its frequency of oscillation will remain constant throughout the descent. When plotted in an *[E X f]* coordinate space, the trajectory of rotational states will define a first order linear relation whose constant of proportionality is a function of the initial energy condition, and whose trajectory, on extension, will intercept (ideally) the origin *(0,0)*. A small negative energy-intercept (and positive frequency-intercept) may result that is associated with air resistance. The resistance, however, is negligible owing to the mass of the cat in the gravitational force field. The small energy-intercept identifies the energy lost to heat dissipation (surface air resistance) averaged (integrated) over a cycle. The small frequency intercept identifies the frequency (velocity) at which rotation would be stopped as a function of the air resistance.

While the interior of the cat is dominated by many nonconservative processes (force structures) the exterior force structure associated with rotational motion is essentially a conservative force structure. The conservative force structure is isolated from dissipative forces as long as there is no surface friction contact (air resistance assumed to be negligible) between the rotating system and its orienting system (center of mass of the earth).

10.5.2. Adiabatic Transformation in an Open, Nonconservative (Entropy Producing) System. The two preceding examples, though instances of animal movement, remain specialized. They highlight the independence of adiabatic transformability from rate restrictions and extend, thereby, the range of processes in which adiabatic invariants might play a role. They are, nevertheless, not sufficiently general. What limits the two preceding examples is their conservative nature. In both cases the force structure of the rotational system was conservative, as the immediately preceding paragraph underlines. Frictional, nonconservative forces were absent and as a result the rotational motion per se incurred no entropy production. The examples, therefore, are continuous in this respect with the examples of adiabatic transformability presented in the earlier sections of this chapter. An adiabatic transformation is defined thermodynamically when the heat production remains invariant with respect to time (Section 10.4.2). A conservative system, by definition, does not produce entropy. Its heat production, therefore, is time invariant, satisfying the thermodynamic criterion.

The production of entropy, of course, characterizes nonconservative systems and contrasts them with the conservative systems (Chapter 3). Despite the difference, as long as the nonconservative system's positive heat production is constant (that is, independent of time) it can, like the conservative system, fulfill the thermodynamic requirement for an adiabatic transformation. The following example illustrates this point. It is an example of a rotational system that includes nonconservative forces by virtue of being in contact with a surface.

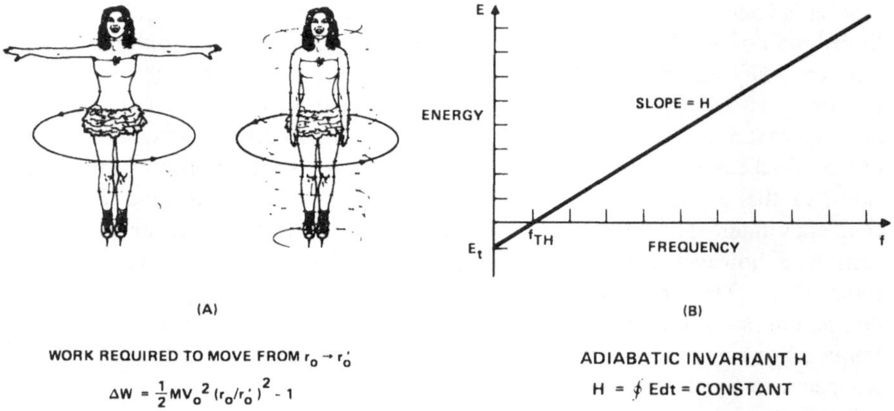

FIGURE 10.14. EXAMPLE OF AN ADIABATIC TRANSFORMATION IN A NONCONSERVATIVE SYSTEM.

The example brings out a difference between conservative and nonconservative systems that was foreshadowed in the falling cat example and which will figure prominently in the subsequent theoretical developments: The trajectory of energy by frequency coordinates of an adiabaticaly transformed nonconservative system has a nonzero, positive frequency intercept.

An ice skater, freely spinning on an ice surface, conserves angular momentum, that is, $I\omega = constant$. If the skater's arms are pulled in close to the body, the moment of inertia I decreases (as a function of changes in the radius of rotation) resulting in an increase in spin rate ω (Figure 10.14). The "pulling-in" of the arm requires a contribution of energy from the skater. This energy appears as kinetic energy associated with the motion. As the arms are pulled in, the energy and frequency of oscillation will increase; however, the ratio of the energy change to frequency change ($H = \Delta E/\Delta f$) will remain constant. The transformation in periodic timing is adiabatic, H = *adiabatic invariant*. When plotted in an energy by frequency space *[E X f]* the trajectory through the space would be a first order linear trajectory with a positive frequency-intercept and a negative energy-intercept. The negative value of the *E-intercept* identifies the amount of energy that is dissipated every cycle by the sliding friction between the ice blade and the ice's surface. The *f-intercept* identifies the frequency of oscillation at which the skater's motion will stop if the skater puts no energy into the cycle to offset the dissipative losses. The slope of the line (which determines the *f-intercept*) is a function of the initial energy put into the cycle before the arms are pulled in. The periodic transformation associated with the the pulling in of the arms is isothermal and adiabatic.

The generation of internal torques is associated with the production of work W that is generated from on-board (interior) metabolic energy reservoirs. The generation of this internal energy in proportion to changes in the frequency of rotation accounts for the constancy of angular momentum with changes in the angular speed:

$$J = \sum m_i r_i^2 \omega = \text{constant} . \tag{10.69}$$

The following adiabatic invariant can be defined:

$$H = \frac{E - E'}{f_o - f_o'} = \text{constant} \tag{10.70}$$

where

$$E - E' = \Delta E . \tag{10.71}$$

ΔE is the work generated from internal energy reservoirs.

The conclusion to be drawn from the various examples presented thus far is that the thermodynamic nature of an adiabatic transformation may apply widely to physical processes (conservative and nonconservative) at all physical scales (macro and micro) within physical limits (such as velocity limits associated with the speed of light and size limits associated with Planck's constant). The preceding three examples identify quantization of action at the macro scale of analysis. The following final example is celebrated. It identifies the quantization of action at the micro scale.

10.5.3. The Photoelectric Effect: An Example of Quantization of Action with a Time-Independent Dissipative ''Work Function'' at the Micro Scale. Electromagnetic radiation interacts with a metal surface in a vacuum. Energy of the photons is transmitted to the electrons of the metal and may result in the release of an electron of the metal into the vacuum. Whether it does so or not depends on whether the energy imparted by the photons exceeds the minimum required for an electron to move from the metal to the vacuum. An electron must do work to get out of the metal; at the surface are dissipative forces that resist its flight. The ''work function'' (A) of a metal refers to this minimum energy value that makes an electron's liberation possible. It depends on the kind of metal and the state of its surface. Electrons liberated by photons are termed photoelectrons. From the conservation of energy, the photoelectric effect is given by:

$$hf = A + E_{K_{max}} , \tag{10.72}$$

where the product hf is the energy of the photons (h is Planck's constant, f is the frequency of the electromagnetic wave) and $E_{K_{max}}$ is the maximum kinetic energy of the photoelectrons in flight in the vacuum. Equation (10.72) can be portrayed geometrically as shown in Figure 10.15). In the geometric portrayal

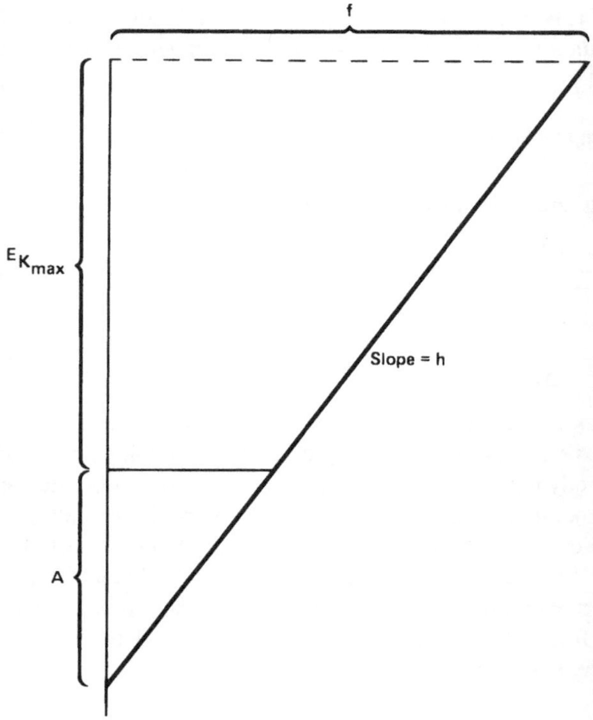

FIGURE 10.15. GEOMETRIC ANALYSIS OF THE PHOTOELECTRIC WORK FUNCTION.

the vertical lengths are magnitudes of energy and the horizontal lengths are magnitudes of frequency (energy and frequency are orthogonal variables). The ratio of total vertical length $(A + E_{K_{max}})$ to horizontal length (f) is the slope of the line connecting the lower extreme of $A + E_{K_{max}}$ to the rightmost extreme of f. This slope is Planck's constant (h). The length of the unlabeled horizontal line dividing the lengths $E_{K_{max}}$ and A identifies the frequency, f_{TH}, below which no photoelectric emission is detectable. From the geometry of Figure 10.15, it is evident that

$$f_{TH} = A/h \ . \qquad (10.73)$$

The parameters f_{TH} and A are characteristics of a metal. Only one of them needs to be determined, however, given that h is a constant of the atomic scale, invariable from metal to metal.

10.6. Exploring the Physical Geometry of the [E X f] Coordinate Space

Equation 10.74 (see Equation 10.30) defines the general form of the law of adiabatic transformability for a conservative system

$$H = \frac{E}{f}, \tag{10.74}$$

where *H* is the adiabatic invariant, *f* is a particular frequency of the pendulum system, and *E* is the respective energy of the pendulum system at *f*. An *[E X f]* coordinate space can be constructed from the oscillator's frequency and respective energy values and a trajectory of states defined relative to the adiabatic constraint (Equation 10.74). The law of action (Equation 10.30) defines a first order linear relationship between frequency and energy. Geometrically, the trajectory of states defined by the constraint is a straight line through the *(E,f)* coordinates that intercepts the origin *(0,0)*. The conservative nature of the adiabatic transformation is geometrically defined by the intercept value: An adiabatic transformation is conservative *if and only if* the trajectory of *(E,f)* points intercepts the *f* and *E* axes at the origin *(0,0)*. If, however, the trajectory of points through the *(E,f)* space is first order and linear and does not intercept the origin *(0,0)*, then the system is nonconservative and the adiabatic tranformation must assume the following form:

$$H = \frac{\Delta E}{\Delta f}. \tag{10.75}$$

The preceding comments are meant to bring to the forefront the physical interpretations of the geometrical properties evident in the coordinate space. Thus, the intercept characterizes a system as conservative or nonconservative: An intercept at the origin *(0,0)* defines conservative systems (see Figure 10.16a),

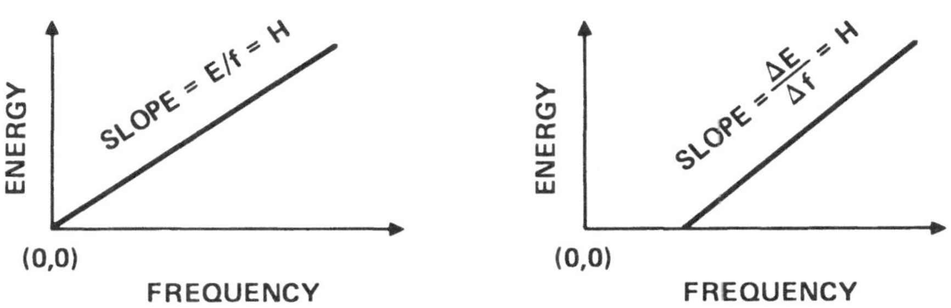

FIGURE 10.16. (a) GRAPHIC REPRESENTATION OF A CONSERVATIVE SYSTEM WITH AN ADIABATIC INVARIANT, $H = E/f$.
(b) GRAPHIC REPRESENTATION OF A NONCONSERVATIVE SYSTEM WITH AN ADIABATIC INVARIANT, $H = \Delta E/\Delta f$.

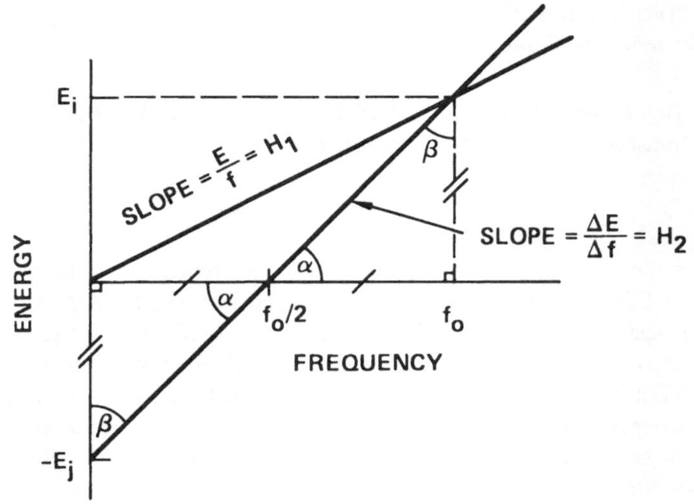

FIGURE 10.17. GEOMETRIC PROOF THAT $|E_i| = |-E_j|$.

intercepts with the positive *f-axis* at points other than *(0,0)* define dissipative systems (see Figure 10.16b). The contrast follows from a consideration of the structure of physical laws: In a conservative system all coupled variables degenerate to zero simultaneously. This is formally captured in descriptive equations of law that are built out of quotients and products among the quantities to which the law refers. This manner of construction guarantees that the quantities are always intrinsically scaled to one another. (There is no such guarantee if the quantities are related by addition and/or subtraction.) Consequently, for conservative transformations it is not possible for one variable of a law to be zero (as E is at the *f-intercept* of Figure 10.16b) without all the quantities to which the law refers being zero: Lawfully related quantities must simultaneously approach the limit in a conservative transformation. In the nonconservative, dissipative case (Figure 10.16b), however, *not all the energy is being seen in the frame of reference in which the quantities are measured* and therefore the energy *(E)* appears to degenerate to zero before the frequency *(f)* degenerates to zero. The point to be appreciated is that the degenerating symmetry around the origin *(0,0)* associated with conservative systems can be used to distinguish graphically (using geometric properties) dissipative (nonconservative) systems from nondissipative (conservative) systems. Figure 10.17 identifies the geometric symmetry that satisfies the conservational (energy) symmetry for nonconservative systems.

Expression (10.75) can be rewritten

$$H_i = \frac{E_2 - E_1}{f_2 - f_1}, \qquad (10.76)$$

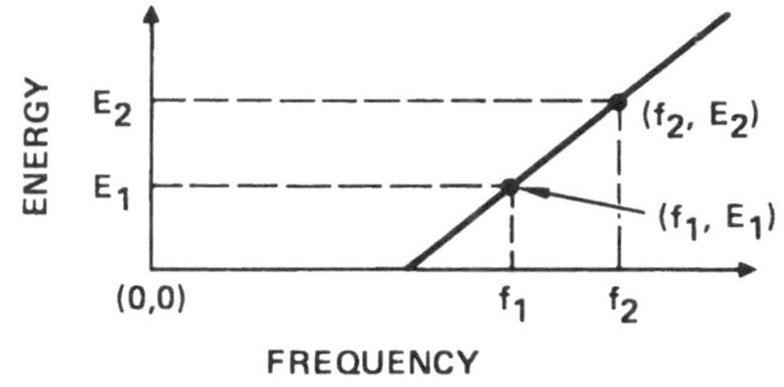

FIGURE 10.18. COMPUTATION OF THE ADIABATIC (ACTION) INVARIANT FOR A DISSIPATIVE SYSTEM: $H = [(E_2 - E_1)/(f_2 - f_1)]$.

where E_1 and E_2 are two energy values, $I_i\omega_1^2/2$ and $I_i\omega_2^2/2$, for the same system (of inertia I_i) at two frequencies, f_1 and f_2, respectively (see Figure 10.18). Substituting for E, with $\omega = \theta/\tau$ (where θ is average angular displacement—*mean free path*—in radians; and τ is average displacement time—*mean relaxation time*—in seconds), gives

$$H_i = \frac{1}{2} I_i \left[\frac{\dfrac{\theta_2^2}{\tau_2^2} - \dfrac{\theta_1^2}{\tau_1^2}}{f_2 - f_1} \right] \tag{10.77}$$

If f_1 is the intercept f_{int}, then $E_1 = 0$, $\theta_1 = 0$ and (10.77) becomes

$$H_i = \frac{1}{2} I_i \left[\frac{\dfrac{\theta_2^2}{\tau_2^2}}{f_2 - f_{int}} \right]. \tag{10.78}$$

Rearranging terms and dropping the subscript on the frequency ($f = 1/\tau$) variable gives

$$\theta = \pm \left[\frac{2H_i\tau^2 (f - f_{int})}{I_i} \right]^{1/2}. \tag{10.79}$$

Transforming frequencies into periods, treating $(2H_i/I_i)^{1/2}$ as a constant, and solving for the positive root of the derivative of θ with respect to τ reveals that $d\theta/d\tau = 0$ when $\tau_{int} = 2\tau_o$. This means that when the amplitude is at a maximum θ_{max}, the associated period τ_o is equal to $(\tau_{int})/2$. Rewriting equation (10.79) for the conditions when $\theta = \theta_{max}$, $\tau = \tau_o$, and $\tau_{int} = 2\tau_o$ yields

$$H_i = I_i \left(\frac{\theta_{max}^2}{\tau_o} \right) . \qquad (10.80)$$

The period of oscillation at which the maximum amplitude is defined identifies a stability point relative to the thermodynamic-mechanical Quality factor of the cycle under the defined boundary conditions (here, adiabatic invariant H_i). At maximum amplitude the Q-factor $= 1$ (see below for a discussion of the Quality factor). The following symmetry defines the relationship between the f and τ intercept values and their respective values at maximum amplitude:

$$\tau_{int} = 2\tau_o \text{ or } f_{int} = f_o / 2 . \qquad (10.81)$$

It follows that, for a given system with $Q = 1$, knowing the 'natural' period (*mean relaxation time*) and its associated amplitude (*mean free path*) permits the determination of the system's constant of action from (10.80). Further, writing Equation (10.80) in the form

$$H_i = I_i \tau_o \left(\frac{\theta_{max}^2}{\tau_o^2} \right) . \qquad (10.82)$$

and then substituting ω^2 for the parenthesized ratio leads to

$$H_i = 2E\tau_o = 2E / f_o , \qquad (10.83)$$

which is the more familiar form of the quantized adiabatic relation among energy, frequency, and action.

10.6.1. Conservative and Nonconservative Systems Compared and Contrasted. Equation (10.74) can be written

$$H_i = \frac{1}{2} \left[\frac{I_i \omega^2}{f} \right] . \qquad (10.84)$$

Substituting θ/τ (θ in radians) for ω and rearranging terms gives

$$\theta = \left[\frac{2H_i \tau}{I_i} \right]^{1/2} , \qquad (10.85)$$

or, more simply, given that H_i/I_i is a constant,

$$\Theta = c\tau^{1/2} . \tag{10.86}$$

There are no degeneracies associated with this function (that is, the derivatives never go to zero), as there are with the corresponding equation for the dissipative case, Equation (10.79). Consequently, the amplitude of the conservative system at its stable period is not the maximum amplitude expressible by the system under the adiabatic constraint ($H = constant$).

Equations (10.82) and (10.83) are written with respect to the stable period of the dissipative system. The corresponding equation for the conservative system with respect to its stable period is:

$$H_i = E\tau_o = E / f_o . \tag{10.87}$$

It is concluded, therefore, from a comparison of (10.83) and (10.87) that at the stable frequency a dissipative system of inertia I_i governed by the constant of action requires exactly *twice* as much energy as a similarly governed conservative system of the same inertial quantity at the same stable frequency.

By way of summary, the two equation sets are (1) $H_i = E/f$ (with intercept $0,0$), and (2) $\theta = c\tau^{1/2}$ for the conservative case; and (a) $H_i = \Delta E/\Delta f$ (with intercept $0,f_{int}$), and (b) $\theta = c\tau(f - f_{int})^{1/2}$ for the dissipative case, where $c = (H_i/I_i)^{1/2}$. Continuous transformations relate (1) and (a), (1) and (2), and (a) and (b); a discontinuous transformation relates (2) and (b). Graphically, the (θ, τ) relationships defined by the two functions (2) and (b) are shown in Figures 11.4a and 10.9b respectively (summarized in Figure 10.19). The monotonic, nonlinear function $\theta = c\tau^{1/2}$ identifies no singular points where the system can stabilize in a steady-state; in comparison, the nonmonotonic, nonlinear function $\theta = c\tau(f - f_{int})^{1/2}$ reveals a singular point at τ_o identifying a potential equilibrium state.

Further, and related, the function $\theta = c\tau^{1/2}$ predicts that amplitude should increase continuously with increase in period; the function $\theta = c\tau(f - f_{int})^{1/2}$ predicts an increasing-decreasing relation. What would be required to transform $\theta = c\tau^{1/2}$ into $\theta = c\tau(f-f_{int})^{1/2}$? Simply, the addition of a dissipative function that reduces amplitude in inverse proportion to period.

If for a given system the link between its amplitude and its period is defined by (b), that is $\theta = c\tau(f - f_{int})^{1/2}$, then it can be concluded that the system under observation is dissipative. That a system is dissipative, however, is not always reason for good cheer. To the degree that some of a system's energy is invisible, the system is not explainable. In the present circumstance this source of disquiet would seem to be removed by the symmetry that relates *at equilibrium* conservative and nonconservative instances of the same system. Whereas the energy of the conservative instance is, by definition, exhibited fully, the energy of the nonconservative instance is exhibited only partially. The comparison of

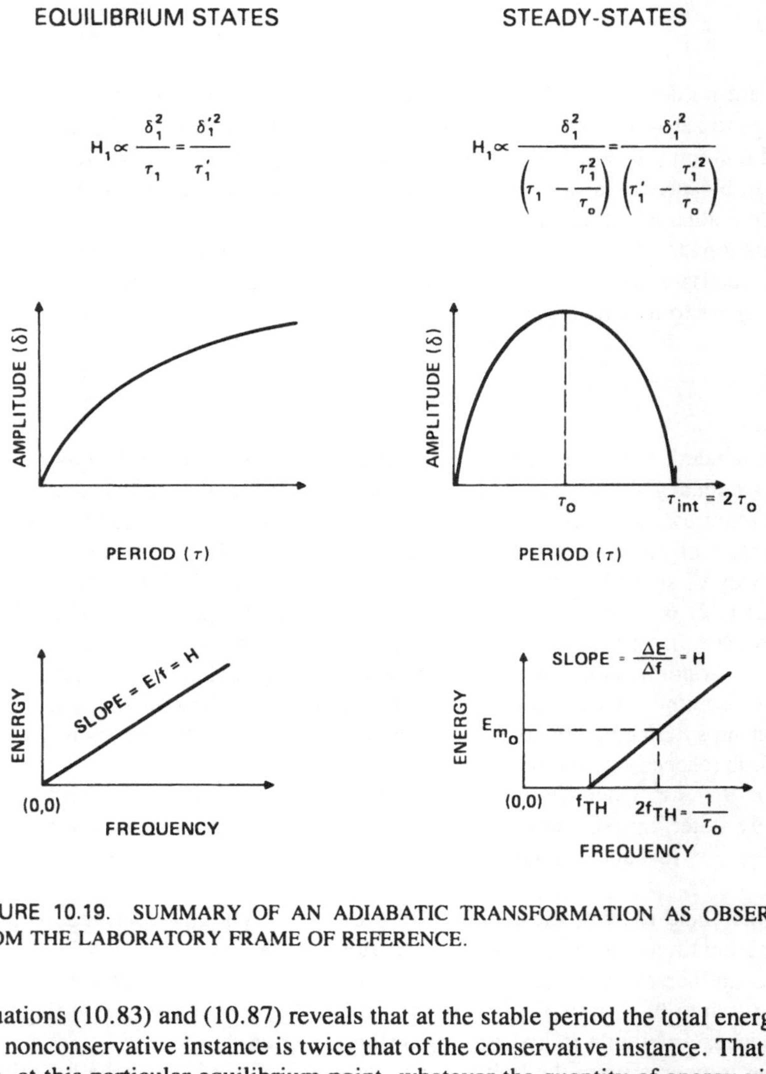

FIGURE 10.19. SUMMARY OF AN ADIABATIC TRANSFORMATION AS OBSERVED FROM THE LABORATORY FRAME OF REFERENCE.

equations (10.83) and (10.87) reveals that at the stable period the total energy of the nonconservative instance is twice that of the conservative instance. That is to say, at this particular equilibrium point, whatever the quantity of energy visible in the laboratory frame of reference, there is an equal quantity of energy not visible (see Figure 10.14 for the geometric proof; Chapter 11 will rationalize this symmetry in the context of biomechanical oscillators). This symmetry is satisfied whenever the Q-factor of the oscillator is equal to unity. An example of an oscillation satisfying this symmetry is Brownian motion.

10.6.2. Brownian Motion: An Example of a Quantized Oscillation with a Quality Factor of Unity. As related by Einstein and Smoluchowski (see Section

3.4.3), a Brownian particle behaves like a gigantic molecule. It is subject to fluctuations in pressure—the result of very many collisions of surrounding molecules with the particle. Because of the distribution of conservations and a critical scale relationship between the macro and micro particles' sizes, an inertial force F wells up from the micro structure that accelerates the macro particle only to be countered by the viscous force F that decelerates the particle. The surrounding medium is responsible simultaneously for initiating and terminating a "hop." On average, a hop of a Brownian particle of mass m covers a distance δ (the mean free path) in a time τ (the mean relaxation time) so that, from the observer's viewpoint, the kinetic energy of one hop is $m(\delta/\tau)^2/2$. However, the (mechanical) work done to move the particle a distance δ is $F_i\delta$ and the (thermodynamic) work done to resist the particle's movement for a distance δ is $F_i\delta$. Therefore the total kinetic energy (mechanical and thermodynamic) associated with the average work $(\Sigma F_i\delta/2)$ done by the medium over one hop is $m(\delta/\tau)^2/2$ *plus* $m(\delta/\tau)^2/2$. That is, for the quantity of energy seen that is associated with the particle's motion from the observer's viewpoint (mechanical

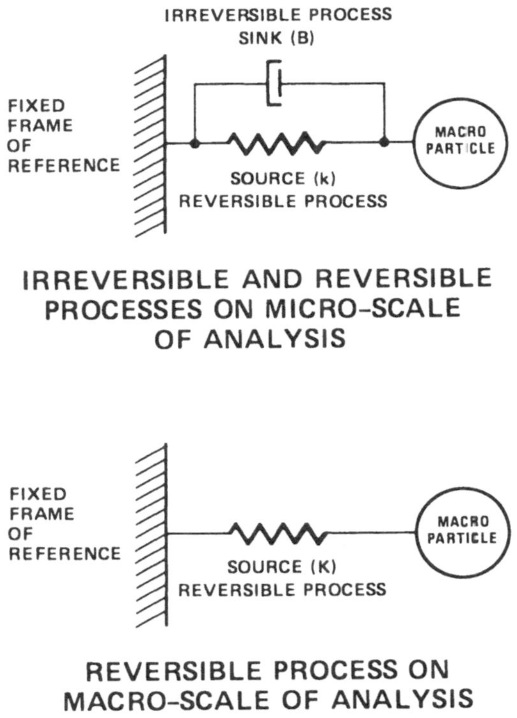

FIGURE 10.20. MACROSCOPIC BROWNIAN PARTICLE VIEWED FROM TWO FRAMES OF REFERENCE (TOP) AND A SINGLE FRAME OF REFERENCE (BOTTOM).

frame of reference) there is an equal quantity of energy not seen (thermodynamic frame of reference). During Brownian motion the conservative (mechanical) forces generated in one hop do not show up in succeeding hops. Oscillators with no conservative force residuals from cycle to cycle have Q-factors ≤ 1. In the case of Brownian motion the Q-factor $= 1$. The result of a $Q > 1$ is the persistence of a "momentum residual" (inertial forces) from hop (cycle) to hop (cycle).

Figure 10.20 depicts a generalized Brownian mechanism, where k identifies the energy source that impels the Brownian particle a step of length δ, and B identifies the energy sink that brakes the particle's motion over a step of length δ. The source and the sink combine in a more macroscopic frame of reference to identify a macroscopic stiffness term K. Arising from the same fluid field, the source and the sink cannot be independent as is the case in artifactual oscillatory processes (see Section 5.4). At equilibrium, the two are related by a nonlinear coupling. It will be argued later that biomechanical oscillatory motion is assembled out of similar nonlinear couplings. A repetitive movement's energy source (macroscopic elastic components that inject energy into the cycle) and energy sink (microscopic viscous components that absorb energy from the cycle) couple nonlinearly because of their common origin in a chemical-thermal-mechanical engine. Whereas the Q-factor of a 'Brownian engine' is necessarily unity, that of a chemical-thermal-mechanical engine is *not* restricted to unity. This difference is of considerable importance and is a major subject of Chapter 11.

REFERENCES

Born, M. (1945). *Atomic physics*. London: Blackie & Son Limited.

d'Abro, A. (1952). *The rise of the new physics*. New York: Dover.

Dugas, R. (1948). *Histoire de la mecanique classique*. Paris: C.N.R.S.

Ehrenfest, P. (1913). A mechanical theorem of Boltzmann and its relation to the theory of energy quanta. *Collected Papers*, No. 13. *Proceedings of the Amsterdam Academy*, *16*, 591-597.

Ehrenfest, P. (1914). Zum Boltzmann entropie-wahrscheinlichkeits-theorem. *Collected Papers*, No. 31. *Physische Zeitschrift*, 657.

Ehrenfest, P. (1916). On adiabatic changes of a system in connection with the quantum theory. *Collected Papers*, No. 37. *Proceedings of the Amsterdam Academy*, *19*, 576-597.

Ehrenfest, P. (1917). Adiabatic invariants and the theory of quanta. *Philosophical Magazine*, *33*, 500.

Ehrenfest, P. (1959). *Collected scientific papers*. Amsterdam: North Holland.

Goldstein, H. (1950). *Classical mechanics*. Reading, MA: Addison Wesley.

Klein, M. J. (1970). *Paul Ehrenfest: The making of a theoretical physicist, Vol. 1*. Amsterdam: North-Holland.

Kramers, H. A. (1957). *Quantum theory*. Amsterdam: North-Holland.

Landau, L. D., and Lifshitz, E. M. (1976). *Mechanics*. New York: Pergamon Press.
MacMahon, T. A. (1975). Using body size to understand the structural design of animals: quadrupedal locomotion. *Journal of Applied Physiology*, *39*, 619-627.
MacMahon, T. A. (1977). Scaling quadrupedal galloping: Frequencies, stresses, and joint angles. In T. J. Pedley (Ed.), *Scale effects in animal locomotion*. London: Academic Press.
Rayleigh, B. (1902). On the pressure of vibrations. *Philosophical Magazine* III, 338-346.
Sommerfeld, A. (1950). *Mechanics*. New York: Academic Press.
ter Haar, D. (1971). *Elements of Hamiltonian Mechanics*. New York: Pergamon Press.
Tolman, R. C. (1938). *The principles of statistical mechanics*. Oxford: Oxford University Press.
Whittaker, E. T. (1937). *Analytical dynamics*. Cambridge: Cambridge University Press.

11 ASSEMBLING AND SUSTAINING ESCAPEMENTS FOR BIOMECHANICAL OSCILLATION: IDENTIFYING THE ENGINE (CHEMICAL-THERMAL-MECHANICAL) PROCESS

ABSTRACT

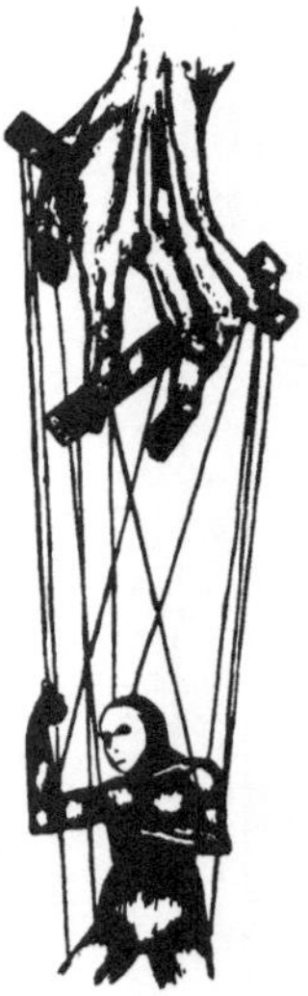

Wrist-pendulum systems are hybrid chemical-thermal-mechanical engines involving the following energy ***converting*** *and* ***partitioning*** *sequence: Chemical energy is first converted to mechanical (kinetic) energy via an elaborate sarcomere cross-bridging; this mechanical energy is then systematically partitioned over external (mechanical work) processes and internal (thermodynamic) processes. The mechanisms underlying these processes are escapements that gate or channel the energy flows. The following questions are addressed: (i) What are the escapement requirements for the* ***assembling*** *processes, those that establish the parameters of oscillation? (ii) What are the escapement requirements for the* ***sustaining*** *processes, those that guarantee the subsistence of oscillations? (iii) What is the relation between mechanical processes and thermodynamic processes? (iv) What properties in the macroscopically observable states can be used to do the bookkeeping on the mechanical and thermodynamic processes? Finally, (v) What is the 'informational' description by which a subject parses the biomechanical work space* ***(E,f)*** *in search of 'preferred' comfort states? A theoretical construct is advanced for the physical basis of 'comfort' states. Central to the construct is an understanding of the lawful mixing of mechanical and thermodynamic processes under*

adiabatic conditions. Under these conditions, mechanical and thermodynamic processes are independent (orthogonal); the thermodynamic processes are invariant with changes in the mechanical processes. It is suggested that these steady-state manifestations in adiabatic conditions constitute a physical basis for 'comfort' that can 'orient' other trajectories of states in the ***(E,f)*** *work space.*

A second orienting property is associated with a ***Q-factor*** *= 1. The Q-factor is a dimensionless scalar built out of the ratio of the (reversible) mechanical energy* $\mathbf{E_m}$ *of a cycle and the (irreversible) thermodynamic energy* $\mathbf{E_t}$*,* $Q = [E_m/E_t]$*. The* ***Q-factor*** *provides a precise measure of the escapement requirements of a cycle. It identifies that portion of the total energy that must be put back into the cycle to restore its initial energy conditions. For example, a high Q-system (*$Q>>1$*) requires very little new escapement energy to return the energy state to its initial condition. For a perfectly conservative system* $Q=\infty$*, indicating that no additional energy is required to recover initial conditions. In contrast, as the Q-factor becomes smaller (*$Q\leq 1$*) the motion of the oscillator becomes dominated by internal thermodynamic processes. At* $Q=1$ *the mechanical and thermodynamic energy flows are symmetric, resulting in a stable steady-state around which oscillatory parameters can orient.*

The ***adiabatic invariant*** *(H) and the* ***quality factor*** *(Q) are independent (orthogonal) variables that share a single common value. It is suggested that this value defines an attractor that orients globally oscillatory trajectories. The global attractor is defined at the intersection of the adiabatic and* $Q=1$ *trajectories. All other states on these two trajectories constitute local "attractive" states that can bias or influence trajectories approaching the global attractor. Collectively, these locally "attractive" regions comprise a* ***slow manifold*** *(the regions are weakly attractive, the characteristic exponents are small). The slow manifold defines an intrinsic set of axes with the global attractor as origin. Information for orienting biomechanical oscillations is defined by vector flow patterns (morphologies) generated by variations in the oscillator's frequency and amplitude (energy). While the externally observed* ***(E,f)*** *work space appears quite complex, the internally-defined* ***(H,Q)*** *control space retains a simplicity that is invariant over variations in the number of mechanical degrees of freedom (muscle/joint complexes), growth, and artificial transformations. The invariance of the intrinsic coordinate space description over subsystem components addresses the problem of* ***motor equivalence****.*

11.1. Identifying the Engine

The wrist-pendulum system, as a muscle/tendon system, is a self-generating, force-transmitting unit composed of elastic and viscous elements. For a given wrist-pendulum system of fixed mass and length, a subject can realize a wide range of stable pendular, clocking organizations. The *assembling* and *sustaining* of these organizations involves the "mixing" of *conservative* (elastic and gravitational) and *nonconservative* (viscous) forces. What are the physical principles governing this mixing of forces and the assembling and sustaining of stable pendular, clocking organizations? The inquiry begins with a review of engine cycles. The search is for an appropriate engine characterization of the pendular clocking mode.

11.1.1. Conservative, Mechanical, Engine Cycles. A mechanical engine consists of a set of physical conditions that sustain a periodic transformation of energy (that is, from potential to kinetic forms, see Section 3.2.1) without any degradation. The physical conditions that define Newton's universe constitute the most celebrated example of a mechanical engine. Whether the "conditions" are constancy of angular momentum (that restricts the periodic motion of a planet around another body) or constancy of energy (that restricts the periodic motion of a pendulum about a fixed point of rotation), the result is the same: An infinitely continuous set of symmetric transformations over which the entity in question subsists. The result of these mechanical transformations is the emergence of a *work cycle*.

A conservative mechanical cycle is *not* stable with respect to perturbation. Perturbation of the oscillating component results in a change in the amount of energy or momentum carried by the cycle. The conservative character of the cycle requires that all perturbations, additions or deletions of energy, result in commensurate changes in the orbit defining the work cycle (see Figure 11.1).

11.1.2. Nonconservative, Thermodynamic, Engine Cycles. Whereas a mechanical engine produces cycles as a function of reversible transformations in the form of the energy (kinetic to potential to kinetic, etc.), a thermodynamic engine produces cycles as a function of irreversible transformations in the organizational grade of the energy (that is, the degradation from a macro distributional characteristic to a micro distributional characteristic, see Figure 11.2; see also, Section 3.2.2 and Figure 3.4). If, and when, the organizational grade of the energy becomes uniform and the distribution functions become time-independent, then the irreversible thermodynamic cycles will cease and reversible mechanical cycles will dominate the field processes (see Figure 11.3).

The transformation of energy from a higher potential source, C_2, to a lower potential sink, C_1 defines a thermodynamic work cycle (see Figure 11.3). The qualitative character of the work cycle is that of the *limit cycle*: In a phase space of displacement and velocity, the work cycle exhibits an essentially closed orbit that is returned to subsequent to a perturbation (see Figure 11.4). The *source→sink* energy degradation is restricted by a set of operational components whose finite storage capabilities result in energy oscillations between well-defined minimum and maximum values. The limits of the cycle are stable with

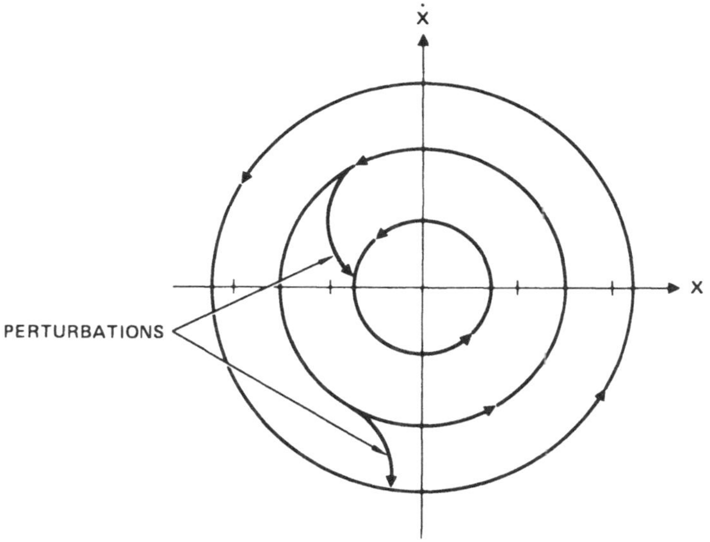

FIGURE 11.1. The response to perturbations of a conservative work cycle.

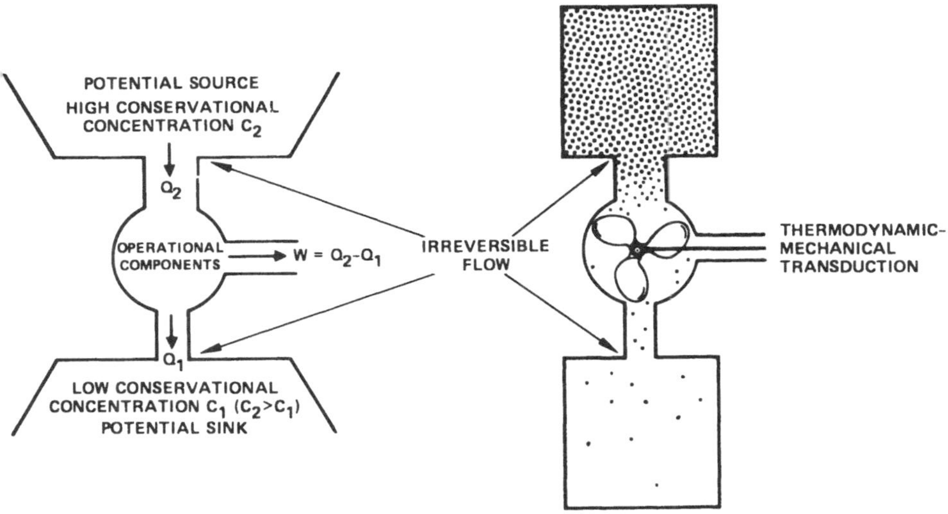

FIGURE 11.2. A generalized nonconservative (thermodynamic) engine.

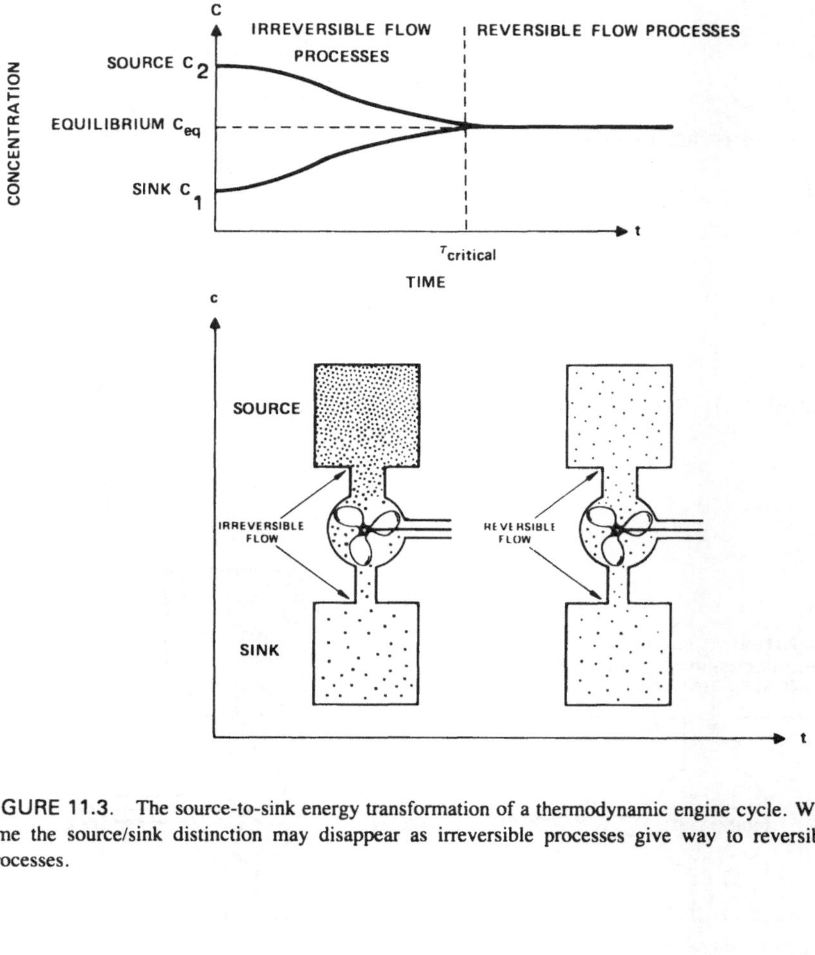

FIGURE 11.3. The source-to-sink energy transformation of a thermodynamic engine cycle. With time the source/sink distinction may disappear as irreversible processes give way to reversible processes.

respect to perturbation since they are scaled by the mechanical and thermodynamic properties of the oscillator, and *not* initial (energy) conditions. Typically, the mechanical properties of the intermediate operational components scale the frequency of oscillation and the thermodynamic properties of the heat sink scale the (maximum) amplitude of oscillation. Perturbations of the orbit are "washed-out" during the degradation phase of the energy cycle. Because of its relative immunity to perturbations, *a thermodynamic limit cycle can be advanced as the basic unit of action of physically autonomous systems* (cf. Kugler, Kelso & Turvey, 1980, 1982).

11.1.3. Carnot Efficiency Limitations on Heat Engines. The ideal (reversible) Carnot engine has been used traditionally to study limitations on maximum engine efficiency. The Carnot efficiency theorem is restricted, however, to engines sustained by *thermal source→sink* flows. Industrial heat engines, driven by high-temperature reservoirs such as boilers or hot gases in cylinders created by burning chemical fuels, exemplify Carnot efficiency-limited engines (see Figure 11.5). The maximum efficiency attainable by these engines is: *Maximum Carnot Efficiency* = $[1-(T_{source}/T_{sink})]$.

The Carnot limitation is a function of a thermal sink T_{sink} restriction associated with the external ambient temperature T_{amb}. If the temperature of the sink is less than the ambient external temperature ($T_{sink}<T_{amb}$), then a refrigerator is required to sustain the temperature differential. The energy required to operate the refrigerator reduces the overall efficiency of the engine, resulting in the

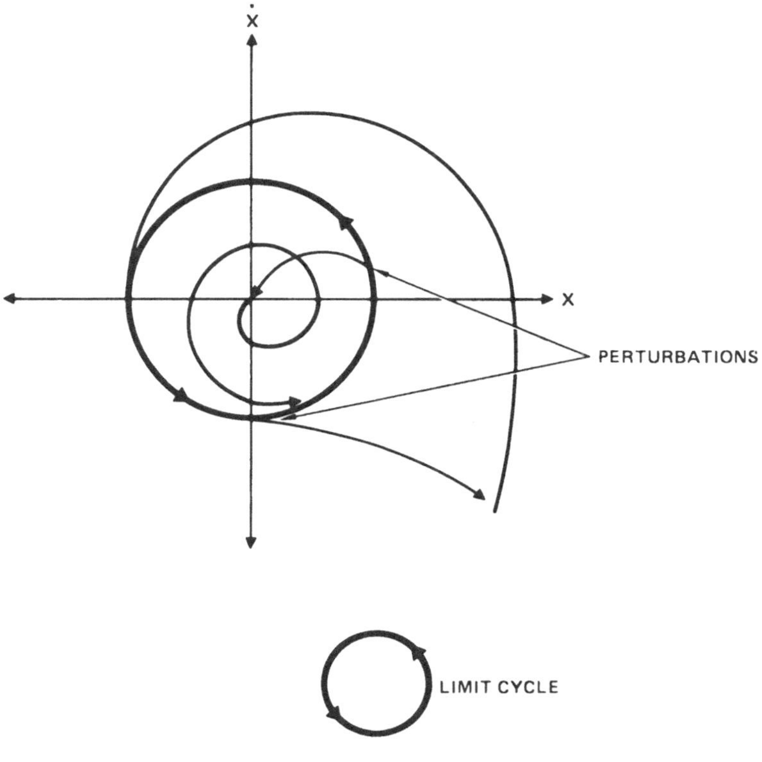

FIGURE 11.4. The response to perturbations of a nonconservative (thermodynamic) work cycle. The qualitative character of this engine process is that of a limit cycle.

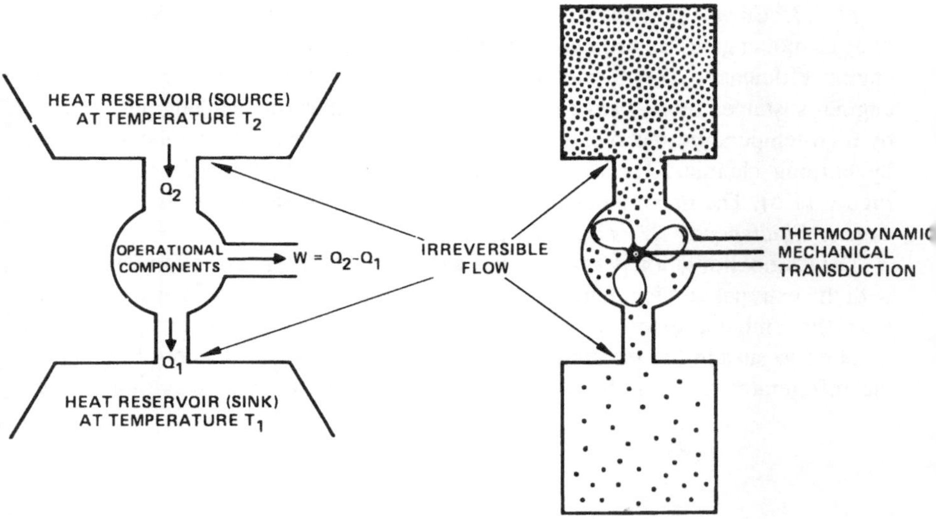

ENERGY TRANSFORMATIONS IN AN IDEAL HEAT ENGINE

Q_2 = ENERGY SUPPLIED TO THE ENGINE (HEAT INPUT)

Q_1 = ENERGY REJECTED BY THE ENGINE TO THE SINK (HEAT OUTPUT)

$W = Q_2 - Q_1$ = WORK PRODUCED BY THE ENGINE (MECHANICAL WORK CYCLE)

$$\text{MAXIMUM CARNOT EFFICIENCY} = \frac{[\text{MECHANICAL WORK CYCLE}]}{[\text{HEAT INPUT}]} = \frac{W}{Q_2} = \frac{Q_2 - Q_1}{Q_2}$$

FIGURE 11.5. The heat engine as paradigmatic of Carnot efficiency-limited engines.

maximum (ideal, reversible) conversion efficiency being defined when $T_{sink} = T_{amb}$. The Carnot efficiency limitation identifies the residual unused energy rejected by the engine into the ambient heat sink. The temperature of the sink is as close to absolute zero T_{abs} as the engine's condenser can attain without the aid of a refrigerator. The remaining temperature drop $T_{sink} \rightarrow T_{abs}$, therefore, is unavailable for conversion into mechanical energy (see Morowitz, 1978).

11.1.4. Muscle Fiber Energy-Converting Systems as Chemical-Thermal-Mechanical Engines not limited by Carnot Efficiency. Muscle fiber energy-converting systems are machines that operate by direct conversion of chemical to mechanical energy—without any intermediate thermal phase—and do so at a relatively constant temperature (Caplan & Essig, 1983; Hoyle, 1983; Morowitz, 1978; Volkenstein, 1970). These machines are not subject to Carnot efficiency limitations because of their lack of thermal dependence (see McClare, 1971). Whereas temperature drops in heat engines are severely limited by the ambient environmental temperature ($T_{source} \rightarrow T_{amb}$), potential drops in chemical engines can, in principle, achieve absolute zero states ($C_{source} \rightarrow C_{abs}$) without additional energy-consuming mechanisms such as refrigerators. There is, therefore, no thermal limitation on the operational efficiency of an engine driven by a chemical gradient.

A chemical engine can, in principle, convert energy with 100% efficiency. At first blush this claim might seem anomalous, given the friction and the resultant energy loss associated with muscle fiber activity. The anomaly is dismissed, however, by recognizing that there are two distinct processes at play. One is the *conversion*, from chemical energy to mechanical energy. The other is a *partitioning* of the mechanical energy into "useful" work and heat. This distinction allows one to see how a chemical engine can be 100% efficient (as an energy converter) and at the same time have a low Q-factor (as a user of the mechanical energy made available by the conversion); the conversion efficiency is *not* a measure of the dissipative character of the system. Conversion and partitioning come about in the following manner.

Biomechanical motions are assembled and sustained by energy generated on the basis of chemical reactions (metabolic breakdown of ATP) in the sarcomere structures (see Figure 11.6). The stored chemical energy E_{chem} is transformed into kinetic energy E_M associated with the mechanical motion of the sliding protein filaments (Carlson & Wilke, 1974; Huxley, 1980; Hoyle, 1983; Volkenstein, 1970; see Figure 11.7). During the cross-bridge conversion, E_{chem}, now in the form E_M, is partitioned over (a) mechanical processes (external work cycles) E_m and (b) thermodynamic processes (internal work cycles) E_t. The energy book-keeping is closed over the entire energy exchange, $E_{chem} \rightarrow E_M = E_m + E_t$ in accordance with the first law of thermodynamics. And, in accordance with the second law of thermodynamics, there are no "free" work cycles, $E_t > 0$ (see Figure 11.6).

11.2. Identifying the E_m and E_t Escapement Laws

11.2.1. Interpreting the Q-Values of Cycles in Terms of Escapement Characteristics. The Q-factor is an index of the thermodynamic processes within a single oscillatory cycle. It is a dimensionless ratio of the mechanical work produced to the energy lost as heat : Q = *[external work cycle/internal work cycle]*. A high Q-factor is associated with oscillations that are dominated by the external (mechanical) work cycles. A completely conservative system is defined when there are no internal work cycles: $Q\text{-}factor = \infty$. For low Q-factors ($Q<1$),

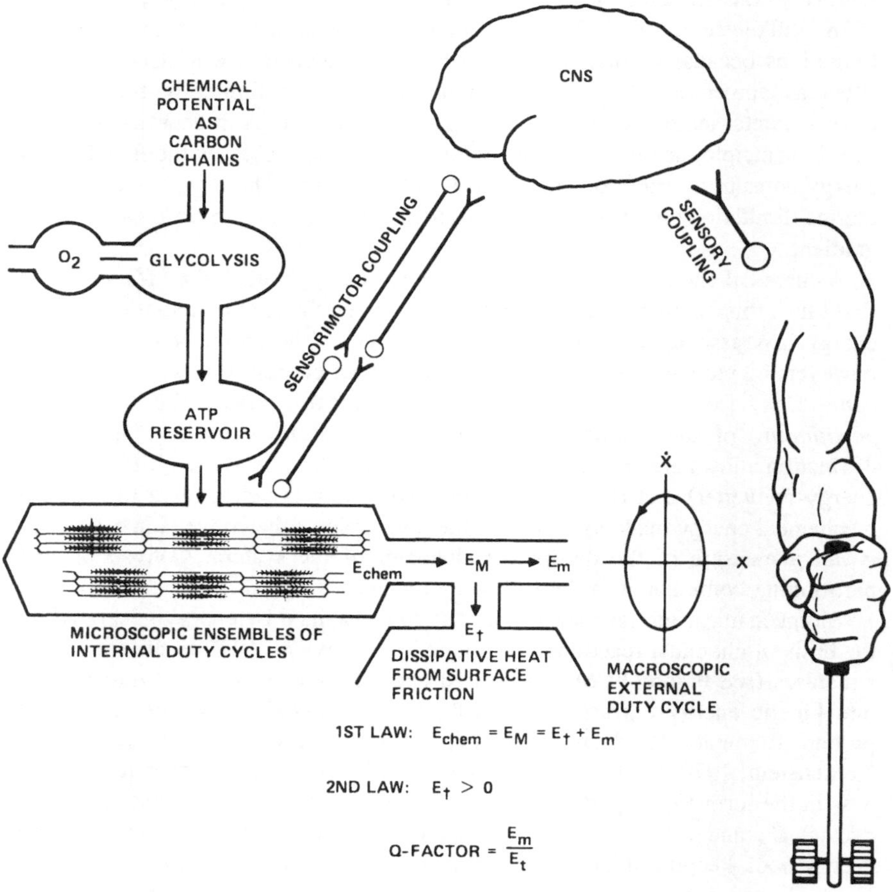

FIGURE 11.6. Energy conversion and energy partitioning in muscular tissues. The paradigmatic example of a chemical-thermal-mechanical engine process.

the oscillation is dominated by internal work cycles. When $Q = 1$, however, the energy associated with (conservative) mechanical work cycles is exactly equal to the (nonconservative) thermodynamic work cycles: $Q = 1$ defines a symmetry relating conservative and nonconservative work cycles.

Over a cycle, muscular forces generate and degenerate the wrist-pendulum kinetic energy observable in the laboratory frame of reference. As noted in Chapter 8, any oscillatory phenomenon requires two properties. One property ensures that the system in question returns to its static steady-state. The other property ensures that the system in question overshoots its static steady-state. A cycle of wrist-pendulum activity requires an inertial force that displaces the system from its static stable position and a restorative force that returns the system to this position. The result is an oscillation about a static equilibrium position.

The overshoots and returns to the static equilibrium position can arise from three sources: (a) the moment of inertia of the wrist-pendulum system; (b) gravity acting on the mass of the pendulum; (c) effective elasticity of neuromuscular machinery and, (d) mechanical energy provided from the neuromuscular

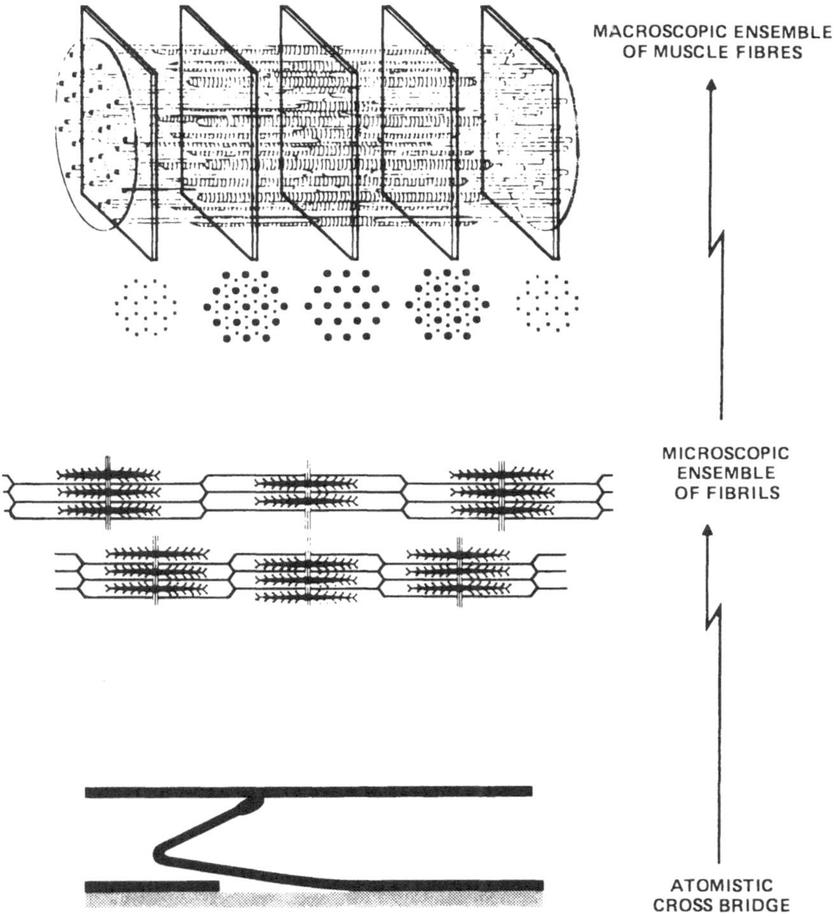

FIGURE 11.7. The mechanical processes in muscular tissues at increasing grains of analysis.

machinery. The first three sources define conservative force structures that vary strictly as a function of the wrist-pendulum's external coordinates of motion—position (height) and velocity, respectively, of the pendulum's center of mass. The fourth source is the neuromuscular escapement energy that is injected into the cycle to offset dissipative losses. The escapement energy is not rigidly coupled to the external coordinates of motion; instead, it varies as a function of internal (thermodynamic) coordinates of motion that index cycle dissipation. The muscular contribution to the overshoot and return can be saltatory and impulsive (some force for some duration), generating and degenerating kinetic energy in segments, with the system's conservative (inertial and gravitational) forces carrying it between segments (cf. Hubbard, 1974). One way to think about this latter point is that not all of the cycle energy need be contributed by the muscles each and every cycle. Some of the energy can be conserved from cycle to cycle as momentum in the external coordinates of motion.

The major implication of this latter point concerning conserved cycle energy is with respect to interpreting the Q-value of a muscle fiber (chemical-thermal-mechanical) engine qua wrist-pendulum system. Let Q be defined by the ratio of the mechanical energy E_m observed in the laboratory frame of reference—external coordinates of motion—to the thermodynamic energy (nonconservative) dissipated through degradation processes E_t. Then:

$$Q = \frac{E_m}{E_t} \,. \qquad (11.1)$$

Which is to say, the muscular energy required to offset the thermal loss is:

$$\text{Muscular Energy} = \frac{E_m}{Q} \,. \qquad (11.2)$$

From Equation (11.2) it follows that the larger the Q-value the smaller is the *proportional* contribution of muscular energy to the energy of the cycle. The muscular forces contribute less and the moment of inertia, elasticity, and gravity contribute more to the return and overshoot properties defining the oscillatory phenomenon. Thus, for example, $Q = 5$ means that the muscles supply on the average only 1/5 of the cycle's energy. To further this idea, consider a Q equal to unity and a Q less than unity. When $Q = 1$, muscular energy is responsible on the average for the entire cycle energy, there is no energy residual conserved from cycle to cycle. The only energy seen in the laboratory frame of reference is that energy added by the neuromuscular escapement each and every cycle. When $Q<1$, the energy seen in the laboratory frame of reference is less than that provided muscularly. Thus, for example, $Q = 0.2$ means that the neuromuscular escapement supplies, on the average, five times the cycle energy observed in the laboratory frame (external coordinates of motion) in order to complete a single cycle. Or, synonymously, five times as much energy as appears in the laboratory frame per cycle is being lost in thermodynamic processes. These odd-sounding statements point to the fact that when $Q<1$ a great deal more energy is being expended in the interior of the chemical-thermal-mechanical engine than in the laboratory frame of reference: Low Q-systems $(Q<1)$ are more dominated by nonconservative internal energy transformations—changes in energy grade—than by conservative external transformations—changes in energy form. As noted above, Q of unity identifies a transitional symmetry condition that separates oscillatory behavior dominated by reversible conservative processes $(Q>1)$ from oscillatory behavior dominated by irreversible nonconservative processes $(Q<1)$. At $Q = 1$ the amount of energy appearing in the mechanical frame (external coordinates of motion) is exactly equal to the amount of energy appearing in the thermodynamic frame (internal coordinates of motion).

11.2.2. Assembling and Sustaining Escapements. The converting/parsing distinction presented in Figure 11.6 leads to a number of important understandings for the escapement mechanisms underlying conditional periodicity. Primarily it brings to the fore two types of escapement stability that can be achieved by the engine processes governing oscillatory movement. Only under certain conditions, however, can both stabilities be satisfied simultaneously. To set the stage for the interplay between the two stable states it will be necessary to distinguish the chemical-thermal-mechanical engine processes by which oscillatory parameters are softly *assembled* from the engine processes by which already assembled oscillatory parameters are *sustained*. The implied differences in these engine processes can be brought to light by contrasting four oscillator designs. These oscillator designs are identified by Figure 11.8 (which is an alternative version of Figure 5.6), by the left panel of Figure 11.9 (which is a reproduction of Figure 10.9), by the right panel of Figure 11.9 and by Figure 11.11. The three major features on which these oscillator designs are to be contrasted are the following: (a) the location of the source of energy for assembling the oscillator's mechanical parameters (an external frame of reference or an internal frame of reference); (b) the dependency of the oscillator's external work cycle on the energy for assembling relative to the energy for sustaining; and (c) the relation of the assembling and sustaining energies. To anticipate, the proposed chemical-thermal-mechanical engine is a cycle assembled from an *internal* frame of reference and characterized by an adiabatic transformation resulting from *orthogonal* assembling and sustaining escapements. *Measurements of this engine's work cycle, made from the external (laboratory) frame of reference, will reveal only the energy contributed by the assembly escapement.* This is because, under an adiabatic transformation, the

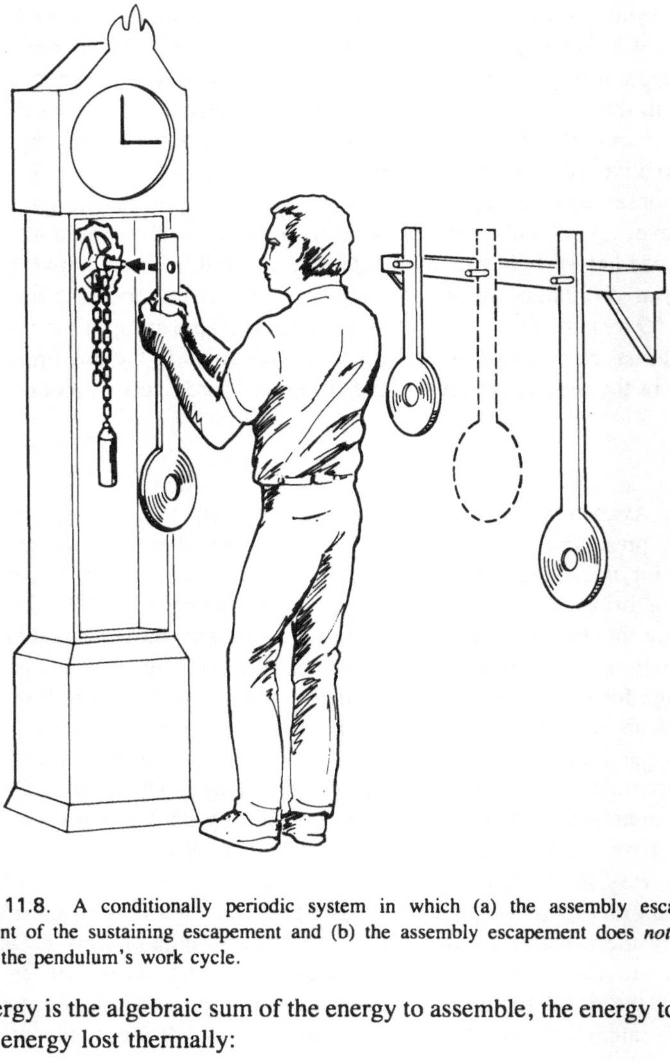

FIGURE 11.8. A conditionally periodic system in which (a) the assembly escapement is independent of the sustaining escapement and (b) the assembly escapement does *not* contribute energy to the pendulum's work cycle.

total energy is the algebraic sum of the energy to assemble, the energy to sustain, and the energy lost thermally:

$$E_{total} = E_{assembly} + (E_{sustaining} - E_{dissipation}) \; ; \tag{11.3}$$

where,

$$E_{sustaining} = E_{dissipation} = |E_t| \; , \tag{11.4}$$

and, therefore,

$$E_{sustaining} - E_{dissipation} = 0 \; . \tag{11.5}$$

Thus, because of the symmetry relating the sustaining escapement energy to the thermal loss, the total energy observed in the laboratory frame reduces to energy contributed by the assembly escapement:

$$\mathrm{E}_{\text{laboratory frame}} = \mathrm{E}_{\text{assembly}} \ . \tag{11.6}$$

The foregoing sequence of equations can be restated in a very general form:

$$\mathrm{E}_{\text{total}} = \mathrm{E}_{\text{conservative}} + \mathrm{E}_{\text{nonconservative}} \ . \tag{11.7}$$

$E_{conservative}$ is a function of mechanical properties and $E_{nonconservative}$ is a function of thermodynamic properties. Equation (11.7) allows for a very succinct statement of the design of an adiabatically transformable system: First, the design of the system is such that manipulation of $E_{conservative}$ can be made independently of $E_{nonconservative}$; and second, the design of the system is such that $E_{nonconservative}$ is invariant over variations in $E_{conservative}$.

Figure 11.8 identifies a nonconservative pendulum clock with a phase-dependent energy escapement. A constant amount of energy is injected into the cycle at a fixed phase of oscillation, such as when the pendulum passes through the static equilibrium position. Operating the phase-dependent oscillator at different frequencies (e.g., exhibiting conditional periodicity) is a four step process: First, the motion of the pendulum is stopped—extracting all energy from the cycle. Second, the pendulum is replaced by a new one that differs in length. Third, the new pendulum is set into motion with a small perturbation so as to activate the escapement. And finally, once activated, the escapement pulses the new pendulum every cycle with new energy to sustain its motion. Steps 1, 2 and 3 are associated with the "assembly" of the new oscillator. The assembly of the new pendular unit is performed at a cost. Assembling or building or modifying requires that energy be put into the system to perform the work. The amount of work done, and the energy dissipated in the process, will vary with the size of the added pendulum—a large pendulum will require more work for its replacement than a small one. This variation in assembly energy, however, does not carry over into the clock's externally observed (laboratory frame of reference) energy cycle. The externally observed energy cycle of the clocking motion is constant, independent of the length of the oscillator (See Figure 5.6). The energy cycle defined by the clock's oscillatory component is independent of the amount of energy required to assemble it. This independence is because the assembly steps (1,2,3) are always completed, and the energy to do them paid for and dissipated, before the sustaining energy comes into the picture. The observed cycle energy is a function only of the sustaining energy, *not* the sustaining energy plus assembly energy. The constancy of the observed energy cycle is, therefore, strictly a function of the constancy of the sustaining escapement. The system depicted in Figures 5.6 and 11.8 is a system whose design characteristics keep the two escapement energies independent. The depicted dissipative pendular clock is an oscillatory system whose *externally observed work cycle* reflects exclusively the *sustaining energy*. It is also a system subject to *non*adiabatic transformations. Oscillator designs in an adiabatic context follow.

The left panel of Figure 11.9 shows a conservative simple pendulum. Unlike the clock of Figures 5.6 and 11.8, with its stops and starts during assembly, the simple pendulum depicted in the left panel of Figure 11.9 maintains continuous motion throughout the assembly process. More precisely, its motion subsists uninterrupted after it has been set into motion with a certain amount of energy. Further, unlike the clock of Figures 5.6 and 11.8, changes in the length of this system—that is, changes in the mechanical property determining its frequency—occur while the system is oscillating. It is because these changes occur so slowly that no external torque is added to the pendular motion and, therefore, the coordinates of motion do not undergo any degeneracy—e.g., the (E/f) ratio is adiabatically invariant. As with the nonconservative clock of Figures 5.6 and 11.8, altering the oscillatory component's length (i.e., assembling a new oscillator) involves work. In the case of the conservative string pendulum, however, the work done in assembling adds energy to, or subtracts energy from, the pendulum's work cycle. Changes in the *(length/frequency)* ratio of the string pendulum incur energy changes ΔE in constant proportion H to frequency changes Δf (see Equation 10.66): $\Delta E = H\Delta f$ *and* $E_{total} = Hf$—Ehrenfest's theorem (see upper left panel of Figure 11.10). These energy changes (ΔE), resulting from the additions and deletions of energy, define the assembly energy of the oscillator. Since the string pendulum is conservative, no sustaining escapement is required to offset dissipative losses. The string pendulum is an oscillatory system whose *externally observed work cycle* reflects exclusively the *assembly energy*.

The right panel of Figure 11.9 depicts a nonconservative pendulum clock with the extraordinary property of continuously malleabale length. It is an intended hybrid of the system depicted in Figures 5.6 and 11.8 (a nonconservative clock with interchangeable pendulum components of discrete lengths) and the system depicted in Figure 11.9, left panel (a conservative simple pendulum with continuously changing lengths). Another extraordinary property of this hybrid oscillator is that the energy dissipated per cycle is frequency independent; that is, it is so constructed so as to exibit an indifference of thermodynamic parameters to mechanical parameters. This hybrid oscillator expresses the fundamental property of mechanical transformability under adiabatic invariance. The mechanical transformation identified in the right panel of Figure 11.9 is depicted in its kinetic *(energy,frequency)* and kinematic *(amplitude,period)* forms in the right upper panel and right lower panel, respectively, of Figure 11.10. The independence of energy dissipation with respect to frequency changes is revealed in Figure 11.10's right upper panel in the negative energy intercept E_t. The negative energy intercept identifies a constant amount of energy degraded per cycle. (For

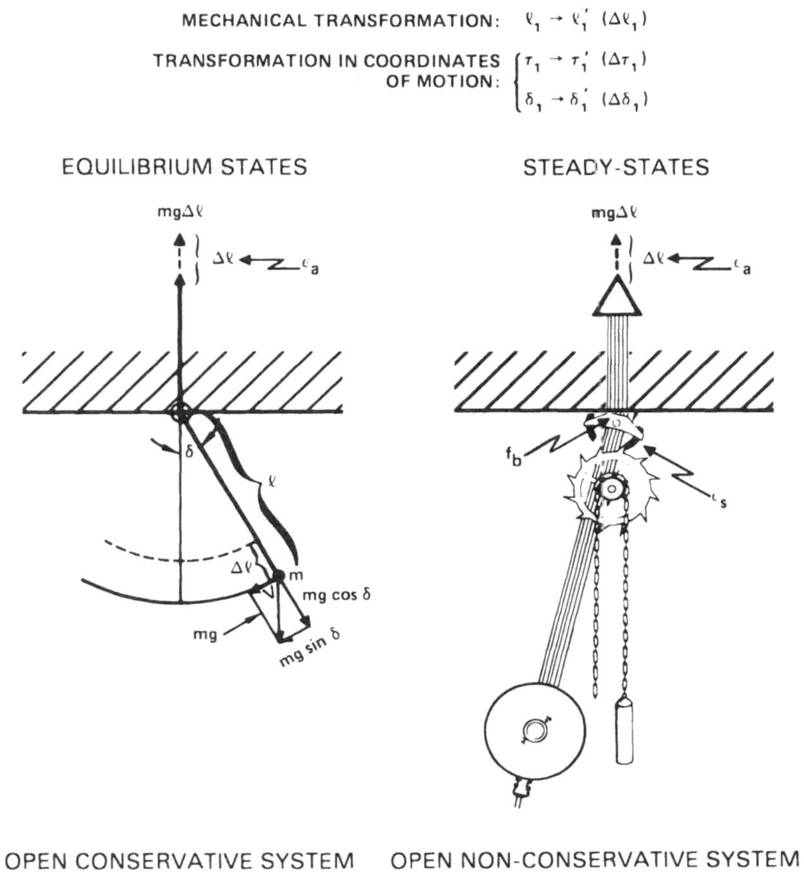

FIGURE 11.9. *Left panel*: An example of a conditionally periodic system in which (a) the assembly escapement contributes *all* of the energy carried reactively in the work cycle, (b) the assembly escapement energy is derived from an *external* energy reservoir, (c) the work cycle degrades *no* energy ($E_t = 0$) and (d) transformations of the equilibrium state coordinates of motion (θ, τ) are *rate-limited*.
Right panel: An example of a conditionally periodic system in which (a) the assembly and sustaining escapements contribute the energy carried reactively in the work cycle, (b) the assembly escapement energy is derived from an *external* energy reservoir, (c) the work cycle degrades a *constant* positive amount of energy per cycle (E_t = *constant* and $E_t > 0$) and (d) transformations of the steady state coordinates of motion (θ, τ) are *rate-limited*

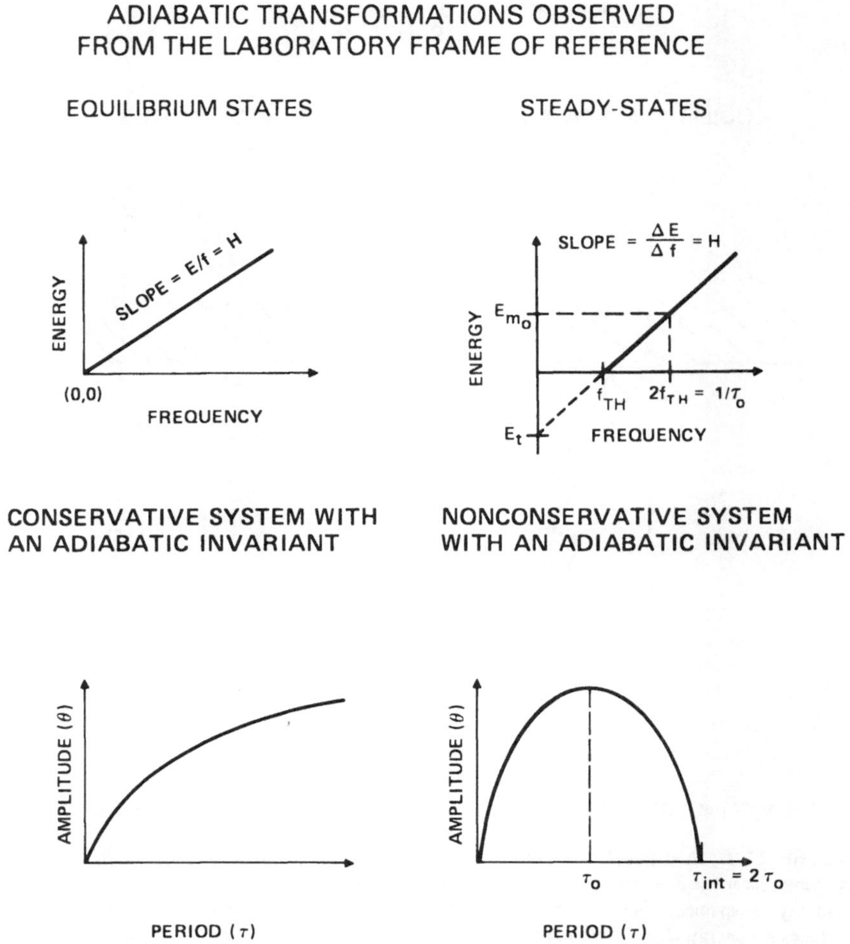

FIGURE 11.10. Analysis of adiabatic transformations for the conservative and nonconservative systems identified in Figure 11.9. The upper panels identify the critical qualitative features in the kinetic *(E,f)* work space. The lower panels identify the critical qualitative features relating the kinematic *(θ,τ)* coordinates.

convenience, Figure 11.10 also presents the corresponding kinetic and kinematic transformations for the conservative, simple pendulum).

As with the conservative, simple pendulum system of Figure 11.9's left panel, the extraordinary hybrid system of the right panel depends upon an external energy source for the assembling of its mechanical parameters, specifically, its length and frequency. Unlike the simple pendulum, the extraordinary clock's cyclic parameters are determined by thermodynamic properties (dissipation) in addition to mechanical properties. And, therefore, unlike the cycle energy of the open and conservative simple pendulum, the cycle energy of the open and nonconservative clock depends on the energy to assemble a length and the energy to sustain a periodic motion in the face of frictional losses. For both kinds of systems, however, the length changes are brought about by energy introduced as a torque from an external frame of reference. To avoid instability and/or degeneracy in the coordinates of motion, external torque must be introduced into the cycle very slowly. In fact, for the extraordinary clock to ride continuously its limit cycle, the length changes would have to be introduced infinitely slowly. That is to say, an adiabatic transformation through energy introduced from an external frame of reference, as depicted in Figure 11.9, is severely *rate-limited* (see Section 10.5.1).

To summarize, the extraordinary oscillatory system of Figure 11.9's right panel is a system in which (a) mechanical and thermodynamic processes contribute to the cycle, (b) the mechanical and thermdoynamic processes are orthogonal, (c) the mechanical parameters (length, frequency) can be changed or assembled in a continuous manner and (d) the continuous mechanical changes constitute an adiabatic transformation if and only if they are executed at an infinitely slow rate. In the final oscillatory design to be discussed, features (a), (b) and (c) are preserved but feature (d), the rate limitation on adiabatic transformability, is removed.

Figure 11.11 presents the fourth and final design. It is meant to express conditional periodicity powered by neuromuscular (chemical-thermal-mechanical engine) processes. More abstractly, it is meant to express conditional periodicity in an external reference frame that is assembled and sustained by an energy reservoir in an *internal reference frame*. A major argument of Chapter 10 is that such conditions, in which energy is introduced as internal torque, circumvent the rate limiting consequences accompanying energy introduced through external torque.

Conditionally periodic oscillations powered by neuromuscluar processes share a number of design similarities with the above two oscillators (Figure 11.9) when operating under adiabatic conditions. First, a neuromuscular oscillator is dissipative and, therefore, requires a sustaining escapement to guarantee persistence of cycle characteristics. Second, when undergoing conditional periodicities that satisfy adiabatic constraints—for example, changing operating frequency while preserving the amount of energy dissipated per cycle constant—the sustaining escapement is invariant over frequency variations (see Chapter 10). Third, the mechanical properties that scale the natural operating frequency, such as muscle stiffness or resting length, are assembled during the oscillatory process, rather than prior to the start-up of oscillation. Fourth, changes in these mechanical properties constitute an assembly escapement that adds or subtracts energy from the externally observed energy cycle. And, fifth, the assembly escapement is powered by an *on-board*, potential-energy source (depicted by the little man

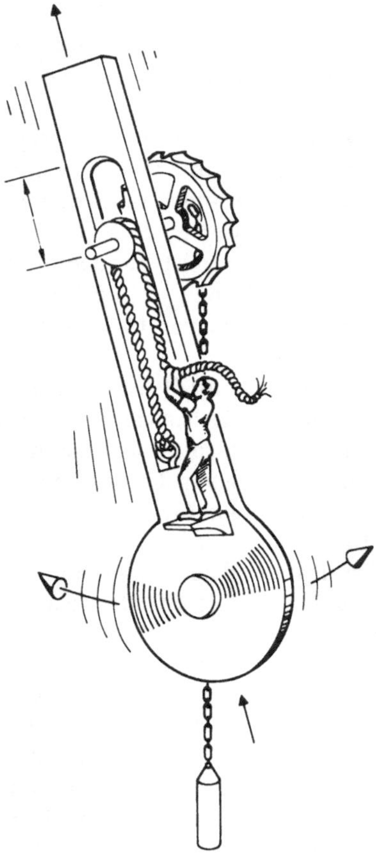

FIGURE 11.11. An example of a conditionally periodic system in which (a) the assembly escapement contributes *all* of the energy carried reactively in the work cycle, (b) the assembly escapement energy is derived from an *internal* energy reservoir, (c) the work cycle degrades a *constant* positive amount of energy per cycle ($E_t = constant$ and $E_t > 0$) and (d) transformations of the steady state coordinates of motion (θ, τ) are *rate-independent*..

'riding' the pendular component in Figure 11.11). That is, the energy for assembling derives from an internal reference frame. This feature eliminates the rate limitation on adiabatic transformability (as argued in Section 10.5.1). In sum, a rhythmically moving neuromuscular complex is a hybrid oscillator with an on-board assembly escapement that is variable and an on-board sustaining escapement that is constant. The hybrid oscillator is designed such that the total energy (conservative plus dissipative portions) of a cycle is the sum of the assembly energy and sustaining energy. The assembly energy contributes to the energy conserved from cycle to cycle (conservative portion) and the sustaining

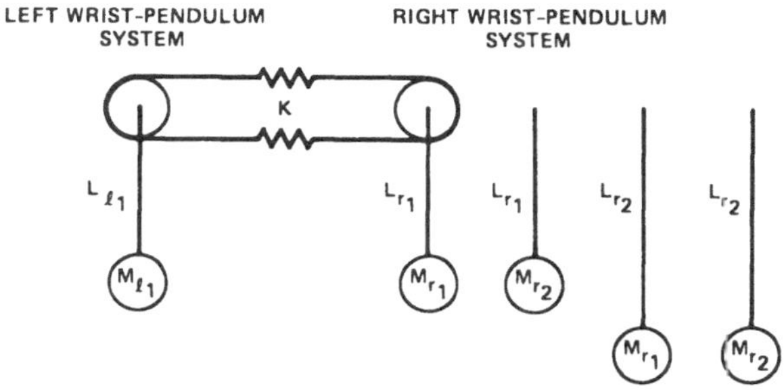

K = MACROSCOPIC STIFFNESS OF VIRTUAL SINGLE SYSTEM

$K = f(M_{\ell_i}, L_{\ell_i}, M_{r_i}, L_{r_i})$

M_ℓ = MASS OF LEFT WRIST-PENDULUM SYSTEM
L_ℓ = LENGTH OF LEFT WRIST-PENDULUM SYSTEM
M_r = MASS OF RIGHT WRIST-PENDULUM SYSTEM
L_r = LENGTH OF RIGHT WRIST-PENDULUM SYSTEM

ΔK IS A MECHANICAL 'ADIABATIC' TRANSFORM THAT HAS ASSOCIATED 'ADIABATIC INVARIANTS', H_i, FOR THE LEFT AND RIGHT WRIST-PENDULUM SYSTEMS:

$H_i = \frac{\Delta E}{\Delta f}$ = CONSTANT (ADIABATIC INVARIANT)

$\Delta K_j \rightarrow (\Delta E_j, \Delta f_j)$ (ADIABATIC TRANSFORM)

$\Delta K_1 \rightarrow (\Delta E_1, \Delta f_1)$

$\Delta K_2 \rightarrow (\Delta E_2, \Delta f_2)$

SUCH THAT

$H_i = \frac{\Delta E_1}{\Delta f_1} = \frac{\Delta E_2}{\Delta f_2}$ = CONSTANT

ACTION IS AN ADIABATIC INVARIANT, H_i, UNDER ADIABATIC TRANSFORMATIONS, ΔK.

FIGURE 11.12. Example of an adiabatic transformation on the external coordinates of motion induced through changes in the the internal stiffness of the oscillator.

energy contributes to the energy dissipated from cycle to cycle (nonconservative portion). This characterization warrants further discussion.

A neuromuscular engine, operating under a given loading (mass and length), is subject to a change in boundary conditions. For example, it is non-mechanically coupled to other, different engine processes—for example, a left limb segment (wrist-pendulum system) is coupled to a right limb segment (wrist-pendulum system). During these coupled (boundary) conditions, exploratory oscillations (see Section 7.2) can be brought about through changes in internal mechanical parameters, such as the stiffness or resting length of the muscle (Fel'dman, 1982). The situation is depicted in Figure 11.12. These mechanical changes are associated with energy flows (additions to or subtractions from the oscillatory component). These flows index an assembling escapement. Chemical energy is converted into the mechanical energy of sliding filaments and some of this energy is put to use in establishing the parameters of oscillation. Arguably there can be, and ordinarily there must be, an energy conversion process oriented specifically to establishing the maximum potential and kinetic energies (observables in the laboratory frame of reference) that will characterize the work cycle. In classical terms, this amounts to the setting of the initial (energy) conditions for the conservative aspect of the cycle. This assembling process is (a) discrete, (b) manifest as either an addition or deletion of energy and (c) scaled to the changes in the mechanical properties. Once a limit cycle has been assembled, it is sustained, in principle, by a second conversion process, an escapement that merely offsets the thermal loss (see Figure 11.6). Under this condition the external work cycle would exhibit an energy equal to the sum of the energies of the two escapements minus the energy dissipated (see Equation 11.3). Equation (11.3) can be rewritten so as to reveal the origin of the escapements' energies in the on-board chemical potential:

$$E_{chem} \xrightarrow{\text{conversion}} E_m \begin{cases} \xrightarrow[\text{escapement}]{\text{assembly}} \epsilon_a \\ \xrightarrow[\text{escapement}]{\text{sustaining}} \epsilon_s - E_t \end{cases} \qquad (11.8)$$

and

$$\int_{t=0}^{t=\tau} E_t\,dt = \text{constant} \qquad (11.9)$$

where ϵ_a is the discrete assembling (and disassembling) escapement and ϵ_s is the continuous sustaining escapement. Only the assembling energy is observed in the laboratory frame of reference. The change in $\epsilon_a + \epsilon_s$ relative to the change in f indexes the action H of the oscillator produced by the chemical-thermal-mechanical engine processes; the ratio of $(\epsilon_a + \epsilon_s)$ to E_t indexes its Q-factor. This sentence provides a glimpse of the two attractor states referred to at the outset.

11.2.3. Adiabatic Trajectories and the Principle of Least Interaction: An "Inward-looking" Stability. Suppose that it was possible to assemble an oscillatory regime such that in running the regime the fullest possible use was made of the reactive phenomena that it engendered. This is Bernstein's (1967, p. 109) "highest stage of coordinational freedom" in which movements are structured so as to "*utilize entirely the reactive phenomena* which arise" (see also Hubbard, 1974). As Bernstein elaborates:

> The secret of coordination lies not only in not wasting superfluous force in extinguishing reactive phenomena but, on the contrary, in employing the latter in such a way as to employ active muscle forces only in the capacity of complementary forces. In this case the same movement (in the final analysis) demands less expenditure of active force
>
> (Bernstein, 1967, p. 109).

Consider Bernstein's "highest stage of coordinational freedom" in the pendular, clocking mode. For a given change in the boundary conditions on a wrist-pendulum system, the change in the amount of chemical fuel degraded to assemble the new oscillatory properties and to sustain the new oscillations is kept to a minimum. The absolute minimum in this case is well defined. If the energy loss per oscillation of a wrist-pendulum system is the same regardless of its frequency, then the absolute minimum is defined by the energy to offset this loss. The question becomes: How can this ideal be attained? The answer lies with ϵ_a. The original cost of ϵ_a, of assembling the oscillatory properties, is carried over into the rhythmic behavior as the conservative part of the total cycle energy. If it can be transferred to a nonmuscular account then the active muscular contribution per cycle need be only with respect to the nonconservative portion, that is, $E_m = \epsilon_s = E_t$. This characterization of the pendular, clocking mode in terms of Bernstein's "highest stage" is important to the conceptions that follow.

Speaking approximately, to function at Bernstein's "highest level of coordinational freedom" is to function vegetatively. Vegetative functions are those of an organism during times of peace, simply conserving bodily resources, disposing of wastes. In Cannon's (1932) classical opposition of autonomous nervous processes, the vegetative functions of the parasympathetic system contrast with the activity functions of the sympathetic system. Vegetative functions comprise the physiological deep structure. They are those states onto which an organism, in whole or in part, relaxes in the absence of external forces imposing contrary demands. They are *time-independent* attractors providing a basis for organizing time-dependent processes.

To off-set heat production, and to do no more, is to function vegetatively. Hence, the ascription of "vegetative" to Bernstein's "highest level". An assembled pendular, clocking mode of organization that subsists merely by compensating frictional loss has relaxed onto (or into) a vegetative function. The same can be said when a succession of organizations in the pendular, clocking mode (defined over a wrist-pendulum system of fixed magnitude) are so assembled and so sustained to yield a trajectory of (E,f) states characterized by constant action. Across the successive organizations, the change in the quantity of chemical fuel degraded into mechanical energy per change in frequency is *minimal* and *constant*. An adiabatic trajectory of states is a vegetative function.

It is so for two reasons. One is that resources are conserved. The other is that the means of conserving is an energy converting regime that is fully autonomous in the formal time-independent sense. An adiabatic trajectory reflects, in short, a time-independent steady-state in which the thermodynamic processes are independent of the mechanical processes. It reflects, more exactly, an "inward looking" stability that directs the *building* and *running* of the pendular, clocking mode. It earns the qualifier "inward looking" from its focus: maintaining the internal reserve of metabolic fuels.

What precisely is entailed by the vegetative function reflected in an adiabatic trajectory of pendular, clocking states? Assembling the properties on the interior (the properties of the chemical-thermal-mechanical engine), to produce a prescribed frequency, requires contact with the current boundary conditions. In the situation under analysis these derive from the magnitudes of the virtual single system (see Figure 11.12). From the perspective of the individual left or right wrist-pendulum system these boundary conditions are imposed externally. To this extent, therefore, the building of a local pendular, clocking organization is mediated through interactions with its immediate surround. The interactions, however, are minimal. In largest part the assembling of the requisite chemical-thermal-mechanical engine processes proceeds noninteractively or autonomously with respect to the sustaining processes.

Returning to Bernstein's optimal strategy. When it is in effect, the sum of the assembling escapement's energy is equal to the energy carried reactively in the external coordinates of motion (that is, nonmuscularly). That is, there is a principled constraint on the minimal quantity of chemical fuel degraded into mechanical energy. It must equal the sum of the energy lost thermally (a constant) *and* the energy returned to the cycle reactively. Consequently, if this minimum is attained then *all* that is required of the chemical-thermal-mechanical engine processes per cycle is the energy to sustain. The major point of the preceding is that, the adiabatic trajectory is a strategy that limits the contribution of active muscular forces during oscillations (of any frequency) to the off-setting of frictional losses. The vegetative function associated with an adiabatic trajectory is, in fact, the basic vegetative function: compensating for heat production.

This fundamental vegetative function promises a physical interpretation of the "comfort state" of biomechanical oscillation. "Comfort state" may be defined by the thermodynamic criterion that cycles are sustained by a time-independent escapement. An important corollary of this definition is that "comfort state" and "autonomy" are linked concepts. Approximately, comfortable oscillations are those in which the system in question interacts minimally with, and largely independently of, its immediate surroundings.

This line of argument follows the tradition of Gel'fand, Gurfinkel, Tsetlin and Shik's (1971) inquiry into nervous system functioning that emphasizes the expediency and autonomy of softly assembled neural subsystems. At its core is a *principle of least interaction*. A system is said to be working expediently in some external environment if it tends to minimize its interaction with that environment. The state of minimal interaction defines maximum stability. In this state, a subsystem functions as if it were independent or autonomous of other subsystems (Gel'fand et al., 1971). It is suggested here that an adiabatic transformation satisfies a principle of least interaction relative to the vegetative escapement processes that off-set dissipative losses.

11.2.4. Q-Factors and Self-Similarity: An "Outward Looking" Stability. The abiabatic stability is with reference to preserving certain local conditions of autonomy on the interior states—keeping the characteristics of the "vegetative" energy conversion largely invariant. In contrast, the Q-factor can be used to define an "outward looking" stability, one that refers to external states of interaction. This stability is with respect to minimizing the effects of external, environmental perturbations on the steady-state limits (coordinates of motion) of the work cycle.

The focus of the Q-factor stability is in the laboratory frame of reference; it identifies the ratio of the total energy carried by the oscillatory component to the energy dissipated. Certain stability benefits arise from low Q-values: Low Q-systems are highly resistant to external perturbations. Whereas the H stability minimizes interactions between subsystems, it can result in organizational states that are not stable with respect to external perturbation. These are the oscillatory states on the adiabatic trajectory exhibiting frequencies that are in excess of the the critical frequency ($f_{critical} = 2f_{int} = 1/\tau_o$). At the critical frequency the quality factor equals *unity:* the conserved energy of the mechanical processes is exactly equal to the nonconserved energy of the thermodynamic processes. Because of the symmetry relating these two energy processes all external perturbations are 'washed out' after a single cycle. The energy book-keeping for both the first and second laws of thermodynamics is closed at the completion of each and every cycle.

To the extent that the Q-value of the auto-oscillator exceeds unity, an external perturbation will play out over more than one cycle. The Q-value identifies the number of cycles required to recover initial energy conditions (prior to perturbation). It is a measure of the oscillator's *autonomy with respect to external perturbation*. This externally oriented autonomy assumes a larger role as the number of coordinate oscillator units is increased. In systems of multiple coupled oscillators the Q-factor is central to the phase resetting characteristics of the system (see Winfree, 1980).

The stability of a $Q = 1$ system amounts to keeping the responsibility for the control of the coordinates of motion anchored in (muscular) forces generated internally—rather than in (reactive) forces generated externally. Where the adiabatic invariant is an internal interactive stability, the Q-factor is an external interactive stability. A system organized around a constant H will not always be stable with respect to changes in the context of external forces; a system organized around the symmetry associated with $Q = 1$ will absorb these changes. Bernstein recognized this "weakness" of the "highest stage of coordinational freedom". He wrote:

> A higher degree of coordination secures a higher reactive conductivity to muscle force impulses at the periphery. We must note that, because this heightening of conductivity is entirely achieved by employing the dynamic transitory phenomena such high conductivity cannot be a stable or constant value.
>
> (Bernstein, 1967, p. 110)

11.3. Building an Intrinsic *(H,Q)* Coordinate Space

The sum of the preceding discussion is this: When a chemical-thermal-mechanical engine generates a variety of movement cycles under a fixed load, the energy of the cycles at a given frequency can be assigned (a) in accordance with either one of two "restrictions", and (b) only in accordance with both simultaneously at a single point in the *(E,f)* work space. It is proposed that these two restrictions orient the engine processes. One restriction is associated primarily with energy conversion and one with energy partitioning. The conversion restriction amounts to adiabatic invariance, in which cycle energy increases in constant proportion to cycle frequency. The change in conservational flow is kept to a minimum. The partitioning restriction amounts to a Q-factor of unity, in which the cycle energy matches the thermal loss. The resistance (autonomy) to perturbation is kept to a maximum. The details of these two restrictions are examined in this section. The goal, ultimately, is to define a set of natural coordinates around which the biomechanical machinery is organized.

11.3.1. Deriving an Expression for Q-Stability in terms of H-Stability. How do these two orienting restrictions relate? Are they independent, orthogonal variables of a machine space or are they linked (linearly or nonlinearly)? The goal of this subsection is to derive an equation set that details the relations between H and Q. To this end it will be helpful to return to the earlier discussion (Section 10.5.3) of the "work functions" that characterize the photoelectric effect.

Recall that adiabatic trajectories in a nonconservative system exhibit a frequency intercept greater than zero and an energy intercept less than zero (see Sections 10.5 and 10.6). With regard to the mechanical pendulum depicted in Figure 10.9 and 11.9 (left panel) there are no restrictions on the frequency or energy values that the system can assume. In the ideal case the *(E,f)* trajectory extends from the origin *(0,0)* to infinity. Whatever the change in length (that is, whatever the amount of work done to transform the length adiabatically) there will be a resultant change in the system's frequency and a fortiori in the system's work cycle. For the conservative system there are no limiting conditions.

It was also noted in Section 10.6 that for any adiabatic trajectory in the (E,f) work space there was a single frequency value, defined at twice the intercept frequency, at which the Q-value $= 1$. Absent from Section 10.6 is a discussion of how Q-values are assigned more generally. More precisely, what is missing is an explicit statement of the relation between H and Q. An explicit statement of this kind follows from generalizing the physical insight embodied in Equations 10.72 and 10.73—that for a given system exhibiting quantized action the magnitude of the action is "determined" completely by the work function A and the threshold frequency f_{TH}. More directly the explicit statement follows from generalizing the observation that the work function is the only hidden energy. Referring again to Figure 10.15, whereas the length of the vertical line labeled $E_{k_{max}}$ is indefinitely variable, the length of the vertical line labeled A is fixed (note $A = E_t$ in Figure 11.13).

Before extending this observation to the problem at hand it is important to note once more that the mechanical and thermodynamic derivations of the adiabatic invariant presented in Chapter 10 point to Q's indifference to adiabatic invariance in periodic (versus conditionally periodic) systems, and vice versa. For both the ideal mechanical oscillator and the ideal thermodynamic ensemble the Q-factor is constant. It is infinity in the former case and unity in the latter case. These values arise because there is no friction, by definition, in the ideal mechanical oscillator and because the distribution of the ideal thermodynamic ensemble is characterizable by the random walk that is paradigmatic of the cycles of Brownian motion (Section 10.6). Consequently, in neither case is there an interdependency of Q and the adiabatic invariant. A chemical-thermal-mechanical engine is unlike these ideal systems in important ways, as detailed above. It will be argued that as a consequence of their special physical conditions, chemical-thermal-mechanical engines sustain a special linkage between quantized action and the quality factor of oscillation. The equations that follow identify the linkage.

Figure 11.13 is Figure 10.15 redrawn with the lengths labelled according to the convention for describing the partitioning of the mechanical energy of the sliding filaments. Thus the symbols E_{m_o} and E_{m_i} designate the energy of the oscillator, assembled through chemical-thermal-mechanical engine processes, at twice its intercept frequency, f_o (see Section 11.1), and at any frequency, f_i, respectively. (In each case the energy is a constant proportion of the frequency.) The symbol E_t designates the thermodynamic energy or "work function". E_t is the minimal energy required to overcome the frictional forces resisting the oscillatory motion (the surface friction unique to each metal retarding the photoelectron emission).

A most important feature of E_t is that it is constant; the metal's surface friction

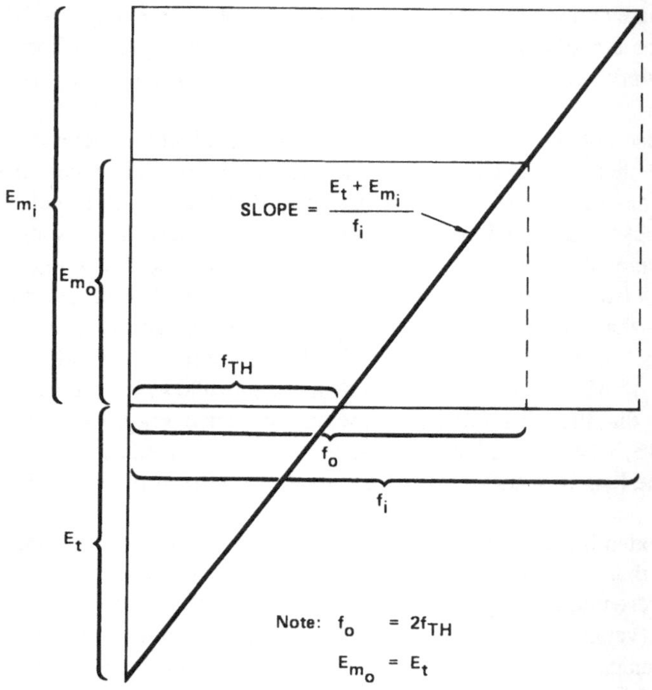

FIGURE 11.13. A geometric portrayal of an adiabatic trajectory of chemical-thermal-mechanical engine processes in an *(E,f)* work space.

does not vary with electron velocity. That is, the amount of energy degradation associated with surface friction is unchanged by changes in the frequency of electron oscillation. Cycle-independent energy degradation is the hallmark of adiabatic transformations. Its mechanistic implication is that the amount of energy required to start-up the oscillation (that is, to overcome the initial static frictional forces) is identical to the amount of energy required to sustain the oscillator at any oscillation within energy restrictions.

The f_{TH} value in Figure 11.13 identifies the start-up frequency. According to the invariance postulate for natural law (see Chapter 3), the geometric structure of adiabatic work cycles is scale-independent within the limits of universal constants. In this regard, the geometrical structure of the law of quantization (of action) is not restricted to any privileged scale. The same physical strategy (law structure) can be applied to adiabatic organizations of thermodynamic state changes as a function of oscillator frequency at any scale of organization (ranging from the photoelectric scale to the biomechanical scale).

In the case of a biomechanical oscillator, such as a wrist-pendulum system, the "work function" (e.g., characteristic energy degradation value) is defined by

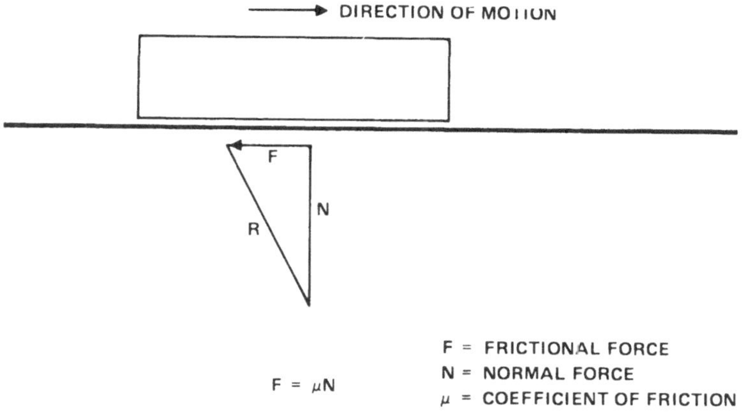

FIGURE 11.14. Sliding friction is a constant proportion of the force normal to the contacting surfaces.

the mass and length characteristics of the hand-held pendulum. Under adiabatic transformation, the work function will be invariant over frequency variations. Simply interpreted, the energy degraded during wrist-pendulum oscillation is independent of the frequency of oscillation. The amount of energy degraded by each (mass-length) oscillator will be indexed directly by the negative energy intercept. More intuitively, an adiabatic transformation, over biomechanical oscillatory states, implies that the energy required initially to overcome the static frictional forces is the same amount of energy (per cycle) required to sustain the oscillatory motion. An example of a velocity-independent nonconservative force is *sliding friction* (see Figure 11.14).

Sliding friction occurs when the surfaces of two ordinary objects, with their ordinary irregularities (e.g., bumps, wrinkles, crevices, etc.), are in contact and one of the surfaces is moving relative to the other. Very approximately, the energy of the sliding surface is converted, in part, into deformations of the other surface's bumps, and so on. Waves result, and the coherent atomic motions that they signify are realized, in turn, as heat. While friction laws are notoriously evasive (see Feynman et al., 1964) the case of sliding friction is reasonably embraced by a law which states that the force (F_B) needed to overcome friction and to drag one object over another is a constant proportion (μ) of the force perpendicular to the contacting surfaces, that is, the normal force (N). In short, $F_B = \mu N$. Sliding friction proves to be largely indifferent to velocity, leading to the realization that the resistance to getting motion started and the resistance to keeping motion going are about the same. If E_t is constant over different (θ, f) combinations (within limits) then it may be conjectured that E_t reflects a form of sliding friction (originating, perhaps, in the creeping action of actin and myosin filaments relative to each other).

The quantity signified by f_{TH} in Figure 11.13 is the lowest frequency at which the mechanical work cycle can be defined. (Stated more generally, it is the lower bound on the frequency range over which constancy of action can be revealed.) From the geometry of Figure 11.13,

$$H = \frac{E_t}{f_{TH}} = \frac{(E_{m_i} + E_t)}{f_i} . \tag{11.10}$$

which leads to an expression for E_{m_i} namely,

$$E_{m_i} = E_t\left(\frac{f_i}{f_{TH}} - 1\right) . \tag{11.11}$$

As already noted, the ratio of E_{m_i} to E_t can be interpreted as the Q-value of a muscle fiber system (a chemical-thermal-mechanical engine). It follows, therefore, from Equations (11.10) and (11.11), that for a system of constant H, the Q values will change with f_i according to:

$$Q = \frac{E_{m_i}}{E_t} = \left(\frac{f_i}{f_{TH}} - 1\right) . \tag{11.12}$$

In Section 10.6 (Equation (10.82), it was shown that $2f_{int} = f_o$ which reads in the current notation as $2f_{TH} = f_o$. Substituting for f_{TH} in Equation (11.12) for the situation in which $f_i = f_o$ gives an expression relating E_{m_o} and E_t, viz.

$$\frac{E_{m_o}}{E_t} = 1 . \tag{11.13}$$

What is the physical interpretation of Equations (11.12) and (11.13)? If the frequency of a chemical-thermal-mechanical engine were to be changed adiabatically, then the chemical energy degraded to fashion an external work cycle would be increased in direct proportion to the increase in the cycle frequency. In compliance with the second law, only a part E_{m_i} of the total mechanical energy E_M produced by chemical degradation will show up in the laboratory frame of reference. The remainder, E_t, will be lost as heat. Because of the constancy of E_t, however, the adiabatic invariant—whose focus is the conversion not the partitioning of energy—can be observed. The mechanical energy E_{m_i} in the laboratory frame of reference should be seen to increase in direct proportion to frequency, f_i. This is where Equations (11.12) and (11.13) apply. If a constant proportionality is to be maintained between E_{m_i} and f_i, then the

Q-value, E_{m_i}/E_t, cannot be constant. Equation (11.12) shows how the Q-value will vary as f_i varies (where f_{TH} is a system constant, given the constancy of H).

Suppose, to the contrary, that the system orients around Q rather than H. Now the constancy of Q determines the degrading of chemical fuel rather than the constancy of H. Now the observable E_{m_i} is constant and indifferent to changes in f_i. For reasons cited above, the Q-value of regimes of a chemical-thermal-mechanical engine that are marginally stable to perturbations across all frequencies is unity. With $Q = 1$, E_{m_i} and E_t are always identical.

11.3.2. (E,f): The Extrinsic Work Space. The major implication of the preceding is that the energy by frequency work space of a chemical-thermal-mechanical engine is populated with a number of possible stable ("attractive") trajectories. A tentative depiction of the structure of the space is given by Figure 11.15. Inspection of Figure 11.15 reveals that as the cycle frequency increases the observed mechanical energy can pursue a number of courses depending on which stability is "dominating". If the dominance is not maintained from one region (f_i range) to another then the trajectory will be more complex (nonlinear) than simple (linear).

An especially important range of f_i values is that in which the adiabatic invariant (H) trajectory lies below E_{m_o}. In this region cyclic engine processes organized around the adiabatic invariant are not only stable with respect to internal (interactive) perturbations but they are also stable with respect to external perturbations. For Q-values at or below unity the forces governing the stability of the coordinates of motion are the sustaining muscular forces. This juncture is the appropriate place to embellish a point only adumbrated above, namely, that Bernstein's "highest stage of coordinated freedom" does not apply to all the rhythmicities of a system operating according to the adiabatic invariant of action. Rather it applies only to those rhythmicities for which the Q-value exceeds unity. The contrast with a system organized constantly around a Q of unity still holds: Such a system never relinquishes responsibility for the energy of a cycle to reactive forces. In the frequency range for which $Q \leq 1$, a system is most strongly biased toward the trajectory of (E_{m_i}, f_i) points determined by adiabatic invariance. In the frequency range for which $Q > 1$ the strength of the adiabatic attraction is less imposing—the $Q = 1$ attractor is more competitive.

11.3.3. (Q,H): The Intrinsic Action Space. To characterize the rhythmic behavior generated by a chemical-thermal-mechanical engine in terms of energy and frequency is to refer that behavior to an extrinsic coordinate space (see Chapters 2 and 3). As noted, this coordinate space is differentiated (Figure 11.15). Some trajectories of *(E,f)* points are highly preferred. It is most unlikely, however, that assembling and sustaining rhythmic behaviors are actually

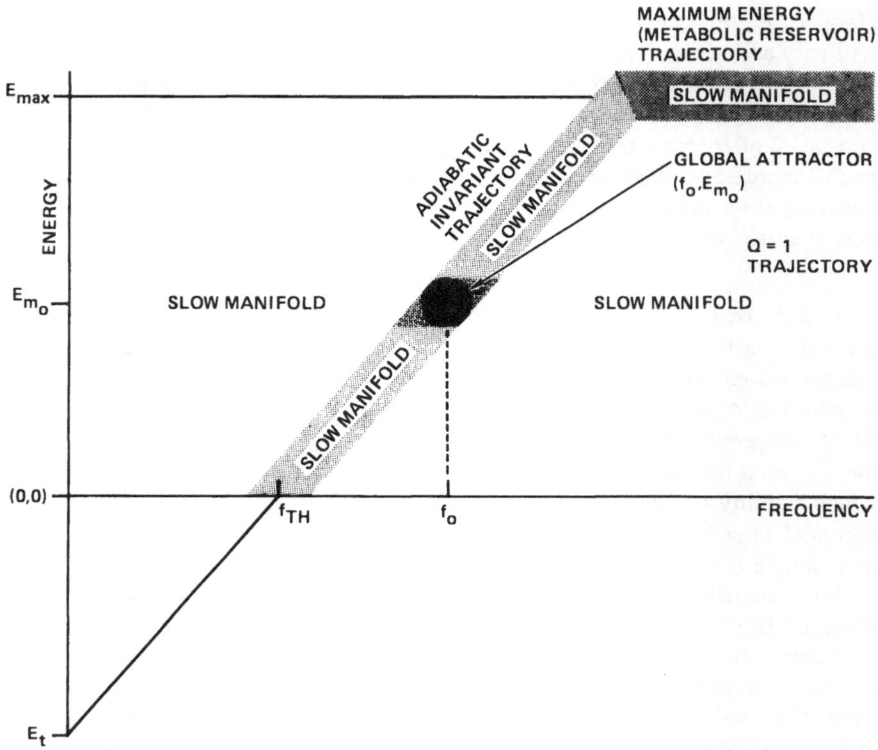

FIGURE 11.15. LAYOUT OF ATTRACTIVE STATES FOR A BIOMECHANICAL OSCILLATOR IN AN *(E,f)* WORK SPACE. The trajectories defined by an adiabatic constant (H_i) and a constant quality factor (Q_i) yield a set of independent (orthogonal) states that share a single common value. At $Q = 1$ the mechanical and thermodynamic energy flows are symmetric resulting in a stable steady-state around which oscillatory parameters can orient. The intersection of the $Q = 1$ trajectory and the adiabatic trajectory defines an attractor that orients globally oscillatory trajectories. All other states on these two trajectories constitute local "attractive" states that can bias or influence trajectories approaching the global attractor. Collectively, these locally "attractive" regions comprise a slow manifold (the regions are weakly attractive, the characteristic exponents are small). The slow manifold defines an intrinsic set of axes with the global attractor as origin. Information for orienting biomechanical oscillations is defined by vector flow patterns (morphologies) generated by variations in the oscillator's frequency and amplitude (energy).

conducted in reference to units of energy and time. The chemical-thermal-mechanical engine processes of converting and parsing energy will define their own metrical structure. That is, there will be one or more sets of *natural* units providing an intrinsic coordinate system to which the assembling and sustaining of oscillations are actually referred.

As underscored in Chapter 3 (and as developed in Chapter 7) a system's equilibrium states or, more generally speaking, a system's attractor states establish the intrinsic metrics. Two attractors, an adiabatic invariant and a unit value of Q, have been identified for the dynamical situation in which stable limit cycles at different frequencies have to be established for the same system. Figure 11.16 shows that there are very many virtual linear relationships between energy and frequency (achieved at E_t with an upper bound due to f_{TH}) and very many virtual constant-Q lines that can be drawn through the energy by frequency space. Let the adiabatic invariant (the only trajectory that can *actually* occur in which energy is a constant proportion of frequency) be designated as H_o; this is unit action. Similarly let $Q = 1$ be designated as Q_o; this is unit Q. Then $[H_i/H_o]$ and $[Q_i/Q_o]$ define distances, respectively, from the internal perturbation stability and the external perturbation stability. Given that H_i is tied to frequency $[\Delta(E_t + E_m)/\Delta f]$ while Q is frequency independent (E_m/E_t), an attractor space can be constructed with orthogonal dimensions H and Q and the origin at $(H_o, Q_o) = (1,1)$. The line of points for which $H_i = 1$ and Q_i is variable comprises

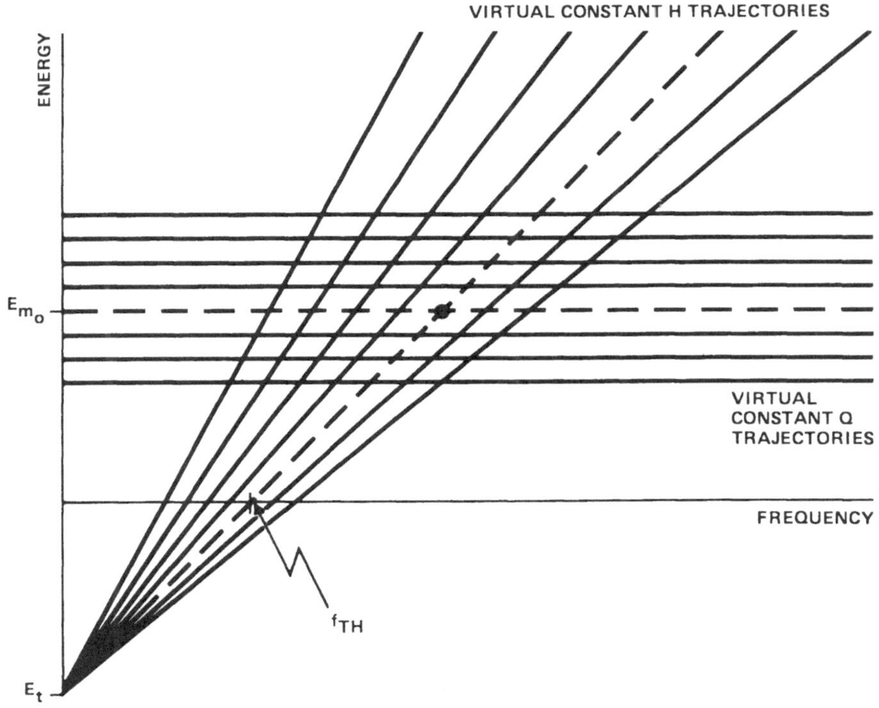

FIGURE 11.16. INTRINSIC *(H,Q)* ACTION SPACE. A macroscopic set of coordinates defined in terms of adiabatic constants and Q-factors.

a one-dimensional attractor set as does the line of points for which H_i is variable and $Q_i = 1$. The origin of the action space (H_o, Q_o) is a two-dimensional attractor.

Any point (E_{m_i}, f_i) in the extrinsic (E,f) work space of Figure 11.16 can be transformed into a point (H_i, Q_i) in the intrinsic (H,Q) action space according to:

$$Q_i = \frac{E_{m_i}}{E_t} \quad \text{and} \quad H_i = \frac{(E_{m_i} + E_t)}{f_i} . \tag{11.14}$$

The intrinsic coordinate system so constructed would be invariant over all engine processes that quantize action with a constant "work function" (E_t). It would, for the same reasons, be invariant under the growth transformation (see Shaw & Pittinger, 1977).

This is an appropriate place to return the discussion briefly to the thesis of haptic information as the abstracted morphology of movement dynamics (Chapter 7). The information question posed here is of the following form: What is the information about the maximum mechanical energy that is to accompany a prescribed frequency? Since this kinetic oscillatory parameter is assembled on the basis of law rather than enforced by law (see Chapter 8), by what lawfully generated nonkinetic properties is it specified? The chemical-thermal-mechanical engine processes that give rise to and sustain different oscillatory regimes over different frequencies are characterized (by the arguments of the present section) in terms of a layout of attractors that populate the local (H,Q) action space. Though kinetic in origin, the (H,Q) action space can be explored using a velocity vector field. Critical flow field properties (minima, maxima, etc.) of the action space are scaled by the *local inertial loadings*. These critical properties provide constraints for the spatial (amplitude) coordinate of motion. These constraints on the spatial coordinate are the thermodynamic counterpart to the mechanical constraints on the temporal coordinate (timing) identified in Chapter 7.

The one-dimensional and two-dimensional orienting properties of the attractors that populate the action space will be revealed only during the execution of the oscillatory activity. The origin and evolution of the attractor space is execution-driven (see Chapter 7). In principle, whatever the initial conditions, the limited choices of values for the preferred comfort orbit will be specified by critical flow regions in the local thermodynamic action space (specifying the spatial coordinate of motion) and the global mechanical work space (specifying the temporal coordinate of motion). Stated differently, the timing (frequency) will be specified by the mechanical attractors and the amplitude (energy) by the thermodynamic attractors.

REFERENCES

Bernstein, N. A. (1967). *The coordination and regulation of movements*. London: Pergamon Press.

Cannon, W. (1932). *The wisdom of the body*. New York: Norton.

Caplan, S. R., & Essig, A. (1983). *Bioenergetics and linear nonequilibrium thermodynamics: The steady state*. Cambridge, MA: Harvard University Press.

Carlson, F. D., & Wilkie, D. R. (1974). *Muscle physiology*. Englewood Cliffs, New Jersey: Prentice-Hall, Inc.

Fel'dman, A. (1980). Superposition of motor programs: I. Rhythmic forearm movements in man; II. Rapid forearm flexion in man. *Neuroscience 5*, 81-95.

Feynman, R. P., Leighton, R. B. & Sands, M. (1964). *The Feynman Lectures on Physics*. New York: Addison-Wesley.

Gel'fand, I. M., Gurfinkel, V. S., Tsetlin, M. L., & Shik, M. L. (1971). Some problems in the analysis of movements. In Gel'fand et al. (Eds.), *Models of the structural-functional organization of certain biological systems*. Cambridge, MA: MIT Press.

Hoyle, G. (1983). *Muscles and their neural control*. New York: Wiley.

Hubbard, A. W. (1974). Homeokinetics: muscular function in human movement. In W. R. Johnson & E. R. Buskirk (Eds.), *Science and medicine of exercise and sport*. New York: Harper and Row.

Huxley, A. F. (1980). *Reflections on muscle*. Princeton, NJ: Princeton University Press.

Kugler, P. N., Kelso, J. A. S., & Turvey, M. T. (1980). On the concept of coordinative structures as dissipative structures: I. Theoretical lines of convergence. In G. E. Stelmach & J. Requin (Eds.), *Tutorials in motor behavior*. New York: North-Holland.

Kugler, P. N., Kelso, J. A. S., & Turvey, M. T. (1982). On the control and co-ordination of naturally developing systems. In J. A. S. Kelso and J. E. Clark (Eds.), *The development of movement control and co-ordination*. New York: John Wiley and Sons.

McClare, C. W. F. (1971). Chemical machines, Maxwell's demon and living organisms. *Journal of Theoretical Biology, 30*, 1-34.

Morowitz, H. J. (1978). *Foundations of bioenergetics*. New York: Academic.

Shaw, R. E. and Pittenger, J. (1977). Perceiving the face of change in changing faces: Implications for a theory of object perception. In R. E. Shaw and J. Bransford (Eds.), *Perceiving, acting and knowing: Towards an ecological psychology*. Hillsdale, NJ: Lawrence Erlbaum Associates.

Volkenstein, M. V. (1970). *Molecules and life*. New York: Plenum Press.

Winfree, A. T. (1980). *The geometry of biological time*. New York: Springer-Verlag (series in Biomathematics, vol. *8*).

12 THE ADIABATIC HYPOTHESIS AND THE PROPORTIONING OF SPACE TO MASS, LENGTH, AND PERIOD: EXPERIMENTAL EVALUATIONS

ABSTRACT

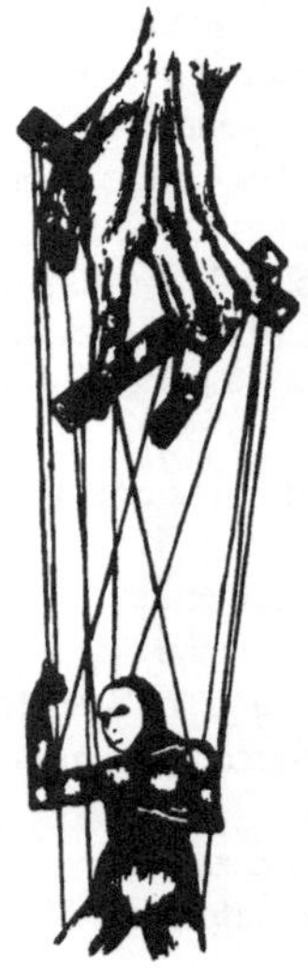

In this chapter the data from the main experiment (presented in Chapter 6) and a third subsidiary experiment are analyzed for the purpose of rationalizing how amplitude is determined as a function of variations in mass, length and period of the oscillator. The following hypotheses, developed in Chapters 10 and 11, direct the analyses: (i) change in the periodic timing of a local wrist-pendulum system of fixed mass and length is brought about through changes in internal mechanical parameters (such as stiffness, resting length, etc.) that preserve the amount of energy degraded constant over varying frequencies of oscillation—that is, the transformation is ***adiabatic****, preserving the ratio of energy of oscillation to frequency of oscillation constant in the laboratory frame of reference; (ii) the constant of proportionality that relates energy to frequency scales as a function of the moment of inertia of the oscillator.*

The following data analyses, bearing on the amplitude-period relation depicted in Figure 10.2, are reported: First, energy by frequency plots are generated for each wrist-pendulum system for each subject. These are characteristically linear (the slope identifying the constant of action) with non-zero frequency and energy intercepts; second, a system's constant of action—as given by the slope of its energy by frequency plot—is used to generate the curve of the system's change in (maximum) amplitude against the system's change in timing (both period and frequency), according to the action-based equation for amplitude developed in Chapter 10. These curves capture the qualitative nature of a system's spacing-time relation—the sometimes rising, the sometimes falling and the sometimes stationary trajectory of amplitudes with increases in period.

The demonstration of constant action under invariable external boundary conditions (fixed moment of inertia) is taken as the departure point for determining amplitude's scaling to mass and length. A proportionality for scaling in terms of action and inertia follows from the constancy of action: **amplitude** $\propto (H_i\tau_o/I_i)^{1/2}$. *The transformation of this action-based proportionality into a scaling relation for spacing in terms of mass and length is brought about by replacing action and periodic timing by their respective moment variables with coordinate space parameters affixed; that is,* $H_i = a_H(ML^2)^{c_H}$ *and* $\tau_o = a_\tau(M^{1/16}L^{1/2})^{c_\tau}$. *Ignoring coefficients, the scaling relation for spacing then has the form:*

$$\text{amplitude} \propto [(ML^2)^{c_H-1}\,(M^{1/16}L^{1/2})^{c_\tau}]^{1/2}$$

The amplitudes of single wrist-pendulum systems, as reported in the main experiment and as reported in a third subsidiary experiment, were used to evaluate the action-derived moment variable for spacing. The fit of predicted and actual mass and length exponents for amplitude was perfected by recognizing that coordinate space inhomogeneities "adjust" the proportioning of dependent quantities (e.g., action, period) to mass and length differentially.

Two major conclusions are drawn highlighting the interplay of laws and coordinate spaces in governing the pendular clocking mode. First, the scaling relation for spacing arises from the nesting of the mechanical law specifying periodic time within the energy degradation law specifying action, with the coordinate spaces of these laws playing the decisive role in determining the signs and magnitudes of amplitude's mass and length exponents. Second, when two local pendular, clocking modes are in isochronous coordination, their common periodic time is specified at the cooperative level by the periodic timing law applied to the virtual mass and length magnitudes and their different amplitudes are specified at the atomisms' level by the locally defined action constants relating energy to frequency.

12.1. Evaluating the Adiabatic Invariant

A wrist-pendulum system of given mass and length, coupled in the comfort state to another wrist-pendulum system, changes its (external) coordinates of motion. Since the inertial properties of the wrist-pendulum system are fixed, any changes in the coordinates of motion of the comfort state are the result of changes in internal parameters. As anticipated in Chapers 10 and 11, the purpose of this chapter is to analyze empirically the physical mechanisms behind these internal parameter changes. Internal parameter changes can be classified as either *thermodynamic* or *mechanical.* If a change in an internal parameter results in a corresponding change in the amount of energy degraded per cycle, then the internal parameter in question is thermodynamic. If a change in an internal parameter does not result in any change in the amount of energy degraded per cycle, then the internal parameter in question is mechanical. Given these definitions the relation between energy degradation and frequency can be determined experimentally. If changes in the external coordinates of motion are due to variations in a thermodynamic parameter, then the quantity of energy degraded per cycle will vary as a function of frequency. In contrast, if changes in the external coordinates of motion are due to variations in a mechanical parameter, then the quantity of energy degraded per cycle will be independent of frequency; transformations so characterized are termed adiabatic. The central hypothesis expressed in Chapters 10 and 11 is that transformations induced internally on comfort state coordinates of motion satisfy the assumptions of adiabatic transformations. The data from the main experiment (Chapter 6) are analyzed to test the adiabatic hypothesis: *Physical changes in the internal coordinates of periodic comfort states involve only changes in mechanical parameters.*

The data of the main experiment provide several measures of support for the adiabatic hypothesis. The measures, identified in Chapters 10 and 11, are reproduced in Figure 12.1 in the context of the main experiment's design. To anticipate, first the data indicate that if the external inertia of the oscillator is fixed, the observed energy varies linearly with the operating frequency. The constant of proportionality is identified with the adiabatic invariant: $H_i = \Delta E/\Delta f$. Second, the adiabatic constant of proportionality is shown to scale directly with variations in the external inertia of the oscillator: $H_i \propto ML^2$. Third, energy by frequency plots identify a frequency intercept that is greater than zero, implying a negative energy intercept, $-E_{int} = H_i f_{int}$. The constancy of H_i over the frequency variations, implies that the amount of energy degraded each cycle is constant and equal to the negative energy intercept: $H_i = \Delta E/\Delta f = |E_{int}|/f_{int}$. Fourth, the parabolic relations that are predicted to relate amplitude and period—if action, $\boldsymbol{H}$, is constant and the system is dissipative, $-E_{int}$—are shown to rationalize the qualitative shifts from increasing amplitudes with increasing periods to unvarying amplitudes with increasing periods to decreasing amplitudes with a further increase in period: $\theta = c\tau_i(f_i - f_{int})^{1/2}$ where $c = (H_i/2\pi^2 I_i)^{1/2}$.

12.1.1. A Graphic Evaluation of (E,f) Coordinates of Motion. The sixteen paired pendulum conditions were produced by combining each $m_i l_i^2$ condition for one hand with the four $m_j l_j^2$ conditions of the other hand. (In the pursuant

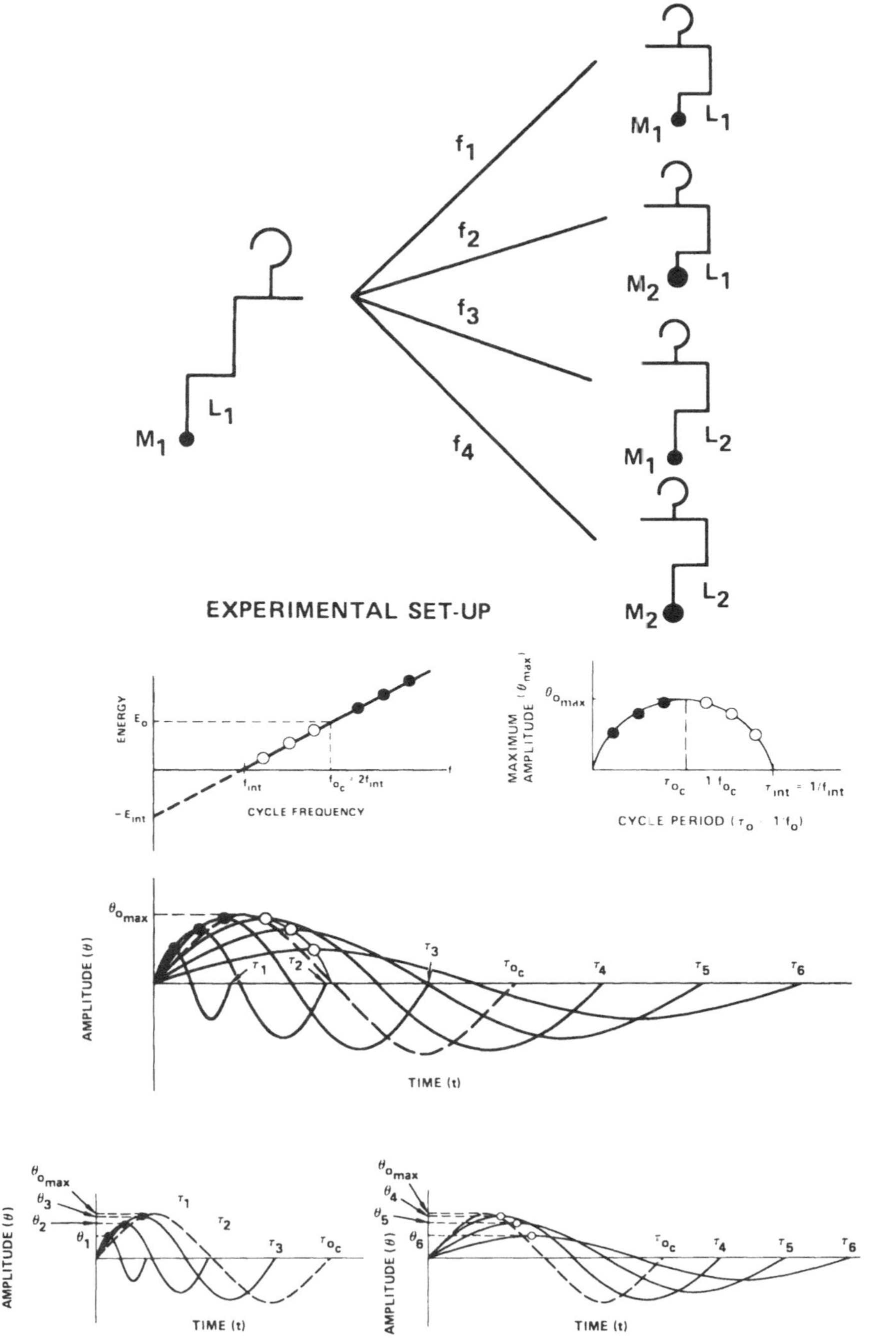

FIGURE 12.1. Summary of the experimental design and critical features in the analysis of an adiabatic invariant.

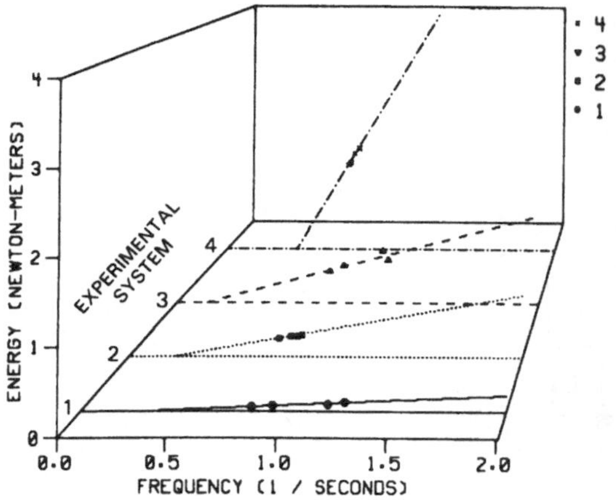

FIGURE 12.2a. *(E,f)* plots: left wrist-pendulum systems of Subject 1.

SUBJECT 1 (left)

Regression Curve Parameter	*Experimental System* 1	2	3	4
I_i (kg−m^2)	0.0065	0.0312	0.0645	0.3125
H_i (kg−m^2/s)	0.1149	0.4577	0.7197	4.6472
f_{int} (l/s)	0.315	0.232	0.162	0.421
E_{int} (kg−m^2/s^2)	−0.0362	−0.1062	−0.1166	−1.9565
Correlation Coefficient	0.996	0.938	0.871	0.963

1		*2*		*3*		*4*	
*Freq.**	*Energy***	*Freq.*	*Energy*	*Freq.*	*Energy*	*Freq.*	*Energy*
1.245	0.109	0.884	0.302	1.159	0.642	0.799	1.743
1.172	0.096	0.858	0.276	1.134	0.779	0.770	1.661
0.909	0.069	0.824	0.281	0.917	0.551	0.752	1.495
0.808	0.056	0.765	0.241	0.842	0.477	0.733	1.461

*(1/s)
**(kg−m^2/s^2)

analyses, depiction of the data is simplified by approximating the moment of inertia I_i of a system with the product of the mass and squared length explicitly manipulated in the experiment as reported in Table 7.1.) For each pairing, a given $m_i l_i^2$ system assumed a frequency determined by the virtual single system that it partially constituted. It is possible, therefore, to observe—for each of the four subjects and for each of his or her two hands—a given wrist-pendulum system operate at four different frequencies with four different energies E_i computed as $I(\omega^2/2)$ (see Figure 12.1).

The plots of energy against frequency are given in Figures 12.2a-h for all the conditions for each subject together with tables indicating moment of inertia I_i,

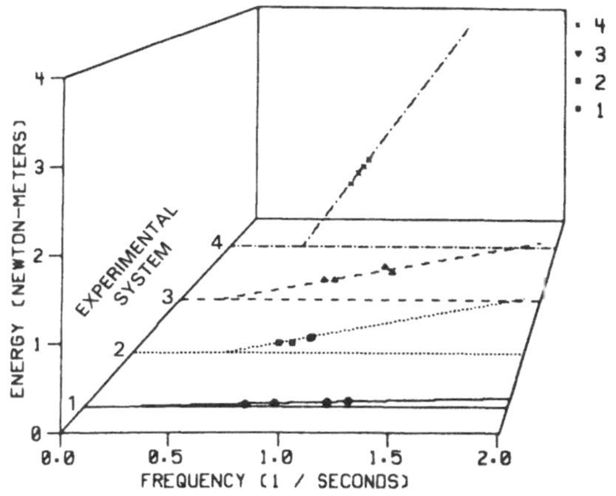

FIGURE 12.2b. *(E,f)* plots: right wrist-pendulum systems of Subject 1.

SUBJECT 1 (right)				
Regression	*Experimental System*			
Curve Parameter	*1*	*2*	*3*	*4*
I_i (kg−m²)	0.0065	0.0312	0.0645	0.3125
H_i (kg−m²/s)	0.0631	0.4852	0.4995	3.7840
f_{int} (l/s)	0.165	0.473	0.226	0.441
E_{int} (kg−m²/s²)	−0.0104	−0.2295	−0.1129	−1.6687
Correlation Coefficient	0.980	0.940	0.916	0.998

1		*2*		*3*		*4*	
*Freq.**	*Energy***	*Freq.*	*Energy*	*Freq.*	*Energy*	*Freq.*	*Energy*
1.248	0.071	0.917	0.227	1.175	0.428	0.842	1.515
1.153	0.059	0.909	0.208	1.133	0.507	0.811	1.400
0.90	0.048	0.820	0.150	0.857	0.301	0.778	1.292
0.770	0.038	0.752	0.146	0.798	0.292	0.733	1.098

*(1/s)
**(kg−m²/s²)

slope H_i, frequency intercept f_{int}, energy intercept E_{int}, and correlation coefficients of the respective linear regressions. Energy might be expected to increase in each case with an increase in frequency, just because frequency appears in the measure of energy. However, whether energy does increase or not depends importantly on how amplitude changes; moreover, if the only basis for the increasing linear relation between energy and frequency was frequency—that is, if amplitude was a constant—then for a given inertial loading (e.g., $I_l = 0.0065 kg\text{-}m^2$) the slopes across hands should not differ. (Different constant amplitudes would lead, however, to different intercept values.) That the slopes do differ across hands for the same inertial loadings means that the variations in amplitude were of

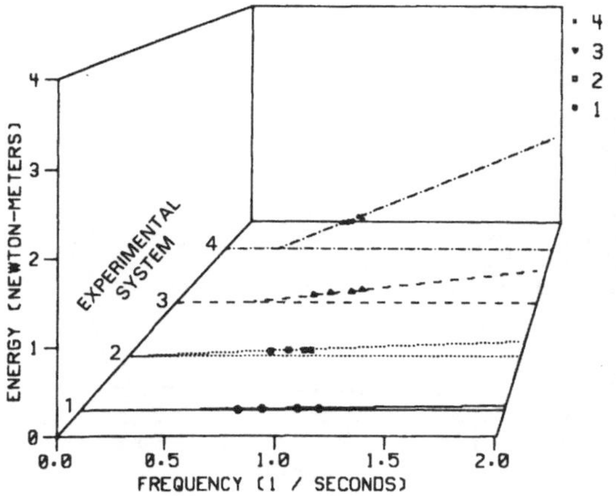

FIGURE 12.2c. *(E,f)* plots: left wrist-pendulum systems of Subject 2.

SUBJECT 2 (left)

Regression Curve Parameter	*Experimental System* 1	2	3	4
I_i (kg−m²)	0.0065	0.0312	0.0645	0.3125
H_i (kg−m²/s)	0.0377	0.1018	0.3091	1.1385
f_{int} (l/s)	0.373	0.031	0.403	0.317
E_{int} (kg−m²/s²)	−0.0141	−0.0032	−0.125	−0.3609
Correlation Coefficient	0.985	0.874	0.983	0.953

1		*2*		*3*		*4*	
*Freq.**	*Energy***	*Freq.*	*Energy*	*Freq.*	*Energy*	*Freq.*	*Energy*
1.135	0.028	0.935	0.084	1.027	0.193	0.831	0.507
1.030	0.026	0.898	0.085	0.970	0.171	0.815	0.575
0.866	0.017	0.821	0.086	0.855	0.149	0.763	0.488
0.753	0.015	0.727	0.068	0.762	0.106	0.724	0.474

*(1/s)
**(kg−m²/s²)

sufficient magnitude to affect the trajectory of points in the energy by frequency coordinates. And because of the strength of the bivariate relation, as measured by the correlation coefficients ($r(2) = .95$, $p<.05$; $r(2) = .99$, $p<.01$), it means that in the majority of cases the variations in amplitude were coupled systematically to variations in frequency so as to maintain action constant.

The observed frequencies f_i of each wrist pendulum system for each subject can be expressed in units of f_o—where $f_o = 2f_{int}$, see Equation 10.81—so as to form a dimensionless unit $\pi_f = f_i/f_o$. Similarly the observed energy values can be expressed in units of the energy value at f_o given by $E_o = H_i/2f_o = |E_{int}|$ (Equation 10.83). A dimensionless coordinate system is thereby con-

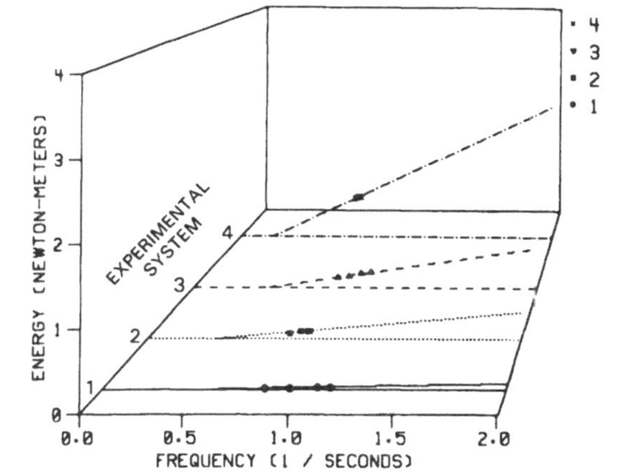

FIGURE 12.2d. *(E,f)* plots: right wrist-pendulum systems of Subject 2.

SUBJECT 2 (right)

Regression Curve Parameter	*Experimental System* 1	2	3	4
I_i (kg−m²)	0.0065	0.0312	0.0645	0.3125
H_i (kg−m²/s)	0.0548	0.2342	0.4150	1.3263
f_{int} (l/s)	0.456	0.379	0.435	0.202
E_{int} (kg−m²/s²)	−0.0250	−0.0888	−0.1805	−1.2679
Correlation Coefficient	0.992	0.956	0.976	0.996

1		*2*		*3*		*4*	
*Freq.**	*Energy***	*Freq.*	*Energy*	*Freq.*	*Energy*	*Freq.*	*Energy*
1.133	0.038	0.868	0.112	1.029	0.248	0.765	0.721
1.067	0.034	0.854	0.111	0.974	0.227	0.754	0.731
0.931	0.025	0.821	0.108	0.898	0.181	0.742	0.719
0.811	0.020	0.756	0.086	0.837	0.173	0.724	0.692

*(1/s)
**(kg−m²/s²)

structed within which the data for all four subjects can be considered simultaneously. Figures 12.3a,b present the dimensionless plots for the left and right systems. The figures give expression to the canonical form of the action law defined by Equation (10.83). The correlation coefficients, relating the observed data values to values predicted by the law, for the left and right system are ρ_L (62) = .997 and ρ_R (62) = .994, respectively.

A useful image for interpreting the data of Figures 12.2 and 12.3 can be sketched using the terminology introduced in Section 9.2. A rising trajectory of *(E,f)* points defines the behavior of a muscle/joint complex that is constrained (a) locally by its atomistic mass and length properties that scale the (local) action and

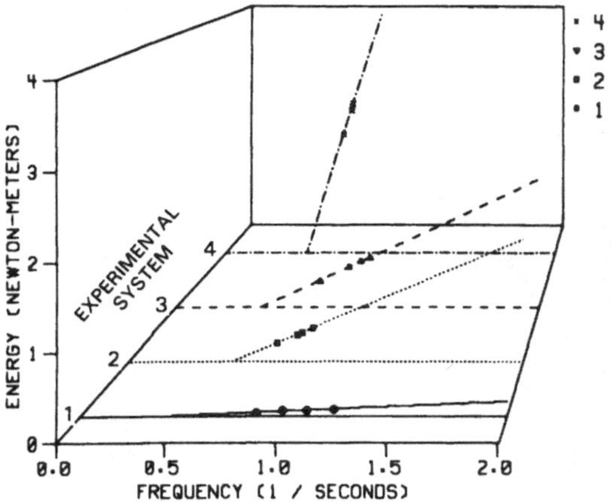

FIGURE 12.2e. *(E,f)* plots: left wrist-pendulum systems of Subject 3.

SUBJECT 3 (left)

Regression Curve Parameter	*Experimental System* 1	2	3	4
I_i (kg−m^2)	0.0065	0.0312	0.0645	0.3125
H_i (kg−m^2/s)	0.1177	1.0963	1.2329	8.9865
f_{int} (l/s)	0.387	0.534	0.474	0.4957
E_{int} (kg−m^2/s^2)	−0.0455	−0.5854	−0.5844	−4.4546
Correlation Coefficient	0.996	0.999	0.999	0.993

1		*2*		*3*		*4*	
*Freq.**	*Energy***	*Freq.*	*Energy*	*Freq.*	*Energy*	*Freq.*	*Energy*
1.199	0.095	0.941	0.449	1.067	0.723	0.775	2.518
1.067	0.079	0.887	0.384	1.016	0.677	0.766	2.395
0.955	0.069	0.860	0.355	0.949	0.588	0.765	2.452
0.831	0.051	0.763	0.253	0.799	0.399	0.717	1.988

*(1/s)
**(kg−m^2/s^2)

(b) globally by its cooperative (virtual) mass and length properties that scale the (global) operating frequency. As the cooperativity (namely, the virtual system) changes its mass and length, its frequency changes in proportion to the product of *[virtual-mass$^{1/16}$ x virtual-length$^{1/2}$]*. How does the atomism behave? The conservative mechanical energy at the atomistic level changes in direct proportion to frequency. Recall that if the transformation is adiabatic, then the amount of energy degraded in thermodynamic processes is *independent* of the amount of energy associated with the mechanical processes. The negative energy intercept of the regression curve identifies this independence. It indexes the constant

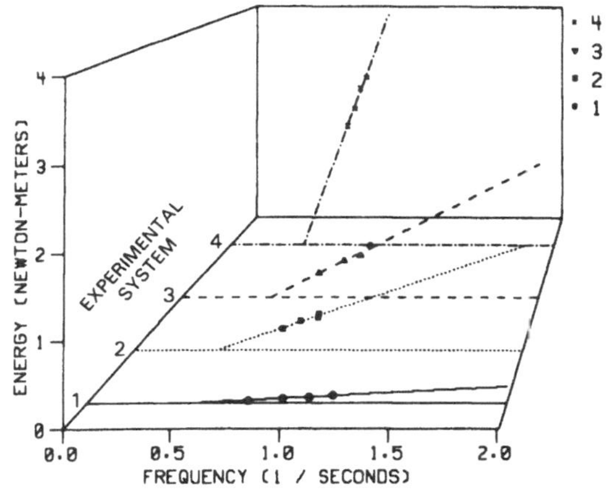

FIGURE 12.2f. *(E,f)* plots: right wrist-pendulum systems of Subject 3.

SUBJECT 3 (right)

Regression Curve Parameter	*Experimental System* 1	2	3	4
I_i (kg−m²)	0.0065	0.0312	0.0645	0.3125
H_i (kg−m²/s)	0.1384	0.8920	1.3566	7.7288
f_{int} (l/s)	0.506	0.432	0.493	0.449
E_{int} (kg−m²/s²)	−0.070	−0.385	−0.569	−3.470
Correlation Coefficient	0.986	0.962	0.984	0.992

1		*2*		*3*		*4*	
*Freq.**	*Energy***	*Freq.*	*Energy*	*Freq.*	*Energy*	*Freq.*	*Energy*
1.181	0.096	0.956	0.485	1.050	0.787	0.831	2.929
1.064	0.072	0.949	0.433	1.002	0.647	0.796	2.749
0.940	0.063	0.857	0.398	0.902	0.564	0.763	2.378
0.773	0.037	0.766	0.289	0.766	0.373	0.716	2.076

*(1/s)
**(kg−m²/s²)

amount of energy that is degraded each cycle by thermodynamic processes. Because of the orthogonality of the thermodynamic work function E_{int} to mechanical processes, transformations at the cooperative level of organization closely approximate the behavior of a rigidly connected, conservative field (see Section 9.2). Thus, under adiabatic conditions, the image is that of an atomism of fixed magnitude cooperating with other atomisms in a gradually dilating (or expanding) conservative energy field that is vibrating at increasingly higher (or lower) frequencies against a backdrop of an invariant thermodynamic energy field (work function).

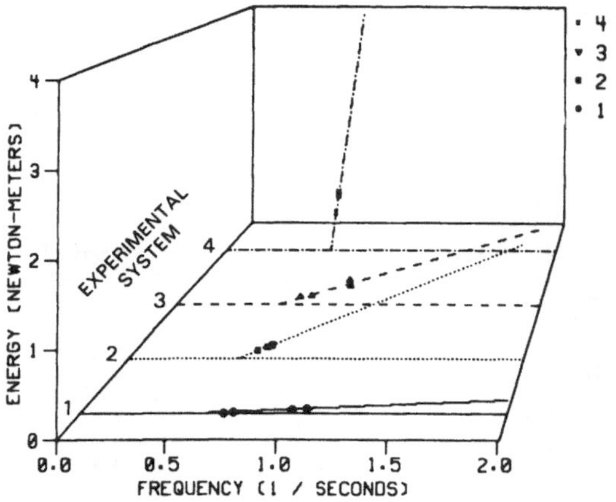

FIGURE 12.2g. *(E,f)* plots: left wrist-pendulum systems of Subject 4.

SUBJECT 4 (left) *Regression Curve Parameter*	*Experimental System* *1*	*2*	*3*	*4*
I_i (kg−m²)	0.0065	0.0312	0.0645	0.3125
H_i (kg−m²/s)	0.1107	1.0477	0.7943	20.7691
f_{int} (l/s)	0.553	0.560	0.560	0.629
E_{int} (kg−m²/s²)	−0.061	−0.587	−0.445	−13.064
Correlation Coefficient	0.996	0.984	0.951	0.981

1 *Freq.**	 *Energy***	*2* *Freq.*	 *Energy*	*3* *Freq.*	 *Energy*	*4* *Freq.*	 *Energy*
1.072	0.059	0.730	0.185	0.959	0.273	0.678	1.021
1.003	0.048	0.727	0.167	0.949	0.354	0.676	0.953
0.728	0.018	0.705	0.152	0.743	0.142	0.674	0.979
0.684	0.016	0.661	0.106	0.681	0.097	0.661	0.658

*(1/s)
**(kg−m²/s²)

12.1.2. Graphic Evaluations of (θ,τ) and (θ,f) Coordinates of Motion. Turning now to the H_i-based equation relating amplitude and period for a dissipative system (Equation 10.79), the (qualitative) prediction is that amplitude should follow a parabolic curve as a function of period. The exact form of the relation (increasing, invariant, decreasing) will depend on where the θ, τ points under consideration are with respect to the singularity of the function defined by Equation (10.79).

For each wrist-pendulum system (e.g., $I_l = 0.0065 kg\text{-}m^2$, right hand) for each subject (e.g., Subject 1) the curve relating θ to τ was computed from Equation (10.79) using the H_i and f_{int} of the system as estimated from the regression

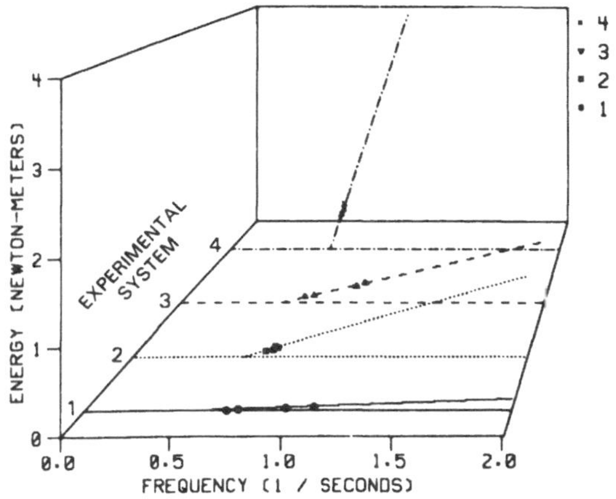

FIGURE 12.2h. *(E,f)* plots: right wrist-pendulum systems of Subject 4.

SUBJECT 4 (right)

Regression Curve Parameter	*Experimental System* 1	2	3	4
I_i (kg−m^2)	0.0065	0.0312	0.0645	0.3125
H_i (kg−m^2/s)	0.0900	0.7486	0.6387	8.5251
f_{int} (l/s)	0.534	0.562	0.550	0.599
E_{int} (kg−m^2/s^2)	−0.048	−0.421	−0.351	−5.107
Correlation Coefficient	0.994	0.802	0.990	0.916

1		*2*		*3*		*4*	
*Freq.**	*Energy***	*Freq.*	*Energy*	*Freq.*	*Energy*	*Freq.*	*Energy*
1.078	0.051	0.741	0.126	1.010	0.306	0.687	0.695
0.946	0.035	0.726	0.144	0.962	0.253	0.685	0.787
0.727	0.017	0.711	0.096	0.732	0.100	0.671	0.609
0.673	0.014	0.674	0.085	0.677	0.096	0.661	0.522

*(1/s)

**(kg−m^2/s^2)

analysis of Section 12.1.1. These curves are shown in Figures 12.4a-h. The experimental coorodinates of motion (amplitude and period) are plotted for left and right systems for each subject. The qualitative character of the dependency of amplitude on period is captured by the parabolic trajectory of amplitude and period coordinates of motion defined by the H_i-based equation. Comparing the observed amplitude (identified by discrete markers) with the predicted amplitude (identified by continuous parabolic trajectories) reveals the subtlety of the variation in amplitude with variation in period reflected in the H_i-based equation.

The qualitative shift in the scaling of period to amplitude can be observed in the comparison of the left System 1 of Subject 1 and the left System 1 of Subject

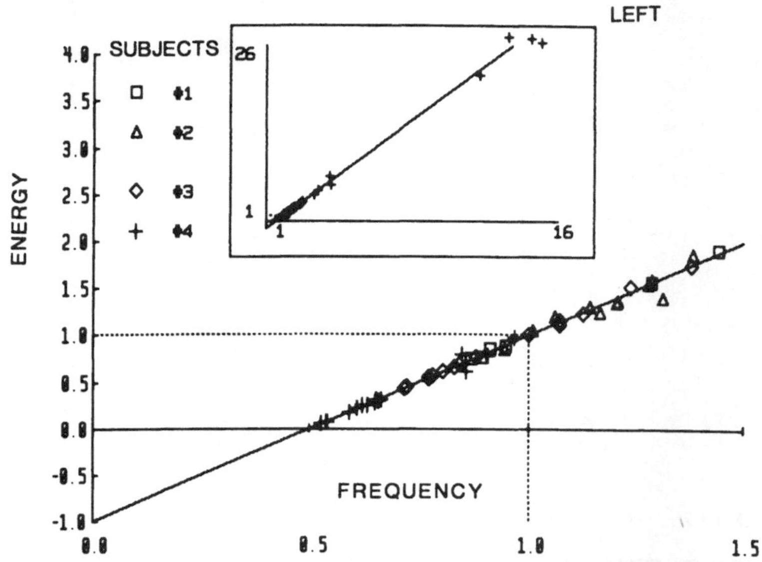

FIGURE 12.3a. *(E,f)* plot of all four subjects in dimensionless coordinates $(E_i/E_o, f_i/f_o)$, where $f=2f_t$ and $E_o=|E_t|$: *Left systems.* (The numerical values of the energy axis identify the Q-value of the oscillator.)

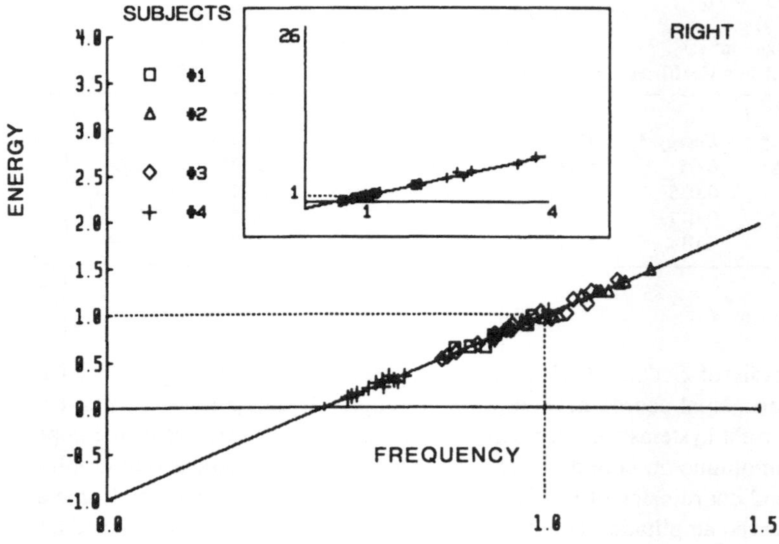

FIGURE 12.3b. *(E,f)* plot of all four subjects in dimensionless coordinates $(E_i/E_o, f_i/f_o)$, where $f=2f_t$ and $E_o=|E_t|$: *Right systems.* (The numerical values of the energy axis identify the Q-value of the oscillator.)

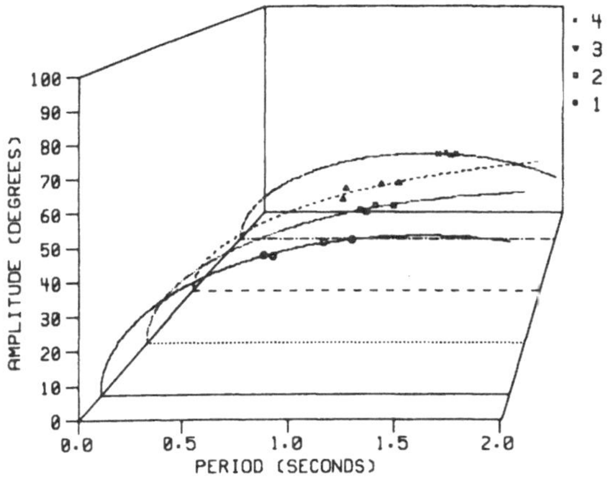

FIGURE 12.4a. (θ,τ) plots: left wrist-pendulum systems of Subject 1.

SUBJECT 1 (left)

	Experimental System							
	1		*2*		*3*		*4*	
Point	*Period*	*Amp*	*Period*	*Amp*	*Period*	*Amp*	*Period*	*Amp*
1	0.803	42.6	1.131	45.3	0.863	35.1	1.252	38.1
2	0.853	42.4	1.165	44.7	0.882	39.5	1.299	38.6
3	1.100	46.4	1.214	47.0	1.090	41.1	1.329	37.5
4	1.237	47.2	1.308	46.8	1.187	41.6	1.364	38.0

4. These two systems exhibit similar action constants, viz., 0.1149*joule-sec* and 0.1107*joule-sec*. In the former case the increase in period from 0.803*sec* to 1.237*sec* was accompanied by an increase in amplitude from 42.6*deg* to 47.2*deg*; in the latter case, however, the increase in period from 0.933*sec* to 1.463*sec* was accompanied by a *decrease* in amplitude from 36.5*deg* to 29.6*deg*. The f_{int}s were -0.315sec^{-1} and -0.553sec^{-1}, respectively. In the foregoing contrast, not only do the individual amplitudes and periods differ quantitatively in the two instances, but their qualitative relationship differs in that the direction of the change in the ratios is increasing in the one instance and decreasing in the other instance. The point to be underscored is that for dissipative systems the same constant of action can be defined over qualitatively distinct sets of (θ,τ) coordinates.

A dimensionless coordinate system can be constructed for the amplitude and period data. Period is expressed as a dimensionless unit scaled to the unit length defined at the singularity, $\pi_\tau = \tau_i/\tau_o$, (where $\tau_o = \tau_{int}/2$, Equation 10.81). Amplitude is similarly expressed as a dimensionless unit scaled to unit length defined at the singularity, $\pi_\theta = \theta_i/\theta_{max}$ (see Equations 10.79 and 10.80). Figures 12.5a,b

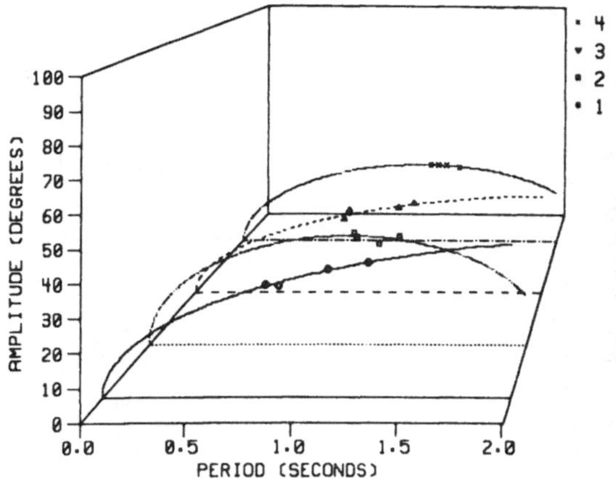

FIGURE 12.4b. (θ,τ) plots: right wrist-pendulum systems of Subject 1.

SUBJECT 1 (right)

	Experimental System							
	1		*2*		*3*		*4*	
Point	*Period*	*Amp*	*Period*	*Amp*	*Period*	*Amp*	*Period*	*Amp*
1	0.801	34.3	1.090	37.9	0.851	28.3	1.187	33.7
2	0.867	33.7	1.100	36.6	0.883	31.9	1.233	33.7
3	1.108	38.9	1.220	34.5	1.167	32.5	1.285	33.7
4	1.299	40.7	1.330	37.1	1.253	34.4	1.364	40.0

present the data for the left and right systems of all four subjects in the dimensionless coordinate system. The parabolic curve defines the predicted trajectory of adiabatic states. It identifies a thermodynamic branch of equilibrium states. Whereas the entire parabolic curve satisfies thermodynamic equilibrium requirements the only state satisfying mechanical equilibrium requirements is located at the singular point. The singular point defines the only state that satisfies *both* mechanical and thermodynamic equilibrium requirements. The singular point identifies a global attractor that can be used to orient movements within the biomechanical work space (see Chapter 11). The data presented in Figures 12.5a,b conform to the predicted action-based law for amplitude (Equation 10.79).

Figures 12.6a-h, presents (θ,f) plots. These figures serves two purposes. They reveal the intercepts that are not always seen in the (θ,τ) plots. And, they provide a basis for comparison with other kinds of oscillatory systems that are more commonly described using (θ,f) profiles (see Section 12.1.3). Figures 12.7a,b plot the data in dimensionless coordinates using the same conventions identified for Figures 12.5a,b. The curve defines the trajectory of states *predicted* by the action law.

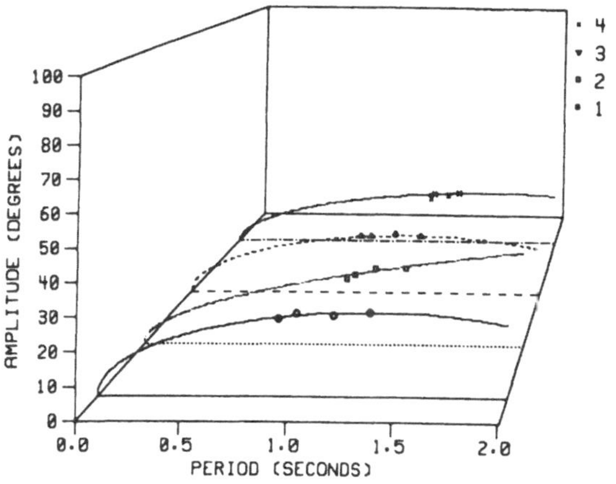

FIGURE 12.4c. *(θ,τ)* plots: left wrist-pendulum systems of Subject 2.

SUBJECT 2 (left)								
			Experimental System					
	1		*2*		*3*		*4*	
Point	*Period*	*Amp*	*Period*	*Amp*	*Period*	*Amp*	*Period*	*Amp*
1	0.881	23.7	1.070	22.6	0.974	21.7	1.203	19.8
2	0.971	25.2	1.113	23.7	1.031	21.6	1.227	21.5
3	1.155	24.5	1.218	26.1	1.169	22.9	1.311	21.1
4	1.328	25.9	1.376	26.2	1.313	21.7	1.382	21.9

Comparison of Figures 12.7a,b with Figures 12.5a,b highlights the lack of reflective (geometric) symmetry in the power function around f_o and the presence of the symmetry around τ_o, respectively.

Finally, Figures 12.8a,b,c present data sets formed by *averaging* the four subjects' amplitudes, their periods and their energies. These data sets are plotted using the conventions for forming dimensionless metrics identified for Figures 12.3, 12.5, and 12.7. The averaging of nonlinear data, such as the *(θ,τ)* and *(θ,f)*, results in a convergence of the averaged data set onto regions near the singularity. In these regions the plotted variables exhibit very little correlation; as the data points approach the singular region, the two variables approach orthogonality, becoming orthogonal at the singularity. The result of averaging, therefore, is an illusory independence of amplitude and period. Extreme departures from the singular region can only be observed in unaveraged data sets. And it is only in these extreme regions that small changes in the period of oscillation can produce large changes in amplitude.

12.1.3. Resonance Curves and Q-factors. By convention, *(θ,f)* defines a

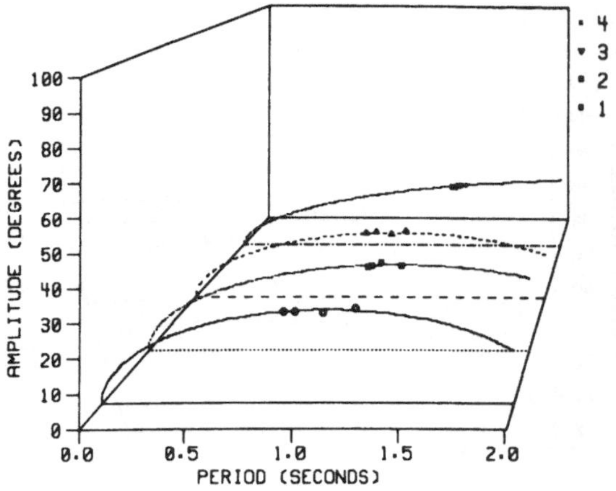

FIGURE 12.4d. *(θ,τ)* plots: right wrist-pendulum systems of Subject 2.

SUBJECT 2 (right)

	Experimental System							
	1		*2*		*3*		*4*	
Point	*Period*	*Amp*	*Period*	*Amp*	*Period*	*Amp*	*Period*	*Amp*
1	0.883	27.5	1.152	28.1	0.972	24.6	1.308	25.6
2	0.937	27.6	1.171	28.5	1.027	24.8	1.327	26.2
3	1.074	27.1	1.218	29.3	1.113	24.1	1.347	26.3
4	1.233	28.3	1.322	28.3	1.195	25.3	1.381	26.5

resonance curve. The type of curve that occurs most frequently in Figure 12.6 is of the broad resonance type—that which typifies a system designed to track faithfully periodic driving influences (e.g., forces). A resonance curve marked by a sharp peak identifies a system designed for selectivity to periodic influences rather than fidelity. The shape of a system's resonance curve is determined by the system's Q-factor: A low Q-factor defines systems of broad resonance; a high Q-factor defines systems of sharp resonance. On the basis of their broad resonance curves wrist-pendulum systems are low Q-factor systems. (This conclusion is buttressed by inspection of Figures 12.3a-b. With the exception of the left systems of Subject 4, the systems in general exhibit Q-values near unity).

Selectivity is defined quantitatively by the half-power width of the resonance curve given by $|f_o - f^*|$ where f^* is that frequency at which the system's energy is exactly one-half its energy at the natural frequency f_o. From the symmetries in the H_i-based equation for amplitude, f^* for a wrist-pendulum system is $\frac{3}{2}f_{int}$, which

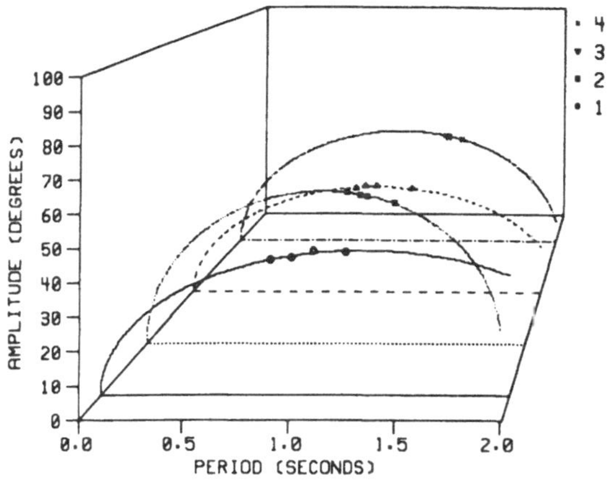

FIGURE 12.4e. (θ,τ) plots: left wrist-pendulum systems of Subject 3.

SUBJECT 3 (left)

	Experimental System							
	1		*2*		*3*		*4*	
Point	*Period*	*Amp*	*Period*	*Amp*	*Period*	*Amp*	*Period*	*Amp*
1	0.834	41.3	1.063	52.0	0.937	40.5	1.290	47.2
2	0.937	42.4	1.128	51.0	0.984	41.1	1.305	46.6
3	1.047	44.2	1.163	50.5	1.054	41.0	1.307	47.2
4	1.203	43.6	1.310	48.1	1.251	40.1	1.395	45.4

equals $\frac{3}{4}f_o$, and the half-power width of a wrist-pendulum system is, therefore, $|f_o - \frac{3}{2}f_{int}|$, which is equal to $\frac{1}{2}f_{int}$. The smaller the value of f_{int} the higher the selectivity of a wrist-pendulum system and, conversely, the larger the value of f_{int} the greater is the system's fidelity. Inspection of Figure 12.6 reveals that where f_{int} is (comparatively) small the resonance curve is peaked.

12.2. A Nesting of Natural Laws

Recall the three-level analysis of a functional unit of movement that was advanced in Section 9.2.1. A further comment can now be added with respect to the relation between the cooperative and the atomistic levels. The system produced by the isochronous coupling of a pair of wrist-pendulum systems is characterized by single quantities of mass and length. To reiterate, it is a virtual single system assembled by the conservations. The scaling law for periodic time applies equally to a virtual single wrist-pendulum system and to a single wrist-pendulum system (Chapter 8). Consequently, given the virtual mass and

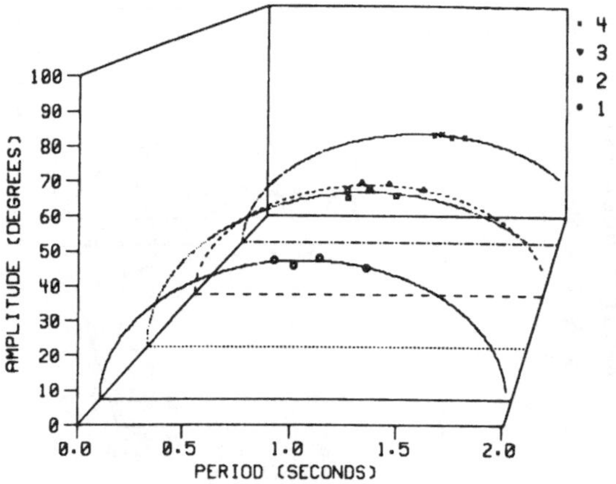

FIGURE 12.4f. (θ,τ) plots: right wrist-pendulum systems of Subject 3.

SUBJECT 3 (right)

	Experimental System							
	1		*2*		*3*		*4*	
Point	*Period*	*Amp*	*Period*	*Amp*	*Period*	*Amp*	*Period*	*Amp*
1	0.847	42.2	1.046	53.1	0.952	42.9	1.203	47.5
2	0.940	40.4	1.054	50.6	0.998	40.8	1.256	48.0
3	1.064	42.8	1.167	53.7	1.109	42.3	1.311	46.6
4	1.293	39.9	1.305	51.2	1.305	40.5	1.396	46.4

virtual length of the cooperative level a periodic timing is defined for that level (but, of course, not determined—see Chapters 1 and 7). The atomistic level must follow suit. That is, the periodic time for the cooperative level is necessarily the periodic time of the atomisms. Now the question is: what specifies the *amplitudes* of the atomisms? The proposed answer is a law at the atomistic level, namely, the law of constant action that relates frequency to energy (see Figure 12.9). In sum, *a nesting of natural laws specifies the rhythmic parameters of the individual wrist-pendulum systems*: The frequency f_{o_v} specified at the cooperative level CS_v (see Figure 9.1) by

$$f_{o_v} = a(M_v^{-1/16} L_v^{-1/2})^c \, , \qquad (12.1)$$

specifies energy quantities E_i and E_j—and, therefore, θ_i and θ_j (given the fixed inertias)—for the atomisms CS_i and CS_j, respectively, through the lawful relations

$$f_{o_v} H_i = 2E_i \qquad (12.2)$$

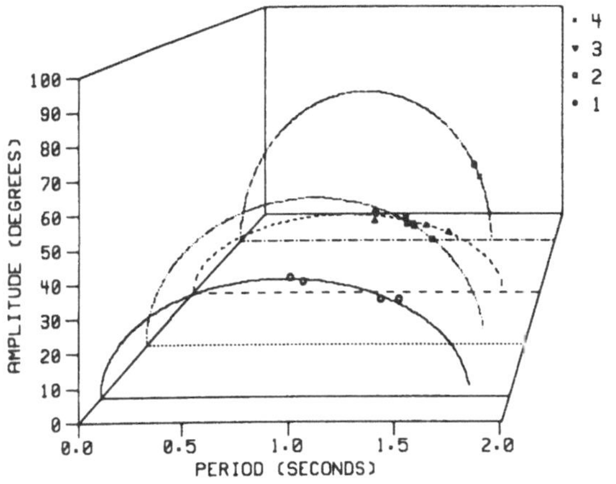

FIGURE 12.4g. (θ,τ) plots: left wrist-pendulum systems of Subject 4.

SUBJECT 4 (left)

	Experimental System							
	1		*2*		*3*		*4*	
Point	*Period*	*Amp*	*Period*	*Amp*	*Period*	*Amp*	*Period*	*Amp*
1	0.933	36.5	1.370	43.0	1.043	27.7	1.474	34.4
2	0.997	35.0	1.375	41.0	1.054	31.9	1.479	33.3
3	1.374	29.7	1.418	40.3	1.345	25.7	1.484	33.9
4	1.463	29.6	1.513	35.9	1.468	23.2	1.513	28.3

and

$$f_{o_v} H_j = 2E_j \ , \tag{12.3}$$

where H_i and H_j are the locally defined constants of action.

12.3. The Mass, Length Scaling of Amplitude in ('Comfort') Steady-State Pendular Clocking Behavior

The focus of Section 12.1 was the relation of amplitude to period when the boundary conditions—the mass and length magnitudes of the wrist-pendulum system—were fixed. The amplitude data to be explained were those produced in the double wrist-pendulum conditions of the main experiment. In the present section, the focus is on the amplitudes of the eight single wrist-pendulum systems (see Table 6.3). The goal is to define the scaling relation of amplitude to mass and length.

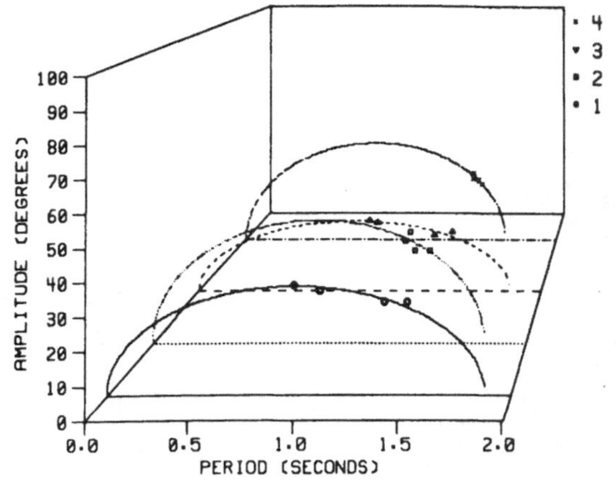

FIGURE 12.4h. *(θ,τ)* plots: right wrist-pendulum systems of Subject 4.

SUBJECT 4 (right)

	Experimental System							
	1		*2*		*3*		*4*	
Point	*Period*	*Amp*	*Period*	*Amp*	*Period*	*Amp*	*Period*	*Amp*
1	0.928	33.5	1.350	34.9	0.990	27.8	1.455	28.0
2	1.057	31.7	1.378	38.1	1.039	26.5	1.460	29.9
3	1.375	28.4	1.406	31.8	1.367	22.0	1.491	26.8
4	1.486	28.2	1.484	31.6	1.478	23.2	1.512	25.2

12.3.1. The Relation of Action to Moment of Inertia. The extension of Ehrenfest's adiabatic hypothesis, that shaped the arguments in Chapters 10 and 11 and motivated the analysis in Section 12.1, serves as the departure point. It will be supposed that single wrist-pendulum systems in the comfort state operate at $Q = 1$. Given this supposition, the law of action (in the form of Equation 10.83) prescribes that the action constant scales to the moment of inertia according to:

$$H_i = 2\left(\frac{E}{f_o}\right) = \frac{I\omega^2}{f_o} . \tag{12.4}$$

For any wrist-pendulum system, action identifies a trajectory in *(E,f)* coordinates for which the quantity given by the average kinetic energy less the average potential energy is a minimum. According to the principle of "least action" the action quantity is the least value of the integral

$$H = \min \int_{t=o}^{t=\tau} (\text{kinetic energy} - \text{potential energy})dt, \tag{12.5}$$

in the external (conservative) frame of reference.

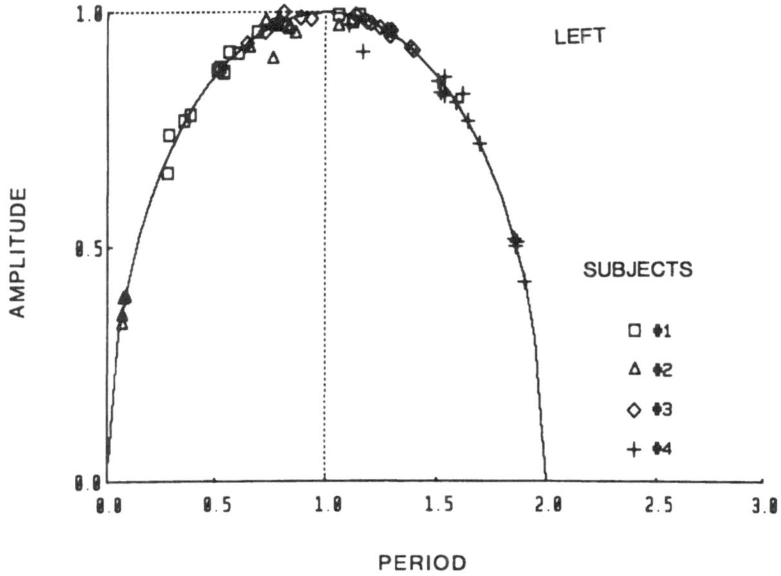

FIGURE 12.5a. (θ,τ) plot of all four subjects in dimensionless coordinates $(\theta_i/\theta_{max},\tau_i/\tau_o)$, where $\tau_o = 1/2f_{int}$: Left systems.

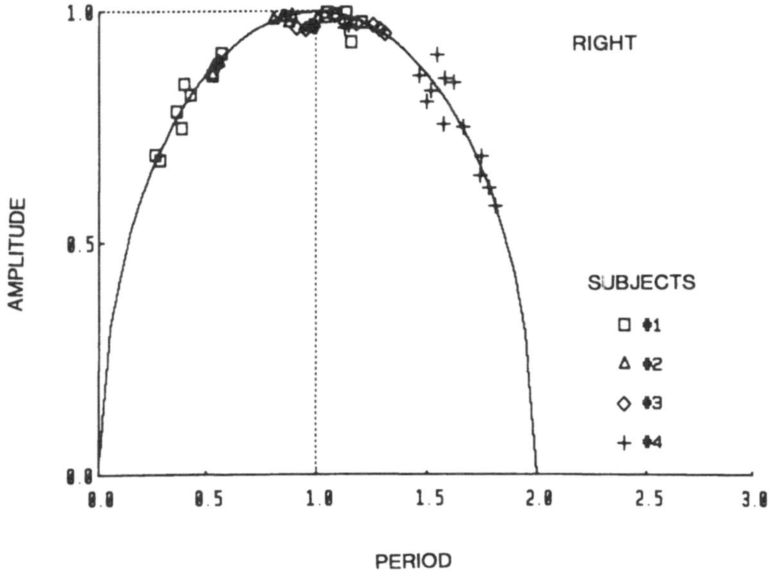

FIGURE 12.5b. (θ,τ) plot of all four subjects in dimensionless coordinates $(\theta_i/\theta_{max},\tau_i/\tau_o)$, where $\tau_o = 1/2f_{int}$: Right systems.

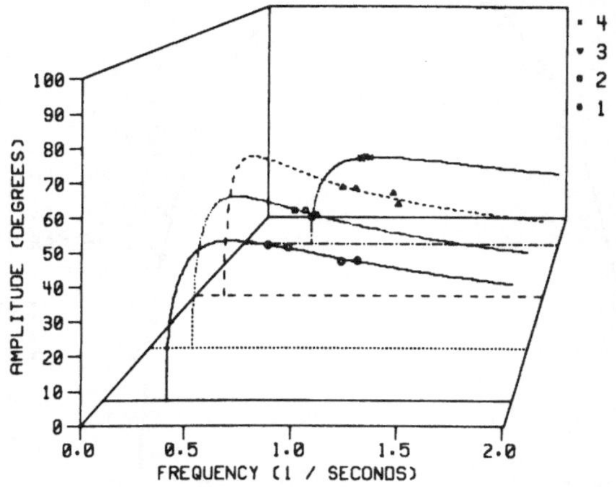

FIGURE 12.6a. *(θ,f)* plots: left wrist-pendulum systems of Subject 1.

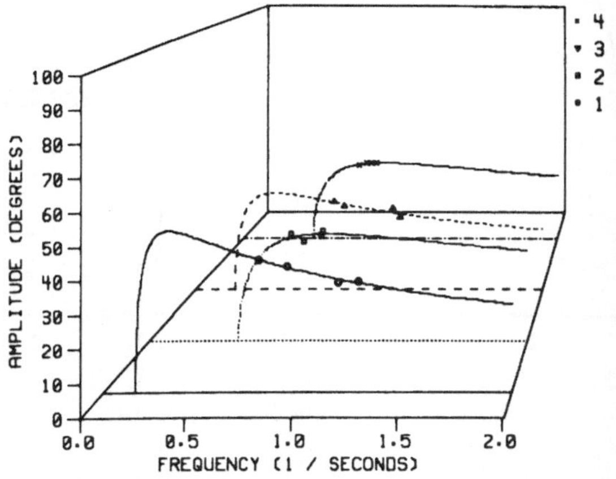

FIGURE 12.6b. *(θ,f)* plots: right wrist-pendulum systems of Subject 1.

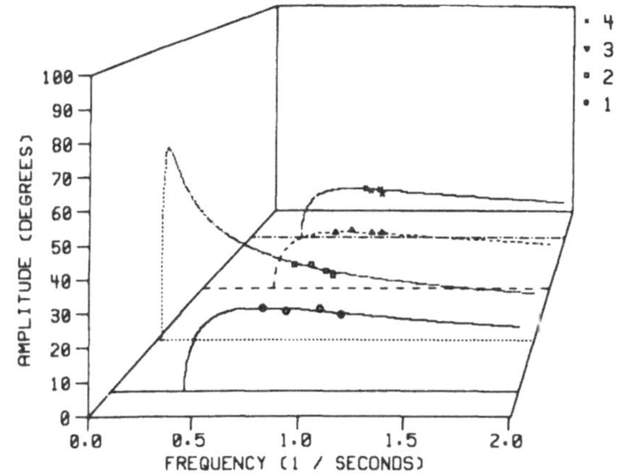

FIGURE 12.6c. (θ,f) plots: left wrist-pendulum systems of Subject 2.

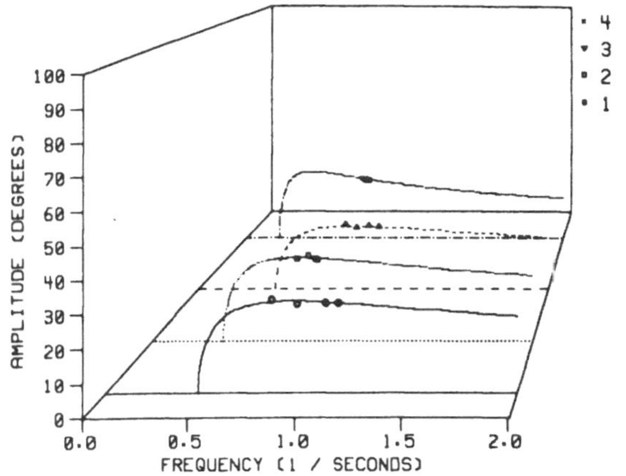

FIGURE 12.6d. (θ,f) plots: right wrist-pendulum systems of Subject 2.

Returning to Equation (12.4), it is expected that for a set of wrist-pendulum systems

$$H_i = a(I_i)^c \ , \tag{12.6}$$

where *a* and *c* are parameters that characterize the coordinate space, that is, the physically-based geometry underwriting the coordination of quantities of action with quantities of moment of inertia. The scaling of log action against log moment of log inertia (from Table 6.1) for the four left and four right single systems of the main experiment is presented in Figure 12.10. Inspection of the

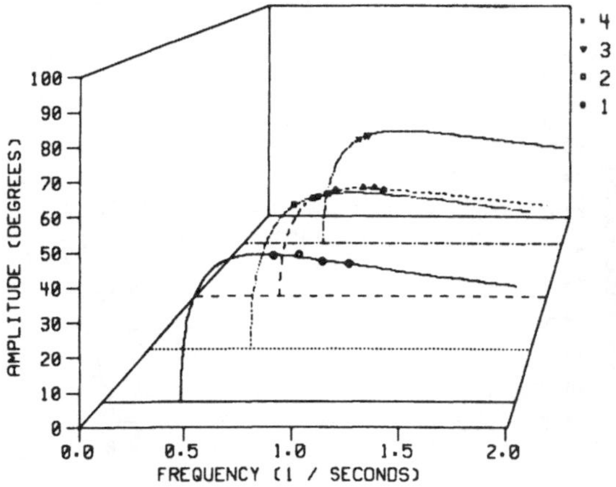

FIGURE 12.6e. (θ,f) plots: left wrist-pendulum systems of Subject 3.

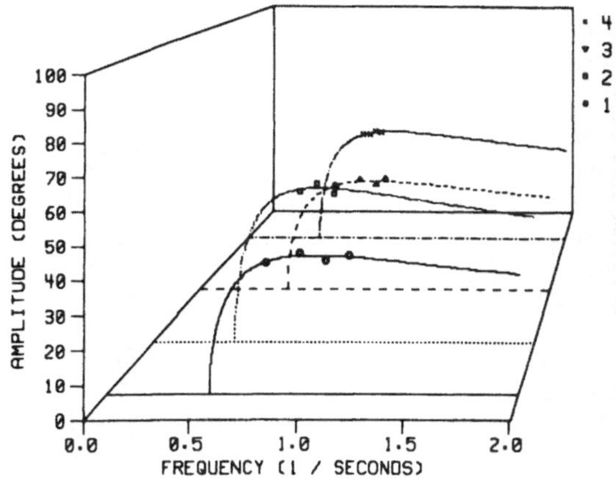

FIGURE 12.6f. (θ,f) plots: right wrist-pendulum systems of Subject 3.

figure reveals a linear dependency of log action on log inertia for the left and right systems of all four subjects. Most important is the fact that each subject's data show a departure from unit slope with the magnitude of the departure roughly equivalent for the left and right systems.

12.3.2. The Moment Variable for Spacing. From the law of action a scaling of amplitude to the mass and length of a wrist-pendulum system can be derived.

Dropping numerical constants from Equation (12.4), rearranging terms and substituting "proportional" for "equals" gives:

$$\theta \propto \left[\frac{H_i \tau_o}{I_i}\right]^{1/2} . \tag{12.7}$$

The two scaling equations for H_i and τ_o are given by Equations (12.6) and (8.32), viz.,

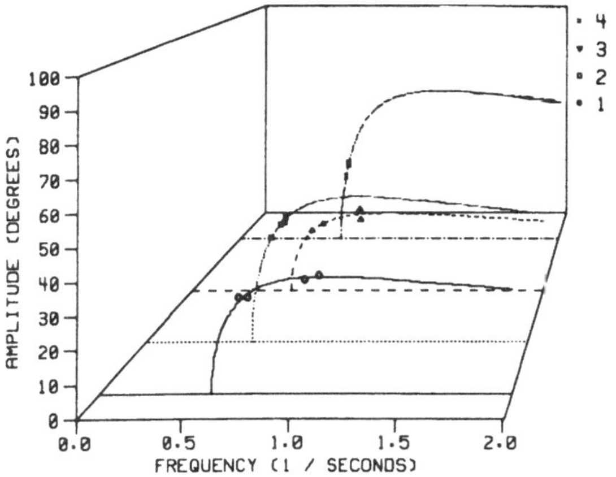

FIGURE 12.6g. (θ,f) plots: left wrist-pendulum systems of Subject 4.

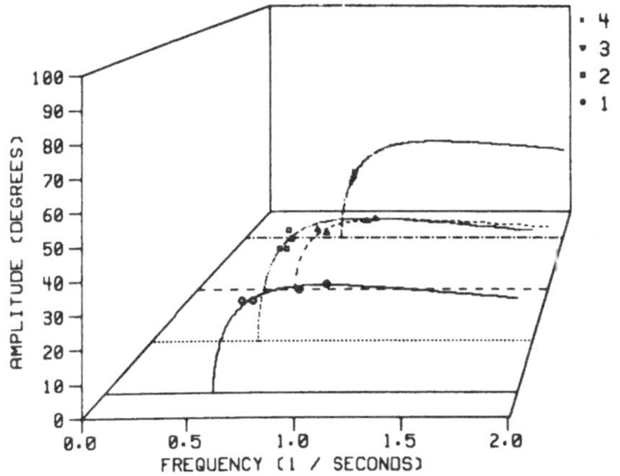

FIGURE 12.6h. (θ,f) plots: right wrist-pendulum systems of Subject 4.

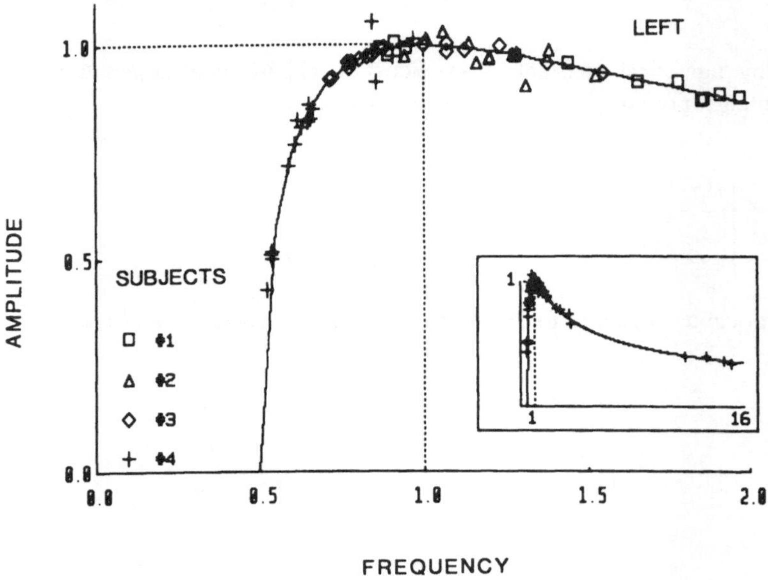

FIGURE 12.7a. (θ, f) plot of all four subjects in dimensionless coordinates $(\theta_i/\theta_{max}, f_i/2f_{int})$: Left systems.

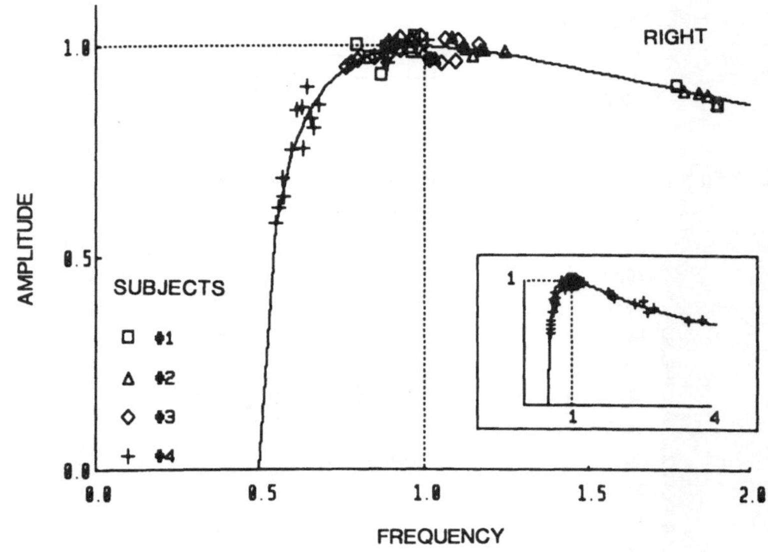

FIGURE 12.7b. (θ, f) plot of all four subjects in dimensionless coordinates $(\theta_i/\theta_{max}, f_i/2f_{int})$: Right systems.

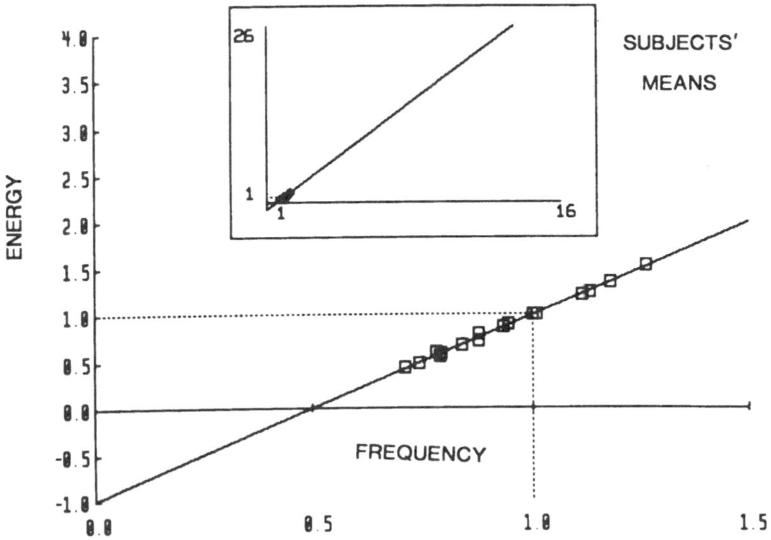

FIGURE 12.8a. (E,f) data averaged across subjects and hands plotted in dimensionless coordinates $(E_i/E_o, f_i/f_o)$.

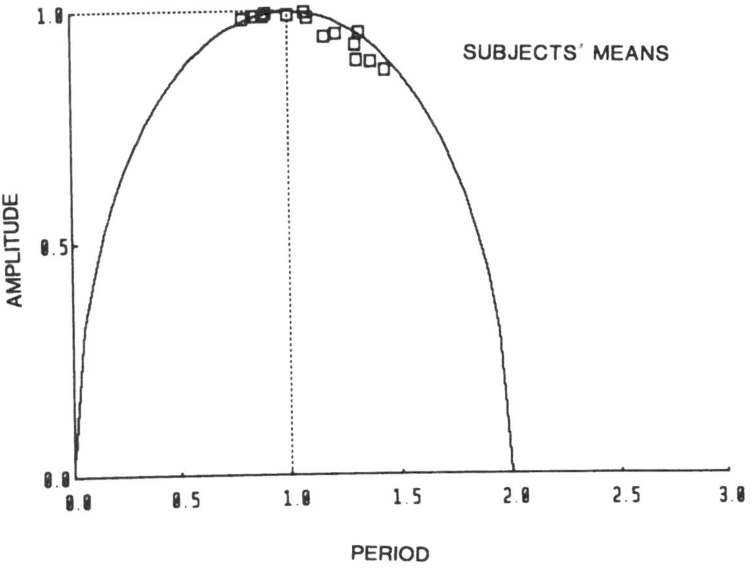

FIGURE 12.8b. (θ,τ) data averaged across subjects and hands plotted in dimensionless coordinates $(\theta_i/\theta_{max}, \tau_i/\tau_o)$.

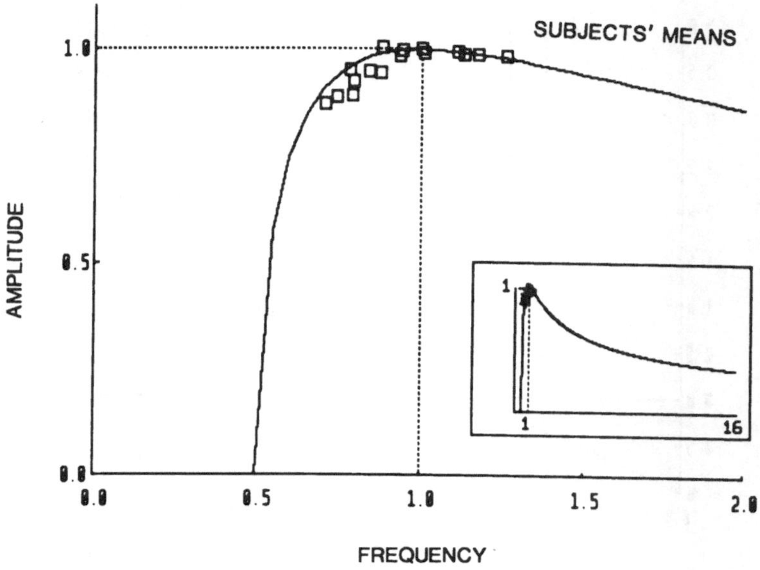

FIGURE 12.8c. (θ,f) data averaged across subjects and hands plotted in dimensionless coordinates $(\theta_i/\theta_{max}, f_i/2f_{int})$.

$$H_i = a_i(ML^2)^{c_i} \tag{12.8}$$

and

$$\tau_o = a_i(M^{1/16}L^{1/2})^{c_i} \ , \tag{12.9}$$

respectively. If the metrical structure of the coordinate spaces were uniform, then the scaling equations for H_i and τ_o are as above with exponents of *unity*. Substituting these ideal scaling equations in (12.7), ignoring their coefficients and expressing inertia (I_i) dimensionally results in:

$$\theta \propto M^{1/32}L^{1/4} \ . \tag{12.10}$$

The monomial identified in (12.10) defines the moment variable for spacing. This moment variable is the square root of the timing moment variable implying a simple, basic relation (for systems with uncorrelated masses and lengths, operating at $Q = 1$) in which space goes as the square root of time.

Inspection of Figure 10.2 reveals that the scaling of amplitude to mass and length is more complex than that expressed in (12.10). As highlighted in Chapters 7 and 10, there is a marked tendency in the data for *amplitude to vary inversely with mass while varying directly with length*. How can the different signs of the two exponents (negative for mass and positive for length) be rationalized? One

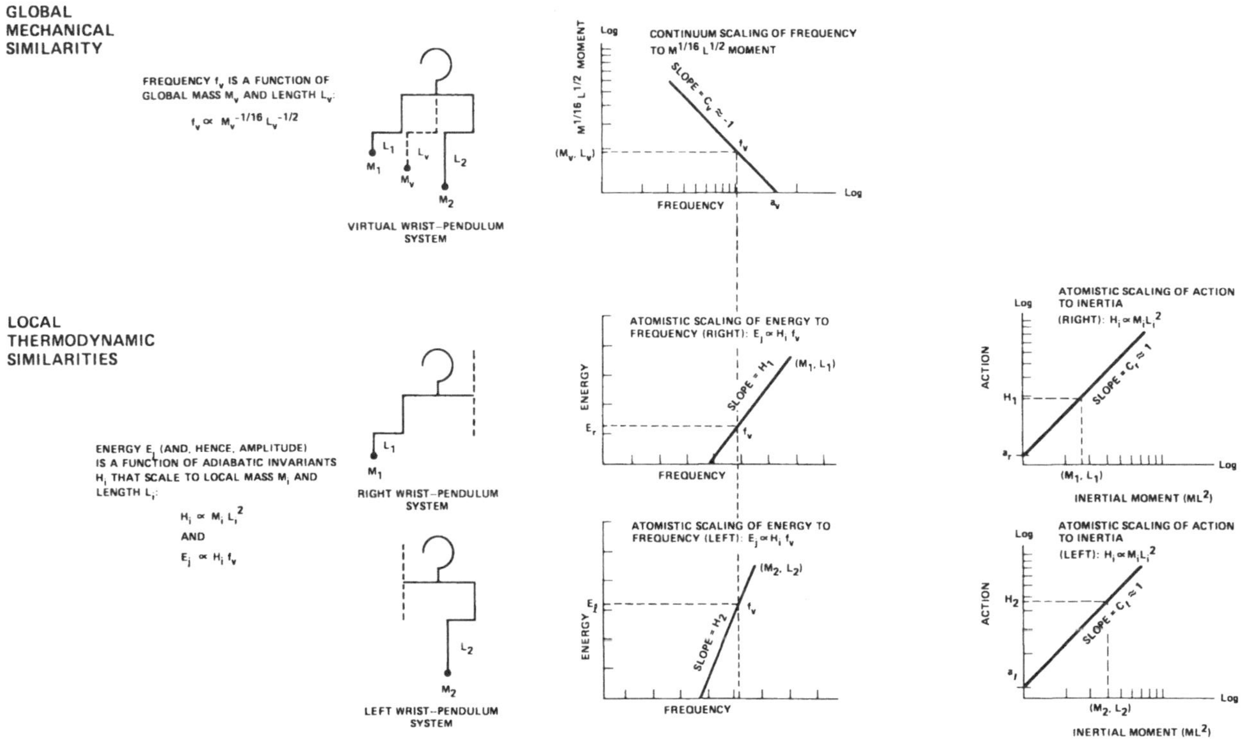

FIGURE 12.9. Nesting of Natural Laws and Boundary Conditions that Determine Global and Local Steady States.

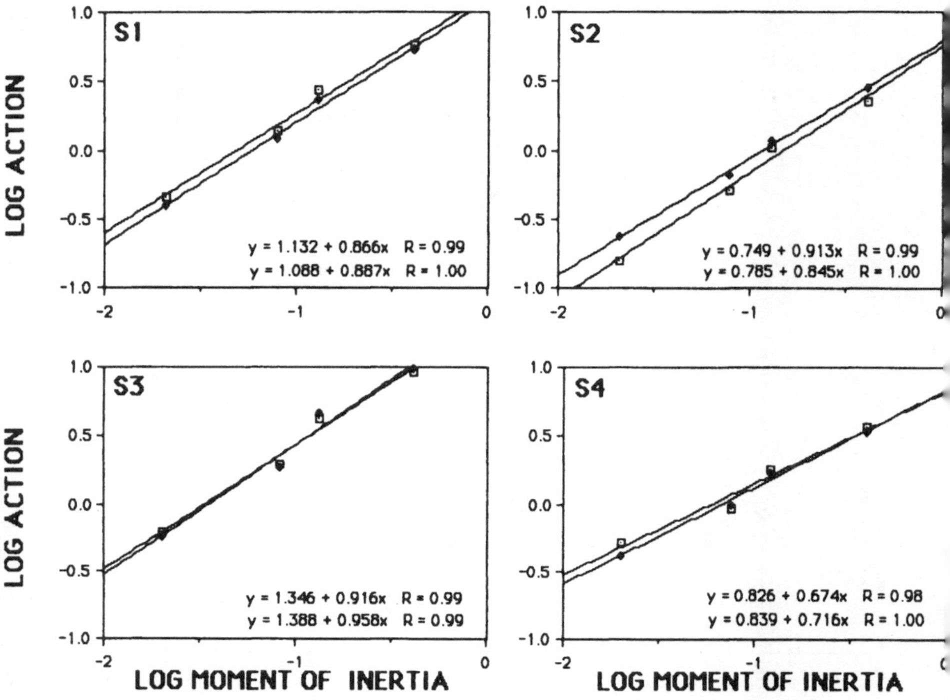

FIGURE 12.10. Log action (kg-m^2/s) x log moment of inertia (kg-m^2) for each subject in the main experiment.

path to an answer, the path taken here, is to recognize explicitly that the amplitude of a system at its natural period is (a) determined jointly by the system's constant of action and the system's natural period, and (b) that the magnitudes of these properties are contingent on the parameters of the coordinate spaces in which their respective scaling laws are embedded. The exponents of the action scaling law, Equation (12.8), and the period scaling law, Equation (12.9), will be designated by C_H and C_τ, respectively. Given these designations, and continuing to ignore the coefficients, the proportionality (12.7) can be written as

$$\theta \propto \left[(ML^2)^{c_H - 1}(M^{1/16}L^{1/2})^{c_\tau}\right]^{1/2} , \tag{12.11}$$

which on rearranging terms can be expressed as

$$\theta \propto M^{\left(\frac{c_H}{2}+\frac{c_\tau}{32}-\frac{1}{2}\right)} L^{\left(c_H+\frac{c_\tau}{4}-1\right)} \tag{12.12}$$

The monomial in mass and length in (12.12) is the moment variable for spacing as manifest in real coordinate spaces. Inspection of the monomial reveals that the mass exponent will be negative when

$$\left(\frac{c_H}{2} + \frac{c_T}{32}\right) < \frac{1}{2} ; \tag{12.13}$$

and the length exponent will remain positive when

$$\left(c_H + \frac{c_T}{4}\right) > 1 . \tag{12.14}$$

Importantly, Scaling Relation (12.12) suggests that *all* combinations of positive and negative exponents are possible; that is, there need be no consistency to the pattern of signs, no consistency to the form of amplitude's dependency on mass and length, and even no dependency (the exponents can be zero). The pattern observed is contingent upon the inherent structure of the coordinate spaces.

The timing and action exponents given in Table 8.1 and Figure 12.10, respectively, were used to construct the expected mass and length exponents according to the Scaling Relation (12.12). The *actual* exponents were determined by regressing angular excursion simultaneously on mass and length for each set (as defined by wrist and subject) of four angular excursions (those associated with systems A, B, C and D; see Table 6.1). (With the limited degrees of freedom available it was not possible for any of these linear regressions with two independent variables to achieve statistical significance. The measured exponents can, nonetheless, be taken as very close approximations for present purposes.) The goodness of fit between predicted and actual exponents for the two quantities is shown in Figure 12.11. The dependence of angular excursion on the moment variable for spacing is shown in Figure 12.12.

In any one panel of Figure 12.12 the eight angular excursions (four left and four right) produced by a subject are related in double logarithm coordinates to the corresponding eight values of the spacing moment variable. Collecting the four regression lines of Figure 12.12 for direct comparison (Figure 12.13) reinforces the belief that a single, common basis underlies the spacings of steady-state pendular clocking behavior exhibited by the four subjects.

Scaling Relation (12.11) which expresses the influence of coordinate spaces on the relation of action to the moment variable (ML^2) and time to the moment variable ($M^{1/16}L^{1/2}$), assumes that a given coordinate space modifies the power of a given moment variable *as a single unit*. A more judicious assumption is that coordinate spaces adjust the proportional relations of action to mass and to length, and the proportional relations of time to mass and to length, differenti-

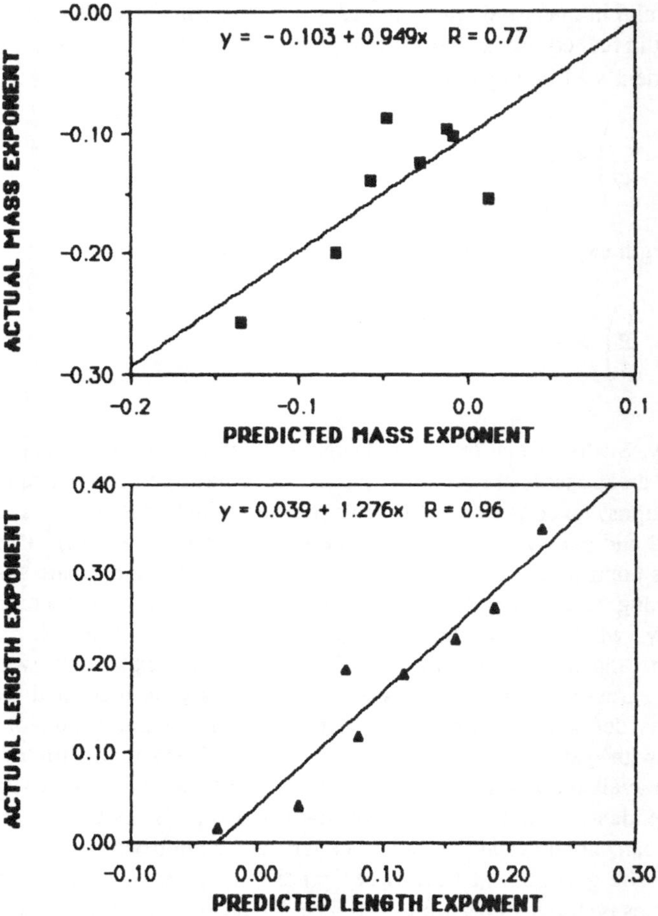

FIGURE 12.11. Comparison of actual mass and length exponents for amplitude scaling with those predicted from Scaling Relation (12.12). The data are those of the left and right single systems of the four subjects in the main experiment.

ally. If this is the case then the predictions manifest in Figure 12.11 can be improved. By hypothesis, Scaling Relation (12.11) incorporates the full complement of influences on the spacing of steady-state pendular clocking behavior. Consequently, the mass and length exponents for amplitude should be predicted precisely. Scaling Relation (12.11) can be modified as follows

$$\theta \propto [(M^{c_{H_l}} L^{c_{H_m}})(M^{-1}L^{-2})(M^{c_{T_m}} L^{c_{T_l}})]^{1/2} \,. \qquad (12.15)$$

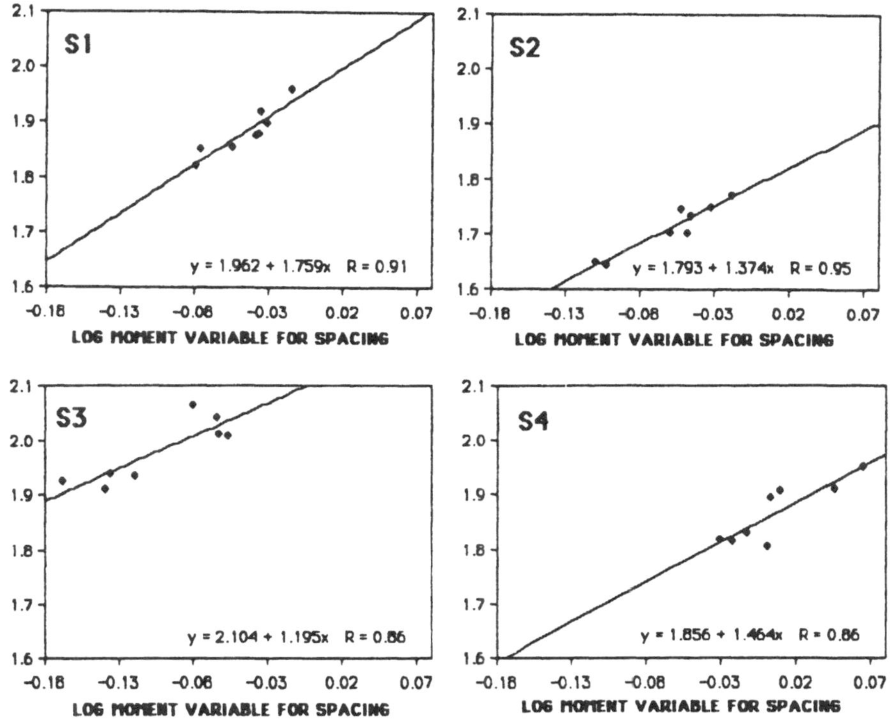

FIGURE 12.12. Log angular excursion (deg) x log moment variable for spacing (see Scaling Relation (12.12)). The data are those of the left and right single systems of the four subjects in the main experiment.

In Scaling Relation (12.15), c_{H_m} and c_{τ_m} are the actual exponents on mass and c_{H_l} and c_{τ_l} are the actual exponents on length in the action (H) and timing (τ) coordinate spaces. Exponents c_{H_m} and c_{H_l} can be obtained by regressing action simultaneously on mass and length; exponents c_{τ_m} and c_{τ_l} can be obtained by regressing periodic timing simultaneously on mass and length. These regressions were performed for the left and right single systems of the four subjects of the main experiment—that is, for the data that yielded the patterns in Figures 12.11 and 12.12. The obtained exponents are reported in Table 12.2 together with (a) the exponents for amplitude assembled through the modified moment variable for spacing (Scaling Relation (12.15) and (b) the actual exponents for amplitude as determined by multiple regression. Inspection of Table 12.1 identifies the fit between the predicted and the actual mass and length exponents for amplitude. To increase the degrees of freedom, each subject's left and right data were collected in the same two-variables regression analysis. Table 12.2 gives the resultant observed and predicted exponents together with the r^2 values.

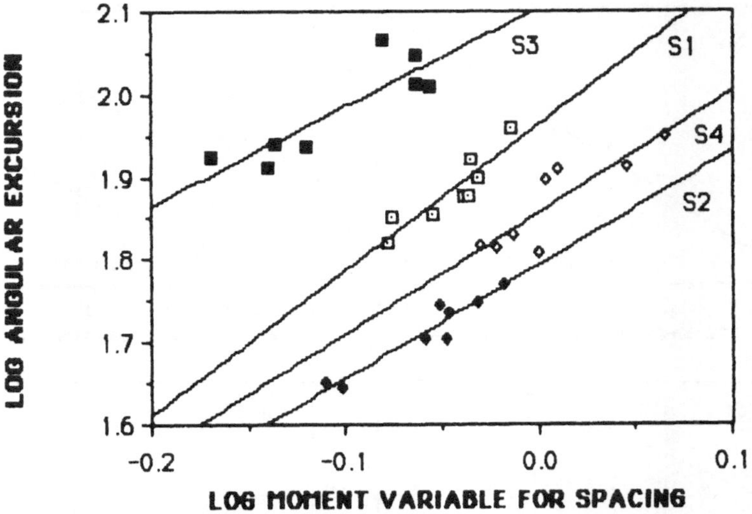

FIGURE 12.13. The four regression lines of Figure 12.12 in a single plot.

TABLE 12.1
Mass and Length Exponents from the Periodic Timing (τ), Action (H) and Amplitude (θ) Multiple Regressions Together with the Predicted Exponents for Amplitude. Data are from the Single Systems of the Main Experiment.

		τ exponents		*H exponents*		*Predicted θ exponents*		*Actual θ exponents*	
S	*Wrist*	C_{τ_m}	C_{τ_l}	C_{H_m}	C_{H_l}	$(M^{C_{\tau_m}+C_{H_m}-1})^{1/2}$	$(L^{C_{\tau_l}+C_{H_l}-2})^{1/2}$	C_{θ_m}	C_{θ_l}
1	left	.045	.504	.684	1.891	−.136	+.198	−.139	+.192
	right	.028	.473	.731	1.913	−.121	+.193	−.124	+.187
2	left	.072	.483	.721	1.956	−.104	+.220	−.097	+.226
	right	.085	.459	.724	1.765	−.096	+.112	−.088	+.118
3	left	.119	.505	.675	2.01	−.103	+.258	−.103	+.260
	right	.083	.521	.609	2.174	−.154	+.348	−.154	+.351
4	left	.037	.468	.454	1.567	−.254	+.018	−.257	+.014
	right	.016	.529	.592	1.557	−.196	+.043	−.199	+.039

The evidence suggests that the observed spacing of steady-state pendular clocking behavior scales to mass and length according to Scaling Relation (12.11). As anticipated (Section 10.1), the basis of the scaling relation for spacing is more complex than that for timing. It arises from the nesting of a mechanical law (that

TABLE 12.2
Mass and Length Exponents from the Periodic Timing (τ), Action (H) and Amplitude (θ) Multiple Regressions Together with the Predicted Exponents for Amplitude. Data are from Combined Left and Right Single Systems of the Main Experiment.

	τ exponents		*H exponents*		*Predicted θ exponents*		*Actual θ exponents*	
Subject	C_{τ_m}	C_{τ_l}	C_{H_m}	C_{H_l}	$(M^{C_{\tau_m}+C_{H_m}-1})^{1/2}$	$(L^{C_{\tau_l}+C_{H_l}-2})^{1/2}$	C_{θ_m}	C_{θ_l}
1	.037	.488	.707	1.902	−.128	+.195	−.132	+.189
	.992[a]		.990				.827	
2	.078	.471	.723	1.861	−.099	+.166	−.092	+.172
	.993		.980				.542	
3	.101	.513	.642	2.092	−.129	+.303	−.129	+.305
	.997		.998				.964	
4	.026	.499	.523	1.562	−.226	+.031	−.228	+.027
	.995		.993				.919	

[a] r^2 of the two variable regression

governing a system's timing) within an energy degradation law (that governing a system's action) and from the coordinate spaces that embed them.

12.3.3. A Third Subsidiary Experiment: Verifying the Moment Variable for Spacing. Rationalizing the relation of spacing in steady-state pendular, clocking behavior to mass and length is the culmination of the ideas developed and evaluated in the preceding pages. Arguments and evidence have been given that a biological pendulum system, auto-oscillating in the steady state, is characterized by two properties: a constancy of action and a natural period. Both properties are linked lawfully to the magnitudes of the pendulum system and information about (in the sense of specific to) both properties is contained in the haptic flow fields engendered by the biological pendular motion. Although the various quantities involved are related lawfully the relationships are not simple. Laws identify possibilities; actual facts come about through the coupling of laws with circumstances (Bunge, 1977). Coordinate spaces are circumstances. These, as observed several times, are physically based geometries (based in the local and momentary distributional character of conservations) that are contexts for the lawful relating of quantities. Consequently, understanding the *actual* behavior of a quantity rests with an appreciation of both the relevant laws and their coordinate spaces. The development and successive modifications of the moment variable for spacing expressed in the Scaling Relations (12.10), (12.11) and (12.15) reflect this joint concern.

A third subsidiary experiment was conducted to test further the scaling relation for spacing. It involved three subjects and 10 pendulums of 10 different

TABLE 12.3
Single Wrist-Pendulum Length (m)/Mass (kg) Conditions in Subsidiary Experiment 3 Ordered According to Length.

	System Number									
Subject	*1*	*2*	*3*	*4*	*5*	*6*	*7*	*8*	*9*	*10*
1	.203[a]	.217	.222	.337	.355	.363	.367	.473	.496	.507
	.772[a]	1.022	1.522	.810	1.060	1.560	2.560	.848	1.098	1.598
2	.199	.213	.219	.335	.353	.361	.365	.472	.494	.506
	.813	1.063	1.563	.851	1.101	1.601	2.601	.888	1.138	1.638
3	.196	.209	.215	.332	.349	.357	.360	.469	.490	.501
	.813	1.063	1.563	.851	1.101	1.601	2.601	.889	1.139	1.639

[a]length (m)
[b]mass (kg)

TABLE 12.4
Mean Periodic Times (s) in Subsidiary Experiment 3

		System Number									
Subject	*Wrist*	*1*	*2*	*3*	*4*	*5*	*6*	*7*	*8*	*9*	*10*
1	left	.769	.827	.898	.884	.920	.986	1.040	1.000	1.038	1.151
	right	.784	.828	.898	.904	.934	.999	1.063	1.004	1.032	1.130
2	left	.803	.858	.914	1.105	1.134	1.194	1.228	1.324	1.370	1.404
	right	.777	.785	.893	1.060	1.116	1.188	1.234	1.269	1.309	1.377
3	left	.754	.819	.865	1.018	1.070	1.109	1.136	1.105	1.306	1.222
	right	.788	.832	.872	1.021	1.102	1.113	1.152	1.135	1.285	1.263

mass/length combinations. Each member of the resulting 10 left wrist-pendulum systems and 10 right wrist-pendulum systems was evaluated independently on eight separate occasions in accordance with the methodology identified in Chapter 6. The mass and length magnitudes of the individual pendulums for each subject (computed in the manner detailed in Chapter 6) are given in Table 12.3. The mean periods and mean angular excursions as a function of the wrist-pendulum magnitudes for each subject are reported in Tables 12.4 and 12.5, respectively.

Figure 12.14 gives the plots in double logarithmic coordinates of the relation between periodic timing and the moment variable ($M^{1/16}L^{1/2}$) for the left and right wrist-pendulum systems of each subject. The double logarithmic plots of the relation between action and the moment variable (ML^2)—the moment of inertia—are given for the left and right systems of each subject in Figure 12.15. Following the logic developed in Chapter 8, to the extent that the r^2 of a regression is not unity and the slope of a regression is not unity then the

TABLE 12.5
Mean Angular Excursions (deg) in Subsidiary Experiment 3

		System Number									
Subject	*Wrist*	*1*	*2*	*3*	*4*	*5*	*6*	*7*	*8*	*9*	*10*
1	left	32.98	33.78	27.92	34.02	34.60	34.41	25.90	35.16	33.78	28.48
	right	29.60	29.85	28.19	34.10	33.52	28.86	26.02	34.11	32.05	28.92
2	left	32.72	31.82	26.32	50.29	46.17	43.54	34.71	57.73	52.93	43.69
	right	35.19	28.14	28.55	49.20	43.19	43.19	36.04	61.45	55.49	46.11
3	left	42.93	54.11	46.77	48.17	52.13	50.65	38.49	44.63	56.65	39.72
	right	45.04	50.34	50.44	48.46	46.54	50.60	47.31	44.74	53.77	47.05

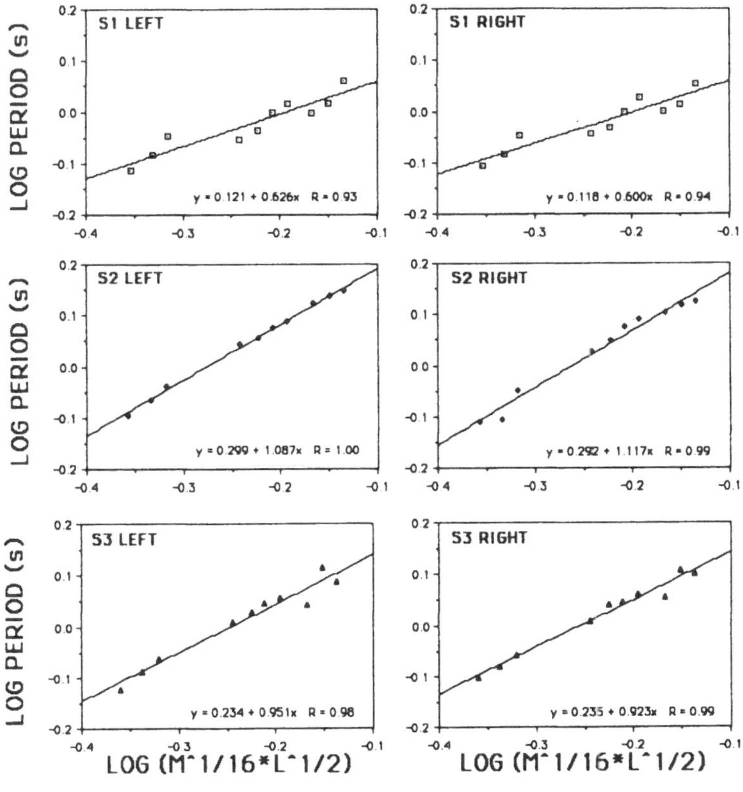

FIGURE 12.14. Log period (s) x log moment variable for timing for the left and right systems of the three subjects in Subsidiary Experiment 3.

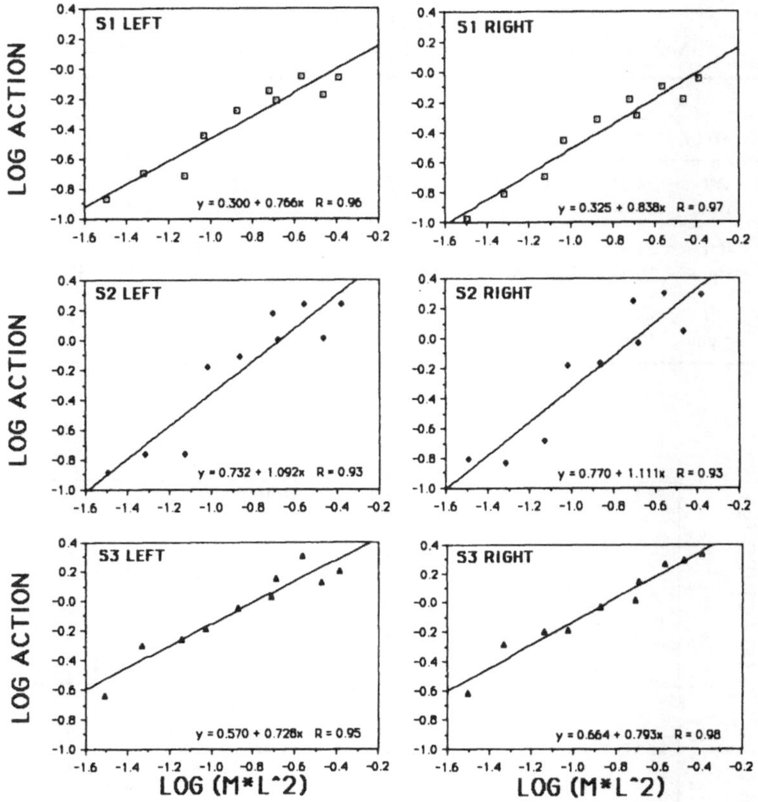

FIGURE 12.15. Log action (kg-m^2/s) x log moment of inertia (kg-m^2) for the left and right systems of the three subjects in Subsidiary Experiment 3.

coordinate space within which the quantities are related is not uniform. Inspection of Figures 12.14 and 12.15 (taking into account both the regression parameters and the visual appearance) suggests that, in particular, the coordinate space housing the periodic timing law in the case of Subject 1 and that housing the action law in the case of Subject 2 are markedly nonuniform. In specific terms, a nonuniformity is realized as an adjustment to the exponents of the governing moment variable. That the mass and length adjustments within a moment variable need to be considered individually is evident in the data of Subjects 1 and 2 when viewed in the context of the wrist-pendulum magnitudes reported in Table 12.3. As Table 12.3 reveals, the wrist-pendulum systems of the third subsidiary experiment fall roughly into three groups distinguished by length (namely, Systems 1-3, 4-7 and 8-10). Within each group the length differences are small relative to the mass differences. Referring to the action plots of Subject

2, it would appear that the coordinate space adjustment is skewed in the direction of amplifying the length exponent and reducing the mass exponent (crossing the regression line in the case of both left and right systems are three distinct, nearly flat, trajectories). In contrast, with respect to the timing plots of Subject 1, the coordinate space adjustment seems to have amplified the mass exponent and reduced the length exponent (crossing the regression line in the case of both left and right systems are three distinct, positively sloped, trajectories).

Both timing and action were subjected to a two variables (mass and length) regression. The actual mass and length exponents thus determined are presented in Table 12.6. (Pronounced adjustments of the exponents in the case of Subject 1's periodic timing and Subject 2's action can be noted). The predicted amplitude scaling based on the actual timing and action scalings for each subject and both wrists are also presented in Table 12.6. The predicted exponents were assembled according to the moment variable for spacing as expressed in Scaling Relation (12.15). For Subjects 1 and 2 the predicted exponents for the scaling of amplitude and those actually obtained were identical, as inspection of Table 12.6 shows. The difference in sign between the mass exponent (negative) and the length exponent (positive) observed in the data of the main experiment (see Tables 12.1 and 12.2) was reproduced in the data of Subjects 1 and 2. A different pattern of signs was predicted for Subject 3. Unlike Subjects 1 and 2, however, the regression of Subject 3's amplitude on mass and length was not significant (r^2 = .177 and r^2 = .028 for the left and right systems, respectively). The reported actual exponents, therefore, are not reliable. It is clear, nonetheless, that the small exponents predicted from the action and timing scalings with the aid of Scaling Relation (12.15) reflect accurately the observed independence of amplitude from the variables of mass and length in the data of Subject 3.

In sum, the signs of the mass and length exponents with respect to amplitude are not fixed. They can be flexible, as implied by the internal structure of Scaling

TABLE 12.6
Mass and Length Exponents from the Periodic Timing (τ), Action (H) and Amplitude (θ) Multiple Regressions Together with the Predicted Exponents for Amplitude. Data are from Subsidiary Experiment 3.

		τ exponents		*H exponents*		*Predicted θ exponents*		*Actual θ exponents*	
S	*Wrist*	C_{τ_m}	C_{τ_l}	C_{H_m}	C_{H_l}	$(M^{C_{\tau_m}+C_{H_m}-1})^{1/2}$	$(L^{C_{\tau_l}+C_{H_l}-2})^{1/2}$	C_{θ_m}	C_{θ_l}
1	left	.137	.279	.341	1.852	−.261	+.066	−.261	+.066
	right	.134	.265	.436	1.980	−.215	+.123	−.215	+.123
2	left	.069	.546	.206	2.795	−.363	+.671	−.363	+.671
	right	.112	.559	.200	2.850	−.344	+.705	−.344	+.705
3	left	.093	.464	.626	1.525	+.141	−.006	+.148	−.001
	right	.086	.454	.955	1.478	+.021	−.034	+.026	+.002

Relations (12.12) and (12.15). Further, significant mass and length dependencies in the timing and action spaces need not be replicated in the amplitude space. (For Subject 2 in the main experiment and for Subject 3 in the third subsidiary experiment, amplitude was orthogonal to mass and length.) Amplitude, though scaled to mass and length in a lawful manner, is orthogonal to mass and length within certain coordinate spaces.

12.4. A Summary

The experimental analyses presented in this chapter have concentrated on the theoretical claim, developed in Chapters 10 and 11, that an adiabatic scalar invariant—action or *H*—figures prominently in the pendular clocking mode of movement organization. This adiabatic magnitude, a hybrid of a conservation (energy) and a coordinate of motion (frequency), scales to a system's moment of inertia. It is one of two scalar quantities, the other being a *Q* of unity, that give direction to the changes in a system's energy accompanying the changes in a system's periodic timing. The competition between these two scalar quantities imposes a structure on the work space (see Figure 11.11).

The *(E,f)* work space and the intrinsically defined *(H,Q)* space warrant careful experimental exploration. Establishing that mechanical transformability is a very general organizing principle of rhythmic movements will require, among other things, a precise determination of (a) the range of conjunctions of inertial and frequency variables within which a constant proportionality of energy and frequency is expected and (b) the movement settings that satisfy the criteria of adiabatic transformations. These two requirements are underscored by a consideration of the first subsidiary experiment reported in Section 8.8.

In that experiment, each of four individual wrist-pendulum systems of one hand assumed three different, stable frequencies as a function of three levels of force produced by the other hand. Table 12.7 summarizes the energy/frequency relations exhibited in this subsidiary experiment. Three (overlapping) observations are called for. First it is evident from inspection of Table 12.7 that the necessary inclusion of frequency in the measure of a wrist-pendulum system's mechanical energy is not of itself sufficient to yield a reliable correlation of energy and frequency in linear coordinates. Second, it is evident that whereas the relation of energy to frequency was consistent (positively monotonic and linearly proportional) in the isochronous coupling data reported in Section 12.1.1, it is inconsistent in the data reported in Table 12.7. Third, it is evident that energy and frequency are not closely related in Table 12.7. Out of the 32 correlations, only one positive correlation was significant at the .05 level. Of the 32 correlations conducted on the isochronous coupling data (Section 12.1.1.), all were positive and 26 were significant at the .05 level and above (see Figures 12.2a-h). One conclusion might be that the two manipulations of a wrist-pendulum system's frequency—isochronous coordination in the main experi-

TABLE 12.7
Energy ($Kg - m^2/s^2$)/Frequency (I/S) Relation Where the Frequency Change is not Produced by Coordination

		Inertia ($kg-m^2$)							
		.0065		*.0312*		*.0645*		*.3125*	
Subject	*Hand*	*Frequency*	*Energy*	*Frequency*	*Energy*	*Frequency*	*Energy*	*Frequency*	*Energy*
		1.242	.0853	.840	.3066	1.130	.6351	.763	2.1694
	Left	1.284	.1075	.870	.4157	1.143	.5838	.776	1.7483
		1.297	.1030	.878	.3920	1.156	.6154	.781	2.1694
		r = .913		r = .918		r = −.399		r = −.715	
1									
		1.148	.0836	.805	.3203	1.073	.4852	.765	2.2637
	Right	1.176	.1038	.838	.2658	1.081	.4476	.777	1.1169
		1.181	.0645	.876	.2412	1.127	.6878	.787	2.5211
		r = −.125		r = −.968		r = .960		r = .120	
		1.580	.1148	.929	.4208	1.318	.7601	.749	1.5190
	Left	1.757	.1691	1.066	.3049	1.418	.8630	.824	1.4317
		1.838	.1692	1.073	.2906	1.495	1.0184	.861	1.7321
		r = .952		r = −.998		r = .982		r = .537	
2									
		1.756	.2138	1.027	.3927	1.397	1.0207	.797	2.0260
	Right	2.079	.1996	1.202	.3721	1.513	1.3800	.937	1.7035
		2.433	.1246	1.328	.4531	1.529	.9080	1.018	2.2843
		r = −.934		r = .649		r = .181		r = .302	
		1.712	.0620	1.206	.1080	1.513	.3132	.957	1.0188
	Left	1.779	.0476	1.299	.0813	1.664	.3650	1.103	1.0011
		2.199	.0719	1.414	.1225	1.848	.5495	1.195	1.195
		r = .708		r = .403		r = .967		r = −.828	
3									
		1.548	.0518	1.045	.1262	1.357	.3790	.898	1.1242
	Right	1.669	.0477	1.191	.1290	1.397	.2380	1.048	.7903
		1.701	.0410	1.212	.1582	1.511	.2688	1.073	.7992
		r = −.895		r = .658		r = −.537		r = −.988	
		1.541	.0809	1.017	.3752	1.328	.5604	.780	2.1800
	Left	1.616	.1218	1.059	.4568	1.385	.6148	.879	2.5450
		1.664	.1529	1.186	.3542	1.408	.6517	.882	2.2306
		r = .999		r = −.460		r = .992		r = .586	
4									
		1.473	.1153	.939	.3561	1.271	1.0012	.773	1.8789
	Right	1.534	.1722	1.042	.5212	1.274	.6976	.840	2.9987
		1.661	.1948	1.103	.4806	1.381	1.154	.876	2.851
		r = .903		r = .817		r = .741		r = .907	

ment and exerting force locally in the first subsidiary experiment—are categorically distinct: the former is an adiabatic (mechanical) transformation, the latter is not.

Although the *experimental* determination of an invariant scaling relation linking energy to frequency is expected to depend on (a) the region of the *(E,f)* work space from which the data are drawn and (b) the nature of the movement conditions inducing changes in frequency, it is nevertheless the case that, by the *theoretical* arguments advanced in Chapters 10 and 11, for any system of magnitude I_i a constant H_i is always defined. It is this assertion that is exploited in Section 12.3 to develop the scaling relation for steady-state spacing. The success of the resultant scaling relation in identifying amplitude's proportional dependencies on mass and length lends further support to the extension (begun in Chapter 10) of Ehrenfest's hypothesis to nonconservative cases: mechanical transformability appears to be a core property of softly-assembled, rhythmic movement systems.

REFERENCE

Bunge, M. (1977). *Treatise on basic philosophy, Volume 3; Ontology 1: The furniture of the world*. Boston: D. Reidel Publishing Company.

13 INTENTIONALITY AND UNITS OF ACTION

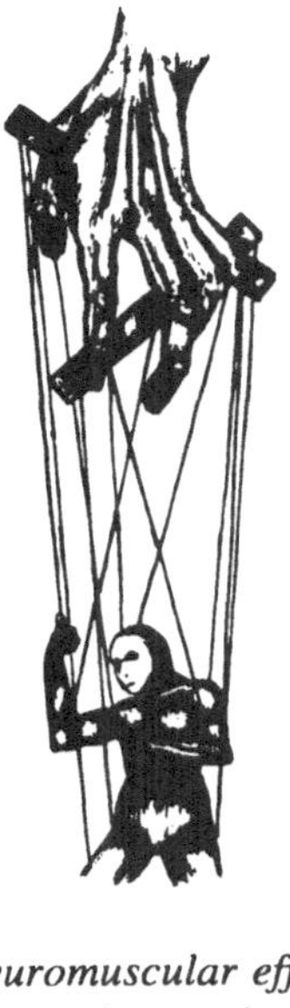

ABSTRACT

It is commonly assumed that everyday movements are constructed by hierarchically arranging elemental movements. The prototypical reflex is such an element. So is the neural oscillator or central pattern generator. The basic idea behind this assumption is that a movement is a complex thing *put together from simpler things. This basic idea is challenged: Movements and their units are not things but* relations. *The pendular clocking mode of behavior is softly assembled through the nesting of lawful relations among properties, namely, action, frequency, energy, and monomials in mass and length. Borrowing an idea of Reed's (1982), the pendular clocking mode of organization is seen to be a specialized usage of "resources," roughly, gravitational, elastic, chemical, and electrical potentials supplemented by information in the specificational sense.*

The foregoing does not completely exhaust the description of the pendular clocking mode, however. A person in the mode has a perspective on his or her various neuromuscular efforts. There is an 'aboutness' or 'directedness' to the activity that points to its intentional character. In very blunt terms, the content of a person's intention (approximately, the desired act) is tantamount to an exceptional boundary condition or constraint that harnesses the lawful relations and fetters the use of resources. This intimated equivalency between the philosoph-

ical/psychological notion of intentional content and the physical notion of (exceptional) constraints is defensible. Closely parallel but independent arguments are offered by philosophers for the physically anomalous nature of intentional content and by physicists for the physically anomalous nature of nonholonomic constraints (as viewed from the conventional standard of physics). The most basic parallels are these: Both intentional content and nonholonomic constraints warrant a descriptive language distinct from that which describes a system's dynamics, and both warrant the status of "causal" despite their failure to instantiate (quasi) universal regularities. These parallels between intentional content and nonholonomic constraints suggest a number of directions for naturalizing intentionality. They also raise the possibility that the notion of causality may need to be "intentionalized" and separated from the notion of regularity. Finally, they suggest—and they do so rather strongly—that physical states with (symbolic) contents that are causally potent are a problem for science in the large and not, as some have implied, a problem that uniquely defines the science of cognition.

13.1. Actions and Their Units

The research reported here has addressed a problem that might be posed variously as follows (following Pattee, 1972): How does a coordinated state arise from a more or less uncoordinated state? How does a more or less uncoordinated aggregate of extreme complexity give rise to a persistent regularity that condenses out simple behavior? What gives rise to the formation of simplicity from complexity? The problem, specifically, was how a person—a very complex dynamical system composed of functionally rich components and very many degrees of freedom—can make of him- or herself a pendular clock—a very simple, dynamical system with few degrees of freedom?

The problem addressed by the preceding questions is not one that is usually posed. More commonly the tendency is to ask how simple things aggregate to form complex things. For example, in traditional theories of animal movement certain simple things (mechanisms such as reflexes, tropisms, central pattern generators, and so on) are thought of as components of actions. The received view is of a number of elemental movements each related to a distinct anatomical structure (e.g., Gallistel, 1980). Activating combinations and permutations of these pieces of the anatomy gives rise to a complex movement. In the received view, actions are *things* that are divisible into other simpler things.

The twin metatheoretical assertions of complexity from simplicity and anatomically specific elemental movements have not been without detractors (Fearing 1930/1970; Weiss, 1941). Both Bernstein (1967) and Gibson (1979; Reed & Jones, 1982), for example, rejected the notion of elemental movements and elemental causes. For Bernstein and Gibson, some movements are *elementary* and some types of informational support are *elementary*, but there are no elements of movement or information from which more complex forms are derived (Kugler, Kelso & Turvey, 1982; Reed, 1982).

The view espoused here is in the spirit of these detractions and it is very close to the view expressed by Reed (1982) that reflects, in turn, ideas of Ghiselin (1981). An action is (a) not a thing but a *relation* (among properties distributed over the acting animal and the surround; see Fowler & Turvey, 1978); (b) not a particular aggregation of elemental anatomical mechanisms but a *specific mode of resource use* (where a resource is interpreted as a potential or a concentration of a conserved quantity); and (c) not categorized by reference to the anatomy that it involves but by reference to the function that it performs, that is, it is *functionally specific*, not anatomically specific.

Because an act is a relation it must be analyzed in terms of nested and interlocking properties rather than in terms of hierarchically organized anatomical parts. Because an act is a specific mode of resource use, its units must also be specific modes of resource use rather than such anatomical structures as reflexes and central pattern generators. And because an act is functionally specific, its variability is in reference to preserving the function it fulfills rather than to preserving any particular aggregation of body parts that it happens to involve.

The preceding points are summarized by Ghiselin's admonition that " . . . units that function should not be treated as units of the function themselves" (1981, p. 200). Phrased in slightly different form, the admonition reads: The structures that support an action should not be confused with the action. To exemplify, consider a hand that grasps. Can the hand be categorized as a unit of grasping? To put the question bluntly, is the hand *a grasp*? The answer is "no" to either form of the question. A hand is a thing, but grasping is a relation realized by the juxtaposition of dispositions (those of the thing that can grasp and those of the thing that can be grasped), which in turn are lawful relations of properties (Shaw & Turvey, 1981; Turvey, Shaw, Reed & Mace, 1981). Categories such as thing and relation provide what may be termed "fundamental ontological units": The units of a thing must be things and the units of a relation must be relations (Ghiselin, 1981). A hand, therefore, is identified correctly as a unit of the body (a thing), but incorrectly as a unit of grasping (a relation).

A simpler example is pendular motion: Imagine a cup with a piece of string tied to its handle. Holding the string at its free end and tapping the cup gives rise to a back-and-forth swinging of the cup. While it is appropriate to say that the string and the cup are things involved in the pendular motion, it is quite unreasonable to say that they are units *of* that style of motion. The very same motion is possible, for example, with a wooden chair replacing the china cup and an iron chain replacing the string. Clearly, what is invariant about, and definitive of, pendular motion is not the things in motion but a relation (of inertial and restorative forces).

As proposed in Chapter 9, locomotion in general, and the rhythmic activity of a pair of wrist-pendulum systems in particular, are cooperative states (produced by the dynamical relation between atomisms from below and boundary conditions from above). In both the general and the particular case the cooperativity is a particular mode of resource use, namely, the pendular, clocking (limit cycle) mode. This mode is a special organization of the uses of the gravitational potential, the body's elastic potential, the body's local depots of chemical fuel, and a (low energy) haptic field. The units of action or the atomisms in the isochronously coupled wrist-pendulum situations are the individual wrist-pendulum systems. They fulfill the requirements of *units of* (as related above) by the fact that they are instances of a specific mode of resource use—again, the pendular clocking mode. The units of action of the individual wrist-pendulum systems are less clear, although a strategy by which their definition might be arrived at has been illustrated by Bloch and Iberall (1982). When defined, these units similarly must be specific ways of using potentials.

To concur with Ghiselin's (1981) stricture, if rhythmic activity is categorized as a relation, then it must be analyzed in terms of relations. In this book an account has been developed of rhythmic biological activity in terms of a nesting of laws that relate mechanical and thermodynamic properties. In addition to its laws, however, there is an aspect of the rhythmic movement that deserves careful comment. From the perspective of the person assembling the pendular clocking mode, the pattern of rhythmic movement has a *meaning* (or a *significance* or a *content*), namely, "I'm trying to move my two wrist joints in opposite directions at what is for me the most comfortable, common tempo." It can be said that the activity that the person performs is directed by this meaning or content. In short, *although rhythmic activity is a relation, it is not fully characterized by nested, lawful relations; it has, in addition, a content.* Customarily, phenomena that involve a relation to a meaning or a directedness to a content are referred to as *intentional.*

Stated very generally, rhythmic movement is an instance of the formation of simplicity from complexity. Like other instances it is largely due to the exceptional boundary conditions through which a system of detailed complexity (an animal or a person) harnesses natural laws to perform specific and reliable functions. These exceptional boundary conditions arise from within the system and its surround. They are, to put it plainly, the contents of an animal's or a person's intentions.

That the content of an intention functions as a field boundary condition of the nonholonomic type is suggested in Chapter 9. In what follows, an effort will be made to uphold this proposed equation of the philosophical/psychological notion of content and the physical notion of constraint.

13.2. Intentional Content and Nonholonomic Constraints

It has generally been supposed in philosophy and psychology that intentionality marks off mental phenomena from physical phenomena. A more moderate supposition is that intentionality entails one description and explanatory language and nonintentionality another (e.g., Dennett, 1969). That is, intentionality can be seen as a mark that divides one kind of description from another kind of description, where the two kinds of descriptions are not reducible one to the other.

On the basis of the instructions from the experimenter one can imagine a subject in the main wrist-pendulum experiment saying (explicity or tacitly) to him- or herself something like, "I intend that I swing the two pendulums comfortably, at the same tempo but in opposite directions." This sentence is an intentional sentence. What the sentence captures is part of the causal context for the resultant act. Just how it can figure causally is part of the problem to be addressed.

Intentional sentences divide into two major parts. One part is the *intentional* (or, more broadly, psychological) idiom. In the target sentence, the intentional idiom is "intends (that)." Additional intentional idioms are "believes (that)," "decides (that)," "understands (that)," and so on. The other part of an intentional sentence is the *intensional context*—roughly, the object of the intentional idiom. Intensional contrasts with extensional and does so in two closely related ways. First, an intensional description *classifies* a physical state of affairs; it ascribes to the state of affairs a *meaning* or a *significance* or a *content*. In contrast, an extensional description is mute on what a physical state of affairs is about. An extensional description is of the details of the physical state pure and simple. Second, an intensional description is *referentially opaque*. This means, technically, that replacing terms in the intensional context of an intentional sentence by different terms that refer to exactly the same thing or things can turn truth into falsehood, and vice versa. For example, it seems most unlikely that any subject in the experiments intended to make of him- or herself "a virtual, single, pendular (limit cycle) clock," even though this latter state is the equivalent of the state that the subjects did in fact intend. Although the two intensional contexts, the original and its replacement, refer to precisely the same thing, they do not have the same meaning. If a subject were asked whether he or she intended to become a "virtual single, pendular (limit cycle) clock," the answer would have probably been "no" or "I don't think so." Of course, there is a sense in which a subject did intend to assemble "a virtual single, etc.," but this is not how he or she would have conceived it. In contrast to an intensional description, an extensional description is referentially transparent—its truth or falsehood is unaffected by replacing its terms by other terms that refer to the same thing or things. For example, replacing "wooden shaft" by "ash dowel" in the extensional sentence "he is swinging a wooden shaft with a weight attached at one end" does nothing to the truth value of the sentence.

The thrust of what follows is that it is advisable to keep distinct the two parts of an intentional sentence, the intentional idiom and the intensional context. The intentional mode of explanation (describing and explaining phenomena in intentional sentences) is both a traditional and a contemporary means of treating certain classes of natural phenomena. However, because intentional idioms refer to intelligent agents—it is, after all, animals and humans that are said to want, believe, expect, recognize, decide, intend, and so on—the use of the intentional mode of explanation is of questionable scientific generality. An intentional idiom conceals an intelligent act of an intelligent agent. Therefore, for science, an intentional idiom is a reference point for something to be explained (even explained away) and not something that can function as an explanation. In the picturesque language of Dennett (1969, 1971), using an intentional idiom in the explanation of a phenomenon is like taking out a loan of intelligence that must eventually be repaid.

Many scientists are inclined to believe that with respect to biological phenomena such as locomotion or visual perception an explanation shot full of intentional idioms (such as depicted in Figure 1.1) is not such a bad thing, given the promise of the Turing reducibility of intellectual processes. Because "recognizing (that)," "expecting (that)," "deciding (that)," and so on, can be reduced, in principle, to the unintelligent syntactical operations of a computational device, then it is supposed that the intelligence loans incurred by the intentional idioms are immediately repayable. This supposition, however, is invalid because it neglects the fact that the operations of computational devices are formatted by intelligent agents (human or their artifacts); they do not self-assemble. Turing reproducibility is a counterfeit payment. Traditionally, intentional idioms and the animal analogues that they conceal have been discharged by the careful, painstaking extension of natural law. This traditionally tried and true response to intentional idioms has been the one taken in the present monograph. How broad is its scope? *At the very least* it must—for matters of biology, physiology, and psychology—encompass all proposed intentional idioms below the scale of the individual animal or human.

In a nutshell, the traditional scientific disdain for intentional explanation is because of its question-begging appeals to intelligence and rationality. These appeals are clued primarily by intentional idioms and it is to them that the disdain has principally been directed. On the other hand, scientific displeasure with intentionality's other aspect, its *intensionalism*, is rarely, if ever, expressed. Intensions, as noted, are classifications (interpretations, meanings, significances, contents, etc.). They are, basically, alternative descriptions of extensionally described states of affairs that selectively ignore many details and enhance others in order to achieve a simplification in prediction or explanation. Intensions are familiar to physicists classical and modern. In analytical mechanics they have paraded under the title of auxiliary conditions or constraints (Lanczos, 1949). Together with initial conditions and laws, constraints comprise the causal context for final conditions or effects. The physical status of constraints, however, is somewhat peculiar. In detailing the physical peculiarity of constraints, insights are gained into the content of intentional sentences. As will become apparent, the problems that philosophers trot out with respect to the content of intentional sentences and explanations of behavior are problems that physicists could trot out with respect to constraints and explanations of motion.

13.2.1. Neither Constraints Nor Intentional Content Meet Conventional Standards for Fundamental Properties of Matter. It is assumed that the laws of motion specify as completely as possible the detailed behavior of matter. At the microscopic level of descriptions there is no room for any further limitations on motion, the laws say it all, as it were. So where do the physical constraints of analytical mechanics come in? The answer is that they come in by way of human invention. The physicist identifies a way of treating a problem that proves to be more simple and more useful than tracking all the degrees of freedom equally. Note how he or she replaces the microscopic description of particles at the surface of a solid by the equation of a geometric surface. The equation is said to define a boundary condition but it is, of course, just an alternative description of the detailed behavior of a huge number of molecules. The equation could also be said to identify the limiting case of the average positions of the particles over an extended period. However, there can be no *explicit* process of averaging within the ensemble of particles that yields the equation.

The latter point is well made by the collective properties of molecules that are termed "temperature" and "pressure." These properties allow simple and useful measures of a gas in a container, yet there is no way that their special significance can be predicted from the laws of motion. "Temperature" and "pressure" result from outside observers classifying the detailed microscopic degrees of freedom so as to result in fewer degrees of freedom at the new macroscopic level. Any classification to simplify details is a constraint; it is a rule of what is to be selectively ignored and it stands over and above the laws of motion. Talking in terms of "selectively ignoring" underscores the inherent subjectivity of many constraints in physics. To summarize these observations: "Constraints cannot be treated in physics as fundamental properties of matter" (Pattee, 1971, p. 200).

A similar sentiment is sometimes expressed by philosophers with regard to the content of intentional sentences: The objects of intentional idioms refer to things that do not exist in the ordinary way that the objects of physical actions (e.g., swinging, lifting) exist. For example, neither the intentional sentence under analysis ("I intend that I swing the two pendulums comfortably, at the same tempo but in opposite directions") nor its contradiction implies that there is or that there is not a something to which the clause " . . . I swing the two pendulums comfortably, etc., etc." truly applies (Chisholm, 1957).

13.2.2. Neither Constraints Nor Intentional Contents Can Be Reduced to Or Paraphrased By the Lower Level Language Describing the Legal Details. Constraints in analytical mechanics are referred to as auxiliary conditions because they are not derivable from the basic axioms or laws of mechanics. Not only are the descriptions of constraints and the descriptions of microscopic matter in motion written in two different languages (one language that is geometrical/kinematical and of low detail, and one language that is kinetic and of high detail), but no simple synonymy exists between them. The translation of a constraint description into the laws of motion is thwarted by an inability to capture the flavor (one might say, colloquially, the information) of the constraint and to render it in a form consistent with the logic of the laws of motion: A constraint classifies degrees of freedom (some are more significant than others) while the laws of motion are blind to differences of significance.

Rather than being reducible, one to the other, the upper level constraint language and the lower level motion language are complementary: The former classifies what is significant, the latter gives the legal details (cf. Pattee, 1970). As already noted, philosophers tell a similar story with respect to intensional and extensional sentences. To reiterate, intensional sentences, sensitive as their truth values are to distinctions of meaning, are not reducible in any obvious way to extensional sentences that ignore such distinctions. Yet both sentence kinds are needed to give an intelligible account of an act. With respect to a subject in the wrist-pendulum situation, the extensional answer to the question of *how* the subject is doing what he or she is doing must be complemented by the intensional answer to the question of *why* the subject is doing what he or she is doing or, more crudely, *what* the subject's movements are about.

13.2.3. Neither Constraints Nor Intentional Content Admit of Alternative Characterizations. Constraints are classified according to whether (a) their mathematical form is time-independent (scleronomic) or time-dependent (rheonomic), and (b) they relate coordinates of the system (holonomic) or they are nonintegrable differentials of the coordinates of the system (nonholonomic). The following equations constrain the motion of a skate on ice, construed as two particles connected by a rod of length l moving on a plane in such a manner that the velocity of the middle of the rod is in the direction of the rod:

$$z_1 = 0, z_2 = 0 \tag{13.1}$$

$$(x_1 - x_2)^2 + (y_1 - y_2)^2 - \ell^2 = 0 \tag{13.2}$$

$$\frac{dx_1/dt + dx_2/dt}{x_1 - x_2} = \frac{dy_1/dt + dy_2/dt}{y_1 - y_2} \tag{13.3}$$

The constraint Equations (13.1) and (13.2) are scleronomic and holonomic, whereas the constraint Equation (13.3) is scleronomic and nonholonomic. The motion of the skate is not prescribed by the laws of motion alone; it is prescribed by the laws of motion with Equations (13.1), (13.2), and (13.3) appended. If the equations of constraint were different, the motion would be different even though the laws of motion are invariant. Speaking more generally, for any given system, its motions are individuated by the equations of constraint. An alternative constraint equation means simply a different constraint and, resultantly, a different motion of the system. Constraints that restrain the middle of the rod to certain trajectories in the x, y plane cannot serve as an alternative for the constraints that keep the velocity of the middle of the rod in the direction of the rod. Constraints are uniquely characterized.

A similar claim can be made for the content of intentional sentences. A different intensional context means a different phenomenon (although it need not result in a completely different behavior as extensionally characterized). As Dennett (1969) remarks, the intention that I leave the room is not the intention that I open the door and so it cannot serve as an alternative characterization of the intentional conditions antecedent to my opening the door. The fundamental principle of intentionality is that intentional phenomena are individuated by their characterizations—to change the characterization is to change the phenomenon (Dennett, 1969).

13.2.4. Neither Constraints Nor Intentional Content Are Causes That Can Be Identified Independently From Their Effects. In words, the foregoing set of constraint equations specify that "the skate will move so as to keep the velocity of its midpoint in the direction of the skate." What does the skate do? The skate moves such that the velocity of its midpoint is in the direction of the skate. There is nothing unusual about the preceding. It is typical of constraints that their characterization is conceptually indistinguishable from the characterization of the effects that they explain.

Propriety requires that proper causes be identifiable independently from their said effects—that the relation of effect to cause be at least "contingent upon" rather than "conceptually the same as" or "logically connected to." Of course, if a constraint could be characterized by the laws of motion or replaced by another constraint equation that makes no reference to the effect, then a constraint would look like a conventionally acceptable cause. As discussed, however, neither move is possible.

Intensional contexts are improper causes for the same reasons. They identify conceptually with the effects that they are said to give rise to and they resist both extensional reduction or paraphrase and intensional substitutions. The failure of intentional content to meet traditional criteria for causes (and, therefore, to fall within the purview of normal standards of scientific explanation) is compounded by the fact that it is vacuous to ask why, given a particular intentional content, a particular action should follow. If one's intention is to swing two pendulums out of phase at a common, comfortable frequency then it makes no sense to ask why that activity occurred rather than some other activity (such as extending, in sequence, the two ankle joints).

There is, therefore, a striking difference between intentional explanation and nonintentional explanation. For the latter, the question of why one property (e.g., amplitude of oscillation) should depend on another (e.g., moment of inertia of a wrist-pendulum system) is meaningful, and it calls for further study to determine the lawful processes on which the regularity is founded. For intentional explanations, however, the antecedent fully explains the consequent; there is no more explaining to be done, no laws to be sought that would rationalize the pairing of these events (Dennett, 1969; Searle, 1983). In sum, the constraints of physics and the content of intentional sentences suffer from the same prejudice: As putative causes of motion and behavior, respectively, they are not part and parcel of the larger explanatory enterprise of science.

13.2.5. Neither Constraints Nor Intentional Content Admit of Rate-Dependent Characterization. Central to the explanation and description of the continuous motion of unconstrained masses is the concept of *rate* of a continuous process. The core notions of dynamics are tied to the derivatives of motion. For example, momentum, kinetic energy, and friction are proportional to velocity; inertial force is proportional to acceleration. How quickly or how slowly things happen matters immensely to dynamical processes. (For a simple example, whether or not two particles of different mass move in opposite directions following a head-on collision depends on their velocities before the collision.) The effects of dynamical laws are rate dependent. However, the effect of a constraint—that it classifies variables in a certain way or that it selects one trajectory rather than another—is indifferent to rate. The *kind* of skate motion identified by the preceding set of constraints remains invariant, by definition, over variations in rate. To put the argument very roughly but plainly, a constraint is like an injunction (e.g., "raise your right arm"): Whether uttered slowly or quickly, the requested act is the same. Or, a constraint is like a classification (e.g., male versus female): It makes no sense to describe the classification per se as slow or quick, although the application of a characterization may occur more or less rapidly.

In the same way that constraints contrast with the laws of motion by being rate-independent and resistant to rate-dependent characterization, so it is with intentional content. The intentional content on which most of the present discussion has focused—"(that) I swing the two pendulums comfortably at the same tempo, but in opposite directions"—can be given a localization and a date, as it were, but it cannot be qualified by rate referential terms such as quick or slow. Paraphrasing Dennett (1969, p. 81), it is more appropriate to refer to intentions, beliefs, hopes, wishes, and so on, as states (that can sometimes be dated) rather than as happenings characterizable in rate-dependent ways.

13.2.6. Neither Nonholonomic Constraints Nor Intentional Content Permanently Eliminate Degrees of Freedom. The intention (that) *x* means that behavior *x* will occur rather than some other possible behavior *y*. The intention (that) *x* selects from among action possibilities. Patently, it is not a permanent selection (in the sense that *y* is forever disallowed) but it is, rather, a flexible selection: The intention (that) *x* *temporarily* rules out relations among animal-environment properties that would thwart the intention rather than identifying those that fulfil the intention. It will be characteristic, therefore, of intentional systems that more degrees of freedom (or more alternative configurations) can be accredited to them in their static state description than can be accredited to them in the performance of any given action.

This latter property is the hallmark of nonholonomic physical systems, that is, systems governed wholly or in part by nonholonomic constraints. Strong forces create relatively permanent linkages among the atomisms of an ensemble, giving the ensemble a structure. If the strong forces are not too dominant—the ensemble is not fully rigid—then there is room for weak forces to contribute to the ensemble's internal dynamics. This means that dynamically the ensemble will appear to be more restricted than is evident from its state description. The state description counts degrees of freedom according to the formula *ND-C*, where *N* is the number of atomisms, *D* is the number of coordinates needed to pinpoint their positions, and *C* is the number of holonomic constraints. The constraints imposed by weak forces constitute a temporary overlay on the state description. They are time dependent constraints whose effects are revealed in the motions of the ensemble, reducing the number of possible trajectories of individual atomisms without reducing the number of degrees of freedom. These are nonholonomic (nonintegrable) constraints. Their effects are tantamount to modifying the laws of motion themselves.

It is not entirely clear how nonholonomic constraints arise (but see 13.3.4, following). Even the origins of a simple, significant example such as the specific catalyst or enzyme are obscure. The enzyme, nevertheless, is recognized as an eminently real constraint comprising a significant component of the causal context for cellular processes. As an example of a nonholonomic constraint, the enzyme is particularly illuminating with respect to the insinuation that the constraints of analytical mechanics are subjective—urged by the expediencies of simpler description and explanation. In analytical mechanics the simpler macroscopic descriptions in the language of constraints seems to be necessitated by the fact that a system is too complicated to follow in microscopic detail. In living systems, however, the simplifications expressed macroscopically by nonholonomic constraints are more than mere convenience. They have to be reckoned with as real. They arise from within the system (rather than being imposed upon it by an outside agent) and they direct or cause the organization of the system's dynamical processes. What analytical mechanics and living things have in common is that whether the stratification of a system into a simpler macroscopic level and a detailed microscopic level are convenient fictions or reality, the stratification is achieved only at the expense of creating distinct languages for each level.

In reckoning with the reality of nonholonomic constraints, one must heed all the attendant conceptual difficulties addressed in points 13.2.1 to 13.2.5. It is especially important to come to terms with the apparently enigmatic fact that the (symbolic) content of a more macroscopic state can figure causally in the behavior of a more microscopic state. Some scholars regard this enigmatic fact as the unique concern of cognitive psychology, setting that science apart from the other sciences (e.g., Fodor, 1981). The feeling expressed here is quite to the contrary. The fact in question is a problem for science most broadly conceived, and it has been since constraints first were appended to laws to ease explanation.

13.3. Directions For Naturalizing Intentionality

13.3.1. Indentifying Intentional Contents With Physical States. Nonholonomic constraints are produced and maintained by kinetic processes in the form of weak forces not derivable from a scalar quantity (Lanczos, 1949). Nonholonomic constraints do not involve any energy and matter over and above (or worse, different from) these kinetic processes. They just involve these kinetic processes under an alternative description, one that is at a different level of abstraction. Paraphrasing previous remarks, a nonholonomic constraint is a "decision," in the sense of a classification of alternatives according to a rule. The alternatives are the dynamical states generated by the atomistic activity of a given scale. The alternatives must be physically possible, otherwise the selection of one rather than the others would be vacuous. However, it is in the nature of nonholonomic constraints that some of these alternatives are rendered impossible, or highly improbable. If the laws that make the alternative dynamical states possible and the constraints that make them impossible operate at the same descriptive level, then a given dynamical state is simultaneously possible and impossible. It has been proposed that natural systems avert this conundrum by having the classificatory and kinetic processes operate at different descriptive levels (Pattee, 1971).

It seems quite proper to say, therefore, that nonholonomic constraints are *produced and maintained by* kinetic processes and *realized in* these same kinetic processes. This conclusion bears on a traditional barrier to identifying intentional contents with physical states. Traditionally it is held that if intentional contents are identical with physical states (say, states of the nervous system) then it cannot be the case that intentional contents are sustained by physical states or vice versa. The argument is that two things or two states, X and Y, cannot be both identical *and* causally correlated (Smart, 1959). But given the equation of intentional content and nonholonomic constraints, there is no irregularity in the claim that intentional content can be realized in (read: identical with) and produced by (read: causally correlated with) physical processes (those of the nervous system) *provided that the realization and the production are phenomena at different descriptive levels*. It is this point that Sperry (1965, 1976), Searle (1983) and others have emphasized in their efforts to demythologize the mental-physical identity problem. Unfortunately it goes a small way in the direction of naturalizing intentional causation, as the remarks that follow reveal.

13.3.2. The Anomaly of Intentional Causation. Return to the intentional content having to do with the pendular clocking mode of organization that is the focus of the present inquiry. If the arguments of 13.3.1 stand, then there are no conceptual barriers to recognizing that a participant's intention to swing the pendulums comfortably, isochronously, and 180 degrees out of phase can be produced by and realized in the structure of the nervous system. There are conceptual barriers, however, to recognizing that this intentional content can play back causally upon the nervous system, in particular the neuromuscular processes that generate the rhythmic movement of wrist-pendulum systems.

One conceptual barrier is the hegemony of the mass-based descriptions through which the interactions between physical states are normally said to occur. The causal efficacy of intentional content is ill-expressed by the Newtonian paradigm for interactions, as the following wry derision highlights: "Are we supposed to think that the mental event pushes against the axons or the dendrites or that somehow it sneaks inside the cell wall and attacks the cell nucleus?" (Searle, 1983, p. 264). Ostensibly, if intentional content is causal, it cannot be causal in the same way that physical states, as classically conceived, are causal. The intimation is that the causality of intentional content must be in the putatively softer sense of nonkinetic interactions (see Chapter 4).

A second conceptual barrier to the causal efficacy of intentional content is the thesis that whenever two events relate regularly as cause and effect—as given initial conditions that lead inexorably to given final conditions—a causal *law* is entailed. The nonintegrable characteristic of nonholonomic constraints (see Appendix A) refers to the fact that they are not in the form of differential equations that admit to integration. That the causal laws governing universal regularities of motion are stated in integrable differential form, and that nonholonomic constraints (by definition) cannot be expressed in this form, suggests that nonholonomic constraints do not instantiate universal regularities. Consequently causal explanations that involve nonholonomic constraints do not sit comfortably with the standard law-based account of causation.

The conceptual barrier of the regularity thesis surfaces in other ways. It is popular in the philosophy of science to view natural laws as universal generalizations or truths of the form:

$$(x)\,(Fx \subset Gx)\,. \qquad (13.4)$$

This is read as: "For any x, if x is an instance of F, then x is an instance of G." (Or, alternatively, "Any x that is in the extension of F must also be in the extension of G.") Accordingly, Newton's law would be read as saying "For any property x, if x is an instance of the quantity given by the product of mass and acceleration, then x is an instance of force."

In this orthodox view, a natural law is an intensional relation between extensions. If this truly is the form of laws, then one has difficulty imagining any law that can cover situations such as that depicted by the sentence: "The person's intention to swing the two pendulums comfortably at the same tempo but in opposite directions caused the pendulums to swing isochronously, 180 degrees out of phase at the periodic time of the coupled wrist-pendulum systems' preferred steady state." That is to say, it is highly improbable that the validity of the preceding sentence and its counterfactual ("if the person did not intend X, then Y would not have occurred") entail a universal truth correlating types of events A and B, of which the intentional content and the behavior of the wrist pendulum systems are respective tokens.

A third barrier to intentional causation is the predilection to regard causal regularities as not merely logical regularities. The connection between the intentional content under inquiry (that I swing . . .) and its effect *is* merely logical or internal—the cause and its effect are conceptually identical. This, of course, is the property common to nonholonomic constraints and intentional contents identified in 13.2.4. In both the cause is self referential; more exactly, nonholonomic constraints/intentional contents present *and* cause their conditions of satisfaction (Searle, 1983).

13.3.3. Intentionalizing Causation. To summarize, intentional causation is not through kinetic observables, it does not instantiate any particular universal generalization, and it does not involve independently defined events. Taken at face value, these anomalies would appear to undercut the reality status of intentional causes. Preserving that status may be achieved only at the expense of rejecting the traditional equation between causation and universal regularities defined over mass dependent quantities. In brief, it may have to be conjectured that causation (the bringing about of events, the inducing of a dynamics) is not necessarily tantamount to a universally regular relation of kinetic events. This conjecture amounts to an intentionalizing of the notion of causation (Searle, 1983), tailoring it to the physically heterodox but commonplace fact (at the ecological scale) of the directedness of biological systems. This conjectured division of the notions of causality and regularity should not be read, however, as a claim that no relation exists between them. The notion of causation must presuppose regularities among kinetic observables even though it may not, in all instances, be reducible to them (see Chapter 4).

13.3.4. Laws as Singular Statements Relating Properties. The development of the notion of causation befitting the phenomena of the ecological scale might benefit from a rejection of the standard construal of natural law (as a universal statement that relates extensions). One promising alternative is that laws are singular statements of fact describing the relationship between properties or magnitudes that are "universal" within a given context (Dretske, 1977, 1978; Turvey et al., 1981). Physical conditions that are far from equilibrium (see Chapters 3 and 4) allow an appreciation of what Bohm has called "the inexhaustible *depth* in the properties and qualities of matter" (1958, p. 134). As the contexts of physical processes vary, so do the characteristic properties and event correlations. Nevertheless, in these narrower contexts, as in the broader context, there will be generalizations of fact possessing explanatory efficacy (Bohm, 1958). Ideally, a formal expression of these generalizations should: (1) explain their explanatory efficacy (given something, something else necessarily follows), and (2) their limited scope. The "universal truth" formulation does not satisfy (2) and it is doubtful that it satisfies (1) (see Dretske, 1977, 1978). A singular statement of the following kind appears to satisfy both requirements:

$$\text{F-ness} \rightarrow \text{G-ness}, \tag{13.5}$$

to be interpreted, roughly, as one property following necessarily from another. Adjunctively (see Appendix C) this statement reads: Since *F-ness* is the case, then *G-ness* will be, or is, necessarily the case. Formulation (13.5), which relates intensions rather than extensions, suggests why it is that laws support counterfactuals, why it is that they successfully explain their instances, and why it is that they exhibit an odd opacity (they are falsified by coextensive substitution). Moreover, this formulation allows for gradations in the symmetry of laws. Not all laws are highly symmetrical in the way of the laws of general relativistic mechanics, but all laws are symmetrical in the sense that for any law there are transformations over which the law is invariant.

The issues raised here deserve careful and separate commentary. For present purposes let it suffice to say that it may be plausible to view both the causality of intentional content and causality in general within a common perspective. In technical terms, this perspective might be provided by the treatment of laws as extensional relations between intensions.

13.3.5. Comparing the Indicational/Injunctional and Specificational Senses of Information. It might be said that nonholonomic constraints/intentional contents define a kind of information. For any dynamical system with many alternative trajectories, a nonholonomic constraint is an optative: It identifies what trajectories are "desirable" (or, more accurately, which ones are "undesirable"). This sense of optative information is referred to in Chapter 1 as indicational/injunctional and distinguished from information in Gibson's specificational sense. The distinction is based, in part, on the fact that information in the indicational/injunctional sense underconstrains the activity that it directs, and that its efficacy depends on the availability of activity-relevant information in the specificational sense.

In the earlier example of the nest-building activity of flying insects (see Chapter 4), the indicational/injunctional information is relatively sparse. In words: "so move as to increase the intensity of the smell; stop and deposit where the smell is strongest." The elaborate architecture of the nest arises from the juxtaposition of these injunctions (multiply instantiated by the very many insects) with the large-scale relational properties of the pheromone diffusion field. The kinetics of nest building are controlled and coordinated through the juxtaposition of the two kinds of information, the indicational/injunctional and the specificational. Neither kind alone will suffice to get the job done.

A central thesis of this monograph is that specificational information is to be equated with kinematic, geometric, and spectral abstractions of kinetic field processes. Can the same equation hold for information in the indicational/injunctional sense? Can intentional content be causal yet force free, as previously conjectured?

Presumably, the injunctions that bias the insects to trajectories on pheromone gradients are structurally embodied. Custom would ascribe them to the individual insect's genetic constitution and it would be strongly inclined to interpret them as rate-independent symbol strings causally independent from the rate-dependent processes of dynamics (see Chapter 1). The upshot is that information in the indicational/injunctional sense and information in the specificational sense may not only differ in the degree to which they constrain activity but they may also differ in their causal relation to dynamics.

To exemplify this latter point, consider the DNA molecule that is advanced as an example of a symbol vehicle. Patently, the DNA molecule is a very particular inhomogeneity of matter and energy. The question to be asked is whether the symbol vehicle, generally construed, is a kind of inhomogeneity ontologically distinct from the inhomogeneities wrought by the dynamical processes of open systems (see Chapter 3). If this question is answered positively, then it follows that dynamically fashioned inhomogeneities cannot perform symbol functions and, further, that there must be an alternative, though unlikely, process fashioning the type of inhomogeneities that can. (The positive answer is consonant with the view of evolution proceeding via randomly occurring combinations of highly improbable symbols tailored by natural and sexual selection.) In contrast, a negative answer to the question is equivalent to the claim that dynamics creates symbols. The negative answer is pursued here for several reasons (cf. Carello, Turvey Kugler & Shaw, 1984; Kugler, 1985), one of which is the promise of a unified view of information and causation.

Recall that a site in a continuum where at least some of the field quantities become discontinuous defines a (physical) singularity. A singularity is nonintegrable, and if it were to function as a constraint it would be, by definition, nonholonomic. By the arguments of Chapters 4 and 7, singularities of kinematic flow fields (pheromone, optic, haptic) *can* function to constrain activity. And by these same arguments, singularities of kinematic flow fields can constitute information for biological systems that the flow envelops and/or connects. Collecting these ideas, the singularities of kinematic flow fields are very close to the intuitive notion of a symbol: They are discrete and nonintegrable and are potentially information. The contention, however, is that a critical property of symbols is specific structural embodiment (Pattee, 1973)—the existence of symbol vehicles (as exemplified by the structure of DNA).

At first blush the singularities of kinematic flow fields would not seem to be structurally embodied in the required sense. They emerge in the course of the flow and they evaporate when it ceases (see Chapter 4). Consider, however, two kinematic flow fields and their respective kinetic bases. If the scale (in terms of relaxation time and mean free path) of one kinetic field process is suitably larger relative to the other, then the singularities condensed out at the larger (slower) scale might appear as rate-independent structurally embodied entities from the perspective of the smaller (faster) scale. Symbols and symbol functions would be, under this view, scale-dependent phenomena dissolving into dynamic processes at extended scales of analyses (Kugler, 1985).

Return again to the nest-building activity. On the basis of the preceding, information in the indicational/injunctional sense and information in the specificational sense share a common origin. Both are (kinematic, geometric, spectral) abstractions of kinetic processes, albeit at different space-time scales. In addition, given that both are lawfully generated, both are specific to their kinetic underpinnings. Only Gibson's sense of information, however, is specific to the properties of the biological system and its surround that bear directly on the system's activity, that is, to those kinetic properties that materialize in the space-time scale of the act. The macroscopic patternings of pheromone diffusion fields are specific to the material layout (actual and possible) with respect to which the insect activities are conducted. In contrast, the injunctions "so move as to increase the intensity of the scent; stop and deposit where the scent is strongest" are specific to properties of the slowly transpiring kinetic processes that distinguish the evolution and persistence of these particular insects as a biological kind.

These ideas can be extended to the pendular clocking mode of organization defined over the two wrist-pendulum systems. A person's intentional content, tantamount to indicational/injunctional information, is realized as a state of the nervous system (Section 13.3.1). This state persists during the course of repeatedly assembling the coupled cyclic movements and is largely independent of the kinetics of these movements. In contrast, the specificational information required to elect reliably the parameters of these movements (the common tempo, the amplitudes) arises during the movements, causally tied to their dynamics. The point is that the kinetic processes sustaining these two senses of information differ; they differ in their relaxation times (the kinetic processes connected with the indicational/injunctional sense relax more slowly) and in their causal dependency on the ongoing movements.

To summarize briefly: The self-assembly of a pendular clocking mode of organization defined over two wrist-pendulum systems is grounded in two senses of information. One sense is closely allied to the notions of intentional context and intentional causation. It is introduced in Chapter 1 as information in the sense of indications and/or injunctions. This information provides (exceptional) boundary conditions for harnessing natural laws to perform specific and reliable functions. The other sense of information is closely allied to the pendular-like, clocking movements, particularly to the dynamic states that they lawfully engender. It provides the basis by which the person *qua* coupled wrist-pendulum systems can be apprised reliably of current and pending kinetic properties that bear on the successful assembly of the required organizational mode. Plainly, these two senses of information complement each other in a most important way. In the soft construction of the pendular clocking mode of behavior, neither sense of information alone will suffice and neither sense of information is reducible to the other. It remains to be underscored that this important complementarity between the two senses of information is based on a functional, not an ontological, distinction. By the arguments advanced here, both senses of information are rooted in dynamics.

REFERENCES

Bernstein, N. A. (1967). *The coordination and regulation of movements*. London: Pergamon Press.

Bloch, E. H., & Iberall, A. S. (1982). Toward a concept of the functional unit of mammalian skeletal muscle. *American Journal of Physiology: Regulatory, Integrative, Comparative Physiology 11*, R411-R420.

Bohm, D. (1958). *Causality and chance in modern physics*. London: Routledge.

Carello, C., Turvey, M. T., Kugler, P. N., & Shaw, R. E. (1984). Inadequacies of the computer metaphor. In M. S. Gazzaniga (Ed.), *Handbook of cognitive neuroscience*. New York: Plenum Press.

Chisholm, R. (1957). *Perceiving: A philosophical study*. Ithaca, NY: Cornell University Press.

Dennet, D. C. (1969). *Content and consciousness*. London: Routledge.

Dennet, D. C. (1971). Intentional systems. *Journal of Philosophy*, *68*, 87-106.

Dretske, F. I. (1977). Laws of nature. *Philosophy of Science*, *44*, 148-268.

Dretske, F. I. (1978). Reply to Niiniluoto. *Philosophy of Science*, *45*, 440-444.

Fearing, F. (1930 and 1970). *Reflex action: A study in the history of physiological psychology*. Cambridge: MIT Press.

Fodor, J. A. (1981). *Representations*. Cambridge, MA: MIT Press.

Fowler, C. A., & Turvey, M. T. (1978). Skill acquisition: an event approach with special reference to searching for the optimum of a function of several variables. In: G. E. Stelmach (Ed.), *Information processing in motor control and learning*. New York: Academic.

Gallistel, C. R. (1980). *The organization of action: A new synthesis*. Hillsdale, NJ: Lawrence Erlbaum Associates.

Ghiselin, M. T. (1981). Categories, life and thinking. *The Behavioral and Brain Sciences*, *4*, 269-313.

Gibson, J. J. (1979). *The ecological approach to visual perception*. Boston: Houghton Mifflin.

Kugler, P. N. (1985). Information as the morphology of dynamics. In W. H. Warren and R. E. Shaw (Eds.), *Persistence and change: Proceedings of the first international conference on event perception*. Hillsdale, NJ: Lawrence Erlbaum Associates.

Kugler, P. N., Kelso, J. A. S., & Turvey, M. T. (1982). On the control and coordination of naturally developing systems. In J. A. S. Kelso and J. E. Clark (Eds.), *The development of movement control and coordination*. New York: John Wiley.

Lanczos, C. (1949). *The variational principles of mechanics*. Toronto: University of Toronto Press.

Pattee, H. H. (1970). Physical problems of heredity and evolution. In C. H. Waddington (Ed.), *Towards a theoretical biology, 3*. Chicago, IL: Aldine.

Pattee, H. H. (1971). Physical theories of biological coordination. *Quarterly Review of Biophysics 4*, *243*, 255-276.

Pattee, H. H. (1972). Laws and constraints, symbols and languages. In C. H. Waddington (Ed), *Towards a theoretical biology, 4*. Chicago: Aldine.

Pattee, H. H. (1973). Physical problems of the origin of natural controls. In A. Locker (Ed.), *Biogenesis, evolution, homeostasis*. Heidelberg: Springer-Verlag.

Reed, E. S. (1982). An outline of a theory of action systems. *Journal of Motor Behavior*, *14*, 98–134.

Reed, E., & Jones, R. (1982). *Reasons for realism*. Hillsdale, NJ: Lawrence Erlbaum Associates.

Searle, J. R. (1983). *Intentionality*. New York: Cambridge University Press.

Shaw, R. E., & Turvey, M. T. (1981). Coalitions as models for ecosystems: A realist perspective on perceptual organization. In M. Kubovy and J. Pomeranz (Eds.), *Perceptual organization*. Hillsdale, NJ: Lawrence Erlbaum Associates.

Smart, J. J. (1959). Sensation and brain process *Philosophical Review*, *LXIII*, 141-156.

Sperry, R. (1965). Mind, brain, and humanistic values. In J. R. Platt (Ed.), *New views on the nature of man*. Chicago: University of Chicago Press 71-92.

Sperry, R. (1976). Mental phenomena as causal determinants in brain function. In G. G. Globus, G. Maxwell & I. Savodnik (Eds.), *Consciousness and the brain*. New York: Plenum Press.

Turvey, M. T., Shaw, R. E., Reed, E. S., & Mace, W. M. (1981). Ecological laws of perceiving and acting: In reply to Fodor and Pylyshyn. *Cognition*, *9*, 237-304.

Weiss, P. (1941). Self-differentiation of the basic pattern of coordination. *Comparative Psychology Monographs 17*, 29-96.

EPILOGUE

In terrestrial locomotion a large number of degrees of freedom assemble cooperatively in a pendular, clocking mode of organization. Although many joints may be involved, the movement organization exhibits a unitary character. What is in the design of biological systems that makes a unitary, pendular clocking mode possible? The question, more generally, is: how is the *unitizing of activity*—the transformations of an aggregate of muscle-joint complexes into a single functional unit—achieved? An answer in terms of neural networks and neural constructs is unlikely to be forthcoming. This skepticism does not derive from an inadequate understanding of, or any recalcitrance of, the architecture of the nervous system and the method by which it conducts and distributes its impulse currency. Rather the skepticism stems from the impression that nervous system properties per se are the properties of a (plastic-elastic-gel-like) medium in which a rich variety of natural laws can play their games. The explanation of unitizing is to be found in these laws.

Laws in general are but "secret ways" of talking about the conservations (Feynman, 1965), and wherever geometries are coupled to conservations, laws arise. Nervous systems can be counted among the multivarious geometries (or boundary conditions) made available at the ecological scale. Some nervous systems bring more geometries, and more varied geometries, to bear than others and, therefore, condense out a larger variety of lawful regularities. This characterization of nervous systems as boundary conditions is not to denigrate nervous systems. Rather, it is meant to put into perspective their contribution to biological phenomena (such as limb coordination) and, concomitantly, to highlight the inadequacy of an explanation couched solely in neurally-based constructs. As some authors (Shaw & McIntyre, 1974; Shaw & Turvey, 1981) have openly suggested, what is to be emphasized is the *transparency* of nervous systems to lawful regularities at the ecological scale. The present manuscript has made some modest inroads into understanding the physical laws and principles that unitize human movements and to which, it might be said, human nervous systems are transparent. Among the modest inroads there are some that bear on the concept of information. These are highlighted, by way of a summary.

1. A rhythmic unit of action is assembled out of generalized reversible (gradient-independent) and irreversible (gradient-dependent) transport processes that compete for spatial and temporal boundaries—that is, the unit oscillates at well-defined spatial and temporal scales.

2. The haptic neural network provides a compact low-energy, low-mass medium dominated by irreversible transports—that is, haptic stimulation radiates irreversibly, forming field-like flow patterns;

3. Competition between and within haptic field processes can be described by a finite number of converging and diverging flow regions, originating and terminating in singular states—that is, haptic flow fields give rise to low-dimensional topological portraits;

4. Haptic flow patterns or portraits can be classified according to the number and layout of singular states and the dimension of the space in which the flow occurs—that is, neural flow patterns can be classified according to theorems;

5. The singular set provides a low-dimension, control space description that can be mapped into a high-dimension machine space description to form an efficient, controlable system—that is, a low-dimension information description constrains a high-dimension neuro-muscle-joint description;

6. The nonlinear thresholds that characterize perceiving-acting couplings are symmetry-breaking devices that partition (categorize) control spaces into regions of reversibility and regions of irreversibility—that is, symmetry-breaking mechanisms categorize qualitatively the field processes;

7. A perceiving-acting system is an execution-driven nonlinear, open flow system—that is, the flow field characterisitics can qualitatively change as a function of internal changes in boundary conditions;

8. New informational properties emerge centrally from the creation and annihilation of singular states; these map back onto the system as constraints for local, muscle-joint, kinetic spaces—that is, the movement system is internally self-organizing.

REFERENCES

Feynman, R. P. (1965). *The character of physical law.* Cambridge, MA: M.I.T. Press.
Shaw, R. E. & M. McIntyre (1974). Algoristic foundations to cognitive psychology. In

W. Weimer & D. Palermo (Eds.), *Cognition and the symbolic processes*. Hillsdale, NJ: Lawrence Erlbaum Associates.

Shaw, R. E. & M. T. Turvey (1981). Coalitions as models for ecosystems: A realist perspective on perceptual organization. In M. Kubovy & J. Pomeranz (Eds.), *Perceptual organization*. Hillsdale, NJ: Lawrence Erlbaum Associates.

APPENDIX A HOLONOMIC AND NON-HOLONOMIC CONSTRAINTS

Constraints that are expressed in the form of an equation relating coordinates are called holonomic. But often a constraint is specified by a restriction on velocities rather than on coordinates. Consider a cylinder rotating (rolling) on a slanted surface of incline α without slipping (see Figure A.1). The velocities ds/dt and $d\theta/dt$ must relate as:

$$\frac{ds}{dt} = r\,\frac{d\theta}{dt}\,. \tag{A.1}$$

Integrating gives:

$$s - r\theta = c, \tag{A.2}$$

where c is a constant. Equation (A.2) is holonomic—it is of the same type as:

$$l = [(X_2 - X_1)^2 + (Y_2 - Y_1)^2 + (Z_2 - Z_1)^2]^{1/2}, \tag{A.3}$$

which requires a fixed distance between pairs of particles (say) in a rigid body. If a constraint on velocity can be integrated to give an equation like (A.2), *then the constraint is still holonomic*.

Sometimes, however, equations of constraint on velocities cannot be integrated to produce an equation like (A.2) or (A.3). A frequently used example (see Symon, 1971) is a disk rolling on a horizontal table under the restriction that it cannot tip over (like the wheel of a wheel-barrow); that is, the diameter touching the table is always vertical. Consider Figure A.2. The position of the disk is specified by three facts: x,y locates the disk's contact with the plane; ϕ determines the orientation of the plane of the disk relative to the x-$axis$; θ is the

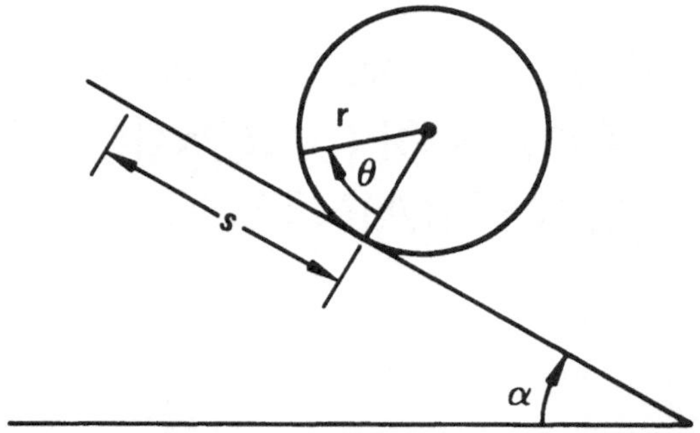

FIGURE A.1. Example of a holonomic constraint that restricts velocity.

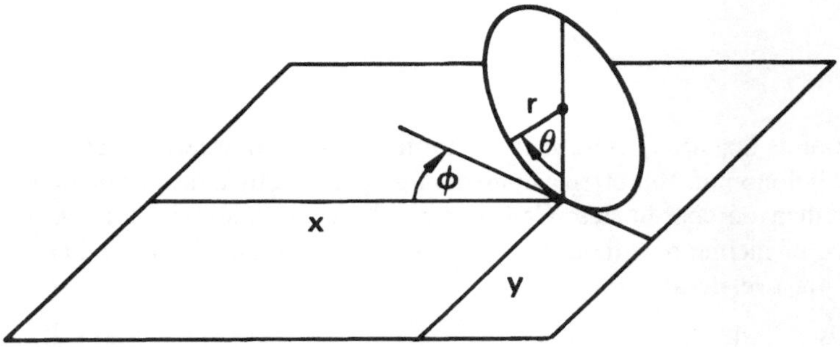

FIGURE A.2. Example of a non-nolonomic constraint that is velocity independent.

angle between a fixed radius and the vertical. Two constraints must be specified as well. First, the velocity of the point of contact must be zero in the perpendicular plane; that is,

$$\frac{dx}{dt}\sin\phi + \frac{dy}{dt}\cos\phi = 0. \tag{A.4}$$

Second the velocity parallel to the plane of the disk must be:

$$\frac{dx}{dt}\cos\phi - \frac{dy}{dt}\sin\phi = r\frac{d\theta}{dt}. \tag{A.5}$$

Equations (A.4) and (A.5) cannot be integrated to get two relations among the coordinates x, y, θ, ϕ. In words, by rolling the disk without slipping and by rotating it about a vertical axis, the disk can be brought to *any* θ. In sum, the four conditions are independent of one another and there cannot be any relation among them. The constraints in this case are said to be non-holonomic (non-integrable).

REFERENCE

Symon, K. R. (1971). *Mechanics*. Reading, MA: Addison-Wesley.

APPENDIX B COLLISIONS AND FRACTURE

B.1. The Theory of Collisions

The concept of collision refers to forces applied to and removed from an object in a very short period of time. The classical theory of collision, based primarily on the impulse-momentum law for rigid bodies, regards the colliding objects as single mass points. All elements of each object are assumed to be rigidly connected and to be subjected instantaneously to one and the same change of motion as the result of the collision. In reality, the forces initiate stress waves which travel at a finite velocity away from the region of contact and through the object. These waves reflect from boundaries of the object and interact with stress waves still being generated at the region of contact to create a complex pattern of stresses and strains in the interior. In short, all regions of an object subjected to a collision are not exposed simultaneously to the same force conditions (Goldsmith, 1960).

The classical theory is most suited to ideal atomisms whose degrees of freedom are exhausted by the three axes of translation. Atomism is a term suggested by Iberall (1977/8) for an entity of any magnitude that is atom-like at the scale of the ensemble to which it belongs. It is conventional to say that ideal atomisms have no internal degrees of freedom, where ''internal'' has the uncommon meaning of ''extratranslational.'' Atomisms of gases such as helium are closest to this ideal. They are single atoms each free to move on the three spatial dimensions. The total energy imparted by collision to a helium atomism is distributed over the translational motion. In terms of the equipartition theorem, the energy received is divided evenly and completely among the atomism's degrees of freedom, which are all translational.

The atomisms of another gas, oxygen, introduce a measure of internal complexity. These atomisms (molecules) consist of two linked atoms. To define the position of each of the atoms of oxygen requires three degrees of freedom for a total of six. However, the linkage between the atoms eliminates a coordinate choice, thereby reducing the degrees of freedom of the oxygen atomism to five. Because translation of the oxygen atomism's center of mass consumes only three of the five degrees of freedom, the two degrees of freedom that remain are "internal." The equipartition theorem would assign three fifths of the energy of collision to the translation of the atomism and two fifths of the energy to the internal bond. The conservation of energy does not hold if only the energy carried by the translation degrees of freedom is taken into account. It is for this reason that collisions of atomisms with internal degrees of freedom are said to be inelastic and that the conservation of momentum (rather than of energy) is the dominant constraint on their equations of collision.

Consideration of the collisions of diatomic atomisms is a small step toward the collisions of *systems*. In a statistical mechanical sense, a system is an ensemble of interacting atomisms with a boundary that prohibits the ensemble from dissolving into the surround. The atomisms of a system may be internally barren (like the helium atomism) or internally complex (of a kind hinted at by the oxygen atomism). As noted, internal complexity is associated with ways of absorbing the energy applied to a unitary thing other than through the translation of its center of mass.

B.2. The Theory of Fracture

The first major advance beyond the classical theory of collisions (viz., the one-dimensional vibrational treatment of colliding objects) recognized the significant proportion of energy converted into oscillations when the system's natural frequency is long compared to the duration of contact. Subsequent analyses of the multidimensional aspect of wave propagation consequent to collision, and of the stress distribution at the region of contact, were made possible by developments in the theory of elasticity (Timoshenko & Goodier, 1951). Elasticity refers to the fact that the internally generated forces of restoration are comparable to the externally applied forces of deformation so that there is a return to the status quo ante on removal of the external forces.

In many collisions, however, the conditions of impact are such that the entire cross-section of one or both of the colliding objects will exhibit a final permanent strain of significant magnitude, or one or both of the objects may fracture. Such nonreversible phenomena result from the conversion of kinetic energy into permanent distortion or fracturing of the structure of the object and the eventual dissipation of this energy in the form of heat. The analysis of the irreversible deformations wrought by the propagation of stresses that exceed the elastic limit (so called plastic flows or plastic waves) is a more recent and less developed aspect of collision theory (Goldsmith, 1960).

While the responses of an internally complex system to collision is difficult to follow, it is possible, nevertheless, to obtain insights into the collision process by considering (a) the behavior of a system under statically imposed forces; and (b) the relation between impact parameters and system failure. The deformation resulting from loading a system statically can be treated as a series of equilibrium states requiring no consideration of the acceleration effects of wave propagations. Of interest is the response to static loading by systems that exhibit a degree of rigidity, that is, systems which preserve their form in the face of perturbations. The requirement is that the system be elastic through some range of perturbation. Solids have an elastic domain as do multiphase systems that are solid or gel in part (for example, living things that are dominated by elastic-plastic-fluid [liquid and gel] processes [Yates, 1982]).

The interior of a solid system can respond in one of three ways to an applied force: (1) the linked atomisms can be forced further apart or closer together than the equilibrium (minimal potential) distance; (2) atomisms can hop into adjacent vacant lattice sites; or (3) the bonds between the atomisms can be broken (Freudenthal, 1950; Nadai, 1950; Walton, 1976). If response (1) is sufficient to absorb the energy of loading, then the solid is operating strictly within its elastic domain. Suppose that a static loading is realized as a force applied along an axis (a stress) so as to stretch or compress (more generally, to strain) the system. Then response (1) means that the system as a whole undergoes a coordinate transformation that changes the distances between all the atomisms but not the topology of the system's internal configuration. This response to static loading is reversible. It is, however, a response of finite capacity. At some point the potential energy stored up within the excessively strained bonds reaches a limit (the elastic yield) and new mechanisms for accommodating the applied energy must be found (that is, a new "escapement" arises). One escapement mechanism is the breaking of some bonds between some atomisms (response 3), another escapement is diffusion (response 2) which is enhanced considerably by the structural changes resulting from bond breaking. (At the elastic limit in a multiphase system, there is a structural change in at least one phase; for example, it is in the continuous solid phase of a two phase solid-fluid system such as gel, or in the more rigid phase of a polyphase solid-solid system such as polycrystalline metal or a polyphase solid-fluid phase system such as a high polymer.)

A brittle system (a physical ideal which is an engineering myth) would be infirmed at the elastic limit. There are no plastic deformations (flow processes) in a brittle system and microscopic bond breaking becomes, immediately, macroscopic fracture. For real, ductile systems, however, the yield point only identifies that loading at which fracturing begins on the atomistic level. Once the yield point is reached in a ductile system, the mutually reinforcing processes of bond breaking and diffusion can continue to accommodate excessive energy

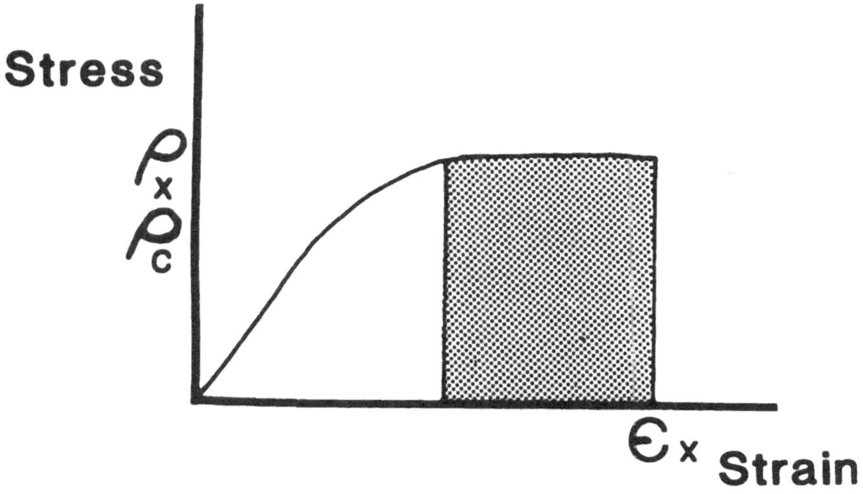

FIGURE B.1. The energy absorption per unit volume of a material is given by the shaded area of its stress-strain curve.

brought in by the static loading. The dissociating of some of the atomisms makes it easier for other bound atomisms to migrate to locations that are more stable than the locations that they currently occupy. This flow process is irreversible. Less energy is required for an atomism to hop from a high to a low potential site than vice versa. Moreover, the consequent relaxing of some bonds brought about by diffusion increases the strain on other, already overstrained, bonds, disposing them to further fracture.

The microfracturing that begins at and proceeds beyond the yield point reduces the long range order or cooperativity of the system (interpreted as bonds that repeat regularly over many thousands of atomic distances). The long range order is replaced by short range order or local cooperatives, not unlike the ''flow unit'' of a liquid. The diffusion occurs at the surface of these local clusters because the atomisms located there are thermally less stable than their partners in the interior. Clearly, the larger the number of local cooperatives and, therefore, internal surfaces, the greater the diffusion. And the greater the diffusion the more disposed to breaking are the already strained bonds at places in the system where diffusion of atomisms is not possible. In sum, fracturing of the bonds between atomisms is a chain reaction process and eventually a ductile system will fracture at the macroscopic scale.

The emphasis of the foregoing has been the gradual progression of macroscopic fracture, or system failure, as might occur under the repeated or prolonged application of static forces that exceed the system's elastic limit. In the range between the initiation of bond breaking on the microscale and the occurrence of system failure on the macroscale, the system gradually loses its ability to absorb

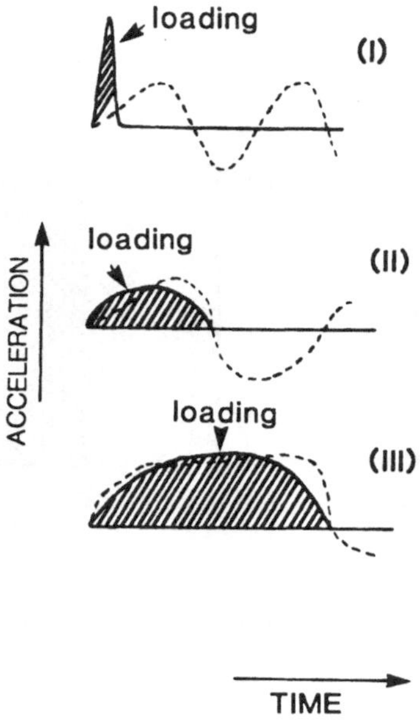

FIGURE B.2. *Acceleration x time* profiles of collisions under three loading durations (after Kornhauser, 1964)

the applied energy. A measure of the energy absorption of a material is given by its stress-strain curve, which relates force per unit area to proportional change in length. The energy per unit volume is approximately equal to the shaded area of Figure B.1. Consequently, the strain energy to failure may be approximated as follows: *[energy/unit] volume* = $1/2(\rho_x+\varepsilon_x)\rho_c$. Where ρ_c is the stress at the yield point and ρ_x and ε_x are the ultimate stress and ultimate strain, respectively, that mark the collapse of the system.

Of course, loss of ability to absorb energy could be quick given a collision. The microscopic processes leading to failure from a single brief loading must be a rapid chain reaction of bond breaking associated with elastic and plastic waves propagating from the point of contact and multiply reflecting from the system's boundary. As noted, however, broad conclusions relating failure to the conditions of collision are possible without considering the complex intermediary processes.

A collision will have an *acceleration X time* profile (where acceleration refers to the system). Three examples of single loadings are given in Figure B.2; to achieve a given response amplitude, shorter durations of loading must be compensated by greater accelerations. Two parameters are of special significance: (a) the change in velocity, and (b) the average acceleration (in units of

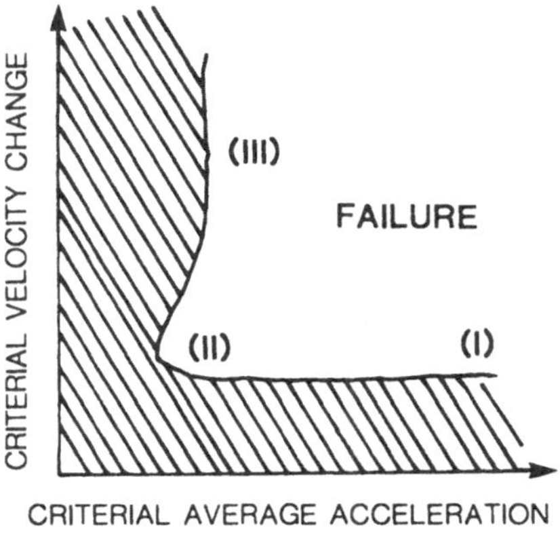

FIGURE B.3. Collision sensitivity plot shows where system failure will occur (after Kornhauser, 1964).

gravity) that is *just sufficient to produce structural failure*. In Figure B.2 the cross-hatched areas express the velocity changes. The average acceleration of any collision is equal to the velocity change divided by duration. A collision sensitivity curve can be generated by plotting criterial velocity change (where fracture occurs) against criterial average acceleration (where fracture occurs) (Kornhauser, 1964). A prototypical collision sensitivity plot is given in Figure B.3. The vertical asymptote is related to acceleration pulses that are steady or of long duration. It implies that no failure occurs unless a certain average acceleration is exceeded, regardless of the change in velocity of the system and the duration of the collision. The horizontal asymptote is related to acceleration pulses of short duration. It implies that system failure does not occur unless a certain velocity change is exceeded regardless of the average acceleration value (Kornhauser, 1964).

The location of the vertical asymptote in Figure B.3 is a function of the shape of the collision (its *acceleration x time* profile). In contrast, the horizontal asymptote is independent of the shape of the loading and is fully characterized by a unique value of velocity change: Collision durations that are short enough to be on the short duration asymptote (marked by (I) and (II)) for a given system will result in the structural failure of that system. There is some evidence (Kornhauser, 1964) to suggest that the collision velocity change required for irreversible damage to mammals is relatively indifferent to species and size (25 feet per second, 7.62 meters per second, is a reasonable approximation). The criterial average acceleration, however, differs markedly with species and size (roughly, 20g for man and 650g for mice).

A simple rule of thumb (from Kornhauser, 1964) relates the criterial velocity change (V_c) and criterial average acceleration (G_c) to the system's natural frequency (ω): $G_c = \omega V_c$. If most collisions between systems and their surrounds are of sufficiently short duration to place the systems on the horizontal asymptote of their collision sensitivity function, then V_c is constant. (For mammals, as noted previously, $V_c = 25f/s$.) In other words, the higher the value of a system's natural frequency, the greater is the system's tolerance to collision (measured in multiples of the gravitational constant).

REFERENCES

Freudenthal, A. M. (1950). *The inelastic behavior of engineering materials and structures*. New York: Wiley.

Goldsmith, W. (1960). *Impact: The theory and physical behavior of colliding solids*. London: Edward Arnold Ltd.

Iberall, A. S. (1977/8). A field and circuit thermodynamics for integrative physiology. I, II and III. *American Journal of Physiology: Regulatory, Integrative Comparative Physiology* 2: R171-R180, 1977; *3*: R3-R19, 1978; *3*: R85-R97, 1978.

Kornhauser, M. (1964). *Structural effects of impact*. Baltimore, MD: Spartan Books Inc.

Nadai, A. (1950). *Theory of flow and fracture of solids, Vol. II*. New York: McGraw-Hill.

Timoshenko, S. & J. N. Goodier (1951). *Theory of elasticity*. New York: McGraw-Hill.

Walton, A. J. (1976). *Three phases of matter*. New York: McGraw-Hill.

Yates, F. E. (1982). Outline of a physical theory of physiological systems. *Canadian Journal of Physiology and Pharmacology 60(3)*: 217-248.

APPENDIX C HUYGENS' DERIVATION OF THE CENTER OF OSCILLATION

When a physical pendulum oscillates, all of its parts interact mutually to determine a natural period. Mass particles closest to the point of suspension tend to oscillate faster than those mass particles positioned further away. The natural period emerges as a result of the competition among the internal tendencies of the motions of the many particles. Huygens' theory of oscillation (1673; see Bell, 1950, for an English translation) showed that the natural period of a compound pendulum depends on the form of the pendulous body (i.e., the distribution of local mass and length parameters) relative to the position of the axis of suspension. The theory determined this dependence by establishing the length L_o of the simple pendulum that oscillated isochronously with the compound pendulum. In determining the center of oscillation Huygens proceeded from three assumptions: (a) the center of gravity of a system, under the influence of gravity only, cannot rise; (b) the center of gravity of a system will rise again to its initial height if, in the absence of friction, the component parts are directed upward after a descent; and, most importantly, (c) assumption (b) applies even if the links connecting the component parts are severed during the movement.

Huygens' derivation of the center of oscillation starts by identifying kinematic similarities associated with Galileo's law of rigidly connected falling bodies. These similarity relations are then related to the ascending motions of the same bodies under disconnected conditions according to kinetic similarities that conserve the height of the center of gravity—an assertion constituting one of the first formal statements of the conservation of energy law. Using the equality relating the kinematic similarities (on the downward swing) with the kinetic

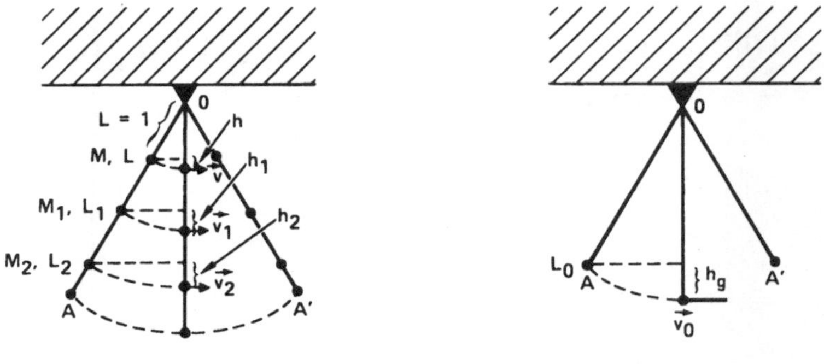

FIGURE C.1. *Left panel*: Compound pendulum.
Right panel: Isochronous simple pendulum.

similarities (on the upward swing), Huygens formulated two expressions—one for the maximum height of the center of gravity and another for the maximum height of the center of oscillation—which when combined yielded an explicit relation defining the center of oscillation strictly in terms of mass and length quantities.

C.1. Huygens' Proof

Figure C.1 (left panel) identifies a rigid pendulum *OA* composed of a large number of masses M_1, M_2, . . . , at distances L_1, L_2, . . . , *from a point of suspension O*. Set free to oscillate at *O*, the pendulum will swing through *B* (*OB* defines the axis of suspension) to *OA'*, where *AB* = *BA'*. In moving from the initial position *OA* to the lowest position *OB* the center of gravity G_c descends a height h_g. Consonant with assumption (b), G_c will rise again if the component parts are directed upward. But consonant with assumption (a), G_c will not rise beyond the height from which descent was initiated. If the individual masses are released from their rigid connections at position *OB* (see Figure C.2), then the masses, by virtue of the velocities derived from their connections, can only attain a collective height with respect to G_c that is equal to the height achieved by the connected system. Observing the uncoupled masses at their maximum heights reveals that the shorter pendulums will be found below the line *OA'*, the longer pendulums will be found above *OA'*, but the collective center of gravity will rise to a height h_g that remains on *OA'*.

According to Galileo's laws of motion the resulting velocities are proportional to the distances from the axis:

$$\frac{\text{speed of } M_1}{\text{speed of } M_2} = \frac{V_1}{V_2} = \frac{\text{arc } A_1B_1}{\text{arc } A_2B_2} = \frac{L_1}{L_2}, \tag{C.1}$$

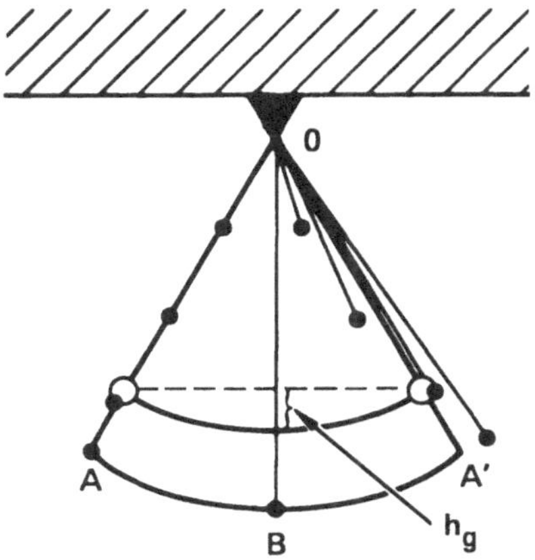

FIGURE C.2. Unconstrained compound pendulum.

where V_1 and V_2 are the respective velocities of masses M_1 and M_2 measured at their lowest point of descent (position B). Because of the preceding constraint (i.e., all pendulums are linked together), if one of the aforementioned sets of relations is known then all of the remaining relations can be determined by kinematic similarity. An additional geometric similarity can be derived relating maximum heights h_i of the centers of gravity for the respective individual pendulums of length L_i:

$$h = L - x$$

$$\cos\theta = \frac{x}{L}$$

$$x = L\cos\theta$$

$$h = L - L\cos\theta$$

$$h = L\,(1 - \cos\theta)$$

$$h_2 = L_2\,(1 - \cos\theta) \tag{C.2}$$

$$h_1 = L_1\,(1 - \cos\theta). \tag{C.3}$$

The geometric similarity is derived by dividing Equation C.2 by Equation C.3:

$$\frac{h_2}{h_1} = \frac{L_2}{L_1} \tag{C.4}$$

and for the particular case of the simple isochronous pendulum Equation (C.4) becomes

$$h_2 = h_1\,\frac{L_2}{L_1}. \tag{C.5}$$

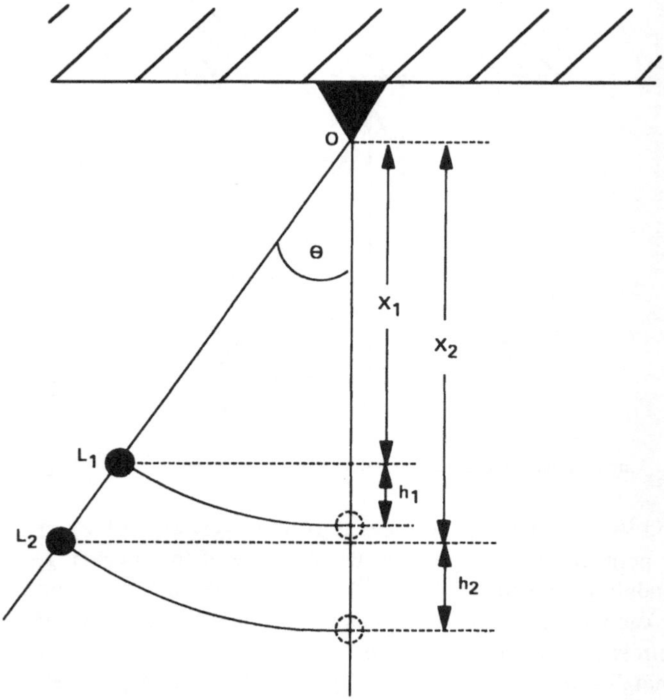

FIGURE C.3. Geometric similarity relating length L_i to height h_i.

Galileo's similarity relations can be used for the case of the simple isochronous pendulum (see Figure C.1, right panel) to give Equation (C.6).

$$\frac{L_2}{L_1} = \frac{V_2}{V_1}, \quad V_2 = V_1 \frac{L_2}{L_1} \tag{C.6}$$

The similarity can be made more transparent by marking off a pendulum of unit length $L_u = 1$ (Mach, 1893/1960; see Figure C.1, left panel) such that $\pi_{L_u} = (L_u/L_u) = (1/1) = 1$, $\pi_{L_1} = (L_1/L_u)$, $\pi_{L_2} = (L_2/L_u)$, ... If the compound pendulum moves from its position of greatest excursion to the position of equilibrium, the point at L_u will fall through the height h_u where V_u is the maximum velocity of the unit pendulum. Equations (C.5) and (C.6) can be rewritten as scaling relations in terms of the unit values for length, velocity, and height as

$$\frac{L_1}{L_u} = \frac{h_1}{h_u}, \quad h_1 = \pi_{L_1} h_u \tag{C.7}$$

$$\frac{L_1}{L_u} = \frac{V_1}{V_u}, \quad V_1 = \pi_{L_1} V_u . \tag{C.8}$$

The masses M_u, M_1, M_2, . . . , at the distances L_1, L_2, . . . , will fall in this case the distances $L_u h$, $L_1 h$, $L_2 h$, ... Expressing the lengths L_i in dimensionless form, using the unit length as the scaling metric, results in the following expression for the descent of the center of gravity of the rigidly connected pendulums:

Ensemble descent of center of gravity h_g

$$h_g = \frac{M_u \pi_{L_u} h_u + M_1 \pi_{L_1} h_u + M_2 \pi_{L_2} h_u + \ldots}{M_u + M_1 + M_2 + \ldots} = h_u \frac{\Sigma M_i \pi_{L_i}}{\Sigma M_i} . \tag{C.9}$$

Let the point at unit distance $L_u = 1$ from the axis acquire, on passing through the lowest (equilibrium) position, the velocity V_u. At the moment of passing the lowest position, all the linkages between the parts are severed such that the individual parts have velocities V_i, which, when directed upward, move the masses M_i to heights h_i. (Recall that the equivalent simple pendulum oscillates with the same period as the compound pendulum.) Under the constraints defined by the severed pendulum condition, the kinematic similarity of Equation (C.6) no longer holds—that is, the amplitudes of the individual pendulums are no longer constrained such that the individual pendulums now assume the timing characteristics that are proportional to their individual lengths, instead of the length of the ensemble's center of oscillation. Operating as independent pendulums, the motions of each pendulum relate to each other in terms of a kinetic similarity in accordance with Galileo's law of falling bodies where V_i^2 is proportional to h_i

$$\frac{h_u}{h_o} = \frac{V_u^2}{V_o^2} = \frac{\pi_{L_u}^2}{\pi_{L_o}^2} \tag{C.10}$$

The height of ascent of the unit pendulum, after the connections have been severed, will be $h_u = (\pi_{L_u} V_u)^2/2g$. The corresponding heights of ascent of the other material particles will then be $(\pi_{L_1} V_u)^2/2g$, $(\pi_{L_2} V_u)^2/2g$, ... *The ensemble* height of ascent of the center of gravity h_g of the *uncoupled masses* will be:

Ensemble ascent of center of gravity h_g

$$h_g = \frac{M_u(\pi_{L_u} V_u)^2/2g + M_1(\pi_{L_1} V_u)^2/2g + M_2(\pi_{L_2} V_u)^2/2g + \ldots}{M_u + M_1 + M_2 + \ldots} = \frac{V_u^2 \, \Sigma M_i \pi_{L_i}^2}{2g \, \Sigma M_i} \tag{C.11}$$

Note that the equation describing the decent of the connected pendulums (Equation C.9) is based on a kinematic similarity, while the equation describing the ascent of the disconnected pendulums is based on a kinetic similarity

(Equation C.11). This is the essence of Huygens' fundamental principle which relates the kinematic similarities of the descending motions to the kinetic similarities of the ascending motions in terms of the height of the center of gravity of the ensemble system. Huygens' principle (Equation C.12) states that the ensemble height of the center of gravity of the connected pendulums—derived according to kinematic similarities—is equal to the ensemble height of the center of gravity of the unconnected pendulums—derived according to kinetic similarities.

Hugyens' Principle:

$$h_g = \underbrace{h_u \frac{\Sigma M_i \pi_{L_i}}{\Sigma M_i}}_{\text{Kinematic Similarity}} = \underbrace{\frac{V_u^2}{2g} \frac{\Sigma M_i \pi_{L_i}^2}{\Sigma M_i}}_{\text{Kinetic Similarity}} . \tag{C.12}$$

From Huygens' principle it is possible to deduce a relationship between h_u and V_u. Given that all the pendulum motions of equal amplitude are temporally similar as in the connected pendulum condition, if one motion is known, then all the motions are completely determined through the kinematic similarity relations.

The final step in Huygens' derivation is to find an expression equating the height of descent of the center of gravity of an isochronous simple pendulum (see Figure C.1, right panel) in terms of the same kinematic and kinetic similarities as Equation C.12 provided for the ensemble system. Using the fact that the center of oscillation and center of gravity are coincident in a simple pendulum, it is then possible to derive an explicit expression for the length of the center of oscillation: If π_{L_o} is the length of the isochronous simple pendulum (in dimensionless units), then $h\pi_{L_o}$ is the distance of its descent (in dimensionless units) according to kinematic similarity and $(V\pi_{L_o})^2/2g$ is the distance of descent according to kinetic similarity:

Descent of center of oscillation (and gravity) h_o of isochronous simple pendulum

$$h_o = h_u \, \pi_{L_o} = \frac{(V_u \pi_{L_o})^2}{2g} . \tag{C.13}$$

Or, expressed in terms of the height h_u of the unit pendulum

Descent of center of gravity of unit pendulum h_u

$$h_u = \pi_{L_o} \frac{V_u^2}{2g} . \tag{C.14}$$

Multiplying equation (C.12) by equation (C.14) and expressing π_{L_o} as L_o and π_{L_i} as L_i results in an explicit relation defining the length of the center of oscillation L_o strictly in terms of mass and length characteristics

Huygens' law relating masses and lengths to center of oscillation

$$L_o = \frac{\Sigma M_i L_i^2}{\Sigma M_i L_i}. \tag{C.15}$$

REFERENCES

Bell, A. E. (1950). *Christian Huygens and development of science in the seventeenth century*. London: Arnold.

Huygens, C. (1673). *Horologium oscillatorium*.

Mach, E. (1960). *The science of mechanics* (6th ed.). T. J. McCormack, Trans. LaSalle, IL: Open Court.

APPENDIX D

PHYSICAL LAWS, STRUCTURAL STABILITY AND DIMENSIONAL HOMOGENEITY

D.1. Qualitative and Quantitative Properties of Physical Laws

From the perspective of physical science events are to be understood in terms of conservations that are geometrically distributed in space and time. A conservation has both *qualitative* (type of conservation, mass, energy, momentum, etc.) and *quantitative* (summational) properties. Both of these properties are preserved invariant under scale and coordinate transformations.

The second law identifies the natural tendency for conserved entities to become distributed uniformly in space and time as $t \rightarrow \infty$. If the distribution function changes with time the process is referred to as *irreversible*. If the distribution function is constant in time the process is referred to as *reversible*. The equipartition principle is the formal descritpion of this ordering tendency. The formal expression of the distribution functions constitute the *equations of state*. Time dependencies of the distribution functions are captured by *equations of change*.

The invariance of the relations defining the state equations under metrical and coordinate space transformations has been referred to by Luce (1978) as an *automorphism on a measurement structure*: The measurement structure is defined by the physical law and the automorphism is defined by the set of transformations that preserve the qualitative form of the law invariant.

D.2. Dimensional Analysis and the Equation of State

The dimensional structure of physical variables and, ultimately, laws is built up from products of powers of a basic set of numerical scales such as mass, length, and time. Formal study of dimensional structure of these variables and equations is the subject of dimensional analysis (see Bridgman, 1922; Duncan, 1953; Kline, 1965; Langhaar, 1967; Rosen, 1978). Dimensional analysis provides a methodology for deriving an estimate of the form of an equation describing a physical process with a limited amount of knowledge about the actual quantitative details of the process. The only knowledge required is an identification of the dimensionality associated with the physical quantities and constants involved. For example, suppose the goal is to describe a physical process that depends on a certain number of physical quantities, or observables, then the law(s) governing the process must be expressible as some functional relation between the observables that are involved in it; that is, it must be of the form

$$\phi(x_1,x_2, \ldots ,x_n) = 0, \tag{D.1}$$

where the x_i terms are the observables in question. This relation is called the *equation of state* of the process.

D.3. Fundamemtal and Derived Dimensions

The modern formulation of equation of states from dimensional considerations came from Fourier in 1822. He emphasized that the measurable quantities appearing in the mathematical representation of physical laws usually are not pure numbers; most of them have associated dimensions, such as velocity (*length/time*), momentum (*Mass x length/time*), force (*mass x length/time*^{2}). These can be expressed in terms of a smaller set of fundamental dimensions, which he took to be mass M, length L and time T. That is, every quantity x_i of the equation of state has a dimension that can be expressed as a monomial of the fundamental ones:

$$x_i = M^{\alpha_i} L^{\beta_i} T^{\gamma_i}, \tag{D.2}$$

or, more generally,

$$x_i = x_1^{\alpha_i} x_2^{\beta_i} x_3^{\gamma_i}, \tag{D.3}$$

where each dimension, fundamental or derived, has a scale of units specified by the measurement situation.

Magnitudes of different dimensions or units cannot be combined arbitrarily; they may be multiplied but they cannot be added. Fourier's principle of *dimensional homogeneity* required that any mathematical expression which purports to describe a physical system must accord with the rules for manipulating dimensional qualities. This principle restricts severely the mathematical expressions that can describe physical relationships. These ideas provided the conceptual backdrop for the development of the π-theorem by Buckingham (1915).

D.4. The Buckingham π-Theorem

Within the framework of dimensional analysis it may be postulated that any law of nature satisfies the property of being *dimensionally invariant*. (For an exact treatment of this issue see Krantz, Luce, Suppes, & Tversky, 1971; Rosen, 1978.) A law of nature satisfies this requirement if the qualitative form of the law does not rest on the choice of units for the several measures involved in its statement. The key fact about laws being dimensionally invariant is that the dimensions involved can be grouped into several products of powers that are *dimensionless* and the law depends only on these dimensionless quantities. Equation (D.3) can be rewritten as a ratio

$$\pi_i = x_i\, M^{-\alpha_i}\, L^{-\beta_i}\, T^{-\gamma_i}, \tag{D.4}$$

where π_i is dimensionless; that is, π_i is a pure number. The equation of state (Equation D.1) governing the process can be written in the following dimensionless form

$$\phi(\pi_4, \pi_5, \ldots, \pi_n) = 0. \tag{D.5}$$

The number p of dimensionless coefficient groups π_i that define the equation of state in this form can be derived from the π-theorem of Buckingham (1915) which states that:

> If there are *m derived dimensions* required to define the system of interest, and these require *n* independent *fundamental dimensions* for their specification (where, $n<m$), then there are $m-n$ independent dimensionless groups required for the equation of state.

$\phi\,[f_1, f_2, \ldots, f_n; d_1, d_2, \ldots, d_m] = 0$
Fundamental Parameters: $[f_1, f_2, \ldots, f_n]$
Derived Parameters: $[d_1, d_2, \ldots, d_m]$
π-Theorem: $m-n = p$.

The critical requirement for the success of the π-theorem concerns the property of *independence*. A fundamental dimension is independent of other fundamental

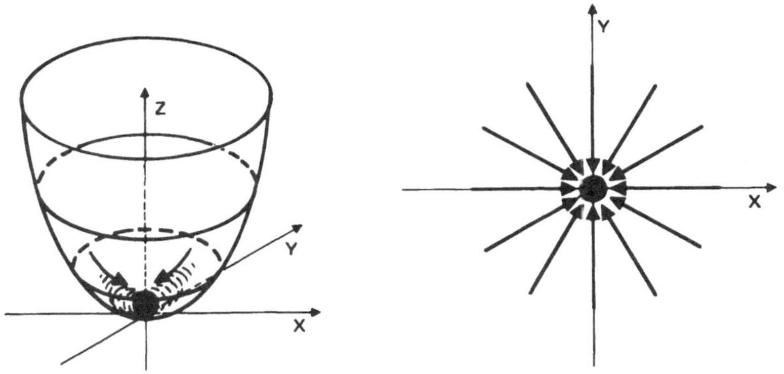

FIGURE D.1 . POINT ATTRACTOR
Left panel: Stable equilibrium point in a potential well.
Right panel: Motion in x-y plane.

dimensions if it cannot be formed from the sum, difference, or product of any two or more of the fundamental quantities. Likewise, a dimensionless product group π_i is not independent of other groups if it can be formed by taking the sums, products, or powers of other groups.

Dimensionless numbers are important in the description of degenerate behavior, that is, transition regions where the system's behavior suddenly changes from one stable mode to another. The π-numbers identify quantitative values that specify precisely where a qualitative transition will occur. The π-theorem allows for prediction of the number of possible qualitative transitions a system can potentially realize if the number of fundamental and derived dimensions are known. π-numbers do not, however, address the question of the qualitative nature of these transitions.

D.5. Classical Stability Analysis

Classical stability analysis is concerned with understanding the dynamical properties associated with certain systems of differential equations. The stability associated with these systems is often represented by the analogy of a marble rolling on a surface. Equilibrium states are defined when the net resultant of the horizontal forces is zero. Displacements in any direction create a force resultant that tends to return the ball to the equilibrium position. Conventionally these changes are associated with a potential function whose values are assigned by the z-axis. Potential functions can come in a variety of *forms*. The particular shape of these forms determines the stability characteristics of the system.

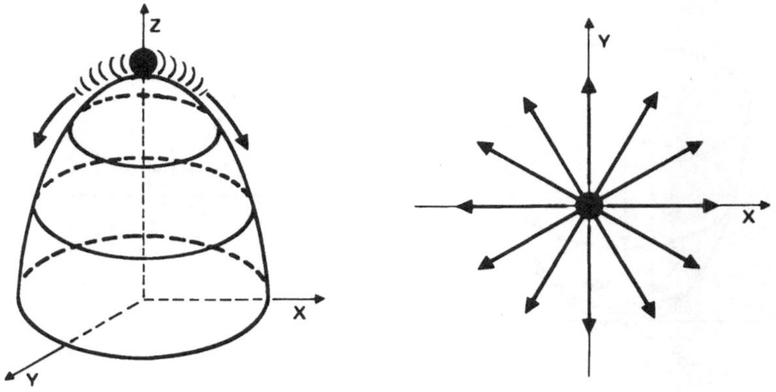

FIGURE D.2. POINT REPELLOR
Left panel: Unstable equilibrium point in a potential well.
Right panel: Motion in x-y plane.

Figure D.1 identifies a bowl-shaped potential function. A single equilibrium point occurs at the bottom of the bowl where the potential function is minimum. At this location the net horizontal forces sum to zero. Displacements of the ball from the minimum result in the development of a net force which tends to return the ball to the minimum—at the equilibrium point the net force returns to zero. The equilibrium point is referred to as *stable* if small perturbations are counteracted by resultant forces that tend to return the ball to the equilibrium condition. Such equilibria are referred to as *attractors*.

Figure D.2 identifies an inverted bowl-shaped potential function. The top of the inverted bowl identifies an equilibrium point. At this point there are no forces that tend to displace the ball. If, however, the ball is slightly perturbed away from this location, the resultant force that develops tends to move the ball away from the equilibrium point. This equilibrium is *unstable* and is referred to as a *repellor*.

Figure D.3 identifies a 'saddle-shaped' potential function. The point defined at the seat of the saddle is a combination attractor and point repellor. Perturbations of the ball in the x-plane confront ascending gradients resulting in forces that tend to return the ball to the seat of the saddle. Motion of the ball is stable in the x-plane. Perturbations of the ball in the y-plane confront descending gradients resulting in forces that tend to move the ball away from the seat. Motion of the ball is *unstable* in the y-plane. This combination equilibrium point is termed a *saddlepoint*. It is only realizable in potential functions that have at least two degrees of freedom relative to the potential plane.

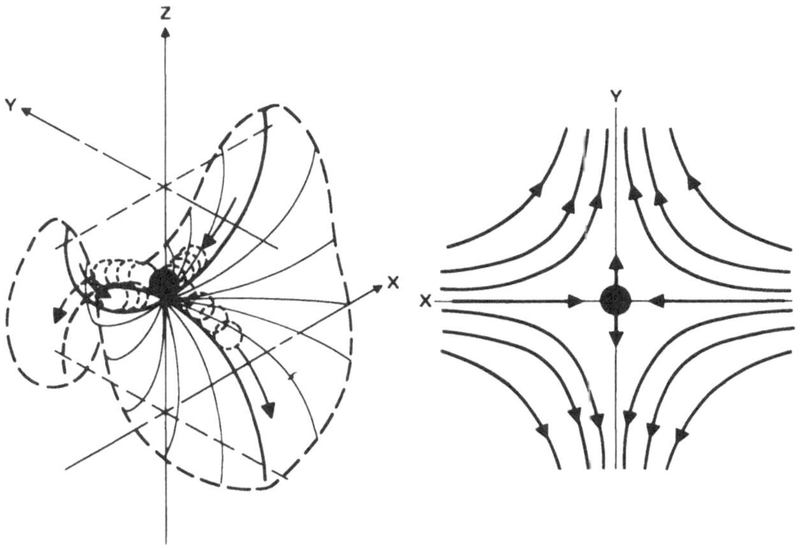

FIGURE D.3. SADDLEPOINT
Left panel: Stable attractor in x-plane and unstable attractor in y-plane.
Right panel: Motion in x-y plane.

D.6. Structural Stability Analysis

The theory of structural stability analysis extends beyond the simple hilltops, valley bottoms and saddlepoints exhibited in conventional stability theory, and offers a richer set of building materials for dynamical landscapes[1]. Structural stability analysis provides a methodology for exploring the *qualitative* nature of change in a system's dynamical character as various independent parameters are changed. Whereas conventional stability theory deals with local response patterns to perturbations, structural stability theory deals with global changes in the dynamical landscapes.

A system of differential equations is referred to as *structurally stable* if the dynamical landscape (i.e., tendencies for numerical solutions to converge or diverge from particular solutions) remains qualitatively unaltered in the course of changes in the independent parameters. Put differently, the system is structurally stable if the numerical solutions of its trajectories change, but the overall pattern of its stabilities remains unchanged. The system is *structurally unstable* if any of

[1] The 'form' of the output states (potential values) has the appearance of hills, valleys, and intermediate gradient basins and therefore is sometimes referred to as the geometric 'layout' or 'landscape.' These terms refer to qualitative characteristics of the potential function's surface manifold.

the stability characteristics change. If a structural instability occurs, it usually occupies only a small region of parameter space—this small submanifold region constitutes the *bifurcation set*. In these critical regions continuous changes in independent parameters may suddenly result in dramatic 'jumps' in the output organization. The sudden transition or jump appears to be discontinuous not because there are no intervening pathways or states, but because none of them is stable; that is, there are no stable equilibrium points in the intervening region. Because of the lack of intervening equilibrium points the passage through this region is likely to be brief in comparison to the time spent in the stable (equilibrium) states. Thom (1975) has named this type of discontinuous bifurcation a 'catastrophe'.

Figure D.4 identifies an example of a catastrophe. Beginning on the left side, the 'form' of the potential surface for a fixed y-value is flat. There are no equilibrium regions associated with changes in x-values. Moving to the right along the y-parameter results in the emergence of a minimum. The ball moves to the bottom of the 'valley' defined by the potential minimum. Further changes along the y-parameter result in the appearance of a small 'bump' along the ascending gradient defining the right wall of the valley. The bump eventually evolves into an *inflection point*. With further evolution the inflection point is annihilated and two new equilibrium points (maximum and minimum) emerge. Throughout this phase of the evolution the ball adheres to the location of its local minimum, prohibited from the new minimum by intervening gradients. With further changes in the y-parameter the new maximum ('peak') coalesces with the original minimum ('valley') resulting in their mutual annihilation. The loss of the ball's local minimum results in the ball becoming unstable. Instability continues until the ball comes to rest in the new minimum. The trajectory of the ball during the instability is marked by the dotted line which identifies that there are no equilibrium points defined in this unstable transition region. Thom (1975) has referred to evolutionary sequences in which the state of the system has to 'jump' across regions in which no equilibrium points are defined as *catastrophes*. (The dashed line identifies the evolutionary path the ball would take if the process proceeded in the opposite direction. A hysteresis loop can result.)

A catastrophic bifurcation is distinct from a *subtle bifurcation*. In a subtle bifurcation the ball's trajectory (on the potential surface) is in contact continuously with an equilibrium point, even during the bifurcation. An example of a subtle bifurcation is the Hopf bifurcation identified in Figure 7.10. As the point attractor evolves with changes in the control parameter a bifurcation occurs as the limit point attractor evolves into a periodic limit cycle attractor. During the evolution of the bifurcation, the state of the system is never in a region that is undefined with reference to an equilibrium point—that is, there is no change of basin. This is not the case for a catastrophic bifurcation. In a catastrophe the state of the system *always* traverses through a region in which *no equilibrium points are defined*.

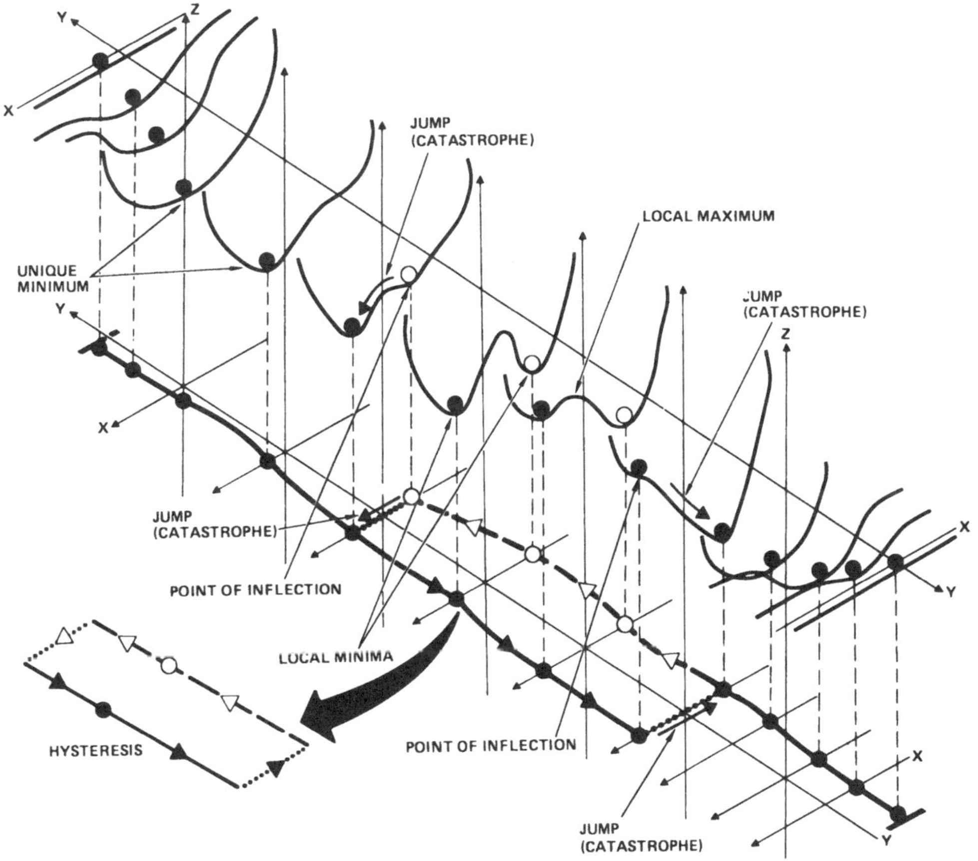

FIGURE D.4. The structural stability of the potential surface changes as a function of variations in the y-parameter.

The most comprehensive analysis of bifurcations (subtle and catastrophic) in one-potential systems with four or less control dimensions (independent parameters) is Thom's (1975) theorem of elementary catastrophes. For example, the theorem states that there are seven distinct catastrophes or subtle bifurcations associated with one-potential systems composed of four dimensions of control. The theorem identifies how many qualitatively different ways solutions can approach and exit equilibrium points and what 'forms' the solutions assume.

REFERENCES

Bridgman, P. W. (1922). *Dimensional analysis*. New Haven, CT: Yale Univiversity Press.

Buckingham, E. (1915). On physically similar systems. *Physical Review*, *4*, 345-370.

Duncan, W. J. (1953). *Physical similarity and dimensional analysis*. London: Edwald.

Kline, S. J. (1965). *Similitude and approximation theory*. New York: McGraw-Hill.

Krantz, D., Luce, R., Suppes, P., & Tversky, A. (1971). *Foundations of measurement, Vol. I*. New York: Academic Press.

Langhaar, H. (1967). *Dimensional analysis and theory of models*. New York: Wiley.

Luce, D. (1978). Dimensionally invariant numerical laws correspond to meaningful qualitative relations. *Philosophy of Science, 45*, 1-16.

Rosen, R. (1978). *Fundamentals of measurement and representation of natural systems*. New York: North-Holland.

Thom, R. (1975). *Structural stability and morphogenesis*. Reading, MA: Benjamin.

APPENDIX E NATURAL LAW AND ADJUNCTIVE LOGIC

The term natural law (or similar terms such as scientific law, law of nature, or simply law) is not a technical term defined in a particular empirical science; rather, it is an ontological term that implicates a fundamental commitment towards a position on existence (i.e., a position on the nature of being). In a physical science the commitment is to the existence of conservations or summational invariants, a commitment which leads, on extension, to the invariance postulate of natural laws. The evocation of the *explanatory* construct of 'law,' however, implies more than a commitment to conservations and the invariance postulate: It further implies a commitment to a formal logic relating to a theory of causal mechanism. That is, a logistic commitment must be introduced to fulfill the requirements of necessity and sufficiency for an explanatory physical science versus simply an ontological physical science. The prima facie difference between an ontological physical science and an explanatory physical science concerns the role of a conditional logic. An ontological physical science does not require the act of interpretation, whereas an explanatory physical science requires an interpretation. The nature of the interpretation is a function of the type of formal logic that relates the initial and auxiliary conditions of a physical event. Different logics necessarily lead to different interpretations. Of the many types of formal logic that have been proposed, three forms—hypothetical, causal, and adjunctive—will be considered in terms of their explanatory relevance for relating natural laws to physical events. The hypothetical or conditional proposition takes the form, "*If* x, then y." This can be contrasted with the causal proposition, "*Because* x, then y." and the adjunctive proposition, "*Since* x, then y." The arguments outlined below follow directly from Turvey and Shaw (1979).

E.1. Hypothetical Logic

The hypothetical proposition is invalid whenever the premise is true and the conclusion is false, and valid otherwise. It can be said to be conditionally true or correct if the opposite of its conclusion contradicts its premise (i.e., by *modus tollens*). However, the opposite of the conclusion (e.g., not *y*) is not necessarily inconsistent with the premise (e.g., *x*). For instance, "If this is Monday, then I go to work" is not inconsistent with the fact that it may indeed be Monday but I do not go to work simply because it is a holiday. In a hypothetical proposition the consequent does not disconfirm the premise.

For an explanatory physical science the hypothetical propositions constitute a poor model for natural laws. The inadequacy is owing to the fact that (a) there are too many ways in which the propositions can be valid, and (b) the propositions fail to provide a necessary relationship between premises and conclusions. The reason that these limitations make the hypothetical propositions an inappropriate model for natural laws becomes apparent as soon as one attempts to fit these laws to this propositional schema.

Let the initial and auxiliary conditions that define the domain of application of the natural law correspond to the premise of the proposition. The valid outcome predicted by the natural law will then correspond to the consequent of the proposition. A relation is considered to be a natural law (a) if the denial of its prediction or consequent is necessarily inconsistent with the premise and (b) when its premise can be shown to be true (i.e., when its initial and auxiliary conditions can be shown to be satisfied). In brief, a natural law is a relation that predicts true outcomes whenever it can be shown to be validly applied.

It clearly violates what is meant by a natural law to say that the law validly applies but does not predict the outcome. In such a case, either the relation is not acceptable in its stated form as being a valid law, or else it would be concluded that the conditions for its application had not actually been satisfied. A law cannot validly be applied to any situation where its premises are not true, but a hypothetical proposition is valid, by definition, even when its premises are false. It is not difficult to see how attempts to interpret laws as hypothetical propositions lead to paradoxes. In particular, what is conventionally and intuitively accepted as the function of scientific theory evades formal confirmation. Thus, Popper (1959) and others have shown that when stated in hypothetico-deductive form neither natural laws nor theories are logically verified simply because their predictions are confirmed. Affirming the consequent of a hypothetical proposition does not affirm the premise. Hence, if law *x* predicts outcome *y*, given that *y* is true does not imply that *x* is true.

Unfortunately, Popper's own attempt at an alternative evaluation of scientific laws and theories fails in like fashion. His argument is that even though laws or theories cannot be verified directly, they can be evaluated by showing that their predictions do not hold. The falsifiability procedure is based on a time-honored and valid argument schema (*modus tollens*): If law x, then outcome y; but not y, therefore, not x.

It is quite unlikely, however, that falsifiability is ever achieved in practice (Lakatos, 1970), because the premises for a law or theory are such a complex of variables that it often proves impossible to determine which one has been falsified. The onus, therefore, falls upon the theorist to decide whether a major or a minor assumption of the theory is at fault. Given the potential arbitrariness of this choice, it would be imprudent to reject a law or the core of a theory that had been developed carefully and arduously.

The validation of theories and laws appears to be logically impossible. Thus it is important to note that it is not that logic makes no difference in theory evaluation (and a law is, of course, just an accepted consequent of a theory), but rather that logic makes so little difference. The fruitfulness of a theory in explaining anomalies and bringing general consistency into science is more important than either logical verification or falsification. In other words, the degree to which a new law applies symmetrically across a wide domain of natural phenomena and relates other laws is the highest criterion of its worth. It is also a realistic measure of the resistance scientists will show in abdicating it.

At all events, logical forms other than the hypothetical proposition should be pursued. What is sought is a logical form that is consonant with the aforementioned facts about theory evaluation. In this perspective, consider the Stoics' conception of causal propositions.

E.2. Causal Logic

A causal proposition begins with a true premise and ends with a necessary consequent, for example, "Because it is day, then it is light." The causal schema more appropriately expresses the form of the law; by contrast, recall that the hypothetical proposition is still valid if its premises are assumed false. For instance, consider the following hypothetical proposition: "If it is night, then it is light." Whether it is night or day, it follows that "It is light"—a valid logical argument but scientifically false.

The causal proposition, however, does not permit such a trick to be performed on scientific intuition. A causal proposition is incorrect (by definition) if it begins with a false premise or ends with a conclusion that does not follow from it. Thus, unlike hypothetical forms, the causal interpretation demands that the premise and conclusion correspond.

Although the schema for natural laws seems to be satisfied by causal propositions, the intuitive notion of causal relation cannot be captured effectively in formal statements. Moreover, it can be argued that invariance laws (Wigner, 1970), those that characterize the symmetry relations existing among natural laws, do not seem to fit the schema for causal propositions (see Shaw & McIntyre, 1974).

E.3. Adjunctive Logic

An adjunctive proposition begins with a true premise and ends with a necessary consequent, for example, "Since it is day, then the sun is shining." This proposition is incorrect when it either begins with a false premise or ends with a consequent that need not follow. The adjunctive proposition professes both that the second member follows from the first and that the first member is true. Adjunctive logic requires a commitment to one's premises, and it is this propositional schema that is consonant with the concept of natural law.

Analyzing the adjunctive proposition "Since *x*, then *y*" in terms of truth tables reveals that in order for the adjunctive relation to hold, both *x* and *y* must be true. The adjunctive relation is false otherwise. On the surface this looks like the truth functional definition of the conjunctive relation (e.g., *x* and *y*). It differs, however, in one important way: Where conjunctive relations are commutative (i.e., "*x* and *y*" is equivalent to "*y* and *x*"), adjunctive relations are not. Hence "Since *x*, then *y*" does not imply "Since *y*, then *x*." (This asymmetry in commutativity is consistent with the time asymmetry associated with the initial and auxiliary conditions of irreversible flow processes.)

The adjunctive formulation seems to capture the sense that laws of nature apply in an inexorable manner to grind out reality. This is expressed simply as the adjunctive proposition that "Since the law applies, then the observed outcome must follow (necessarily)." If the outcome does not follow invariably upon correct application of the law, then there are grounds for falsifying the whole proposition, because in an adjunctive proposition the truth of the conclusion follows necessarily from the truth of the premises. Thus, in this special sense, the falsifiability criterion is preserved.

The verification criterion, however, does not hold for adjunctive propositions at the level of natural law. The observation (*y*) that some event (*x*) occurs as predicted by physical law *x* does not verify that *x* is a law. Thus, given *y*, it is fallacious to affirm *x*. Although natural laws can be applied to predict outcomes, no number of observed outcomes can be used logically to verify the law, a fact that is expressed in the non-commutativity of the adjunctive propositional form of natural law.

One might even question the utility of the verification procedure, because natural laws are postulated on more grounds than observation. The major grounds for accepting or rejecting principles as natural laws are whether or not they fit into the invariance structure of a science, resolving anomalies and relating other principles (Wigner, 1970).

The accuracy with which natural laws predict effects is not as important as the degree to which they contribute to the coherence of explanation for natural phenomena. Since one might predict what one does not understand, prediction alone is an insufficient criterion for the explanatory worth of theories, hypotheses, or laws. Scientific theories or natural laws that help simplify a field will never be rejected solely on the grounds that they are not predictive. Indeed, they should not be, since the conditional logic of verification does not apply.

REFERENCES

Popper, K. (1959). *The logic of scientific discovery*. New York: Harper and Row.

Lakatos, I. (1970). Falsification and the methodology of scientific research programmes. In I. Lakatos & A. Musgrave (Eds.), *Criticism and the growth of knowledge*. Cambridge, England: Cambridge University Press.

Shaw, R. E. and M. McIntyre (1974). Algoristic foundations to cognitive psychology. In W. Weimer & D. Palermo (Eds.), *Cognition and the symbolic Processes*. Hillsdale, NJ: Lawrence Erlbaum Associates.

Turvey, M. T. & R. E. Shaw (1979). The primacy of perceiving: An ecological reformulation of perception for understanding memory. In L-G. Nilsson (Ed.), *Perspectives on memory research: Essays in honor of Uppsala University's 500th anniversary*. Hillsdale, NJ: Lawrence Erlbaum Associates.

Wigner, E. P. (1970). *Symmetries and reflections*. Cambridge, MA: MIT Press.

Author Index

Subject Index

Page numbers followed by "n" indicate footnotes.

B

C

E

F

K

L

N

Q

R

S

T